Polar Microbiology

The Ecology, Biodiversity and Bioremediation Potential of Microorganisms in Extremely Cold Environments

Edited by
Asim K. Bej
Jackie Aislabie
Ronald M. Atlas

CRC Press
Taylor & Francis Group
Boca Raton London New York

CRC Press is an imprint of the
Taylor & Francis Group, an **informa** business

Cover: Frozen lakes in Schirmacher Oasis and Lake Untersee in East Antarctica. Photo Courtesy: Asim Bej from 2008 Tawani/NASA Antarctic Scientific Expedition.

CRC Press
Taylor & Francis Group
6000 Broken Sound Parkway NW, Suite 300
Boca Raton, FL 33487-2742

Printed in the United States of America on acid-free paper
10 9 8 7 6 5 4 3 2 1

International Standard Book Number: 978-1-4200-8384-2 (Hardback)

Library of Congress Cataloging-in-Publication Data

Polar microbiology : the ecology, biodiversity, and bioremediation potential of microorganisms in extremely cold environments / editors, Asim K. Bej, Jackie Aislabie, Ronald M. Atlas.
p. ; cm.
Includes bibliographical references and index.
ISBN 978-1-4200-8384-2 (hardcover : alk. paper)
1. Extreme environments--Microbiology. 2. Polar regions. I. Bej, Asim K. II. Aislabie, Jackie. III. Atlas, Ronald M., 1946-
[DNLM: 1. Microbiological Phenomena--Antarctic Regions. 2. Microbiological Phenomena--Arctic Regions. 3. Biodegradation, Environmental--Antarctic Regions. 4. Biodegradation, Environmental--Arctic Regions. 5. Cold Climate--Antarctic Regions. 6. Cold Climate--Arctic Regions. QW 4 P762 2010]

QR100.9.P65 2010
616.9'041--dc22

2009041209

Visit the Taylor & Francis Web site at
http://www.taylorandfrancis.com

and the CRC Press Web site at
http://www.crcpress.com

Contents

Preface

Polar Microbiology: The Ecology, Biodiversity, and Bioremediation Potential of Microorganisms in Extremely Cold Environments explores the field of polar microbiology and, in particular, the environmental impact of hydrocarbon contamination and the potential for microbial bioremediation. The polar regions of this planet include the Antarctic continent and the area north of the tree line of the Arctic Circle. Ice dominates much of the Antarctic continent and the Arctic Ocean. In fact, the polar environment is the coldest on Earth and poses significant challenges to life in the Arctic and the Antarctic. Yet, in this extremely cold environment, populations of microorganisms thrive and carry out critical ecological functions. These microorganisms are adapted to growing at low temperatures and have diverse and often unique metabolic capabilities.

Although they have had only limited exposure to pollutants, many microorganisms that are indigenous to Arctic and Antarctic ecosystems are able to degrade pollutants. It is this metabolic capacity that forms the basis for bioremediation as a potential treatment for hydrocarbons that may contaminate polar environments. Contamination may happen as a result of oil extraction and transport in the Arctic and human activities in both the northern and southern polar regions.

Currently, much of the polar environment is relatively undisturbed by direct human activities. However, global warming is causing the melting of ice, and the depletion of the atmospheric ozone is exposing polar environments to increased ultraviolet radiation. It is evident that the polar environment is changing, and increasing human activities threaten to bring about significant pollution in the polar terrestrial and aquatic ecosystems. In particular, human activities pose serious risks of environmental contamination with petroleum hydrocarbons, especially in the Arctic due to oil exploration, marine transportation, and human habitation, and in the Antarctic due to tourism and scientific research activities.

It is critical that we exercise all prudent measures to prevent the environmental contamination of polar regions. However, if we fail, we must have contingency plans for abating pollution levels thereby restoring the ecosystem. It is within this context that this book examines the diversity of polar microorganisms and their abilities to degrade petroleum hydrocarbons that may contaminate ice, soil, and both marine and freshwater aquatic environments in the Arctic and the Antarctic. In the Antarctic, increased scientific knowledge about microbes is being generated as scientific research activities have taken on a priority. In the Arctic, much is known about the practical problems of dealing with pollutants.

Pollution with petroleum hydrocarbons has accompanied polar exploration from its inception and now seems to be an unavoidable consequence of oil exploration and extraction in the Arctic and scientific research activities in the Antarctic. Since the first landing on the Antarctic continent by Captain John Davis in 1821 (in the Cecilia islands at Hughes Bay in the Antarctic Peninsula), numerous expeditions have been completed resulting in the establishment of permanent research stations and bases representing over 30 nations.

Even the earliest polar expeditions by Scott, Shackleton, and Amundsen polluted the environment as they carried out their historic explorations. These early explorers brought large quantities of supplies, including petroleum fuels, to support their activities as they sought to reach the North and South Poles. When the expeditions ended, materials that contaminated the environment—including stored fuels—were left behind. Corrosion and weathering of wooden structures led to leakages into the environment. Consequently, high concentrations of polyaromatic hydrocarbons, such as anthracene, benzo[b]fluoranthene, benzo[k]fluoranthene, chrysene, fluorene, and pyrene, as well as benzo[a]anthracene, benzo[a]pyrene, and fluoranthene, which are recognized carcinogens, have contaminated Antarctic soils under and around the historic fuel depots.* In the Arctic, many early exploration wells were not properly capped and many are now surrounded by oozing oil, resulting in the contamination of these historic sites, which need to be remediated.

Since 1960, human presence in the Antarctic continent has increased significantly, thereby requiring a greater need for energy to power the research stations and for aviation and ground transportation refueling. It has therefore become necessary to have year-round storage of hydrocarbon-based fuels, such as diesel and petroleum, at bases and locations where temporary research stations have been established. Such activities have all too frequently resulted in accidental oil spills causing the contamination of ice, soil, lakes, and coastal sediments and waters. In the Arctic, the development of oil wells and pipelines to transport oil southward has resulted in the oil pollution of many previously pristine environments.

Petroleum hydrocarbon spills in the Antarctic and Arctic environments have created pockets of microbial consortia in which biodegradative microorganisms have been selected and have naturally become enriched over time with the hydrocarbon and diesel fuels acting as the carbon source. This book is a compilation of the state of scientific knowledge about the microbial populations that are present in polar ecosystems and the diversity of their genes and physiological capabilities. It provides an overview of the extremophilic microorganisms that inhabit polar ecosystems and their potential for microbial bioremediation in the cold polar environments. All the chapters have been written by leading investigators in the field of polar microbiology. This is the first book to present a broad overview of the study of the ecology, the diversity, and the bioremediation potential of microorganisms in extremely cold polar environments. It should therefore serve as a very valuable resource for scientists and those charged with environmental protection and the management of polar regions.

Asim K. Bej
Birmingham, Alabama

Jackie Aislabie
Hamilton, New Zealand

Ronald M. Atlas
Louisville, Kentucky

---

* For a discussion of the environmental pollution caused by early Antarctic expeditions see Blanchette, R.A., B.W. Held, J.A. Jurgens, J. Aislabie, S. Duncan, and R.L. Farrell. 2004. Environmental pollutants from the Scott and Shackleton expeditions during the "Heroic Age" of Antarctic exploration. *Polar Record* 40: 143–151.

Editors

Asim K. Bej is currently a professor of biology at the University of Alabama at Birmingham (UAB). He received his PhD in biology with research in microbial molecular genetics from the University of Louisville at Kentucky, where he continued as a postdoctoral fellow in environmental molecular microbiology. He joined UAB as an assistant professor in 1991 and was promoted to professor of biology in 2003. His research interests include the study of microbial extremophiles including biodegradative microorganisms in cold environments, mechanisms of cold adaptation, and the application of microbial metabolites in biomedical science and diagnostics. He is a member of the editorial board of several microbiology journals and has served as a guest journal editor. He has published nearly 100 manuscripts in scientific journals and over 20 book chapters. He is an honorary member of the Golden Key International Honor Society, a recipient of the NASA Faculty Fellowship program, and a member of the Tawani International Antarctic Scientific Expedition.

Jackie Aislabie is a research leader within the Soils and Landscape Team at Landcare Research in Hamilton, New Zealand. She received her BSc (Hons) and PhD from the University of Otago, Dunedin, New Zealand. She was a postdoctoral fellow at the University of Louisville, Kentucky, where she worked on the use of microbes for the selective removal of nitrogen from shale oil. She has also worked as a lecturer in microbiology at the University of Canterbury, Christchurch, New Zealand, and has authored 70 manuscripts. She has been working in Antarctica since 1995 and has added to the knowledge of microbial diversity of Antarctic soils and the potential of bioremediation for hydrocarbon-contaminated Antarctic soils.

Ronald M. Atlas is a professor of biology and public health, and a codirector of the Center for Health Hazards Preparedness at the University of Louisville, Kentucky. He received his BS from the State University at Stony Brook, New York, his MS and PhD from Rutgers University, and his DSc (honoris causa) from the University of Guelph, Ontario, Canada. He was a postdoctoral fellow at the Jet Propulsion Laboratory, Pasadena, California, where he worked on Mars life detection. He has previously served as the president of the American Society for Microbiology; as a member of the NIH Recombinant Advisory Committee; as a member of the DHS Homeland Security Science and Technology Advisory Committee; as the chair of NASA's Planetary Protection Subcommittee; as the chair of the Wellcome Trust Pathogens, Immunology and Population Health Strategy Committee; and as a cochair of the American Society for Microbiology Biodefense Committee. He is a fellow of the American Academy of Microbiology and has received the ASM Award for Applied and Environmental Microbiology, the ASM Founders Award, and the Edmund Youde Lectureship Award in Hong Kong. He has authored nearly 300 manuscripts and 20 books. He has helped pioneer the field of bioremediation and has published extensively on oil biodegradation in the Arctic and sub-Arctic regions.

Contributors

Jackie Aislabie
Landcare Research
Hamilton, New Zealand

Ronald M. Atlas
Department of Biology
University of Louisville
Louisville, Kentucky

Asim K. Bej
Department of Biology
University of Alabama at Birmingham
Birmingham, Alabama

Andrew M. Booth
Department of Marine Environmental Technology
SINTEF Materials and Chemistry
Trondheim, Norway

John P. Bowman
School of Agricultural Sciences
University of Tasmania
Hobart, Tasmania, Australia

Odd G. Brakstad
Department of Marine Environmental Technology
SINTEF Materials and Chemistry
Trondheim, Norway

Paul Bridge
British Antarctic Survey
Natural Environment Research Council
Cambridge, United Kingdom

Vivia Bruni
Department of Animal Biology and Marine Ecology
Italian Collection of Antarctic Bacteria
University of Messina
Messina, Italy

Frédéric Coulon
Sustainable Systems Department
Centre for Resource Management and Efficiency
School of Applied Sciences
Cranfield University
Cranfield, United Kingdom

Donald A. Cowan
Department of Biotechnology
Institute for Microbial Biotechnology and Metagenomics
University of the Western Cape
Cape Town, South Africa

Daniel Delille
Observatoire Océanologique de Banyuls
Université Pierre et Marie Curie
Unité Mixte de Recherche-Centre National de la Recherche Scientifique
Banyuls sur mer, France

Liv-Guri Faksness
Department of Marine Environmental Technology
SINTEF Materials and Chemistry
Trondheim, Norway

Julia M. Foght
Biological Sciences Department
University of Alberta
Edmonton, Alberta, Canada

Perry A. Gerakines
Department of Physics
University of Alabama at Birmingham
Birmingham, Alabama

Angelina Lo Giudice
Department of Animal Biology and Marine Ecology
Italian Collection of Antarctic Bacteria
University of Messina
Messina, Italy

Charles W. Greer
National Research Council Canada
Biotechnology Research Institute
Montreal, Quebec, Canada

Caroline Heath
Department of Biotechnology
Institute for Microbial Biotechnology and Metagenomics
University of the Western Cape
Cape Town, South Africa

Richard B. Hoover
Space Science Office
NASA Marshall Space Flight Center
Huntsville, Alabama

Kevin A. Hughes
British Antarctic Survey
Natural Environment Research Council
Cambridge, United Kingdom

Nuraan Khan
Department of Biotechnology
Institute for Microbial Biotechnology and Metagenomics
University of the Western Cape
Cape Town, South Africa

Emanuele Kuhn
Department of Microbiology
Institute of Biomedical Sciences
University of São Paulo
Sao Paulo, Brazil

Adriana Philippi Luz
Environmental Protection Agency of Santa Catarina
Florianópolis, Santa Catarina, Brazil

Luigi Michaud
Department of Animal Biology and Marine Ecology
Italian Collection of Antarctic Bacteria
University of Messina
Messina, Italy

Nazia Mojib
Department of Biology
University of Alabama at Birmingham
Birmingham, Alabama

Moola Mutondo
Department of Biotechnology
Institute for Microbial Biotechnology and Metagenomics
University of the Western Cape
Cape Town, South Africa

Emilien Pelletier
Institut des Sciences de la Mer de Rimouski
Université du Québec à Rimouski
Rimouski, Quebec, Canada

Vivian Helena Pellizari
Department of Microbiology
Institute of Biomedical Sciences
University of São Paulo
Sao Paulo, Brazil

Elena V. Pikuta
Astrobiology Laboratory
National Space Science and Technology Center
Huntsville, Alabama

Shane M. Powell
Australian Antarctic Division
Department of Environment, Water, Heritage and the Arts
Kingston, Tasmania, Australia

and

Australia and Tasmanian Institute of Agricultural Research
University of Tasmania
Hobart, Tasmania, Australia

G.S.N. Reddy
Centre for Cellular and Molecular Biology
Hyderabad, India

Birgit Sattler
Institute of Ecology
University of Innsbruck
Innsbruck, Austria

Sisinthy Shivaji
Centre for Cellular and Molecular Biology
Hyderabad, India

Michael C. Storrie-Lombardi
Kinohi Institute, Inc.
Pasadena, California

Shannon J. Williamson
Department of Microbial and Environmental Genomics
J. Craig Venter Institute
San Diego, California

1 Microbiology of Antarctic Terrestrial Soils and Rocks

Donald A. Cowan, Nuraan Khan, Caroline Heath, and Moola Mutondo

CONTENTS

1.1 INTRODUCTION

The Antarctic continent offers the coldest and most arid environments on Earth (Onofri et al., 2004). However, climatic conditions are by no means homogenous across the continent and widely differing climatic regions exist; for example, temperatures and water availability differ hugely between the glacial dome, peninsular, and Dry Valleys regions of the continent.

While the vast majority of the continental land surface area is permanently covered by a multi-kilometer thick layer of ice, around 1%–3% is ice-free for at least part of each year (Cowan and Ah Tow, 2004). These ice-free regions mainly comprise mountain ranges that emerge above the ice plateau (nunataks), or coastal desert regions characterized by bare arid soils containing little or no vegetation. Coastal regions that have been the focus of terrestrial microbiological research include the McMurdo Dry Valleys, the Vestfold Hills, the Antarctic Peninsula, and the Mars Oasis.

The extreme nature of the environment, in addition to the complete absence of visible biological structures in some Antarctic desert regions, led to early conclusions that some Antarctic soils were devoid of microbial life. Of course, it is now known that abundant microbial life exists in most "extreme" habitats and microbial diversity and biomass in Antarctic desert soils has proven to be even higher than expected.

1.2 MICROCLIMATE AND GEOCHEMISTRY OF ANTARCTIC SOILS

The deglaciated terrestrial environments of the Antarctic continent (Figure 1.1) offer a unique combination of extreme climatic conditions which include extremely low temperatures, wide temperature fluctuations, low water availability, long periods of darkness, and high periodic incident solar radiation.

The region encompassing the McMurdo Dry Valleys (or Ross Desert) in South Victoria Land constitutes the most extensive ice-free desert in Antarctica (Friedmann and Ocampo, 1976; Onofri et al., 2004) and comprises an area of roughly 4800 km^2. The total ice-free area of Antarctica comprises <0.3% of the continent (Fox and Cooper, 1994). Dry katabatic winds greater than 100 km h^{-1} descend from the Antarctic ice plateau into the valleys and contribute to the maintenance of desert conditions (Friedmann, 1982; Doran et al., 2002). These deserts consist of exposed rock and soil, with no consistent ice or snow cover, although lakes are sometimes present on the valley floors (Figure 1.1) (Friedmann, 1982).

The mean annual air temperature of the Ross Desert ranges from −20°C to −25°C (Wynn-Williams, 1988; Ascaso and Wierzchos, 2002), although summer and winter temperatures vary markedly. The air temperature fluctuates between

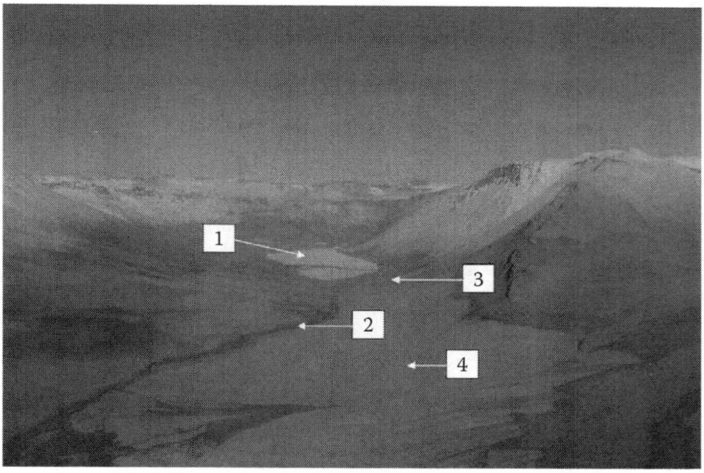

FIGURE 1.1 (See color insert following page 276.) The Miers Valley, Eastern Antarctica. This valley incorporates all the typical terrestrial microbial habitats of the Dry Valleys including lakes (1), lake-derived streams (2), glacial meltwater streams (3), and desert soils (4).

−20°C and −50°C in the winter season (mid-April to mid-October), although values as low as −60°C have been recorded (Friedmann, 1982; Wharton et al., 1986; Onofri et al., 2004). During summer (mid-November to mid-February), the air temperature can fluctuate between −35°C and +3°C depending on the weather and particular geographic location (de la Torre et al., 2003).

Surface ground temperatures in the Ross Desert region can vary significantly from the air temperatures. While the mean summer air temperature hovers around 0°C, the ground surface temperature during periods of direct sunlight frequently exceeds 15°C (Ellis-Evans, 1997). Extreme fluctuations in surface temperature are common and depend on direct sunlight and cloud cover. During the summer months, variations in cloud cover cause significant temperature oscillations that commonly occur over a matter of minutes. Indeed, observations of temperature fluctuations from −15°C to +27.5°C within 3 h have been reported for Ross Desert soils (Cameron, 1974). Furthermore, continuous summer sunshine can increase the internal temperature of rocks up to 10°C above the air temperature (de la Torre et al., 2003).

The sole form of precipitation in the Dry Valleys region is snowfall, with a mean annual precipitation of approximately 10 g cm^{-2} $year^{-1}$ (Ugolini, 1970). The infrequent snowfall mostly sublimes due to the extremely low atmospheric humidity or is blown away by the high winds (Friedmann, 1982; Onofri et al., 2004). The rate of evaporation/sublimation exceeds that of precipitation (Marchant and Head, 2007). The soil surface is, therefore almost continually exposed to a desiccating atmosphere, with summer atmospheric humidity levels typically ranging from 20% to 50% RH. Winter humidity levels in the region are extremely low (<10% RH). This, together with the low precipitation rates, leads to extreme desiccation in the upper layers of the soil profile. Average moisture contents ranging from 0.2% to 3.9% wt water during the Antarctic summer have been recorded in Taylor Valley surface soils (Connell et al., 2006).

"Permafrost" refers to the lower part of the soil profile that constantly remains below 0°C. The moisture regime of Dry Valley soils is determined by the proximity of ice-cemented permafrost to the soil surface, where soils with the ice-cemented layer closer to the surface are typically moister due to melting of the ice in summer and the resultant upward movement of moisture across a strong desiccation gradient (Ugolini, 1970). The recession of the ice-cemented layer occurs due to loss of moisture to the atmosphere and this recession results in the ice-cemented surface ultimately resting below the permafrost table, part of which is dry permafrost. The ice-cemented permafrost layer in the Dry Valleys is generally found from a few centimeters to a meter below the active surface layer. However, this is variable at locations within a region. For example, in the Beacon Valley, ground ice is present less than 50 cm below the surface in some locations, but absent down to a few meters in other locations within the Dry Valley (McKay, 2009). The factors affecting the distribution of dry permafrost and ground ice in Antarctica are poorly understood, although some evidence indicates that the recurrence of snow may be linked to the location of ice-cemented ground and the depth of dry permafrost above it (McKay, 2009). The upward capillary transport of melt water from the seasonally thawed ice-cemented permafrost causes the soil layer directly above to be saturated. A clear boundary with the desiccated surface horizon is often observed, in which the steep humidity gradient is evident (Figure 1.2) (Ugolini, 1970; Campbell et al., 1997). However, while it has been suggested that there is insufficient liquid water from the melted permafrost layer to support the growth of microorganisms in the upper soil horizon (Wynn-Williams, 1988), the upward transport of water vapor potentially provides a high concentration of available water, albeit in gaseous form or as a basis for condensation on soil particles.

Biological water availability in desiccated Antarctic soils is possibly further reduced due to high soil salinity levels (Claridge and Campbell, 1977; Vishniac, 1993), a condition generally considered unfavorable for microbial growth in soils (Cameron, 1971). Although unevenly distributed on a microscale, high soil salinity is evident in most Ross Desert soils, with salinity levels decreasing with altitude (Vishniac, 1993). The accumulation of salts in surface soils originates from the substratum (by transport in capillary water from melted permafrost) and from the aeolian transport of sea sprays (Wada et al., 1981; Wynn-Williams, 1990). Halophilic bacteria capable of growth in media containing 15% (Horowitz et al., 1972) and 20% NaCl (James et al., 1990) have been isolated from Ross Desert soils.

The organic content of Dry Valley soils is very low, generally constituting <0.1% of dry soil weight (Campbell et al., 1998) with an average of $0.064 \pm 0.035\%$ total organic carbon (Matsumoto et al., 1983). The organic content is presumed to derive from past primary production, from the deposit of organic material by glacial movement in the region, from the physical abrasion of rocks (and consequential exfoliation of endolithic communities), from erosion of the exposed sediments of ancient lake beds, and from marine and lake-derived matter (Nienow and Friedmann, 1993; Moorhead et al., 1999; Burkins et al., 2000). More recent studies show that

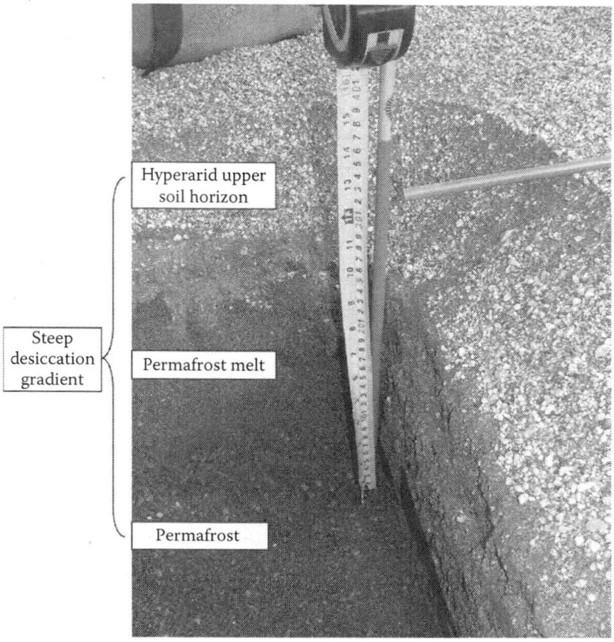

FIGURE 1.2 Depth profile of typical Dry Valley mineral soils indicating a supersaturated permafrost layer, permafrost melt layer, and hyperarid surface soils (Miers Valley, Eastern Antarctica).

some contribution to the organic pools is due to primary production by existing soil microorganisms and, to a greater extent, lithic (rock) communities (Moorhead et al., 1999). The distribution of organic matter is thought to be aided by aeolian factors (Moorhead et al., 1999). Organic carbon concentrations of soils in the Taylor Valley increase with proximity to the Ross Sea coastline, with the 0.01 g C g^{-1} soil average for inland sites increasing to 0.05 g C g^{-1} soil for coastal sites (Burkins et al., 2000, 2001). This observation might suggest a direct marine supplementation or some indirect effect resulting from the more "benevolent" climate in coastal regions.

The mean residence time of organic carbon in the Dry Valleys has been calculated at 23 years (Burkins et al., 2001), which is an extremely low value considering that most of the organic content of these soils is thought to be a legacy of lake sediments deposited during the last glacial maximum (Barrett et al., 2005). This low value may indicate that the simple food webs in the Dry Valley ecosystem actively cycle recently fixed carbon in addition to legacy sources of organic matter (Barrett et al., 2005).

CO_2 efflux from soil is the result of either biological respiration or georespiration (the inorganic equilibration of soil carbonate in the presence of water vapor) or a combination of both factors (Oyama and Berdahl, 1977; Burkins et al., 2001). Due to the low biological activity in Antarctic desert soils, the physical controls on CO_2 flux from soil are of similar magnitude to biological controls (Parsons et al., 2004).

While higher biological activities in other regions probably mask the relatively small contribution of georespiration to CO_2 flux, nonbiological inputs may become significant and should be taken into account in low productivity environments (Parsons et al., 2004). Eliminating the effects of georespiration, the CO_2 efflux of Taylor Valley soils was calculated to be $0.1 \pm 0.08\,\mu$mol CO_2 m^{-2} s^{-1} (Burkins et al., 2001). This is very low in comparison with the CO_2 efflux of hot deserts and the Arctic tundra (Raich and Schlesinger, 1992). Assuming maximum productivity for 60 days per year, the Dry Valley soils could release 6.5 ± 5 g C m^{-2} $year^{-1}$, one of the lowest soil carbon respiration figures reported for terrestrial ecosystems (Burkins et al., 2001).

Antarctica experiences high periodic incident UV irradiation, with long periods of complete darkness (Wynn-Williams, 1990; Cowan and Ah Tow, 2004; Onofri et al., 2004). The effects of stratospheric ozone depletion and increased UV-B irradiation are pronounced on the continent, where over 50% of the ozone column can be depleted during the spring. This, in turn, leads to increased UV-B levels (Madronich et al., 1998), which increases the environmental stress inflicted on living organisms in the region.

Due to the extreme environmental conditions, no vascular plants grow in the Antarctic Dry Valleys. With the absence of plant roots, there is little vertical transport of carbon. The dominant food web is therefore limited to the near surface environment (Virginia and Wall, 1999) where nutrient cycles and trophic interactions are limited to microbial populations and microinvertebrates (Vishniac, 1996). Community structures and dynamics are therefore thought to be principally controlled by two factors: resource availability and environmental conditions (Moorhead et al., 1999). That is, there are strong "bottom up" controls on ecosystem structures (Moorhead et al., 1999) and these are largely abiotic rather than biotic (Connell et al., 2006).

It is widely believed that the community structures in arid Antarctic soils are relatively simple compared to the complexity of soil communities in temperate regions (Wall and Virginia, 1999; Connell et al., 2006). While this is certainly true with respect to trophic complexity and prokaryote–eukaryote interactions, the recent demonstration of high prokaryotic phylogenetic diversity (Aislabie et al., 2006; Smith et al., 2006) and the existence of complex refuge habitats (hypolithons) suggest that assumptions of extreme community simplicity should be treated with caution. Nevertheless, these "simple" communities represent excellent systems in which to identify and define interactions between soil microorganisms and the environment.

The environmental conditions experienced in the Dry Valleys, especially the extreme temperature fluctuations and extreme aridity, mimic those postulated for early Mars. This region of the Antarctic continent has long served as a model for exobiological studies (e.g., Andersen et al., 1990; McKay, 1993).

1.3 ANTARCTIC SOIL MICROBIOLOGY AND MICROBIAL DIVERSITY

Microbial diversity has been analyzed using modern molecular phylogenetic methods for a range of Antarctic habitats (Table 1.1), including open desert soils of the Dry Valleys, desiccated soils at higher altitudes, soil that receives external carbon and nitrogen input (i.e., from beneath seal carcasses) and sublithic communities (Figure 1.3).

TABLE 1.1
List of the Sources of Soils and Their Nature

Location(s)	Soil Characteristics	References
Fossil Bluff, Mars Oasis, Coal Nunatak, Sentinel Range (Ellsworth Mountains)	Bare frost-sorted soils	Yergeau et al. (2007c)
Luther Vale L 1–3, Admiralty Range, Northern Victoria Land	A low productivity, low moisture soil in a dry cirque	Niederberger et al. (2008)
Bratina Island; Penance Pass, Miers Valley Ross Dependency, Eastern Antarctica	Desiccated mineral soils and soils from under seal carcasses	Smith et al. (2006)
Marble Point and Wright Valley, Victoria Land	High and low moisture soils	Aislabie et al. (2006)
Vestfold Hills	Quartz stone sublithic communities	Smith et al. (2000)

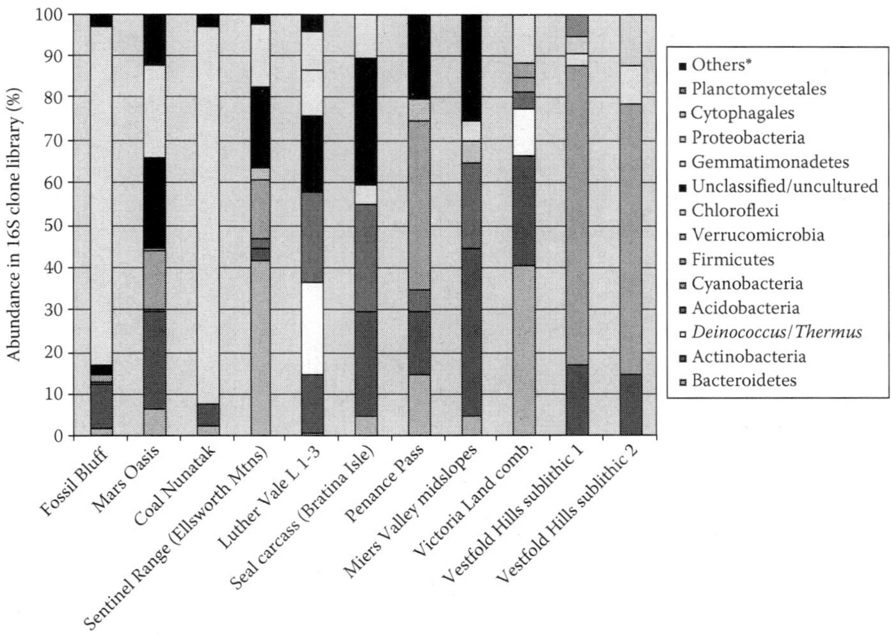

FIGURE 1.3 (See color insert following page 276.) A snapshot of prokaryotic diversity across a range of Antarctic arid soil habitats. *NB: For Mars Oasis, Fossil Bluff, Coal Nunatak, and Sentinel Range, "Others" category includes Chloroflexi, Firmicutes, Fusobacteria, Gemmatimonadetes, Planctomycetes, and the candidate phyla OP10 and TM7. For Luther Vale, "Others" category includes Chloroflexi, Firmicutes, Bacteroidetes, Verrucomicrobia, Nitrospira, and Cyanobacteria.

In general terms, prokaryote community structure and diversity is thought to be largely determined by location and/or associated location-dependent environmental conditions (Yergeau et al., 2007a). On a continental scale, comparisons of the microbial ecology of the Antarctic Peninsula and Dry Valley regions clearly show a decrease in community complexity with increasing latitude (Barrett et al., 2006a; Yergeau et al., 2007b), reflecting the increase in climatic severity. It is thus assumed that the Antarctic continent should contain a high level of site isolation and endemism (Lawley et al., 2004).

Until recently, studies focusing on the distribution, abundance, and diversity of microorganisms in Antarctic Dry Valley habitats were culture-dependent. Estimates of the number of culturable bacteria in Dry Valley soils, based on such studies, have ranged from 0 to 10^7 colony forming units (CFU) g^{-1} soil (Boyd et al., 1966; Cameron and Devaney, 1970). Culture-based analyses of microbial diversity in Dry Valley soils yielded numerous cosmopolitan soil organisms (typically aerobic heterotrophs belonging to genera such as, *Achromobacter*, *Arthrobacter*, *Bacillus*, *Corynebacterium*, *Flavobacterium*, *Micrococcus*, *Planococcus*, *Pseudomonas*, *Streptomyces*, and *Nocardia* (Cameron, 1972). Most were identified (using chemical taxonomic methods) as known species, although new, possibly endemic, species are regularly isolated (Cowan and Ah Tow, 2004).

Some local differences in microbial distribution have been noted; for example, chromogenic (pigmented) bacteria are commonly isolated from soil surface layers whereas nonpigmented bacteria dominate below the soil surface (Cameron and Devaney, 1970). This might be expected as the presence of photoprotective pigments in chromogenic bacteria affords some protection from the high periodic incident radiation in the region (Wynn-Williams et al., 2002). The observation that hyperarid surface soils contained fewer culturable bacteria than the moist subsurface soils above the permafrost (Aislabie et al., 2006) is at odds with the higher ATP titers determined in surface soils (Cowan et al., 2002).

Although culture-based techniques are still widely used and remain vital tools of microbial ecology, it is now widely accepted that such methods cannot provide a realistic estimate of microbial diversity (e.g., Hill et al., 2000). It is therefore reasonable to conclude that all early microbiological studies in the Antarctic soils represent gross underestimates of both microbial loads and diversities. Indeed, recent studies using ATP analysis have estimated viable microbial biomass in the Dry Valley soils to be up to four orders of magnitude higher than previously reported (Cowan et al., 2002).

With the advent of molecular phylogenetic methods, it has become possible to more accurately assess both microbial diversity and population numbers. A number of detailed microbial phylogenetic surveys of Antarctic soils have been published in recent years (Shivaji et al., 2004; Aislabie et al., 2006; Smith et al., 2006; Shravage et al., 2007; Babalola et al., 2008; Wood et al., 2008) (also see Chapter 3). The most striking and consistent observation from these studies is that a substantial proportion of 16S rRNA sequences do not show high homology to the sequences of any cultured organisms, suggesting that Antarctic soils harbor a large number of novel, possibly endemic, genospecies. For example, 50% of sequences in three Miers Valley 16S rRNA clone libraries corresponded to uncultured bacteria (Smith et al.,

2006). The Miers Valley libraries showed a diverse range of prokaryote phylotypes that were assigned to eight broad phylotypic groups: Cyanobacteria, Actinobacteria, Acidobacteria, Verrucomicrobia, α-Proteobacteria, β-Proteobacteria, Chloroflexi, and Bacteroidetes. Indeed, the dominant members of major phylotypic groups, such as the Acidobacteria and Verrucomicrobia, have never been cultured from Antarctic soils (Aislabie et al., 2006). Interestingly, no anaerobic bacterial 16S rRNA gene signals have been identified in Dry Valley mineral soil metagenomic DNA extracts (Smith et al., 2006). While this observation is consistent with the belief that anaerobic niches are not found in surface mineral soils (Tiedje et al., 1984), it represents something of an anomaly with respect to the widely held view that lacustrine biomass (in which anaerobic microorganisms are commonly found) is a major source of carbon in mineral soils (Barrett et al., 2006a,b). Archaeal 16S rRNA gene signals have only recently been detected in soils from the Miers Valley (Khan, 2008) (see the following text; also see Chapter 2).

A comparison of culturable and culture-independent microbial diversity highlights the limitations of both methods. In a comparative assessment of bacterial diversity in soils from the Dry Valley sites in the Wright Valley and at Marble Point (Aislabie et al., 2006), 16S rRNA clone libraries identified 27 dominant phylotypes, with 11 of the 27 phylotypes belonging to the group Bacteroidetes, 8 belonging to the Actinobacteria, and the rest distributed among the Proteobacteria, *Deinococcus/Thermus*, Acidobacteria, Firmicutes, and Cyanobacteria. However, cultured isolates represented only 8 of the 27 dominant phylotypes identified by molecular methods (Aislabie et al., 2006). Some phylotypes, such as those belonging to the Proteobacteria, were present in all sites sampled, while phylotypes belonging to the Actinobacteria and the *Deinococcus/Thermus* group were found only in specific sites (Aislabie et al., 2006). These, and other phyla represented in pristine Antarctic soil 16S rRNA gene libraries (such as, Planctomycetales, and Gemmatimonadetes) represent the common phyla found in many soil communities (Buckley and Schmidt, 2001; Zhou et al., 2003, 2004; Janssen, 2006).

1.4 IMPORTANT MICROBIAL TAXA IN ANTARCTIC SOIL SYSTEMS

1.4.1 *DEINOCOCCUS/THERMUS*

Members of the *Deinococcus/Thermus* group are found in a number of Antarctic desert soil environments (de la Torre et al., 2003; Saul et al., 2005; Aislabie et al., 2006; Shravage et al. 2007; Niederberger et al., 2008), including sublithic communities (Smith et al., 2000). In a comparison of high- and low-productivity soils in Luther Vale, Victoria Land, members of the phylum were found exclusively in the drier, low-productivity environments (Niederberger et al., 2008). Members of this phylum are typically tolerant to both radiation and desiccation, with the genes involved in resistance to radiation also contributing to resistance to desiccation (Mattimore and Battista, 1996). These properties are likely to contribute to the ability of these organisms to survive in open Antarctic soils where they will be exposed to high levels of desiccation and high periodic fluxes of UV irradiation. *Deinococcus/Thermus* phylotypes have also have been identified in hot arid environments (Chanal et al., 2006).

1.4.2 Proteobacteria

Proteobacteria are common in virtually all Antarctic terrestrial habitats. Members of Proteobacteria are represented in clone libraries from Victoria Land Dry Valley environments (Aislabie et al., 2006; Khan, 2008), from habitats across the Antarctic Peninsula (Yergeau et al., 2007c), in sublithic habitats in the Miers Valley and Vestfold Hills (Smith et al., 2000; Khan, 2008), and beneath seal carcasses (Smith et al., 2006). Proteobacterial phylotypes were found to be more dominant in high than in low productivity soils and this nonhomogeneous distribution has led to the suggestion that the phylum might be more strongly selected for in soils that are carbon and nitrogen enriched (Niederberger et al., 2008). These include seal carcass-associated communities, which are seeded with C and N from degrading tissues, sublithic communities and those at higher altitudes, both of which receive C and N supplementation through the autotrophic activities of cyanobacteria. However, the distribution of proteobacterial phylotypes in the region based on organic nutrient availability is not that clear-cut as they have also been found to be prevalent in the low productivity soils in the Wright Valley (Aislabie et al., 2006).

Members of Proteobacteria cultured from hot arid soil (in particular the α-Proteobacteria) show a high degree of resistance to solar radiation (Chanal et al., 2006). Isolates of one of the more unusual genera of the Proteobacteria, *Ramlibacter*, which are members of the dwarf bacteria (Rutz and Kieft, 2004), have been obtained from hot desert sands. It has been suggested that dwarf bacteria have a particular ability to colonize arid environments (Chanal et al., 2006). Although *Ramlibacter* spp. have not yet been detected in Antarctic soils, a closer examination of the uncultured and unclassified bacteria in Antarctic soils may reveal species that possess similar morphological or physiological features, conferring the ability to survive under desiccating conditions.

1.4.3 Actinobacteria

Members of Actinobacteria contribute to a significant proportion of all the libraries from the dry habitats, and a high frequency of novel phylotypic signals has been observed in libraries from Miers Valley soils (Smith et al., 2006). The radiation-resistant *Rubrobacter* sp. of this phylum has been found in hot deserts of the Australian (Holmes et al., 2000) and African continents (Chanal et al., 2006) and is observed in all Antarctic low-productivity, open soil communities (Smith et al., 2000, 2006; Saul et al., 2005; Aislabie et al., 2006; Shravage et al., 2007). Actinobacteria were the second most common phylotype to cyanobacteria in the Vestfold Hills crypotoendolithic clone libraries (Smith et al., 2000). A recent phylogenetic survey of Actinomycete diversity in Dry Valley soils (Babalola et al., 2008) clearly demonstrated that a very large pool of previously undescribed species (and probably even previously undescribed genera) exist in such habitats. The study also demonstrated that standard isolation protocols largely fail to penetrate this uncultured diversity.

1.4.4 Bacteroidetes

Bacteroidetes are widespread in Antarctic desert soil habitats. They are typically decomposers and probably contribute to the turnover of polymeric carbon (Aislabie et al., 2006). The order Sphingobacteriales, and specifically *Chitinophaga*, dominated Ellsworth Mountain clone libraries (Yergeau et al., 2007c). These organisms are able to degrade chitin (Buckley and Schmidt, 2001), and access to insoluble polymeric substrates (e.g., fungal cell wall material) may well confer an advantage in otherwise low-nutrient environments.

1.4.5 Acidobacteria

Acidobacterial phylotypes are widely represented in Antarctic soil libraries, and have been prominent members of clone libraries from some drier, low productivity Antarctic soils (Smith et al., 2006; Niederburger et al., 2008). Acidobacteria are common, and sometimes dominant members of soil, marine, and other microbial communities (Kuske et al., 1997; Dunbar et al., 1999; Chanal et al., 2006). Although their exact physiological role within the community structures is currently unknown, they have been implicated in biogeochemical cycling (Lee et al., 2008).

1.4.6 Cyanobacteria

In terrestrial environments where environmental extremes preclude the survival of the higher eukaryotic phototrophs, the cyanobacteria typically play a pivotal role in the C/N acquisition and the development of biological communities (de la Torre et al., 2003). It would therefore be reasonable to predict that cyanobacteria will be major components of all Antarctic soil phylotypic communities.

Classical surveys of cyanobacterial distribution have demonstrated that a wide variety of cyanobacterial species are found in moist habitats (lakes, meltwater ponds, and streams) where macroscopic growth may be observed (Taton et al., 2003, 2006; de los Rios et al., 2004; Jungblut et al., 2005) and protected (lithic) habitats in Antarctica (de la Torre et al., 2003; Omelon et al., 2007). The genera *Chroococcidiopsis*, *Nostoc*, and *Gloeocapsa*, and members of the *Leptolyngbya-Phormidium-Plectonema* group, are widespread (Nienow and Friedmann, 1993) although an understanding of distribution patterns and relative abundance is still lacking (Adams et al., 2006).

A similar trend has been found in recent phylogenetic studies of soil habitats (Smith et al., 2006; Niederberger et al., 2008). However, cyanobacterial distribution is clearly nonhomogeneous and is probably linked with water availability. Cyanobacterial signals were more prominent in clone libraries at higher, rather than lower altitudes, in soils of higher rather than lower productivity (Niederberger et al., 2008) and associated with lithic habitats rather than open desiccated soils (Smith et al., 2006). This approach has also shown a much higher cyanobacterial diversity (including possible endemic species) than previously known (Taton et al., 2003; Wood et al., 2008). Fewer studies have focused on the comparative distribution of cyanobacteria in Antarctic soils (Cavacini, 2001; Wood et al., 2008).

Miers Valley soils contain cyanobacteria belonging to the orders Nostocales, Chroococcales, and Oscillatoriales as well as a number of as yet uncharacterized cyanobacteria and 16S rRNA sequences with low similarity to any database entries (Wood et al., 2008). It has been suggested that cyanobacterial populations in the soils of the Miers Valley are free-living species, rather than deriving from cryptoendolithic populations, as suitable rock structures to support cryptoendolithic growth are absent in the sampling region. Moreover, the cyanobacterial phylotypes (*Nostoc*, *Phormidium*, and *Oscillatoria*) differed from those typically associated with cryptoendoliths (Smith et al., 2006).

Some apparent anomalies in cyanobacterial distribution exist. These organisms did not constitute a significant proportion of 16S rRNA clone libraries from low nutrient sites at Coal Nunatak and Fossil Bluff (Yergeau et al., 2007c), although higher proportions of cyanobacterial sequences were found at Mars Oasis and Ellsworth Mountain sites.

It has been suggested that cyanobacterial biomass in Dry Valley mineral soils may be largely derived from the aeolian distribution of lake and pond cyanobacterial mat material (Broady, 1996; Moorhead et al., 2003; Aislabie et al., 2006) and that this process is an important carbon input (Adams et al., 2006) with implications for ecosystem functioning. Wood et al. (2008) combined the use of ARISA (Automated Ribosomal Intergenic Spacer Analysis) and 16S rRNA clone library techniques to demonstrate that soils in the Miers Valley contained multiple cyanobacterial phylotypes, a significant proportion of which were also found in the nearby Lake Miers and associated hydroterrestrial cyanobacterial mats. While this apparently confirms that lakes and ponds contribute significantly to the cyanobacterial diversity of Dry Valley soils, a quantitative understanding on this process remains to be established.

1.4.7 Fungi

Yeast, yeast-like organisms, and filamentous fungi are widely dispersed in Antarctic ecosystems, but only in low numbers (Vishniac, 1996) (also see Chapter 12). These lower eukaryotes have been found in cryptoendolithic (Vishniac, 1996), marine (Dearborn and Fell, 1974), air, and soil samples from various locations on and around the Antarctic continent (Sun et al., 1978). Endemic fungi have been found growing on historic hut timbers (Arenz et al., 2006) and a diversity of cosmopolitan fungi have been isolated from Dry Valley lakes and areas affected by human activity. Although fungi are known to be important in soil ecosystems, the distribution, importance, and function of fungi in the Dry Valley and other Antarctic soil systems are largely unknown.

Connell et al. (2006) investigated the distribution and abundance of fungi in soils of different moisture contents (averaging from 0.2% to 3.9% wt) in the Taylor Valley. The distribution and abundance of filamentous and nonfilamentous (yeast and yeast-like) fungi were apparently affected by different parameters. For example, while the abundance of filamentous fungi showed a significant positive correlation with increased distance from marine and glacial water sources, with soil carbon content, with moisture content, and with soil pH (pH 7.2–10.5), no such correlation was established for nonfilamentous fungi. In contrast, yeast and yeast-like fungi were more evenly distributed throughout the valley and showed significant correlations only with low soil salinity and lower elevation. The study also found the distribution of nematodes to be similar to that of filamentous fungi. Both cosmopolitan and endemic fungal species were identified, although the true diversity of fungi in this region is still unknown due to the dependence of this (and other) studies on culture-dependent methods. The presence of both indigenous and cosmopolitan filamentous fungi and yeasts has been confirmed by other studies (e.g., Lawley et al., 2004; Arenz et al., 2006).

1.5 TROPHIC STRUCTURE

The food web of the Antarctic Dry Valleys is much simpler than a contrasting hot desert (Chihuahuan desert, New Mexico) (Wall and Virginia, 1999). Given the absence of higher phototrophs and chemoautrophy, algae and cyanobacteria naturally form the basis of the food web in the Antarctic Dry Valley system (Wall and Virginia, 1999). We also suggest that mosses, which are found in specific lithic habitats (see later) also contribute to carbon input in desert systems. Together these organisms contribute to the total organic load, and support a fairly complex array of heterotrophic organisms, both prokaryotic (principally bacteria) and eukaryotic (fungi). The highest trophic level of the Dry Valley soil ecosystem is composed of three nematode species: a single microbivore (*Scottnema lindsaye*) that feeds on bacteria and yeast, one bacterivore (*Plectus antarcticus*) that feeds on bacteria, and a single omnivore–predator (*Eudorylaimus antarcticus*) that feeds on algae, bacteria, fungi, nematodes, and other small fauna, with no higher invertebrates detected (Wall and Virginia, 1999). This may constitute the simplest trophic structure on Earth, as no other soil systems are known where nematodes represent the pinnacle of the food chain (Virginia and Wall, 1999). With such a low level of complexity at the upper trophic level, a high degree of susceptibility to disturbance is likely (Virginia and Wall, 1999; Wall and Virginia, 1999). The lower trophic levels (cyanobacteria, bacteria, and fungi) are significantly more diverse (e.g., Smith et al., 2006; Wood et al., 2008), with greater functional diversity and therefore stability.

1.6 LITHOBIONTIC HABITATS

The severe climatic conditions of the Antarctic Dry Valleys have resulted in the apparent retreat of life to protected microenvironments offered by stones and rocks. Habitats associated with rock or stone substrates are referred to as lithic or lithobiontic and can be classified into three types depending on whether organisms grow on (epilithic), in (endolithic), or under the rock (hypolithic). The term euendolithic refers to microorganisms that actively bore into lithic substrates and thus create their habitats (Nienow and Friedmann, 1993). Although epilithic growth occurs in the climatically less extreme coastal regions of the continent, rock surfaces are mostly devoid of visible life in the more extreme continental interior. Environments where epilithic growth is inhibited by extreme climatic conditions are more ideally suited to the endolithic and hypolithic lifestyle (Friedmann and Ocampo-Friedmann, 1984).

Such lithic communities are prevalent in the Antarctic desert and their distribution is ubiquitous in the ice-free regions (Nienow and Friedmann, 1993).

The most obvious function that rocks serve is to form an effective shelter from scouring by strong winds. Additionally, the overlying rock acts as a buffer, protecting the microbial habitat from the other elements of the macroenvironment (Friedmann, 1982), such as intense UV irradiation. Although essential for photosynthetic processes, UV irradiation can have lethal effects on organisms, primarily via damage to biological macromolecules and other cellular components, particularly DNA, through direct absorption and UV-induced generation of excited oxygen states (Vincent and Quesada, 1994). While all microorganisms possess various levels of repair processes for UV irradiation damage, lithobiontic organisms are further protected by the overlying rock strata (Cockell et al., 2002). The rocks are also thought to trap moisture, providing a source of liquid water, which is scarce in the desiccating conditions of the Dry Valleys (Cockell and Stokes, 2004).

1.6.1 Establishment of Lithobiontic Communities

Lithic communities from a variety of substrates have been described, including flint (Berner and Evenari, 1978), limestone (Cockell and Stokes, 2004), gypsum (Hughes and Lawley, 2003), and sandstone (Friedmann and Ocampo, 1976). The microbial colonization of rock depends on a combination of factors. In addition to environmental factors (discussed later), physicochemical properties of rock, such as the ultrastructure, mineral compositions, and pH all contribute to the ability of rock to host microbial life (Friedmann and Ocampo-Friedmann, 1984). It has been noted that cyanobacteria-dominated habitats were typically associated with rock substrates having higher pH values and higher levels of calcium and magnesium carbonates (Johnston and Vestal, 1989). Conversely, communities dominated by fungi were found in habitats characterized by higher concentrations of metal oxides. A more recent study of Arctic lithobiontic communities showed similar pH relationships (Omelon et al., 2007).

1.6.1.1 Temperature

The temperature of rock and soil surfaces is closely linked to ambient air temperature and solar radiation (Nienow and Friedmann, 1993). During the winter period, when sunlight is largely or wholly absent, rock surface temperatures are on average 1°C–2°C lower than the ambient air temperature. In summer, however, the surface temperatures increase significantly above the ambient air temperature (Nienow and Friedmann, 1993). Furthermore, wide temperature fluctuations occur on the rock surfaces; for example, temperature oscillations over a range of 7.7°C within 42 min have been recorded, with the temperature shifting from positive to negative figures 14 times within the period tested (Friedmann, 1982). This may be a significant factor in the preferential colonization of the underside of rocks by microorganisms.

1.6.1.2 Water Availability

A major challenge faced by microorganisms in the Dry Valleys and other Antarctic desert areas is thought to be access to water, despite the fact that such communities are often positioned close (10–30 cm) to large water reserves. In the McMurdo Dry Valleys (76°20′–78°20′S; 160°–164°E), the saturated horizon from permafrost melt (see Figure 1.2) may extend to within 10–15 cm of the surface (and lithobiontic communities positioned thereon). Despite the proximity of this saturated layer, liquid water content in the surface profiles (as determined by gravimetric analysis) is typically very low. Values of 0.5%–2% wt are typically recorded, and are of the same order as found in hot deserts.

Lithobiontic communities occupy niches where water is available either as liquid water (i.e., from occasional snow melt), or as vapor. The rock strata may be capable of adsorbing water from the vapor phase using physical mechanisms that are independent of the biological processes of the lithobiontic community. Such processes may increase the amount of water available for microbial growth, particularly in summer months when solar warming of rocks and soil causes substantial permafrost melting (Palmer and Friedmann, 1990). The rock can therefore act as a store of biologically accessible water during cycles of freezing and thawing (Friedmann et al., 1987).

Photosynthesis in lithobiontic communities is thought to be linked to thermodynamic water activity and therefore to ambient humidity (Cowan et al., 1979). In this study, it was noted that photosynthetic CO_2 fixation processes in temperate lichens were highly desiccation sensitive, showing little activity below 86% RH. Active prokaryotic photosynthesis requires higher humidity levels than eukaryotic photosynthesis (Friedmann et al., 1988), which may influence the conditions under which cyanobacteria-dominated or lichen-dominated lithobiontic communities will thrive.

1.7 CRYPTOENDOLITHIC COMMUNITIES

1.7.1 Cryptoendolithic Habitats

Cryptoendolithic microorganisms colonize the interstitial spaces of crystalline or porous rocks. Sandstone is a common rock type of the Dry Valleys and is readily colonized by cryptoendolithic communities, making it a major cryptoendolithic habitat (Friedmann et al., 1988). Cryptoendolithic communities are classified either as lichen-dominated or as cyanobacteria-dominated, based on the dominant organism in the community (Friedmann, 1982).

1.7.2 Physical Characteristics of Cryptoendoliths

Lichen-dominated cryptoendolithic communities are the most prevalent in the Antarctic Dry Valleys (Friedmann, 1982; de la Torre et al., 2003). The communities are visible as distinct colored bands, formed by filamentous fungi and unicellular green algae (Friedmann, 1982). The bands, or "zones," are usually comprised of a 1 mm thick upper black zone, a 2–4 mm thick middle white zone, and a lower green zone of similar dimensions (Friedmann, 1982).

1.7.3 Microbiology of Cryptoendoliths

Studies of cryptoendolithic communities in the Dry Valleys have revealed that cyanobacteria and lichens are the primary producers of these systems (Friedmann et al., 1988). In lichenized cryptoendolithic communities, molecular methods showed a single green algal species, *Trebouxia jamesii*, to be dominant (de la Torre et al., 2003). *T. jamesii* is a known component of lichen associations, such as the lichenized fungal genus *Letharia* (Friedl and Rokitta, 1997; Kroken and Taylor, 2000). This observation is not consistent with microscopic analyses, which have indicated the presence of morphologically distinct photobionts. However, it is possible that some photobionts can exist as morphological variants (as is the case in some fungal associations), which would make identification based on morphology difficult. Alternatively, other photobionts may be present in the community but at levels that are below the detection limit of the molecular method used (de la Torre et al., 2003).

The analysis of the prokaryotic members of the community showed a diverse range of phylotypes related to Actinobacterial, α-Proteobacterial, γ-Proteobacterial, and Planctomycetales species, and an abundance of an unidentified bacterium, sharing only 95% identity to a Cytophagales species (de la Torre et al., 2003).

Cyanobacteria-dominated communities have been shown, by a combination of microscopic and molecular methods, to be comprised of several species related to the genus *Chroococcidiopsis* (Friedmann, 1982), together with cyanobacteria morphologically similar to *Gloeocapsa*, *Plectonema*, and *Hormathonema* species (de la Torre et al., 2003). Such communities also contain fungal phylotypes (Ascaso and Wierzchos, 2002). Abundant prokaryotes included a member of the *Deinococcus/Thermus* phylogenetic group and a representative of the α-Proteobacteria. Based on clone frequencies in the 16S rRNA gene library, these organisms were thought to be present in similar numbers to the principal cyanobacterium, suggesting that they may be involved in a syntrophic relationship (de la Torre et al., 2003). The clone libraries from the cyanobacteria-dominated community also showed a high degree of diversity, with a total of 16 Actinobacterial phylotypes identified. Other sequences were encountered only once, or were not similar to any known phylogenetic group.

1.8 HYPOLITHIC COMMUNITIES

1.8.1 Hypolithic Habitats

Hypolithic communities colonize the underside of rocks at the interface between rock and soil (Nienow and Friedmann, 1993). They have been found in both hot (Schlesinger et al., 2003; Warren-Rhodes et al., 2006) and cold (Smith et al., 2000; Cockell and Stokes, 2004; Wood et al., 2008) deserts. Both are subject to extreme temperature fluctuations and extreme aridity. Most hypolithic communities characterized to date have been found beneath quartz stones, a common translucent rock type in some (but not all) of the Antarctic Dry Valleys and other desert areas (Schlesinger et al., 2003) (Figures 1.4 and 1.5). Other translucent rock types (e.g., marble) support similar communities.

FIGURE 1.4 (See color insert following page 276.) Filamentous, fungal-dominated hypolithic community loosely attached to the underside of translucent quartz (Miers Valley, Eastern Antarctica).

FIGURE 1.5 (See color insert following page 276.) Unattached, moss-dominated community beneath a quartz rock (Miers Valley, Eastern Antarctica).

1.8.2 Physical Characteristics of Hypoliths

Hypolithic communities are primarily photosynthetic and the overlying rock must be sufficiently translucent for adequate light to penetrate (Cockell and Stokes, 2004). The lower depth limit for the growth of the community is probably dictated by light transmission, but hypolithic communities have been observed at sites where the irradiance levels are less than 0.1% of the incident light (Schlesinger et al., 2003; Thomas, 2005). It is interesting to note that in the Southern Mojave Desert (California), hypolithic communities were able to photosynthesize under stones of up to 25 mm thick that transmit only 0.08% of incident light (Schlesinger et al., 2003).

Microbial colonization beneath thicker rocks is confined to the periphery, as insufficient light to support photosynthesis is transmitted to the base of the rock (Broady, 1981; Schlesinger et al., 2003). Variation in light transmittance through the rock also leads to zonation. This can be observed in some hypoliths where green- or orange-pigmented green algae are found nearer the soil surface, while cyanobacteria are found deeper beneath the rocks (Figure 1.6) (Nienow and Friedmann, 1993). In some cases, thin (0.2–0.5 mm) green biofilms adhere very tightly to the stone surface (Smith et al., 2000). These biofilms are mainly comprised of cyanobacteria and unicellular green algae (Thomas, 2005). In other cases, growth may be either loosely attached or unattached to the underside of the rock (Figures 1.4 and 1.5), depending on the main community components. Fungal-dominated communities (Figure 1.4) appear loosely attached to the rock while moss-dominated communities (Figure 1.5) are visible on the soil beneath, but not adhering to, the rock. In both cases, the perimeter of the community is confined to that of the overlying rock.

Soils directly beneath hypolithic rocks have consistently been found to have higher water contents than the surface layers of the adjacent soil (Broady, 1981). The moisture contents of the basic (pH 7.4–7.7) soils beneath hypoliths sampled from areas of the Vestfold Hills were estimated to fall in the range of 6%–14%, whereas the adjacent soils contained 0.5%–2% soil moisture (Smith et al., 2000). These values may not be typical, however. The mean gravimetric water contents of hypolithic communities in the Miers Valley were 0.85% (±0.48%, $n = 9$), values not statistically different from open soils at the same depth (~2 cm; 0.72±0.40, $n = 9$). Water vapor, entrained upward from the melting ice-permafrost boundary, may be trapped by the overlying rock (Cowan and Ah Tow, 2004).

FIGURE 1.6 (See color insert following page 276.) Zonation visible in a typical cyanobacteria-dominated hypolithon with orange-pigmented organisms around the periphery and green cyanobacteria directly underneath the quartz rock (Miers Valley, Eastern Antarctica).

1.8.3 Microbiology of Hypoliths

Antarctic hypolithic communities are diverse and complex, and are dominated by cyanobacteria. ARISA profiles of hypolithic communities sampled in the Miers Valley suggest the presence of multiple cyanobacterial species within each community (Wood et al., 2008). The direct metagenomic analysis of 16S rRNA genes from two Antarctic hypolithic communities also showed cyanobacteria to be the dominant group (71%–76% of clones) (Smith et al., 2000). The unicellular (*Chroococcidiopsis*) and filamentous (*Plectonema*) cyanobacteria dominate hypolithic communities in the low organic-content mineral soils typical of the Dry Valleys (Nienow and Friedmann, 1993). Communities from the Vestfold Hills showed the presence of oscillatorian cyanobacteria with morphologies typical of the *Lyngbya/Phormidium/Plectonema* group and coccoidal cells morphologically similar to *Chroococcidiopsis* (Smith et al., 2000). Cyanobacterial-specific isolation studies yielded two dominant filamentous morphotypes, both of which grew between 2°C and 20°C but displayed no growth at 25°C (growth temperature behavior typical of "psychrophiles"). Both isolates belonged to the *Phormidium* subgroup according to 16S rRNA gene sequence analysis (Smith et al., 2000). Enrichment isolations yielded *Chroococcidiopsis*-like and *Synechococcus*-like cells and an isolate most closely related to *Lyngbya* species PCC7419 (Smith et al., 2000).

Desmococcus-like algae, along with *Navicula*-like diatoms, some mites, and nematodes (Smith et al., 2000) were also found in hypolithic communities.

Using culture-based methods, Smith et al. (2000) have shown that heterotrophic bacterial isolates from Vestfold Hills hypolithic communities fall into three groups; the Proteobacteria (α and γ subdivisions), the order Cytophagales and the class Actinobacteria, with many isolates representing novel taxa at the species level. While 20%–40% of the viable isolates from hypolithic community samples displayed psychrophilic tolerance, the remainder were mostly non-halophilic and psychrotolerant (Smith et al., 2000). Species such as *Pseudomonas*, *Psychrobacter*, *Stenophomonas*, *Arthrobacter*, *Achromobacter*, *Micrococcus*, *Rhodococcus*, *Janibacter*, and *Gelidibacter* occurred in all samples. Other species, however, were more erratically distributed and were not found in all hypolithic communities.

In the same study, clones generated using prokaryote-specific 16S rRNA gene primers clustered predominantly with the α-Proteobacteria, the γ-Proteobacteria, the order Cytophagales, the order Planctomycetales and the class Actinobacteria (which accounted for 9% of all clones and was thus the second most abundant phylum) (Smith et al., 2000). Interestingly, only a single eukaryote clone, clustering with the class Ulvophyceae and no archaeal clones, were detected using appropriate 16S rRNA gene primers.

1.9 CHASMOENDOLITHIC COMMUNITIES

1.9.1 Chasmoendolithic Habitats

Chasmoendoliths colonize cracks and fissures in lithic substrates and the habitats of these microorganisms are restricted to rock types that are prone to this form of weathering. Fissures that run perpendicular to the rock surface, as well as flakes that run parallel to the rock surface, form ideal habitats for chasmoendoliths. Granite rocks are particularly prone to flaking and chasmoendoliths have been found as predominant colonizers of this rock type (de los Rios et al., 2005). Chasmoendoliths also colonize porous rock made up of mineral grains such as sandstone in the fissures that form along the grain boundaries (Hoppert et al., 2004). In the Dry Valleys, both chasmoendoliths and cryptoendoliths colonize sandstones, although this rock type better supports the cryptoendolithic microbial niche due to the abundance of internal cavities in the rock ultrastructure (Golubic et al., 1981).

1.9.2 Physical Characteristics of Chasmoendoliths

Chasmoendolithic communities often show zonation with lichens and green algae found nearer the surface and cyanobacteria found at the lowest vegetated level where the intensity of transmitted light is lowest (Longton, 1988). Chasmoendolithic biofilms have also been observed, adhered to the underside of cleaved granite rock fragments (de los Rios et al., 2007).

1.9.3 Microbiology of Chasmoendoliths

Little is known about Antarctic chasmoendolithic microbiology, compared with other lithobiontic groups. However, the characterization of structurally different chasmoendolithic granite biofilms by de los Rios and coworkers showed that each of the biofilms was dominated by a single cyanobacterium (de los Rios et al., 2007). The biofilm dominated by a cyanobacterium closely related to *Acaryochloris marina* formed a loose attachment to the rock. The cells were small and were surrounded by a multilayered envelope, a feature of cyanobacteria associated with adaptation to extreme environments (de los Rios et al., 2004, 2007). The tightly attached biofilm was dominated by a cyanobacterium phenotypically similar to *Gloeocapsa* but not as closely related phylogenetically. Unpigmented heterotrophic bacteria were found in association with the biofilms (de los Rios et al., 2007).

Chasmoendolithic mosses are often associated with granites in the continental interior, although little is known about their microbiology (Riffenburgh, 2007).

1.10 SOURCES OF LITHOBIONTIC BIODIVERSITY

The apparent simplicity of lithobiontic communities makes them intriguing targets for studies of community origins and evolutions, trophic functionings, and interspecies relationships. The origin of the consortia members in such communities has been a source of some speculation. Both metagenomic and culture-dependent data from hypolithic samples indicate that many identified organisms cluster with marine taxa (i.e., previously only detected in marine environments) (Smith et al., 2000). Conversely, a relatively high proportion of psychrophilic and halophilic bacterial phylotypes typical of marine environments are represented in lithobiontic libraries. It has therefore been proposed that the colonization of lithobiontic habitats is dependent on the aeolian transport of organisms from marine to terrestrial environments, where they colonize the underside suitable "refuge" habitats (Smith et al., 2000). However, recent phylogenetic studies of open mineral soil and hypolithic bacterial communities do not necessarily support this contention. Firstly, bacterial phylogenetic data from open soils show numerous phylotypes that are more typical of temperate terrestrial soils than marine systems (Smith et al., 2006). Secondly, the multivariate analysis of Miers Valley hypolithic communities confirmed that each hypolith harbored a unique community (Khan, 2008), and the analysis of similarity (ANOSIM) data showed that hypolithic cyanobacterial diversity was not significantly different from that of lake or hydroterrestrial cyanobacterial mats (Wood et al., 2008). This result suggests that fresh-water (lake) systems, including the associated macroscopic mats, may contribute significantly to the terrestrial cyanobacterial populations and, by extension, to other prokaryotic groups in refuge habitats.

The soils underlying the hypolithic communities have been shown to contain most of the heterotrophic bacteria, but lack Cyanobacteria, α-Proteobacteria, and Flavobacteria (Smith et al., 2000), groups, which appear to be exclusively restricted to the sublithic growth that is closely adhered to the rock.

1.11 NUTRIENT CYCLING AND PRODUCTIVITY OF ANTARCTIC DRY VALLEYS

Endolithic communities are dependent upon photosynthesis. The effective colonization and development of such a community therefore requires some level of photosynthetic activity, which in turn is dependent on the translucence of the overlying rock substrate (Nienow and Friedmann, 1993). Incident light flux is also a critical factor. Activity estimates based on microclimate (light) data suggest that endoliths are able to metabolize for less than 1000 h $year^{-1}$ (Friedmann et al., 1987).

As with other low productivity environments, the relative contribution of hypolithic communities to the carbon and nitrogen turnover rates may be significant, despite their low total biomass. Hypolithic growth in the Vestfold Hills has been shown to cover 66 cm^2 m^{-2} and contribute a total biomass of 0.85 mg chlorophyll m^{-2} (Broady, 1981). This represents a significant addition to the overall ecosystem. Similarly, the productivity (in terms of carbon assimilation) of hypolithic communities in the Arctic was estimated at approximately 0.8 ± 0.3 g m^{-2} $year^{-1}$ (Cockell and Stokes, 2004). As the combined mean productivity from plants, lichens, and bryophytes in the region is estimated to be 1.0 ± 0.4 g m^{-2} $year^{-1}$ (Bliss et al., 1984), then hypolithic community productivity almost doubles the productivity estimates for the polar desert environment (Cockell and Stokes, 2004). Although these rates of productivity are extremely low when compared to those of temperate grasslands and prairies (~1 kg m^{-2} $year^{-1}$) (Thomas, 2005), hypolithic community production is a vital source of nutrition for grazing nematodes and protozoans. Hypolithic phototrophs can, therefore, be viewed as the basis for the survival of an entire ecosystem in the extremely cold, dry polar environments (Freckman and Virginia, 1997).

1.12 CONCLUSIONS

The application of modern molecular phylogenetic techniques to Antarctic desert soils has thrown up some surprises. Traditionally labeled as "one of the most extreme habitats on planet Earth," where some locations were thought to be essentially sterile, Antarctic soils have been shown to contain a large diversity of microbial phylotypes. Microbial taxa that typically dominated culture-dependent analyses are often minor components or even completely absent from small to medium clone libraries. It is also clear that the Antarctic soil systems are anything but homogeneous. Depending on the local geology, they contain biological "hotspots" such as refuge lithobiontic communities taking advantage of the physical protection provided by rock strata. There are indications that such localized communities contribute significantly to the total productivity of the system.

There is much we do not yet know about Antarctic desert microbiology. Our knowledge has progressed little beyond an analysis of microbial diversity, and even this knowledge has limits. For example, the phylogenetic analyses reported to date are relatively limited, where the analysis of a few hundred 16S rRNA gene sequences provides little more than a survey of the most dominant organisms. An in-depth analysis would require one to two orders of magnitude greater penetration, but the associated labor and sequencing costs are currently beyond the scope of most laboratories.

Less is known about the diversity of eukaryotes than of prokaryotes in Antarctic desert soils. Information on the diversity of lower eukaryotes (filamentous and non-filamentous fungi) is largely derived from culture-dependent studies, and no comprehensive phylotypic surveys using eukaryotic phylotypic markers have yet been published. Similarly, the diversity and role of archaea in Antarctic soils is virtually unknown. Some archaeal phylotypes assigned to the low temperature Crenarchaeotes have been observed in clone libraries (Khan, 2008), but nothing is yet known about the physiological properties or functions of these organisms. Chapter 2 describes the Archaeal diversity in the Antarctic continent.

Our current understanding of viral (phage) diversity in Antarctic soils is restricted to a few isolation studies. Given the expectation that phage diversity will be at least as large as bacterial diversity, this remains a potentially exciting and greatly under-represented field of study. Chapter 7 elaborates the role of bacteriophages in Antarctica for horizontal gene transfer.

The wide availability of high-volume, high-throughput sequencing capacity may well change this position radically in the very near future, and an Antarctic soil metagenome sequence will reveal much more than merely the diversity of microorganisms. A large (e.g., multi-Gbp) metagenome sequencing program would provide a wealth of information on the phylotypic diversity of all three kingdoms, and could potentially provide details on the putative physiological functions of many of the phylotypes. Even the partial assembly of a prokaryotic genome may allow a linkage between putative function, based on the annotation of genes, operons and pathways, and phylotypic identity (putative taxonomy).

Some issues remain to be resolved. The extent and impact of relict DNA on phylogenetic analyses remains an unknown factor. It is generally assumed that in temperate moist environments the half-life of turnover of exogenous DNA (from dead lysed cells) is relatively short. The role of relict DNA is therefore assumed to be small, and microbial ecologists consistently assume that a phylotypic signal is indicative of a viable member of a microbial community. This assumption is not necessarily valid for Antarctic desert soils, where the microenvironmental conditions are exactly those exploited for DNA stabilization (i.e., low temperature and low water activity). While this issue has not as yet been addressed in any depth, preliminary studies have indicated that exogenous DNA may be stable over relatively long periods in this system (Ah Tow and Cowan, 2005) and a cautionary approach to the interpretation of metagenomic sequence data is therefore advised.

ACKNOWLEDGMENTS

The authors wish to thank the University of the Western Cape, Antarctica NZ, the University of Waikato, and the South African National Research Foundation for supporting field and laboratory research programs that have facilitated their understanding of Antarctic terrestrial microbiology.

REFERENCES

Adams, B. J., Bardgett, R. D., Ayres, E. et al. 2006. Diversity and distribution of Victoria Land biota. *Soil Biology and Biochemistry* 38:3003–3018.

Ah Tow, L. and Cowan, D. A. 2005. Dissemination and survival of non-indigenous bacterial genomes in pristine Antarctic environments. *Extremophiles* 9:385–389.

Aislabie, J. M., Chhour, K.-L., Saul, D. J., Miyauchi, S., Ayton, J., Paetzold, R. F., and Balks, M. R. 2006. Dominant bacteria in soils of Marble Point and Wright Valley, Victoria Land, Antarctica. *Soil Biology and Biochemistry* 38:3041–3056.

Andersen, D. T., McKay, C. P., Wharton, R. A. Jr., and Rummel, J. D. 1990. An Antarctic research outpost as a model for planetary exploration. *Journal of the British Interplanetary Society* 43:499–504.

Arenz, B. E., Held, B. W., Jurgens, J. A., Farrell, R. L., and Blanchette, R. A. 2006. Fungal diversity in soils and historic wood from the Ross Sea Region of Antarctica. *Soil Biology and Biochemistry* 38:3057–3064.

Ascaso, C. and Wierzchos, J. 2002. New approaches to the study of Antarctic lithobiontic microorganisms and their inorganic traces, and their application in the detection of life in Martian rocks. *International Microbiology* 5:215–222.

Babalola, O. O., Kirby, B. M., Le Roes-Hill, M., Cook, A. E., Cary, S. C., Burton, S. G., and Cowan, D. A. 2008. Ohylogenetic analysis of actinobacterial populations associated with Antarctic Dry Valley mineral soils. *Environmental Microbiology*, doi:10.1111/j.1462-2920.2008.01809.

Barrett, J. E., Virginia, R.A., Parsons, A. N., and Wall, D. H. 2005. Potential soil organic matter turnover in Taylor Valley, Antarctica. *Arctic, Antarctic, and Alpine Research* 37:108–117.

Barrett, J. E., Virginia, R. A., Hopkins, D. W., Aislabie, J., Bargagli, R., Bockheim, J. G., Campbell, I. B., Lyons, W. B., Moorhead, D., Seastedt, T., Sletten, R., Steltzer, H., Wall, D. H., and Wallenstein, M. 2006a. Terrestrial ecosystem processes of Victoria Land, Antarctica. *Soil Biology and Biochemistry* 38:3019–3034.

Barrett, J. E., Virginia, R. A., Parsons, A. N., and Wall, D. H. 2006b. A soil carbon turnover model for the McMurdo Dry Valleys, Antarctica. *Soil Biology and Biochemistry* 38:3065–3082.

Berner, T. and Evenari, M. 1978. The influence of temperature and light penetration on the abundance of the hypolithic algae in the Negev Desert of Israel. *Oecologia* 33:255–260.

Bliss, L. C., Svoboda, J., and Bliss, D. I. 1984. Polar deserts, their plant cover and plant production in the Canadian High Arctic. *Holarctic Ecology* 7:305–324.

Boyd, W. L., Staley, J. T., and Boyd, J. W. 1966. Ecology of soil microorganisms in Antarctica. In *Antarctic Soils and Soil Forming Processes*, vol. 8, ed. J. C. F. Tedrow, pp. 125–129. Washington, D.C.: American Geophysical Union.

Broady, P. A. 1981. The ecology of sublithic terrestrial algae at the Vestfold Hills, Antarctica. *British Phycological Journal* 16:231–240.

Broady, P. A. 1996. Diversity, distribution and dispersal of Antarctic terrestrial algae. *Biodiversity and Conservation* 5:1307–1336.

Buckley, D. H. and Schmidt, T. M. 2001. Exploring the biodiversity of soil—A microbial rain forest. In *Biodiversity of Microbial Life*, ed. J. Staley and A. Reysenbach, pp. 183–208. New York: John Wiley & Sons.

Burkins, M. B., Virginia, R. A., Chamberlain, C. P., and Wall, D. H. 2000. The origin of soil organic matter in Taylor Valley, Antarctica. *Ecology* 81:2377–2391.

Burkins, M. B., Virginia, R. A., and Wall, D. H. 2001. Organic carbon cycling in Taylor Valley, Antarctica: Quantifying soil reservoirs and soil respiration. *Global Change Biology* 7:113–125.

Cameron, R. E. 1971. Antarctic soil microbial and ecological investigations. Research in the Antarctic. In *Research in the Antarctic*, eds. L. O. Quam and H. D. Porter, pp. 137–189. Washington, D.C.: American Association for the Advancement of Science.

Cameron, R. E. 1972. Microbial and ecological investigations in Victoria Valley, Southern Victoria Land, Antarctica. *Antarctic Research Series* 20:195–260.

Cameron, R. E. 1974. Application of low latitude microbial ecology to high latitude deserts. In *Polar Deserts and Modern Man*, eds. T. L. Smiley and J. H. Zumberge, pp. 71–90. Tucson: University of Arizona Press.

Cameron, R. E. and Devaney, J. R. 1970. Antarctic soil algal crusts. A scanning electron and optical microscope study. *Transactions of the American Microscopical Society* 80:264–73.

Campbell, I. B., Claridge, G. G. C., Balks, M. R., and Campbell, D. I. 1997. Moisture content in soils of the McMurdo Sound and Dry Valley region of Antarctica. In *Ecosystem Processes in Antarctic Ice-free Landscapes*, eds. W. B. Lyons, C. Howard-Williams, and I. Hawes, pp. 61–76. Rotterdam, the Netherlands: A. A. Balkema.

Campbell, I. B., Claridge, G. G. C., Campbell, D. I., and Balks, M. R. 1998. The soil environment. In *Ecosystem Dynamics in a Polar Desert: The Mcmurdo Dry Valleys, Antarctica*, ed. J. Priscu, pp. 297–322. Washington, DC: American Geophysical Union.

Cavacini, P. 2001. Soil algae from northern Victoria Land (Antarctica). *Polar Bioscience* 14:45–60.

Chanal, A., Chapon, V., Benzerara, K., Barakat, M., Christen, R., Achouak, W., Barras, F., and Heulin, T. 2006. The desert of Tataouine: An extreme environment that hosts a wide diversity of microorganisms and radiotolerant bacteria. *Environmental Microbiology* 8:514–525.

Claridge, G. G. and Campbell, I. B. 1977. The salts in Antarctic soils, their distribution and relationship to soil processes. *Soil Science* 123:337–384.

Cockell, C. S. and Stokes, M. D. 2004. Widespread colonization by polar hypoliths. *Nature* 431:414.

Cockell, C. S., Rettberb, P., Horneck, G., Wynn-Williams, D. D., Scherer, K., and Gugg-Helminger, A. 2002. Influence of ice and snow covers on the UV exposure of terrestrial microbial communities: Dosimetric studies. *Journal of Photochemistry and Photobiology B: Biology* 68:23–32.

Connell, L., Redman, R., Craig, S., and Rodriguez, R. 2006. Distribution and abundance of fungi in the soils of Taylor Valley, Antarctica. *Soil Biology and Biochemistry* 38:3083–3094.

Cowan, D. A. and Ah Tow, L. 2004. Endangered Antarctic environments. *Annual Reviews in Microbiology* 58:649–690.

Cowan, D. A., Green, T. G. A., and Wilson, A. T. 1979. Lichen Metabolism. 1. The use of tritium-labelled water in studies of anhydrobiotic metabolism in *Ramalina celastri* and *Peltigera polydactyla. New Phytologist* 82:489–503.

Cowan, D. A., Russell, N. J., Mamais, A., and Sheppard, D. M. 2002. Antarctic Dry Valley mineral soils contain unexpectedly high levels of microbial biomass. *Extremophiles* 6:431–436.

Dearborn, J. H. and Fell, F. J. 1974. Ecological investigations of echinoderms from the Antarctic Peninsula. *Antarctic Journal US* 9:304–306.

de la Torre, J. R., Goebel, B. M., Friedmann, E. I., and Pace, N. R. 2003. Microbial diversity of cryptoendolithic communities from the McMurdo Dry Valleys, Antarctica. *Applied and Environmental Microbiology* 69:3858–3867.

de los Rios, A., Ascaso, C., Wierzchos, J., Fernandez-Valiente, E., and Quesada, A. 2004. Microstructural characterization of cyanobacterial mats from the McMurdo Ice Shelf, Antarctica. *Applied and Environmental Microbiology* 70:569–580.

de los Rios, A., Grube, M., Sancho, L. G., and Ascaso, C. 2007. Ultrastructural and genetic characteristics of endolithic cyanobacterial biofilms colonizing Antarctic granite rocks. *FEMS Microbial Ecology* 59:386–395.

de los Rios, A., Sancho, L. G., Grube, M., Wierzchos, J., and Ascaso, C. 2005. Endolithic growth of two *Lecidea* lichens in granite from continental Antarctica detected by molecular and microscopy techniques. *New Phytologist* 165:181–190.

Doran, P. T., McKay, C. P., Clow, G. D., Dana, G. L., Fountain, A. G., Nylen, T., and Lyons, W. B. 2002. Valley floor climate observations from the McMurdo dry valleys, Antarctica, 1986–2000. *Journal of Geophysical Research* 107(D24) 4772, doi:10.1029/2001JD002045.

Dunbar, J., Takala, S., Barns, S. M., Davies, J. A., and Kuske, C. R. 1999. Levels of bacterial community in four arid soils compared by cultivation and 16SrRNA gene cloning. *Applied and Environmental Microbiology* 65:1662–1669.

Ellis-Evans, J. C. 1997. Microscale distribution of photoautotrophic micro-organisms in relation to light, temperature and moisture in Antarctic lithosols. In *Ecosystem Processes in Antarctic Ice-free Landscapes*, eds. W. B. Lyons, C. Howard-Williams, and I. Hawes, pp. 89–95. Rotterdam, the Netherlands: A. A. Balkema.

Fox, A. J. and Cooper, P. R. 1994. Measured properties of the Antarctic ice sheet derived from the SCAR digital database. *Polar Record* 30:201–204.

Freckman, D. W. and Virginia, R. A. 1997. Low-diversity Antarctic soil nematode communities: Distribution and response to disturbance. *Ecology* 78:363–369.

Friedl, T. and Rokitta, C. 1997. Species relationships in the lichen alga *Trebouxia* (*Chlorophyta Trebouxiophyceae*): Molecular phylogenetic analyzes of nuclear-encoded large subunit rRNA gene sequences. *Symbiosis* 23:125–148.

Friedmann, E. I. 1982. Endolithic microorganisms in the Antarctic cold desert. *Science* 215:1045–1053.

Friedmann, E. I. and Ocampo, R. 1976. Endolithic blue-green algae in the Dry valleys: Primary producers in the Antarctic desert ecosystem. *Science* 193:1247–1249.

Friedmann, E. I. and Ocampo-Friedmann, R. 1984. Endolithic microorganisms in extreme dry environments: Analysis of a lithobiontic microbial habitat. In *Current Perspectives in Microbial Ecology*, eds. M. J. Klug and C. A. Reddy, pp. 177–185. Washington, D.C.: ASM Press.

Friedmann, E. I., Hua, M., and Ocampo-Friedmann, R. 1988. Cryptoendolithic lichen and cyanobacterial communities of the Ross Desert, Antarctica. *Polarforschung* 58:251–259.

Friedmann, E. I., McKay, C. P., and Nienow, J. A. 1987. The cryptoendolithic microbial environment in the Ross Desert of Antarctica: Nanoclimate data, 1984 to 1986. *Polar Biology* 7:237–287.

Golubic, S., Friedmann, I., and Scheider, J. 1981. The lithobiontic ecological niche, with special reference to microorganisms. *Journal of Sedimentary Petrology* 51:475–478.

Hill, G. T., Mitkowski, N. A., Aldrich-Wolfe, L., Emele, L. R., Jurkonie, D. D., Ficke, A., Maldonado-Ramirez, S., Lynch, S. T., and Nelson, E. B. 2000. Methods for assessing the composition and diversity of soil microbial communities. *Applied Soil Ecology* 15:25–36.

Holmes, A. J., Bowyer, J., Holley, M. P., O'Donoghue, M., Montgomery, M., and Gillings, M. R. 2000. Diverse, yet-to-be-cultured members of the Rubrobacter subdivision of the Actinobacteria are widespread in Australian arid soils. *FEMS Microbiology Ecology* 33:111–120.

Hoppert, M., Flies, C., Pohl, W., Günzl, B., and Schneider, J. 2004. Colonization strategies of lithobiontic microorganisms on carbonate rocks. *Environmental Geology* 46:421–428.

Horowitz, N. H., Cameron, R. E., and Hubbard, J. S. 1972. Microbiology of the Dry Valleys of Antarctica. *Science* 176:242–245.

Hughes, K. A. and Lawley, B. 2003. A novel Antarctic microbial endolithic community within gypsum crusts. *Environmental Microbiology* 5:555–565.

James, S. R., Dobson, S. J., Franzman, P. D., and McMeekin, T. A. 1990. *Halomonas meridiana*, a new species of extremely halotolerant bacteria isolated from Antarctic saline lakes. *Systematic and Applied Microbiology* 13:270–278.

Janssen, P. H. 2006. Identifying the dominant soil bacterial taxa in libraries of 16S rRNA and 16S rRNA genes. *Applied and Environmental Microbiology* 72:1719–1728.

Johnston, C. G. and Vestal, J. R. 1989. Distribution of inorganic species in two Antarctic cryptoendolithic communities. *Geomicrobiological Journal* 7:137–153.

Jungblut, A. D., Hawes, I., Mountfort, D., Hitzfeld, B., Dietrich, D. R., Burns, B. P., and Nellan, B. A. 2005. Diversity within cyanobacterial mat communities in variable salinity meltwater ponds of McMurdo Ice Shelf, Antarctica. *Environmental Microbiology* 7:519–529.

Khan, N. 2008. Microbial diversity of hypolithonsin Miers Valley, Antarctica. Ph.D. dissertation, Bellville, South Africa: University of the Western Cape.

Kroken, S. and Taylor, J. W. 2000. Phylogenetic species, reproductive mode, and specificity of the green alga *Trebouxia* forming lichens with the fungal genus *Letharia. The Bryologist* 103:645–660.

Kuske, C. R., Barns, S. M., and Busch, J. D. 1997. Diverse uncultivated bacterial groups from the soils of the arid south-western United States that are present in many geographic regions. *Applied and Environmental Microbiology* 63:3614–3621.

Lawley, B., Ripley, S., Bridge, P., and Convey, P. 2004. Molecular analysis of geographic patterns of eukaryotic diversity in Antarctic soils. *Applied and Environmental Microbiology* 70:5963–5972.

Lee, S., Ka, J., and Cho, J. 2008 Members of the phylum Acidobacteria are dominant and metabolically active in rhizosphere soil. *FEMS Microbiol Letters* 285:263–269.

Longton, R. E. 1988. The cryptogamic vegetation. In *The Biology of Polar Bryophytes and Lichens*, pp. 32–64. Cambridge, U.K.: British Bryological Society: CUP Archive.

Madronich, S., McKenzie, R. L., Bjorn, L. O., and Caldwell, M. M. 1998. Changes in biologically active ultraviolet radiation reaching the Earth's surface. *Journal of Photochemistry and Photobiology B: Biology* 46:5–19.

Marchant, D. R. and Head, J. W. III. 2007. Antarctic dry valleys: Microclimate zonation, variable geomorphic processes, and implications for assessing climate change on Mars. *Icarus* 192:187–222.

Matsumoto, G., Chikazawa, K., Murayama, H., Torii, T., Fukushima, H., and Hanya, T. 1983. Distribution and correlation of total organic carbon and mercury in Antarctic dry valley soils, sediments and organisms. *Geochemical Journal* 17:241–246.

Mattimore, V. and Battista, J. R. 1996. Radioresistance of *Deinococcus radiodurans*: Functions necessary to survive ionizing radiation are also necessary to survive long prolonged desiccation. *Journal of Bacteriology* 178:633–637.

McKay, C. P. 1993. Relevance of Antarctic microbial ecosystems to exobiology. In *Antarctic Microbiology*, ed. E. I. Friedmann, pp. 593–601. New York: Wiley-Liss, Inc.

McKay, C. P. 2009. Snow recurrence sets the depth of dry permafrost at high elevation in the McMurdo Dry Valleys of Antarctica. *Antarctic Science* 21:89–94.

Moorhead, D. L., Barrett, J. E., Virginia, R. A., Wall, D. H., and Porazinska, D. 2003. Organic matter and soil biota of upland wetlands in Taylor Valley, Antarctica. *Polar Biology* 26:567–576.

Moorhead, D. L., Doran, P. T., Fountain, A. G., Lyons, W. B., McKnight, D. M., Priscu, J. C., Virginia, R. A., and Wall, D. H. 1999. Ecological legacies: Impacts on ecosystems of the McMurdo Dry Valleys. *BioScience* 49:1009–1019.

Niederberger, T. D., McDonald, I. R., Hacker, A. L., Soo, R. M., Barrett, J. E., Wall, D. H., and Cary, S. C. 2008. Microbial community composition in soils of Northern Victoria Land, Antarctica. *Environmental Microbiology* 10:1713–1724.

Nienow, J. A. and Friedmann, E. I. 1993. Terrestrial lithophytic (rock) communities. In *Antarctic Microbiology*, ed. E. I. Friedmann, pp. 343–412. New York: Wiley-Liss, Inc.

Omelon, C. R., Pollard, W. H., and Ferris, F. G. 2007. Inorganic species distribution and microbial diversity within High Arctic cryptoendolithis habitats. *Microbial Ecology* 54:740–752.

Onofri, S., Selbmann, L., Zucconi, L., and Pagano, S. 2004. Antarctic microfungi as models for exobiology. *Planetary and Space Science* 52:229–237.

Oyama, V. I. and Berdahl, B. J. 1977. The Viking gas exchange experiment results from Chryse and Utopia surface samples. *Journal of Geophysical Research* 82:4669–4676.

Palmer, R. J. and Friedmann, E. I. 1990. Water relations and photosynthesis in the cryptoendolithic microbial habitat of hot and cold deserts. *Microbial Ecology* 19:111–118.

Parsons, A. N., Barrett, J. E., Wall, D. H., and Virginia, R. A. 2004. Soil carbon dioxide flux in Antarctic Dry Valley ecosystems. *Ecosystems* 7:286–295.

Raich, J. W. and Schlesinger, W. H. 1992. The global carbon dioxide flux in soil respiration and its relationship to vegetation and climate. *Tellus* 44B:81–99.

Riffenburgh, B. 2007. *Encyclopedia of the Antarctic*, ed. B. Riffenburgh, p. 654. New York: Routledge, Taylor & Francis Group: CRC Press.

Rutz, B. A. and Kieft, T. L. 2004. Phylogenetic characterisation of dwarf archaea and bacteria from a semiarid soil. *Soil Biology and Biochemistry* 36:825–833.

Saul, D. J., Aislabie, J. M., Brown, C. E., Harris, L., and Foght, J. M. 2005. Hydrocarbon contamination changes the bacterial diversity of soil from around Scott Base, Antarctica. *FEMS Microbiology Ecology* 53:141–155.

Schlesinger, W. H., Pippen, J. S., Wallenstein, M. D., Hofmockel, K. S., Klepeis, D. M., and Mahall, B. E. 2003. Community composition and photosynthesis by photoautotrophs under quartz pebbles, Southern Mojave Desert. *Ecology* 84:3222–3231.

Shivaji, S., Reddy, G. S., Aduri, R. P., Kutty, R., and Ravenschlag, K. 2004. Bacterial diversity of a soil sample from Shirmacher Oasis, Antarctica. *Cell Molecular Biology* 50:525–536.

Shravage, B. V., Dayananda, K. M., Patole, M. S., and Shouche, Y. S. 2007. Molecular microbial diversity of a soil sample and detection of ammonia oxidizers from Cape Evans, McMurdo Dry Valley, Antarctica. *Microbiological Research* 162:15–25.

Smith, J. J., Ah Tow, L., Stafford, W., Cary, C., and Cowan, D. A. 2006. Bacterial diversity in three different Antarctic cold desert mineral soils. *Microbial Ecology* 51:413–421.

Smith, M. C., Bowman, J. P., Scott, F. J., and Line, M. A. 2000. Sublithic bacteria associated with Antarctic quartz stones. *Antarctic Science* 12:177–184.

Sun, H. S., Huppert, M., and Cameron, R. E. 1978. Identification of some fungi from soil and air of Antarctica. In *Terrestrial Biology*, vol. III 30, ed. B. C. Parker, pp. 1–26. Washington, D.C.: American Geophysical Union.

Taton, A., Grubisic, S., Brambilla, E., De Wit, R., and Wilmotte, A. 2003. Cyanobacterial diversity in natural and artificial microbial mats of Lake Fryxell (McMurdo Dry Valleys, Antarctica): A morphological and molecular approach. *Applied and Environmental Microbiology* 69:5157–5169.

Taton, A., Grubisic, S., Balthasart, P., Hodgson, D. A., Laybourn-Parry, J., and Wilmotte, A. 2006. Biogeographical distribution and ecological ranges of benthic cyanobacteria in East Antarctic lakes. *FEMS Microbiology Ecology* 57:272–289.

Thomas, D. N. 2005. Photosynthetic microbes in freezing deserts. *TRENDS in Microbiology* 13:87–88.

Tiedje, J. M., Sexstone, A. J., Parkin, T. B., Revsbech, N. P., and Shelton, D. R. 1984. Anaerobic processes in soil. *Plant and Soil* 76:197–212.

Ugolini, F. C. 1970. Antarctic soils and their ecology. In *Antarctic Ecology*, vol. 2, ed. M. W. Holdgate, pp. 673–692. London, U.K.: Academic Press Inc.

Vincent, W. F. and Quesada, A. 1994. Ultraviolet radiation effects on cyanobacteria: Implications for Antarctic microbial ecosystems. In *Ultraviolet Radiation in Antarctica: Measurements and Biological Effects*, vol. 62, eds. C. S. Weiler and P. A. Penhale, pp. 111–124. Washington, DC: American Geophysical Union.

Virginia, R. A. and Wall, D. H. 1999. How soils structure communities in the Antarctic Dry Valleys. *BioScience* 49:973–983.

Vishniac, H. S. 1993. The microbiology of Antarctic soils. In *Antarctic Microbiology*, ed. E. I. Friedmann, pp. 297–341. New York: Wiley-Liss, Inc.

Vishniac, H. S. 1996. Biodiversity of yeasts and filamentous microfungi in terrestrial Antarctic ecosystems. *Biodiversity and Conservation* 5:1365–1378.

Wada, E., Shibata, R., and Torii, T. 1981. ^{15}N abundance in Antarctica: Origin of soil nitrogen and ecological implications. *Nature* 292:327–329.

Wall, D. H. and Virginia, R. A. 1999. Controls on soil biodiversity: Insights from extreme environments. *Applied Soil Ecology* 13:137–150.

Warren-Rhodes, K. A., Rhodes, K. L., Pointing, S. B. et al. 2006. Hypolithic cyanobacteria, dry limit of photosynthesis, and microbial ecology in the hyperarid Atacama Desert. *Microbial Ecology* 52:389–398.

Wharton, R. A. J., McKay, C. P., Simmons, G. M. Jr., and Parker, B. C. 1986. Oxygen budget of a perennially ice-covered Antarctic lake. *Limnology and Oceanography* 31:437–443.

Wood, S. A., Rueckert, A., Cowan, D. A., and Cary, S. C. 2008. Sources of edaphic cyanobacterial diversity in the Dry Valleys of Eastern Antarctica. *The ISME Journal* 2:308–320.

Wynn-Williams, D. D. 1988. Television image analysis of microbial communities in Antarctic fellfields. *Polarforschung* 58:239–249.

Wynn-Williams, D. D. 1990 Ecological aspects of Antarctic microbiology. *Advances in Microbial Ecology* 11:71–146.

Wynn-Williams, D. D., Edwards, H. G. M., Newton, E. M., and Holder, J. M. 2002. Pigmentation as a survival strategy for ancient and modern photosynthetic microbes under high ultraviolet stress on planetary surfaces. *International Journal of Astrobiology* 1:39–49.

Yergeau, E., Bokhorst, S., Huiskes, A. H. L., Boschker, H. T. S., Aerts, R., and Kowalchuk, G. A. 2007a. Size and structure of bacterial, fungal and nematode communities along an Antarctic environmental gradient. *FEMS Microbiology Ecology* 59:436–451.

Yergeau, E., Kang, S., He, Z., Zhou, J., and Kolwachuk, G. A. 2007b. Functional microarray analysis of nitrogen and carbon cycling genes across an Antarctic latidunal transect. *ISME Journal* 1:1723–1734.

Yergeau, E., Newsham, K. K., Pearce, D. A., and Kowalchuk, G. A. 2007c. Patterns of bacterial diversity across a range of Antarctic terrestrial habitats. *Environmental Microbiology* 9:2670–2682.

Zhou, J., Xia, B., Huang, H., Treves, D. S., Hauser, L. J., Mural, R. J., Palumbo, A. V., and Tiedje, J. M. 2003. Bacterial phylogenetic diversity and a novel candidate division of two humid region, sandy surface soils. *Soil Biology and Biochemistry* 35:915–925.

Zhou, J., Xia, B., Huang, H., Palumbo, A. V., and Tiedje, J. M. 2004. Microbial diversity and heterogeneity in sandy subsurface soils. *Applied and Environmental Microbiology* 70:1723–1734.

2 Archaeal Diversity in Antarctic Ecosystems

Jackie Aislabie and John P. Bowman

CONTENTS

2.1 INTRODUCTION

Archaea and bacteria are difficult to distinguish based on their morphology by microscopy because they were historically grouped together as prokaryotes. However, with the advent of molecular phylogenetic analysis of these taxa, we now know that archaea and bacteria are as different from each other as they are from Eukarya (Woese and Fox, 1977).

The archaeal domain is classified into two major phyla, the Euryarchaeota and the Crenarchaeota, and two proposed additional phyla, the Korarchaeota (Barns et al., 1996) and the Nanoarchaeota (Huber et al., 2002). Until recently the Euryarchaeota were considered to consist predominantly of methanogens, thermophiles, and extreme halophiles, whereas the Crenarchaeota and the Korarchaeota were believed to consist of either obligately anaerobic sulfur-dependent extreme thermophiles or hyperthermophiles, respectively (Barns et al., 1996). With the discovery of Crenarchaeota and Euryarchaeota in mesophilic environments, the view that members of the Archaea are only found in "extreme" environments has been revised (DeLong, 1998). Previously unknown, uncultivated Archaea were first detected in 16S rDNA sequences obtained from planktonic assemblages in the Pacific Ocean and from coastal marine environments (DeLong, 1992; Furhman et al., 1992). These organisms fall into two groups, Group I forming a deeply branching lineage within the Crenarchaeota (DeLong, 1992; Furhman et al., 1992), and Group II within the Euryarchaeota (DeLong, 1992). Since their initial detection in 1992, these Archaea have been widely detected in numerous environments, including, marine and freshwater environments; arboreal and agricultural soils; deep subsurface paleosols; and anaerobic digesters (DeLong, 1998). The phylum Nanoarchaeota is currently known from a single organism *Candidatus* "Nanoarchaeum equitans;" a nanosized (0.35–0.5 μm in diameter) hyperthermophilic symbiont that grows attached to the surface of *Ignicoccus hospitalis* species (Huber et al., 2002; Paper et al., 2007).

Despite harsh environmental conditions, prokaryotes occupy a diverse range of habitats in Antarctica. They have been detected, for example, in marine waters and sediment, sea ice, snow, glacial ice, cryoconite holes, lakes, and soils. In this chapter, we review the abundance and diversity of archaea in Antarctic ecosystems and speculate on their roles in ecosystem functioning. Table 2.1 provides a summary of studies that have reported archaeal abundance and/or diversity in Antarctica.

2.2 ARCHAEA IN ANTARCTIC MARINE ECOSYSTEMS

2.2.1 Seawater

The Southern Ocean is generally characterized by oligotrophic conditions in terms of biomass and production. The water column is not strongly stratified. Coastal waters stratify during periods of meltwater formation when the pack ice retreats during spring and summer. The Southern Ocean receives virtually no freshwater inflow, no terrigenous organic matter, and only small amounts of terrestrially derived micronutrients (e.g., iron). West of the Antarctic Peninsula, water temperature and salinity are reasonably stable with temperatures ranging from −1.8°C to +2°C and salinities from 33.5% to 34.7% (Church et al., 2003). Chlorophyll *a* concentrations, however, vary from below detection limits to 14.9 μg L^{-1}, with highest concentrations detected in surface waters in summer.

2.2.1.1 Archaeal Abundance in Seawater

In seawater off the Antarctic Peninsula, archaea exhibit temporal and spatial differences in abundance and diversity (DeLong et al., 1994; Murray et al., 1998; Massana et al., 1998, 2000; Church et al., 2003).

TABLE 2.1
Summary of Investigations of the Abundance and Diversity of Archaea in Antarctic Marine and Terrestrial Habitats

Habitat and Geographic Location	Sample Description	Techniques Used to Measure Archaeal Abundance and/or Diversity	Main Findings	References
		Marine Ecosystems		
Seawater				
Arthur Harbor, Antarctic Peninsula	Concentrated picoplankton from seawater samples collected in winter at the surface and underneath seasonal pack ice	Archaeal abundance estimated by hybridization of group-specific oligonucleotide probes to rRNA extracted from picoplankton Archaeal diversity determined by cloning and sequencing rRNA genes directly from picoplankton DNA	Archaeal abundance was high, comprising up to 34% of the prokaryote biomass Most clones (9 of 14) clustered with marine Crenarchaeota, and the remainder were similar to marine Euryarchaeota	DeLong et al. (1994)
Coastal waters off Anvers Island, Antarctic Peninsula Samples were collected from nearshore waters of Arthur Harbor, coastal waters of the Bismarck Strait and deeper offshore waters of Palmer Basin	Surface seawater samples collected at depths of 3, 40, and 50 m throughout the year	Archaeal abundance estimated by hybridization of group-specific oligonucleotide probes to rRNA extracted from picoplankton Archaeal abundance estimated by fluorescent *in situ* hybridization using group specific probes	During late winter to early spring archaeal rRNA abundance ranged from 1% to 17% of the total picoplankton in nearshore waters. Archaeal rRNA in nearshore waters decreased during summer *In situ* hybridization experiments revealed that 5%–14% of DAPI-stained cells were archaeal corresponding to $0.9–2.7 \times 10^4$ archaeal cells mL^{-1} in late winter samples	Murray et al. (1998)

(*continued*)

TABLE 2.1 (continued)
Summary of Investigations of the Abundance and Diversity of Archaea in Antarctic Marine and Terrestrial Habitats

Habitat and Geographic Location	Sample Description	Techniques Used to Measure Archaeal Abundance and/or Diversity	Main Findings	References
Coastal waters off Anvers Island, Antarctic Peninsula	Samples collected in August 1996	Archaeal diversity determined by cloning and sequencing rRNA genes directly from picoplankton DNA Archaeal lipids were analyzed by gas chromatography–mass spectrometry	Approximately 11%–18% of total rRNA was attributed to Archaea, most of which belonged to Group I Crenarchaeota Antarctic marine plankton samples contained caldarchaeol-derived acyclic and cyclic diphytanes derived from Archaea	DeLong et al. (1998)
Gerlache Strait, Antarctic Peninsula	Coastal seawater samples collected in early spring from 0 to 500 m depth	Archaeal abundance estimated by hybridization of group-specific oligonucleotide probes to rRNA extracted from picoplankton	Decrease in archeal numbers from late winter to spring Archaeal rRNA more abundant at depth than at the surface. Most of the archaeal signal was attributable to Group I crenarchaeotes. Euryarchaeotes were more prevalent in surface waters than at depth	Massana et al. (1998)
Southern Ocean—three sites were analyzed across Drake Passage (Sub-Antarctic) and one in Arthur Harbor (Antarctic) off the Antarctic Peninsula	Seawater samples collected from coastal (20 m depth) and offshore sites (5 and 200 m depths)	16S rRNA gene clone libraries constructed using DNA extracted from plankton and amplified using Archaea-specific primers. Clones were used for membrane hybridization experiments with universal or group-specific archaeal probes. Selected clones were subjected to RFLP analysis and sequencing	Most belonged to Group I Crenarchaeota and members of Group II Euryarchaeota were rare. The clone libraries were dominated by one or two phylotypes with most appearing only once	Massana et al. (2000)

Drake Passage between the Antarctic Peninsula and South America and Gerlache Strait of the Antarctic Peninsula	Seawater collected from 25 to 3000 m depth	16S rRNA gene clone libraries were constructed using archaeal or prokaryote-specific primers. Selected clones were sequenced	In deep sea samples from the 3000 m depth of the Antarctic Polar Front only Euryarchaeota were detected. While most belonged to Groups II and III, a cluster of distinct sequences related to the Halobacteria (so called Group IV) were detected Group IV were never detected in surface waters	López-García et al. (2001a,b)
West of the Antarctic peninsula, including shallow, near shore waters off the Antarctic Peninsula and deep open ocean waters	Seawater samples collected along transects in summer and winter from the entire water column (0–3500 m)	Abundance and distribution of Archaea in plankton was determined using FISH probes targeting Archaea (Crenarchaea Group I or Euryarchaea Group II)	Group I Crenarchaeota varied seasonally in deeper waters. During summer, numbers of Group I crenarchaeota were low in surface waters, accounting for <1% of the picoplankton, but numbers increased with depth and accounted for 9%–39% of picoplankton in deep water. In winter, the Group I Euryarchaeota were more evenly distributed in the water column, averaging 10% of the picoplankton in surface waters and 13% in deeper waters. The abundance of the Group II euryarchaetoa were <2% of the picoplankton throughout the water column in summer and winter	Church et al. (2003)
Coastal waters from Gerlache Strait and Dallman Bay near the Antarctic Peninsula	Sea water samples collected in early spring from Gerlache Strait (depths 5, 50, 125, 250, and 500 m) and summer from Dallman Bay (0 and 150 m depths).	Samples were compared using DGGE	DGGE fingerprints revealed the richness of Archaea assemblages was greater in samples from deeper water than in the upper water column.	Bano et al. (2004)

(continued)

TABLE 2.1 (continued)
Summary of Investigations of the Abundance and Diversity of Archaea in Antarctic Marine and Terrestrial Habitats

Habitat and Geographic Location	Sample Description	Techniques Used to Measure Archaeal Abundance and/or Diversity	Main Findings	References
McMurdo Sound, Ross Sea region.	Marine biofilms were established on glass slides placed 1 cm above sediment surface at 12 m and 18 m three locations and retrieved after 1 year.	The relative abundance of Archaea was determined using FISH.	Archaea were detected at all sites with the highest abundance detected near Scott Base (2.5% of total prokaryotes).	Webster and Negri (2006)
Marine Sediments				
Taynaya Bay, Fjord of the Vestfold Hills, East Antarctica.	Single anoxic sediment samples from 32 m depth.	16S rRNA gene clone libraries were constructed using universal primers. Clones were subjected to RFLP and representatives of unique phylotypes were sequenced.	A single euryarchaeal phylotype was detected which clustered within Group III.	Bowman et al. (2000a)
Continental shelf within the Mertz Glacier Polynya, East Antarctica.	A sediment core (21 by 1 cm diameter) was aseptically extracted from a larger core obtained from a depth 761 m. Sediment core depths 0–0.4 cm, 1.5–2.5 cm, and 20–21 cm were analyzed.	16S rRNA gene clone libraries were constructed using universal primers. Clones (338–369 per sample) were sequenced. 109 clones were archaeal.	Group 1.1a Crenarchaeota predominated in the surface layer. The deeper layers were more diverse. A single abundant Group 1.1a phylotype (found at all depths) was most similar to a hydrothermal vent clone. The CRA8-27/BB6A phylotype cluster was the next most important crenarchaeal group. It includes clones from deep-sea and shelf sediments. The euryarchaeal clones also grouped with those previously found in sediments. They fell into groups represented by Ace-6 and Pendant-33. One clone each belonged to euryarchaeal Group II and III (Euryarchaea).	Bowman and McCuaig (2003)

Continental shelf within the Mertz Glacier Polynya, East Antarctica.	Sediment core and grab samples collected from 16 to 843 m depth across the continental shelf.	Archaeal abundance estimated by hybridization of group-specific oligonucleotide probes to rRNA extracted from sediment. Sediments were analyzed for Archaea-derived isoprenoid glycerol dialkyl glycerol tetraether (GDGT).	The Archaeal signal was greater in surface sediment layers but declined rapidly between 0–1 cm and 1–4 cm. Archaea-derived GDGTs were detected in sediment at all depths.	Bowman et al. (2003)
Brown and O'Brien Bays, Windmill Island, East Antarctica.	Nearshore marine sediments collected.	16S rRNA gene clone libraries were constructed using universal primers and 250 clones per library were sequenced.	Archaea were detected in the sediment belonged to the Euryarchaea and were distantly related to known methanogens.	Powell et al. (2003) and Powell (2004)
Shallow Bay, Signy Island.	Coastal marine sediment	16S rRNA gene clone libraries were constructed using Archaea-specific primers. Clones chosen at random were sequenced. RNA extracted from the sediment was probed with 16S rRNA-targeted oligonucleotide probes including those specific for methanogens.	Archaeal diversity was limited. All clones were most closely related to methanogenic Euryarchaeota including *Methanogenium cariaci*, *M. burtonii*, and *Methanolobus*. The Archaeal community represented only 0.2% of total prokaryotic community.	Purdy et al. (2003)
Weddell Sea.	Three surficial sediment samples obtained from the bathypelagic zone (depth 2–65–3406 m)	16S rRNA gene clone libraries were constructed using Archaea-specific primers. 146 clones were sequenced and 47 phylotypes defined	The majority of the sequences formed three clusters within the Group 1.1a Crenarchaeota. Only one clone belonged to the Euryarchaeota	Gillan and Danis (2007)

(continued)

TABLE 2.1 (continued)
Summary of Investigations of the Abundance and Diversity of Archaea in Antarctic Marine and Terrestrial Habitats

Habitat and Geographic Location	Sample Description	Techniques Used to Measure Archaeal Abundance and/or Diversity	Main Findings	References
Marine Sponges				
McMurdo Sound, Ross Sea	Specimens of the Antarctic sponges *K. varialosa*, *L. apicalis*, *M. acerata*, *H. balfourensis*, and *S. antarcticus* were collected by SCUBA from McMurdo Sound. Four replicates of each sponge species and four replicate seawater samples were collected.	16S rRNA gene clone libraries were constructed using Archaeal-specific primers. Clones were subjected to RFLP, and representatives of unique phylotypes were sequenced.	No archaeal PCR products were detected from *H. balfourensis* or *S. antarcticus*. 150 archaeal clones, comprising four phylotypes, clustered closely together within Crenarchaea Group I.	Webster et al. (2004)
		Terrestrial Ecosystems		
Freshwater to Moderately Saline Lakes				
Ace Lake, Vestfold Hills, East Antarctica	Lake water sample collected from 24 m depth.	Enrichment cultures established in methanogenic growth media inoculated with lake water. Serial dilution technique was used to isolate pure cultures.	Isolation of methanogens identified as *M. burtonii* and *M. frigidum*. *M. burtonii*, a methylotrophic methanogen, utilized methanol and methylamines but not H_2:CO_2, formate or acetate. The optimum temperature for growth was 23.4°C. *M. frigidum* is a psychrophile which grew by CO_2 reduction using H_2 as reductant and required salt.	Franzmann et al. (1992, 1997)

Ace, Pendant, Scale, Burton, and Clear Lakes, Vestfold Hills, East Antarctica	Single anoxic sediment samples from 10 to 59 m depth	16S rRNA gene clone libraries were constructed using universal primers. Clones were subjected to RFLP and representatives of unique phylotypes were sequenced.	Ten euryarchaeal phylotypes were detected, only one of which clustered with a cultivated species, the methanogen *Methanosarcina barkeri*. Two phylotypes, containing clones from Scale and Burton lakes, clustered within Group III and the remaining phylotypes (most of the clones) belonged to a single divergent cluster now known as Pendant-33.	Bowman et al. (2000a)
Lake Heywood, Signy Island, Antarctic Peninsula.	Sediment samples from deepest point in the lake.	16S rRNA gene clone libraries were constructed using Archaea and Crenarchaea-specific primers. Clones chosen at random were sequenced. RNA extracted from the sediment was probed with 16S rRNA-targeted oligonucleotide probes including those specific for methanogens.	Archaeal diversity was limited. No Crenarchaeota were detected. All clones were most closely related to methanogenic Euryarchaeota. Most clones were closely related to *M. concilii*, and relatives of *M. organophilum* and *Methanogenium* spp. were also detected. The archaeal community represented 34% of the total prokaryotic community, a significant proportion of which was detected with a *Methanosaeta*-targeted probe.	Purdy et al. (2003)
Ace Lake, Vestfold Hills, East Antarctica.	Particulate organic matter filtered from water samples from various depths from the oxic mixolimnion (1.65–10.65 m) to anoxic and sulfidic monomolimnion (21.5–22.0 m).	Genes encoding archaeal 16S rRNA were amplified from water or sediment DNA and resolved by DGGE. Selected bands were reamplified and sequenced.	No archaeal PCR products were recovered from the oxic mixolimnion. From the anoxic, sulfidic bottom waters, 15 archaeal sequences were recovered by DGGE. They represented 12 unique phylotypes, three belonging to the Crenarchaeota, and nine to the Euryarchaeota. Many of the phylotypes were most closely related to clones retrieved from Antarctic sources.	Coolen et al. (2004)

(continued)

TABLE 2.1 (continued)
Summary of Investigations of the Abundance and Diversity of Archaea in Antarctic Marine and Terrestrial Habitats

Habitat and Geographic Location	Sample Description	Techniques Used to Measure Archaeal Abundance and/or Diversity	Main Findings	References
	Sediment collected from 25 m depth.		Thirteen unique phylotypes were recovered from the sediment, two belonging to the Crenarchaeota and 11 to the Euryarchaeota. Many were closely related to Antarctic-derived clones. One phylotype was related to *Methanosarcina*.	
Lake Fryxell, Taylor Valley, McMurdo Dry Valleys.	Water samples from 9, 11, 14, and 17 m and surface sediment samples.	Genes encoding archaeal 16S rRNA were amplified from water or sediment DNA and resolved by DGGE. Selected bands were reamplified and sequenced.	Four clusters of Archaea were detected: three euryarchaeotes and one crenarchaeote. Two clusters of methanogens detected in sediment, and another cluster of possible methanogens in anoxic water just above the sediment. One cluster was closely related to *Methanosarcina*, which have been isolated from Lake Fryxell sediment, and the other clustered with *Methanoculleus*. A large cluster in anoxic deep waters was related to marine euryarchaeotes and several orders of methanogens. The crenarchaeotes clustered with Marine Benthic Group C.	Karr et al. (2006)

Hypersaline Lakes				
Deep Lake, Vestfold Hills, East Antarctica.	Diphasic water-sediment sample.	Sample was incubated in the light for 3 months then streaked on to solid media prepared with Deep Lake water amended with yeast extract and vitamins.	*Halorubrum* (formerly *Halobacterium*) *lacusprofundi* was isolated; it required salt for growth and utilized sugars, alcohol, and organic acids as carbon source and complex nitrogen compounds, not ammonium, for growth.	Franzmann et al. (1988)
Deep, Organic and Ekho Lakes, Vestfold Hills, East Antarctica.	Sediment from the deepest point of each lake.	16S rRNA gene clone libraries were constructed using universal primers. Clones were subjected to RFLP and representatives of unique phylotypes were sequenced.	Archaea were prevalent in sediments from Deep Lake and absent from Ekho Lake. Archaeal clones belonged to the halophilic Archaea (order Halobacteria) were detected. The predominant Deep Lake phylotype grouped closely with *H. lacusprofundi*.	Bowman et al. (2000b)
Lake Bonney, Taylor Valley, McMurdo Dry Valleys.	Water samples collected from five depths from the East (16, 19, 25 m) and West (13 and 16 m) lobes of Lake Bonney.	16S rRNA gene clone libraries were constructed using archaeal-specific primers or via a two-step nested PCR. Clones were subjected to RFLP and representatives of unique phylotypes were sequenced.	Archaeal 16S rRNA gene products were obtained from two hypersaline samples from the 16 m depth of the West lobe and 25 m depth of the East lobe. A single phylotype comprising >90% of clones belonged to the Euryarchaeota and was most closely related to clones from various hypersaline environments.	Glatz et al. (2006)
Microbial Mats				
Lake Fryxell, Taylor Valley, McMurdo Dry Valleys	Mat sample from shallow moated area of the lake.	16S rRNA gene clone libraries were constructed using archaeal-specific primers. Clones were subjected to RFLP and representatives of unique phylotypes were sequenced.	The clone library was dominated by clones distantly related to Methanoculleus palmolei.	Brambilla et al. (2001)

(continued)

TABLE 2.1 (continued)
Summary of Investigations of the Abundance and Diversity of Archaea in Antarctic Marine and Terrestrial Habitats

Habitat and Geographic Location	Sample Description	Techniques Used to Measure Archaeal Abundance and/or Diversity	Main Findings	References
Soil				
Cape Bird, Cape Evans and near Scott Base on Ross Island, Marble Point, Minna Bluff, Bull Pass, and near Lake Vanda in the Wright Valley, Mt Fleming, Victoria Valley, Granite Harbor, Beacon Valley, and Cape Hallett, Ross Sea region.	Soil pits were dug at each site and samples collected from multiple depths from the surface to the permafrost.	Archaeal abundance determined using FISH 16S rRNA gene clone libraries were constructed using archaeal-specific primers. Clones were subjected to RFLP and representatives of unique phylotypes were sequenced.	Archaeal abundance was below detection limits in most samples analyzed. Archaeal DNA sufficient to produce clone libraries was obtained from 18 to 51 samples. Archaea were restricted to coastal mineral soils (Scott Base, Marble Point, Granite Harbor), except for inland soil samples from Victoria Valley. They were absent from ornithogenic soils. Most soil clones belonged to Group 1.1b Crenarchaeota. One or two phylotypes dominated each clone library.	Ayton et al. (submitted)
Mount Erebus, Ross Island	Soil sample collected from 2 to 4 cm depth.	16S rRNA gene clone libraries were constructed using archaeal-specific primers. Clones were subjected to RFLP and representatives of unique phylotypes were sequenced.	Five archaeal phylotypes were obtained from the sample and they exhibited high levels of sequence similarity with Crenarchaeaota sourced from deep-subsurface environments.	Soo et al. (2009)

Archaeal abundance in picoplankton smaller than 1 μm was high in surface coastal waters in the late winter and early spring (DeLong et al., 1994; Murray et al., 1998). Archaeal rRNA abundance ranged from 21% to 34% of the total prokaryotic rRNA in late winter samples (DeLong et al., 1994) but declined to below detection levels during summer (Murray et al., 1998; Church et al., 2003). *In situ* hybridization experiments revealed that 5%–14% of DAPI-stained cells were archaeal, corresponding to $0.9–2.7 \times 10^4$ archaeal cells mL^{-1} in late winter samples (Murray et al., 1998). In the Austral spring, rRNA was more abundant at depth (500 m) than at the surface, much of which was attributed to Group I Crenarchaeota. In contrast, Group II Euryarchaeota were more prevalent in surface waters (Massana et al., 1998).

Using fluorescent *in situ* hybridization (FISH) probes targeting Group I Crenarchaeota and Group II Euryarchaeota, Church et al. (2003) revealed that Group I abundance varied seasonally and with depth. During summer, numbers of Group I in surface waters accounted for <1% of the picoplankton, but numbers increased with depth to account for 9%–39% of picoplankton in deep water. In winter, Group I members were more evenly distributed in the water column, averaging 10% of the picoplankton in surface waters and 13% in deeper waters. Group II represented <2% of the picoplankton abundance throughout the water column in summer and winter (Church et al., 2003).

2.2.1.2 Archaeal Diversity in Seawater

To assess archaeal diversity in marine Antarctic waters, clone libraries have been constructed using amplified DNA extracted from picoplankton from coastal and offshore sites off East Antarctica (DeLong et al., 1994, 1998; Massana et al., 2000; López-García et al., 2001a,b). Using this methodology, Group I Crenarchaeota belonging to lineage 1.1a (or Group 1.1a) (Figure 2.1) appeared generally to be the most abundant archaeal group in marine waters (DeLong et al., 1994; Massana et al., 2000). Group II Euryarchaeota clones were less frequent but were still relatively common in surface water (Massana et al., 2000). In contrast, Group III Euryarchaeota were more common in deeper waters but were less widely distributed and abundant than either Group 1.1a or Group II (López-García et al., 2001a,b). The clone libraries were generally characterized by dominance of one or two phylotypes, with some phylotypes represented by very few clones (Massana et al., 2000). Furthermore the comparison of the phylotypes from different libraries revealed that the dominant types were present in most samples examined. This pattern suggests the presence of a limited number of ubiquitous taxa (Massana et al., 2000). In deep-sea samples from the 3000 m depth of the Antarctic Polar Front, only Euryarchaeota were detected (López-García et al., 2001b). While most belonged to Groups II and III, others were assigned to Group IV, a basal lineage related to the Halobacteria (López-García et al., 2001a,b) (Figures 2.2 and 2.3). The Group IV Euryarchaeota have never been detected in surface waters (López-García et al., 2001b).

In Antarctica, most investigations of archaeal diversity in marine waters have focused on the Antarctic Peninsula. Webster and Negri (2006), however, have detected archaea in marine biofilms on glass slides incubated in the coastal waters of the Ross Sea region.

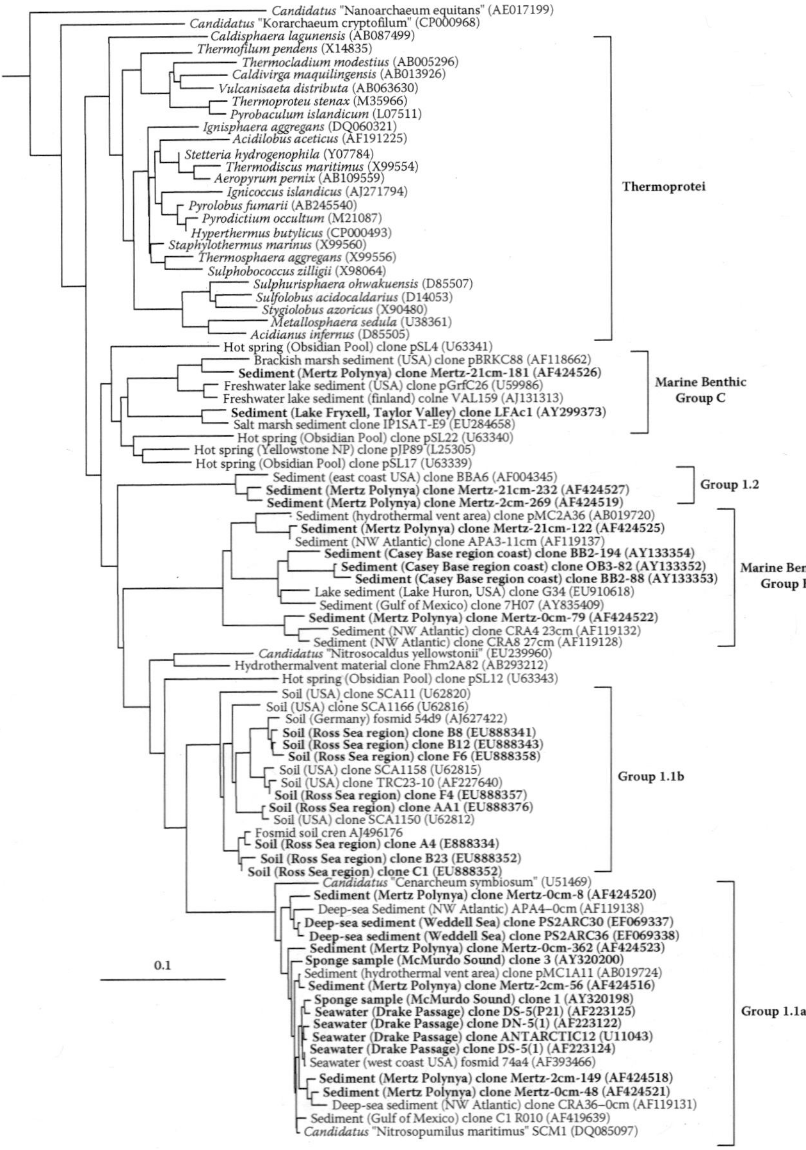

FIGURE 2.1 16S rRNA gene-based phylogenetic tree showing the relationship of cloned archaeal Antarctic sample-derived sequences (indicated in bold type) to the phylum Crenarchaeota. Sequences of cultured species represented are from type strains and for most genera only show the type species. Group labels were adapted from Schleper et al. (2005). Uncultured clones were derived from sediments of the continental shelf, deep sea and a lake (Mertz Polynya, Weddell Sea, Lake Fryxell), coastal areas (Casey Base region), and seawater (Drake Passage); sponge samples (McMurdo Sound) and soil samples (Ross Sea region). The tree is based on sequences clustered by Maximum Likelihood and Neighborhood joining algorithms.

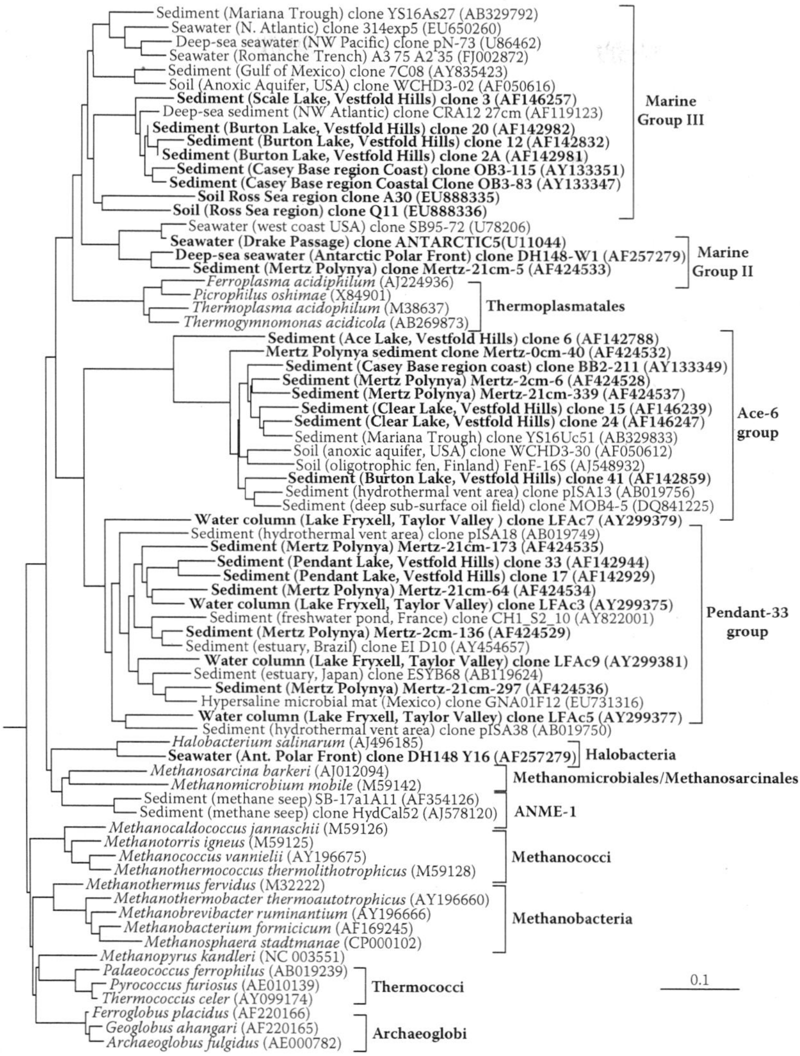

FIGURE 2.2 16S rRNA gene-based phylogenetic tree showing the relationship of cloned archaeal Antarctic sample-derived sequences (indicated in bold type) to the phylum Euryarchaeota. Sequences of cultured species represented are from type strains and for most genera only show the type species. Group labels were adapted from Schleper et al. (2005). The widely distributed Ace-6 group is equivalent to the DHVE6 clade shown in Schleper et al. (2005). Uncultured clones were derived from seawater (Drake Passage, Antarctic Polar Front); soil (Ross Sea region); and the sediment of freshwater to moderately saline lakes (Taylor Valley, Vestfold Hills), continental shelf sediments (Mertz Polynya), and coastal areas (Casey Base region). The tree is based on sequences clustered by Maximum Likelihood and Neighborhood joining algorithms.

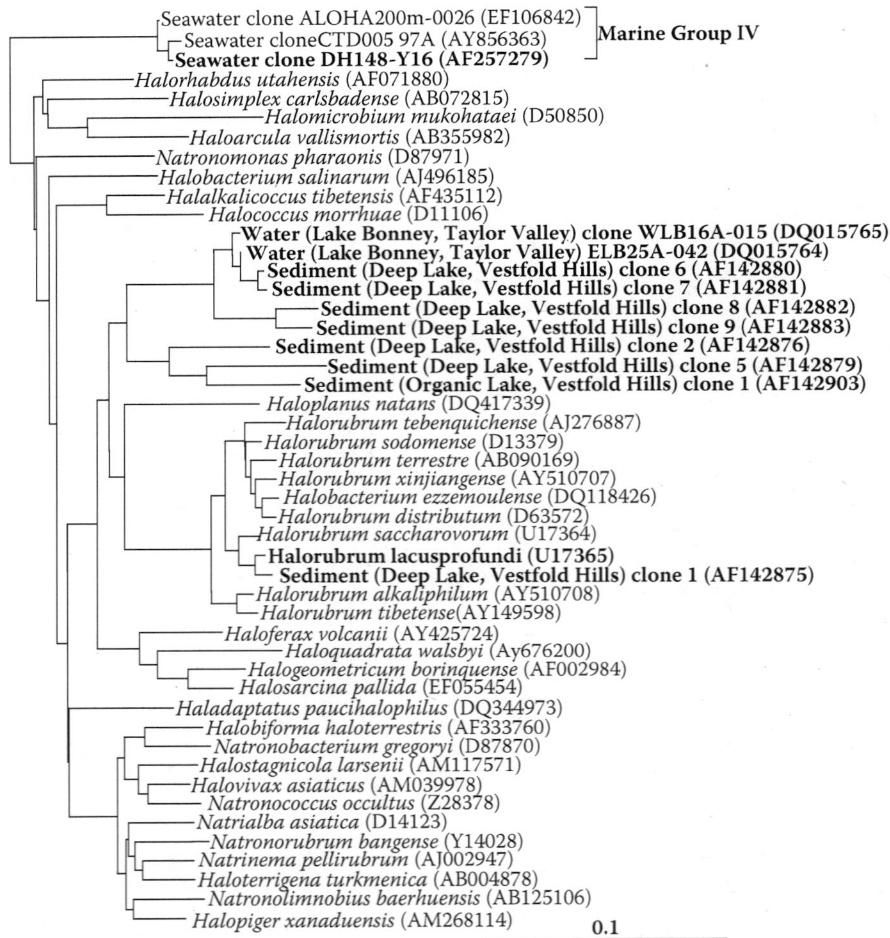

FIGURE 2.3 16S rRNA gene-based phylogenetic tree showing the relationship of cloned and cultured archaeal Antarctic sample-derived sequences (indicated in bold type) to the class Halobacteria. Sequences of cultured species represented are from type strains and for most genera only show the type species. Marine Group IV represents a related but a separate lineage of uncultured taxa obtained mainly from deep-sea seawater samples. Cultured halobacteria include *H. lacusprofundi*. Uncultured clones were derived from the sediment of hypersaline and saline lakes of the Vestfold Hills region and Taylor Valley. The tree is based on sequences clustered by Maximum Likelihood and Neighborhood joining algorithms.

2.2.1.3 Putative Archaeal Function in Seawater

The abundance of archaea in Antarctic marine waters suggests they play a key role in globally significant biogeochemical processes (DeLong et al., 1994). In a metagenomic study of the Saragasso Sea, Venter et al. (2004) noted the presence of genes encoding the ammonia monoxygenase (*amoA*) enzyme complex associated with archaeal 16S rRNA genes, indicating the potential role of archaea in nitrification in the ocean. The physiological role of these putative genes was confirmed by the isolation of the crenarchaeote *Candidatus* "Nitrosopumilus maritimus" SCM1 (belonging to Group 1.1a) from a temperate marine aquarium. This organism grows chemolithotrophically with ammonia as sole energy source (Könneke et al., 2005), but its growth was inhibited by organic compounds even at low concentrations. This may explain the observed negative correlation between archaeal rRNA levels and phytoplankton in surface coastal waters in Antarctica in summer (Murray et al., 1998). The release of organic material by phytoplankton, combined with low concentrations of ammonia could limit the abundance of Group 1.1a Crenarchaea. Archaeal ammonia-oxidizing genes have subsequently been shown to be widely distributed in marine environments (Francis et al., 2005; Wuchter et al., 2006), although it is not the case in the Southern Ocean. In addition to nitrification, marine archaea may contribute to carbon cycling. Both Crenarchaeota and Euryarchaeota in Antarctic waters from 200 to 3000 m depth were reported by Herndl et al. (2005) to assimilate ^{14}C-bicarbonate and ^{3}H-leucine. These authors concluded that archaea, particularly at depth, might play a significant role in carbon cycling in ocean ecosystems.

2.2.2 Marine Sediment

It has been reported that archaea are neither abundant nor diverse in coastal sediments, with most belonging to the Euryarchaeota in either Group III or the methanogens (Bowman et al., 2000a; Purdy et al., 2003). In contrast, archaea are abundant in continental and deep-sea sediment, with Group 1.1a Crenarchaeota predominating (Bowman and McCuaig, 2003; Gillan and Danis, 2007).

In coastal sediments from Shallow Bay off Signy Island, the archaeal community represented <1% of the total prokaryotic community (Purdy et al., 2003). All clones constructed from sediment DNA and sequenced were Euryarchaeota closely related to methanogens belonging to the genera *Methanogenium*, *Methanococcoides*, and *Methanolobus* (Figure 2.4). In Shallow Bay sediment, methanogenesis represented only a small proportion of the total carbon flow (~2%) through the system (Purdy et al., 2003). However, when sulfate reduction was inhibited, methanogenesis increased, suggesting that methanogens were not able to compete with sulfate reducing bacteria for available growth substrates. Methanogens, most closely related to *Methanobacterium*, were also detected in coastal sediments from near Casey Base in East Antarctica (Powell, 2004). Other euryarchaeal clones detected in East Antarctic sediments from near Casey Base (Powell et al., 2003; Powell, 2004) or Tayana Bay (Bowman et al., 2000a) belonged to Group III, or Ace-6 (Figure 2.2). Crenarchaeal clones in sediments from near Casey Base belonged to Marine Benthic Group B and were related to those from continental sediments (Figure 2.1).

In continental shelf sediments collected off East Antarctica, archaea were abundant in surface layers of sediment but declined significantly below 1 cm (Bowman et al., 2003). Subsequent investigations revealed that Group 1.1a Crenarchaeota predominated in the surface layer of a sediment core collected from 761 m depth within the Mertz Glacier Polynya off East Antarctica. The deeper layers were more diverse in archaeal composition (Bowman and McCuaig, 2003). A single Group 1.1a

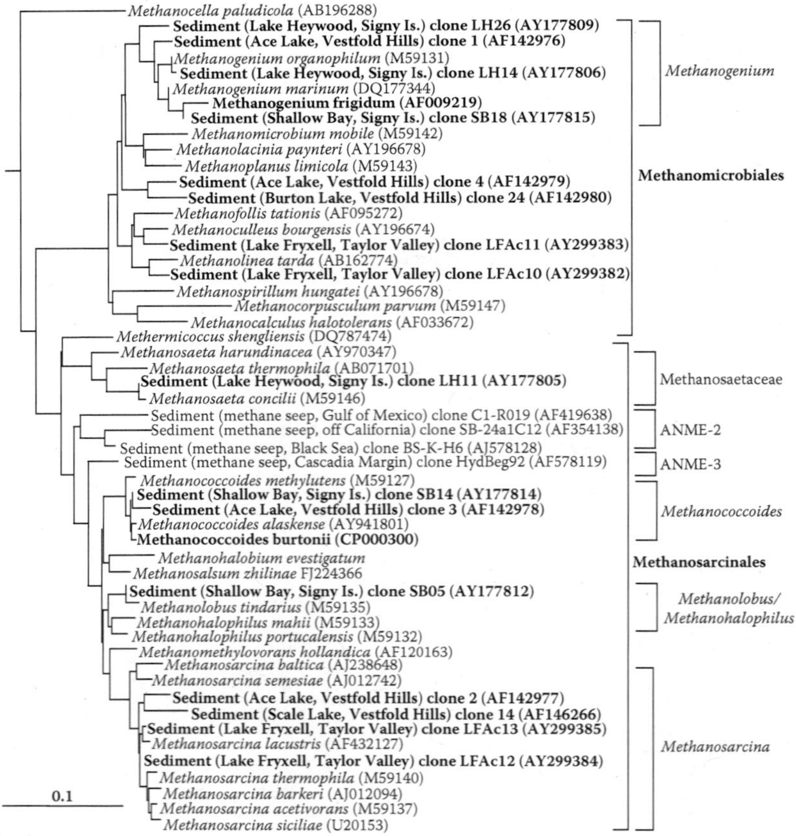

FIGURE 2.4 16S rRNA gene-based phylogenetic tree showing the relationship of cloned and cultured archaeal Antarctic sample-derived sequences (indicated in bold type) to methanogens of the orders Methanomicrobiales and Methanosarcinales. Sequences of cultured species represented are from type strains and for most genera only show the type species. *Methanocella paludicola* represents a related but separate lineage. Cultured Antarctic methanogens include *Methanococcoides burtonii* and *M. frigidum*. Uncultured clones were derived from sediment from lakes and coastal areas of Antarctica including Signy Is., Taylor Valley, and Vestfold Hills. The tree is based on sequences clustered by Maximum Likelihood and Neighborhood joining algorithms.

phylotype (clone Mertz-2cm-56), representing 80% of all archaeal clones (found at all depths), was most similar to a hydrothermal vent clone (Figure 2.1). Other crenarchaeal clones clustered with organisms from deep-sea and shelf sediments and belonged to Marine Benthic Groups B and C, and Group 1.2 (Figure 2.1). The Euryarchaeota clones were phylogenetically similar to those previously found in brackish or saline sediment, and belonged to Ace-6, Pendant-33 and Groups II and III (Figure 2.2). In this study it was noted that the relative abundance of archaea may be underestimated in clone libraries constructed with universal primers targeting the 16S rRNA genes due to PCR bias.

Similar to the observations by Bowman and McCuaig (2003), Gillan and Danis (2007) found that Group 1.1a Crenarchaeota predominate in surface sediment samples from the bathypelagic zone (2165–3406 m) in the Weddell Sea. All but one of the 146 sequenced clones belonged to the Group 1.1a Crenarchaeota (Figure 2.1). The crenarchaeal sequences formed three phylogenetic clusters, two of them containing known nitrifiers (Gillan and Danis, 2007). One cluster (78.8% of the clones) comprised the nitrifier *N. maritimus* and clones from deep-sea sediments. A second cluster (7.5% of the clones) contained two symbionts of marine sponges: *Candidatus* "Cenarchaeum symbiosum" (host: *Axinella mexicana*) and clone A7G11 (host: *Axinella polycapella*). The environmental genomic analysis of these symbiotic crenarchaeotes indicates their ability to use autotrophic pathways for energy generation derived from the oxidation of ammonia (Hallam et al., 2006). The closest relatives of the third group (13% of clones) were clones retrieved from near the mid-Atlantic ridge, a tropical seawater tank, and the Atlantic Ocean abyssal plain. These data indicate Group 1.1a Crenarchaeota are abundant in Antarctic bathypelagic sediments and phylogenetic inference indicates these communities may be autotrophs that oxidize ammonia.

2.2.3 Sea Ice

Sea ice in the Southern Ocean provides an extensive habitat for the growth of microorganisms. Sea ice is temporally and spatially variable. Most Antarctic sea ice is annual. At winter, maximum (September) ice cover may reach 20×10^6 km^2 but by midsummer (February) is usually reduced to less than 4×10^6 km^2. The internal temperature of the sea ice ranges from −1°C to −15°C and the salinity of sea ice brine inclusions from <10‰ to >150‰. Microbial assemblages form in these brine inclusions within ice floes, and are concentrated near the sea ice seawater interface. The assemblages consist mainly of diatoms, as well as autrotrophic nanoflagellates, autotrophic and heterotrophic dinoflagellates, and heterotrophic flagellates. Bacteria occur throughout the ice, particularly at the surface and associate with algae. In contrast, numbers of archaea are below detection levels (<1% of total numbers) during the spring and summer months (Brown and Bowman, 2001). Whether numbers of archaea are higher in sea ice in winter, as has been observed for these organisms in marine picoplankton below the sea ice, has yet to be determined (DeLong et al., 1994). The support for a possible seasonal pattern is indicated by a study in Arctic sea ice, where archaea were at detectable levels (up to 3.4%) in winter samples only (Junge et al., 2004). Archaea were not detected in accretion ice from under the Amery ice shelf (Twin, personal communication, unpublished).

2.2.4 Sponge Symbionts

There has been little consideration of archaea associated with Antarctic plants and animals. A notable exception is the study of Webster et al. (2004) who assessed the archaeal communities associated with the Antarctic sponges *Kirkpatrickia varialosa*, *Latrunculia apicalis*, *Homaxinella balfourensis*, *Sphaerotylus antarcticus*, and *Mycale acerata* from McMurdo Sound in the Ross Sea. One hundred and fifty archaeal clones were screened by restriction fragment length polymorphism and four phylotyes were obtained. All belonged to Group 1.1a Crenarchaeota (Figure 2.1). Phylogenetic analysis indicated close relationships to clones of the tropical sponge *Rhopaloeides odorabile* and, more distantly, *Candidatus* "C. symbiosum" from the temperate sponge *A. mexicana* (Webster et al., 2004). No archaeal PCR products were detected from the sponges *H. balfourensis* or *S. antarcticus*.

2.3 ANTARCTIC TERRESTRIAL ECOSYSTEMS

2.3.1 Antarctic Soil

Soils are located mainly along the continental coastline and adjacent areas, particularly on the Antarctic Peninsula and in the Ross Sea region. These are cold desert soils, characterized by low soil temperatures and moistures. The soils comprise a surface pavement and a seasonally thawed active layer over permafrost. The surface pavement is a layer of gravel, stones, and/or boulders. The depth of the active layer ranges from <5 to >70 cm. Permafrost underlies all exposed ground surfaces except those heated by volcanism. The soils typically have low levels of organic carbon and nitrogen and are low in clay. Soil pH may range from weakly acidic (pH 6) in inland soils to highly alkaline in coastal regions (pH 9). To our knowledge, archaea have only been investigated in soils of the Ross Sea region.

2.3.1.1 Archaeal Diversity and Abundance in Antarctic Soils

The abundance and diversity of archaea was assessed in mineral and ornithogenic soils from 12 locations across the Ross Sea region (Ayton et al., submitted). Archaea were not abundant. Cells labelled with archaeal-specific or crenarchaeal-specific FISH probes were observed by microscopy in one surface soil sample only, from near Scott Base on Ross Island, but at numbers too low to enumerate. Archaeal DNA sufficient for producing 16S rRNA gene clone libraries was extracted from 18 of 51 soil samples from four locations. Between 67 and 89 clones per sample were analyzed by restriction fragment length polymorphism and assigned to 43 phylotypes. Representatives of each phylotype were sequenced. More than 99% of the archaeal soil clones belonged to Group 1.1b Crenarchaeota and clustered together with 16S rRNA genes from fosmid 54d9 (GenBank accession number AJ496176) that contains the *amoA* gene (Treusch et al., 2005) (Figure 2.1). Typically the crenarchaeal clone sequences were most similar (>98%) to those retrieved from soils, including those from agricultural fields, deglaciated sites, and alkaline–saline environments. The few Euryarchaeota clones obtained belonged to Group III (Figure 2.2) and were most similar to clones from alkaline–saline soil or cold sulfidic springs.

Ayton et al. (submitted) found archaea to be most prevalent in pristine coastal mineral soils from Scott base, Marble Point, Granite Harbor, and Minna Bluff, but they were also detected in two inland soil samples from Victoria Valley. Archaea were absent from ornithogenic soils from Cape Bird and Cape Hallett. Similar to pristine soils archaea in oil-contaminated soil from Scott Base and Marble Point were dominated by Group 1.1b crenarchaeota (Ayton et al., submitted). Some clones were more prevalent in soils at depth, associated with lower values of pH, electrical conductivity, organic C, NH_4^+–N, and total P. In contrast, different clones tended to occur in surface soils with high total N and NO_3–N (Ayton et al., submitted). Archaea were distributed throughout the active layer, but sparse in the surface layer at some sites.

The analysis of archaeal diversity in a single thermally heated soil sample from Mount Erebus on Ross Island revealed the presence of Crenarchaeota affiliated with organisms from deep-subsurface environments (Soo et al., 2009).

2.3.1.2 Putative Archaeal Function in Soil

Recent investigations of soil Crenarchaeota have suggested that members of Group1.1b are nitrifiers. Metagenomic studies have revealed that soil Crenarchaeota belonging to Group 1.1b contain and express genes encoding the ammonia monoxygenase (*amoA*) enzyme complex (Nicol and Schleper, 2006). The significance of archaeal nitrification in soil was supported by the report that archaeal *amoA* genes were more abundant than bacterial *amoA* genes in a range of temperate soils, particularly at depth (Leininger et al., 2006). Furthermore, Nicol et al. (2008) observed remarkable phylogenetic congruence between both 16S rRNA genes and the functional gene *amoA* that encodes the ammonia monoxygenase enzyme complex in archaea. Potential nitrification activity has been measured in Antarctic soils (Hopkins et al., 2006; Yergeau et al., 2007), and bacterial, but not archaeal, *amoA* genes have been investigated (Yergeau et al., 2007). A gene that encodes the methanogen-specific methyl coenzyme reductase has been detected in Dry Valley soils (Hopkins et al., 2005). It was thought to originate from methanogens in wet, organic-rich lake margins. The rates of *in situ* methanogenesis within 1 m of lake margins in the Garwood Valley were comparable to those from temperate and sub-Arctic wetlands; however, Ayton et al. (submitted) found Euryarchaeota not to be abundant in soil under a cyanobacterial mat on the margins of a lake below the Victoria Upper Glacier.

2.3.2 Antarctic Lakes

Antarctic lakes occur in ice-free regions and in subglacial environments. Here we consider only surface lakes, derived from either freshwater systems or progressively saline through evaporation or release of salts from sediments of marine origin, or by uplift of the land trapping a body of sea water. In many of the saline systems there has been an influx of fresh melt-water that has overlain the saline water. The exceptional stability provided by the saline bottom waters and the presence of a thick ice cover, often year round, makes the stratification permanent (a meromictic system). Microbial mats that develop on the upper surface of lake sediments, are formed by filamentous matrices of cyanobacteria including *Phormidium*, *Oscillatoria*, and *Nostoc* species. Heterotrophic microbes are also present in the mats (Brambilla et al., 2001). A significant cycling of carbon, nitrogen, and methane has been measured in the lake sediments and anoxic bottom waters. Archaeal diversity assessments have been conducted on lakes of the Vestfold Hills in East Antarctica, those of the Dry Valleys in the Ross Sea region, and Signy Island on the Antarctic Peninsula.

2.3.2.1 Freshwater to Moderately Saline Lakes—Archaeal Abundance and Diversity

Archaea in freshwater to moderately saline lakes appear be more abundant in anoxic bottom waters or sediments. Methanogens are common and have been isolated from or detected in these environments (Franzmann et al., 1992, 1997; Bowman et al. 2000a; Purdy et al., 2003; Coolen et al., 2004; Karr et al., 2006).

Ace Lake is the most studied marine meromictic lake in Antarctica. Saline stratification in the lake maintains zones that are supersaturated with oxygen, through to anoxic bottom waters. Conditions in the lake are cold (maximum of 5°C) with sulfidic, sulfate-depleted and methane-saturated bottom waters. The lake is about 25 m deep, is ice-covered for 11 months of the year, and the oxic/anoxic interface occurs between 11 and 12 m depth (Coolen et al., 2004).

Archaeal PCR products were obtained from particulate organic matter (POM) in the anoxic, sulfidic bottom waters and the sediment of Ace Lake but not from the oxic mixolimnion, and were subjected to denaturing gradient gel electrophoresis (DGGE). Coolen et al. (2004) detected 15 POM archaeal sequences by DGGE, 13 of which represented unique phylotypes. Ten phylotypes belonged to the phylum Euryarchaeota, all of which were most closely related to clones previously known from Antarctic lakes or continental shelf sediments and appear to belong to the lineages Pendant-33 and Ace-6 (Bowman et al., 2000a; Bowman and McCuaig, 2003). Three Crenarchaeota phylotypes were obtained from the 14–19 m depth and similarly clustered phylogenetically with clones from Antarctic continental shelf sediments (Bowman and McCuaig, 2003). From the lake sediment, 16 phylotypes were recovered, with 14 belonging to the Euryarchaeota, and two to the Crenarchaeota. Of the Euryarchaeota, two were phylogenetically related to the genus *Methanosarcina*, one was distantly related to clones from a freshwater lake, and the remainder were distantly related to clones from various marine environments, including members of Group III.

Bowman et al. (2000a) examined prokaryotic diversity in the anoxic sediments from brackish to marine salinity meromictic lakes in the Vestfold Hills. All the archaeal clones in the 16S rRNA gene libraries belonged to the Euryarchaeota. Some of the clones were phylogenetically related to methanogens (Figure 2.4). Clones from Ace and Scale Lakes were related to *Methanosarcina*, members of which have been cultivated in the laboratory, whereas others from Ace Lake and Burton Lake clustered within the Methanomicrobiales but typically had no close cultured relatives. Many of the clones formed a divergent cluster within the euryarchaeal "sediment Archaea" group. Two clades within this cluster have since been renamed as Pendant-33 group and Ace-6 group (Figure 2.2). Because they were affiliated with clones from anoxic environments, the authors suggested that this group is capable of anaerobic metabolism. Burton and Scale Lakes contained clones that belonged to Group III and were affiliated with clones from marine sediments.

Methanococcoides burtonii and *Methanogenium frigidum* were isolated from anoxic water from Ace Lake (Franzmann et al., 1992, 1997). *M. frigidum* is a non-motile, salt-requiring (350 mM Na^+) psychrophile that uses H_2 and CO_2 for growth. In contrast *M. burtonii* uses methyl substrates (trimethylamine and methanol), is flagellated and motile, and psychrotolerant as it grows at low temperatures but has an optimum temperature for growth at 23.4°C. Because *M. burtonii* is a methylotrophic methanogen, it does not compete with hydrogen-utilizing, sulfate-reducing bacteria in the environment. *M. burtonii* may be cosmopolitan in cold environments, as it has a close relative (*Methanococcoides alaskense*, 99.8% 16S rRNA identity) isolated from Alaska (Singh et al., 2005) and is closely related to clones retrieved from the sediment of Shallow Bay, Signy Island (>99% 16S rRNA gene identity) (Purdy et al., 2003) (Figure 2.4). The genomes of *M. burtonii* and *M. frigidum* have been sequenced and investigations of their molecular mechanisms for cold adaptation are proceeding (Cavicchioli, 2006).

Lake Fryxell in the McMurdo Dry Valleys region is a freshwater lake 18 m deep that exhibits significant sulfur and methane cycling. Genes encoding archaeal 16S rRNA genes were amplified from DNA from Lake Fryxell water and sediment and subjected to DGGE (Karr et al., 2006). Selected DGGE bands from the sediment and 11, 14, and 17 m water depths were reamplified and sequenced. Phylogenetic analysis showed that at least four clades of Archaea inhabited Lake Fryxell: three Euryarchaeota and one Crenarchaeota (Karr et al., 2006). Two clades of methanogens were detected in the sediment, one of which was most closely related to *Methanosarcina* species (Figure 2.4). The presence of *Methanosarcina*-like methanogens in Lake Fryxell sediments was confirmed by the isolation of strain FRX-1, a methylotrophic *Methanosarcina* species (Kendall et al., 2007). The second clade of methanogens in Lake Fryxell sediments was related to species of *Methanoculleus,* and *Methanolinea*. *Methanoculleus*-like 16S rRNA gene sequences have also been reported in a cyanobacterial mat sample from the shallow, moated area of Lake Fryxell. Of the 72 clones obtained from the mat, >90% was distantly related to *M. palmolei* (Brambilla et al., 2001). A large clade of Euryarchaeota (clones LFAc2 to 9), belonging to Pendant-33 group (Figure 2.2) and related to marine clones, was found in anoxic waters of Lake Fryxell between 14 and 17 m depth. Between 12 m depth and the surface sediment, methane concentrations declined and sulfide increased. This led the authors to propose that the Euryarchaeota represented by clones LFAc2 to 9 were involved in anoxic methanotrophy using sulfide as an electron acceptor since alternative electron acceptors such as NO_3^-, Mn^{4+}, and Fe^{3+} were undetectable in the anoxic waters. Clones LFAc2 to 9 were only distantly related to the anaerobic methane-oxidizing (ANME) groups that catalyze anoxic methanotrophy in marine sediments (Hallam et al., 2003). The closest relatives of the sole crenarchaeota phylotype recovered from Lake Fryxell were obtained from water above the oxycline (11 m). The sequence was most closely related to crenarchaeotal Marine Benthic Group C (Figure 2.1). Because both oxygen and sulfide were present at 11 m, the authors speculated that the Crenarchaeota may grow by utilizing sulfide as either an electron donor or an electron acceptor (Karr et al., 2006).

Lake Heywood is a freshwater lake on Signy Island. It is up to 7 m deep and is ice-covered for most of the year. When ice-covered, the lake is stratified and has anoxic bottom waters. Both methanogenesis and sulfate reduction were measured in the sediments in the laboratory, though methanogenesis was dominant (Purdy et al., 2003). Archaea were abundant in the sediment of Lake Heywood (~34% of the total prokaryotic community), but their diversity was limited. Three groups of clones were detected, all members of phylum Euryarchaeota. Most of the clones (e.g., clone LH11) were related to the known obligate acetate-utilizing *Methanosaeta concilii* and others were related to hydrogen-utilizing *Methanogenium* spp. including *Methanogenium organophilum* (Figure 2.4).

2.3.2.2 Hypersaline Lakes—Archaeal Abundance and Diversity

From the few investigations of hypersaline lakes, it appears that archaeal diversity is limited, with most clones assigned to the Halobacteria. Furthermore, difficulties experienced in amplifying archaeal DNA suggest that the overall abundance of archaea is low (Bowman et al., 2000b; Glatz et al., 2006).

Archaeal diversity has been assessed in sediments from three hypersaline lakes: Deep Lake, Organic Lake, and Ekho Lake in the Vestfold Hills in East Antarctica (Bowman et al., 2000b). These lakes, formed by the evaporation of seawater, are nearly or completely perennially ice-free. Deep and Organic Lakes have sediments with subzero temperatures. Deep Lake is 36 m deep, and monomictic, with salinity 10 times that of seawater. Organic Lake is a relatively shallow (7.5 m deep) and meromictic lake. 16S rRNA gene clone libraries were constructed using universal primers. None of the clones detected in Ekho Lake sediment were archaeal, whereas 7.8% of the clones retrieved from Organic Lake sediment and most of the clones from Deep Lake sediment were archaeal. The archaeal clones were assigned to eight distinct phylotypes affiliated with the Halobacteria. The predominant phylotype represented by Deep Lake sediment clone 1 grouped closely with the psychrotolerant *Halorubrum lacusprofundi* originally isolated from the same lake (Franzmann et al., 1988) (Figure 2.3). *H. lacusprofundi* has been shown to require 1.5 M NaCl for growth and utilizes sugars, alcohol and organic acids as carbon source and complex nitrogen compounds (not ammonium), for growth. The remaining phylotypes, common to both Deep Lake and Organic Lake, clustered together and were affiliated with a clone from a salt marsh environment (Bowman et al., 2000b).

Lake Bonney is a permanently ice-covered lake in Taylor Valley. The lake consists of two lobes, both approximately 40 m deep, connected by a narrow shallow passage (12–13 m deep). The bottom waters are suboxic to anoxic, have subzero temperatures and salinities 6–10 times that of seawater. DNA was extracted from water collected from 5 depths (13 and 16 m in the West lobe and 16, 19, and 25 m in the East lobe) and amplified with archaeal specific primers either directly or using nested PCR (Glatz et al., 2006). The restriction fragment length polymorphism analysis of 190 archaeal clones retrieved from only two samples (East lobe 25 m and West lobe 16 m) revealed a dominant clone (>90% of all clones) affiliated with Euryarchaeota clones obtained from other hypersaline environments including sediment from Deep Lake and Organic Lake (Figure 2.3) (Glatz et al., 2006).

2.3.3 Archaeal Abundance and Diversity of Other Terrestrial Habitats

Archaea were not detected in some studies of selected Antarctic terrestrial habitats. These include investigations of cyanobacterial biofilms in quartz stone subliths (Smith et al., 2000), cryptoendolithic environments (de la Torre et al., 2003), alkaline epiglacial lakes, glacial ice cores (Lee et al., 2005; Twin, personal communication, unpublished), cryoconite holes (Christner et al., 2003), and accretion ice from Lake Vostok (Christner et al., 2001). As far as we know, archaea have not been sought in transient rivers and streams or their hyporheic zone.

2.4 SUMMARY

In the absence of pure cultures, understanding the habitat preferences, probable carbon and energy sources and physiological properties of Archaea in Antarctica is a

TABLE 2.2
Distribution of Archaeal Groups within Antarctic Habitats

Habitat	Archaeal Phylum	Taxonomic Affiliation[a]	References
Marine waters	Crenarchaeota Euryarchaeota	Group I[b] Group II, Group III, Group IV	DeLong et al. (2004), Massana et al. (2000), and López-García et al. (2001a,b)
Marine coastal sediments	Crenarchaeota	Marine Benthic Group B	Bowman et al. (2000a)
	Euryarchaeota	Group III, Ace-6, *Methanomicrobiales*, *Methanosarcinales*	Purdy et al. (2003), Powell et al. (2003), and Powell (2004)
Marine off-shore sediments	Crenarchaeota	Group 1.1a, Group 1.2, Marine Benthic Group B, Marine Benthic Group C	Bowman and McCuaig (2003)
	Euryarchaeota	Group II, Group III, Pendant-33, Ace-6	
Sponge symbionts	Crenarchaeota	Group 1.1a	Webster et al. (2004)
Soil	Crenarchaeota Euryarchaeota	Group 1.1b Group III	Ayton et al. (submitted)
Microbial mats	Euryarchaeota	Methanomicrobiales	Brambilla et al. (2001)
Freshwater/ moderately saline lakes	Crenarchaeota Euryarchaeota	Marine Benthic Group C Group III, *Methanomicrobiales*, *Methanosarcinales*, Pendant-33, Ace-6	Franzmann et al. (1992, 1997), Bowman et al. (2000a), Purdy et al. (2003), Karr et al. (2006)
Saline lakes	Euryarchaeota	Halobacteria	Franzmann et al. (1982), Bowman et al. (2000b), and Glatz et al. (2006)

[a] Per Cavicchioli (2006).

[b] The Group I Crenarchaeota contains a number of subclusters that appear to be specific to particular environments. Group I Crenarchaeota from marine plankton fall into subcluster Group 1.1a; whereas Group I Crenarchaeota from soil, lake sediments and marine snow tend to belong to subcluster Group 1.1b.

challenge. From the few studies conducted so far it appears that Archaea are however widely distributed in Antarctic ecosystems and may play important ecological roles. Where archaea have been detected in Antarctica, Group 1.1a Crenarchaeota dominate marine waters and off-shore sediments (particularly at depth), and symbiotic communities associated with marine sponges. In coastal mineral soil Group 1.1b Crenarchaeaota predominate (Table 2.2). There is increasing evidence that marine and soil crenarchaeota, belonging to lineages Group 1.1a and 1.1b, respectively, are nitrifiers. Euryarchaea are prevalent in coastal sediments and lakes. Diverse methanogens, some of which have been cultured, have been detected in coastal sediments and anaerobic bottom waters and sediments of lakes. The affiliation of clones belonging to Pendant-33 and Ace-6 with anoxic environments has led to speculation that they may be capable of anaerobic metabolism. Halobacteria dominate hypersaline lakes, and have been detected in deep sea samples. Predicted roles for archaea in Antarctic ecosystems include heterotrophy and nitrification.

ACKNOWLEDGMENT

This research was supported by funding from the Foundation of Research, Science and Technology, New Zealand (C09X0307).

REFERENCES

Ayton, J., Aislabie, J., Barker, G. M., Turner, S., and D. Saul. Archaeal diversity in soils from the Ross Sea Region of Antarctica. *Environmental Microbiology* (submitted).

Bano, N., Ruffin, S., Ransom, B., and J. T. Hollibaugh. 2004. Phylogenetic composition of Arctic Ocean archaeal assemblages and comparison with Antarctic assemblages. *Applied and Environmental Microbiology* 70:781–789.

Barns, S. M., Delwiche, C. F., Palmer, J. D., and N. R. Pace. 1996. Perspectives on archaeal diversity, thermophily and monophyly from environmental rRNA sequences. *Proceedings of the National Academy of Sciences* 93:9188–9193.

Bowman, J. P. and R. D. McCuaig. 2003. Biodiversity, community structural shifts, and biogeography of prokaryotes within Antarctic continental shelf sediments. *Applied and Environmental Microbiology* 69:2463–2483.

Bowman, J. P., Rea, S. M., McCammon, S. A., and T. A. McMeekin. 2000a. Diversity and community structure within anoxic sediment from marine salinity meromictic lakes and a coastal meromictic marine basin, Vestfold Hills, Eastern Antarctica. *Environmental Microbiology* 2:227–237.

Bowman, J. P., McCammon, S. A., Rea, S. M., and T. A. McMeekin. 2000b. The microbial composition of three limnologically disparate hypersaline Antarctic lakes. *FEMS Microbiology Letters* 183:81–88.

Bowman, J. P., McCammon, S. A., Gibson, J. A. E., Robertson, L., and P. D. Nichols. 2003. Prokaryotic metabolic activity and community structure in Antarctic continental shelf sediments. *Applied and Environmental Microbiology* 69:2448–2462.

Brambilla, E., Hippe, H., Hagelstein, A., Tindall. B. J., and E. Stackebrandt. 2001. 16S rDNA diversity of cultured and uncultured prokaryotes of a mat sample from Lake Fryxell, McMurdo Dry Valleys, Antarctica. *Extremophiles* 5:23–33.

Brown, M. V. and J. P. Bowman. 2001. A molecular phylogenetic survey of sea-ice microbial communities (SIMCO). *FEMS Microbiology Ecology* 35:267–275.

Cavicchioli, R. 2006. Cold-adapted archaea. *Nature Reviews Microbiology* 4:331–343.

Christner, B. C., Kvitko, B. H. II, and J. N. Reeve. 2003. Molecular identification of bacteria and eukarya inhabiting an Antarctic cryoconite hole. *Extremophiles* 7:177–183.

Christner, B. C., Mosely-Thompson, E., Thompson, L. G., and J. N. Reeve. 2001. Isolation of bacteria and 16S rDNAs from Lake Vostok accretion ice. *Environmental Microbiology* 3:570–577.

Church, M. J., DeLong, E. F., Ducklow, H. W., Karner, M. B., Preston, C. M., and D. M. Karl. 2003. Abundance and distribution of planktonic Archaea and Bacteria in the waters west of the Antarctic Peninsula. *Limnology and Oceanography* 48:1893–1902.

Coolen, M. J. L., Hopmans, E. C., Rijpstra, W. I. C. et al. 2004. Evolution of the methane cycle in Ace Lake (Antarctica) during the Holocene: Response of methanogens and methanotrophs to environmental change. *Organic Geochemistry* 35:1151–1167.

de la Torre, J. R., Goebel, B. M., Friedmann, E. I., and N. R. Pace. 2003. Microbial diversity of cryptoendolithic communities from the McMurdo Dry Valleys, Antarctica. *Applied and Environmental Microbiology* 69:3858–3867.

DeLong, E. F. 1992. Archaea in coastal marine environments. *Proceedings of the National Academy of Sciences USA* 89:5685–5689.

DeLong, E. F. 1998. Everything in moderation: Archaea as "non-extremophiles". *Current Opinion in Genetics & Development* 8:649–654.

DeLong, E. F., King, L. L., Massana, R. et al. 1998. Dibiphytanyl ether lipids in nonthermophilic Crenarchaeotes. *Applied and Environmental Microbiology* 64:1133–1138.

DeLong, E. F., Wu, K. Y., Prézelin, B. B., and R. V. M. Jovine. 1994. High abundance of Archaea in Antarctic marine picoplankton. *Nature* 371:695–697.

Francis, C. A., Roberts, K. J., Beman, J. M., Santero, A. E., and B. B. Oakley. 2005. Ubiquity and diversity of ammonia-oxidising archaea in water columns and sediments of the ocean. *Proceedings of the National Academy of Sciences USA* 102:14683–14688.

Franzmann, P. D., Liu, Y., Balkwill, D. L., Aldrich, H. C., Conway de Macario, E., and D. R. Boone. 1997. *Methanogenium frigidum* sp. nov., a H_2-using methanogen from Ace Lake, Antarctica. *International Journal of Systematic Bacteriology* 47:1068–1072.

Franzmann, P. D., Springer, N., Ludwig, W., Conway de Macario, E., and M. Rohde. 1992. A methanogenic archaeon from Ace lake, Antarctica: *Methanococcoides burtonii* sp. nov. *Systematic and Applied Microbiology* 5:573–581.

Franzmann, P. D., Stackebrandt, E., Sanderson, K. et al. 1988. *Halobacterium lacusprofundi* sp. nov., a halophilic bacterium isolated from Deep Lake, Antarctica. *Systematic and Applied Microbiology* 11:20–27.

Furhman, J. A., McCallum, K., and Davis, A. A. 1992. Novel major archaebacterial group from marine plankton. *Nature* 356:148–149.

Gillan, D. C. and B. Danis. 2007. The archaebacterial communities in Antarctic bathypelagic sediments. *Deep-Sea Research II* 54:1682–1690.

Glatz, R. E., Lepp, P. W., Ward, B. B., and C. A. Francis. 2006. Planktonic microbial community composition across steep physical/chemical gradients in permanently ice-covered Lake Bonney, Antarctica. *Geobiology* 4:53–67.

Hallam, S. J., Girguis, P. R., Preston, C. M., Richardson, P. M., and E. F. DeLong. 2003. Identification of methyl coenzyme M reductase (*mcrA*) genes associated with methane-oxidising Archaea. *Applied and Environmental Microbiology* 69:5483–5491.

Hallam, S. J., Mincer, T. J., Schleper, C. et al. 2006. Pathways of carbon assimilation and ammonia oxidation suggested by environmental genomic analyses of marine Crenarchaeota. *PloS Biology* 4(4):e95.

Herndl, G. J., Reinthaler, T., Teira, E. et al. 2005. Contribution of archaea to total prokaryotic production in the deep Atlantic ocean. *Applied and Environmental Microbiology* 71:2303–2309.

Hopkins, D. W., Eberling, B., Greenfield, L. G. et al. 2005. Soil micro-organisms in Antarctic dry valleys: Resource supply and utilization. In: *Microorganisms and Earth Systems: Advances in Geomicrobiology, SGM Symposium 65G*, eds. M. Gadd, K. T. Semple, and H. M. Lappin-Scott, pp. 71–84. Cambridge, U.K.: Cambridge University Press.

Hopkins, D. W., Sparrow, A. B., Eberling, B. et al. 2006. Carbon, nitrogen and temperature controls on microbial activity in soils from an Antarctic dry valley. *Soil Biology Biochemistry* 38:3130–3140.

Huber, H., Hohn, M. J., Rachel, R., Fuchs, T., Wimmer, V. C., and K. O. Stetter. 2002. A new phylum of Archaea represented by a nanosized hyperthermophilic symbiont. *Nature* 417:63–67.

Junge, K., Eicken, H., and J. W. Deming. 2004. Bacterial activity at −2 to −20°C in Arctic wintertime sea ice. *Applied and Environmental Microbiology* 70:550–557.

Karr, E. A., Ng, J. M., Belchik, S. M. et al. 2006. Biodiversity of methanogenic and other Archaea in the permanently frozen Lake Fryxell, Antarctica. *Applied and Environmental Microbiology* 72:1663–1666.

Kendall, M. M., Wardlaw, G. D., Tang, C. F., Bonin, A. S., Liu, Y., and D. L. Valentine. 2007. Diversity of Archaea in marine sediments in Skan Bay, Alaska, including methanogens, and description of *Methanogenium boonei* sp. nov. *Applied and Environmental Microbiology* 73:407–414.

Könneke, M., Bernhard, A. E., de la Torre, J., Walker, C. B., Waterbury, J. B., and D. A. Stahl. 2005. Isolation of an autotrophic ammonia-oxidising marine archaeon. *Nature* 437:543–546.

Lee, S. H., Bidle, K., Falkowski, P., and D. Marchant. 2005. Genomic DNA extracted from ancient glaciers ice of molecular analyses on the indigenous microbial communities. *Ocean and Polar Research* 27:205–214.

Leininger, S., Urich, T., Schloter, M. et al. 2006. Archaea predominate among ammonia-oxidising procaryotes in soils. *Nature* 442:806–809.

López-García, P., López-López, A., Moreira, D., and F. Rodríguez-Valera. 2001a. Diversity of free-living prokaryotes from a deep-sea site at the Antarctic Polar Front. *FEMS Microbiology Ecology* 36:193–202.

López-García, P., Moreira, D., López-López, A., and F. Rodríguez-Valera. 2001b. A novel haloarchaeal-related lineage is widely distributed in deep oceanic regions. *Environmental Microbiology* 3:72–78.

Massana, R., DeLong, E. F., and C. Pedrós-Alió. 2000. A few cosmopolitan phylotypes dominate planktonic archaeal assemblages in widely different oceanic provinces. *Applied and Environmental Microbiology* 66:1777–1787.

Massana, R., Taylor, L. T., Murray, A. E., Wu, K. Y., Jeffrey, W. H., and E. F. DeLong. 1998. Vertical distribution and temporal variation of marine planktonic Archaea in the Gerlache Strait, Antarctica, during early spring. *Limnology & Oceanography* 43:607–617.

Murray, A. E., Preston, C. M., Massana, R. et al. 1998. Seasonal and spatial variability of bacterial and archaeal assemblages in the coastal waters near Anvers Island, Antarctica. *Applied and Environmental Microbiology* 64:2585–2595.

Nicol, G. W. and C. Schleper. 2006. Ammonia-oxidising Crenarchaeota: Important players in the nitrogen cycle? *Trends in Microbiology* 14:207–212.

Nicol, G. W., Leninger, S., Schleper, C., and J. I. Prosser. 2008. The influence of pH on the diversity, abundance and transcriptional activity of ammonia oxidizing archaea and bacteria. *Environmental Microbiology* 10:2955–2978.

Paper, W., Jahn, U., Hohn, M. J., Kronner, M., Nather, D. J., Burghardt, T., Rachel, R., Stetter, K. O., and H. Huber. 2007. *Ignicoccus hospitalis* sp. nov., the host of "*Nanoarchaeum equitans*". *International Journal of Systematic and Evolutionary Microbiology* 57:803–808.

Powell, S. A. 2004. Molecular microbial ecology of contaminated marine sediment near Casey Station, Antarctica. Ph.D. dissertation, University of Tasmania, Tasmania, Australia.

Powell, S. A., Bowman, J. P., Snape, I., and J. S. Stark. 2003. Microbial community variation in pristine and polluted nearshore Antarctic sediments. *FEMS Microbiology Ecology* 45:135–145.

Purdy, K. J., Nedwell, D. B., and T. M. Embley. 2003. Analysis of the sulphate-reducing and methanogenic archaeal populations in contrasting Antarctic sediments. *Applied and Environmental Microbiology* 69:3181–3191.

Schleper, C., Jurgens, G., and M. Jonuscheit. 2005. Genomic studies of uncultivated Archaea. *Nature Reviews Microbiology* 3:479–488.

Singh, N., Kendall, M. M., Liu, Y., and D. R. Boone. 2005. Isolation and characterization of methylotrophic methanogens from anoxic marine sediments in Skan Bay, Alaska: Description of *Methanococcoides alaskenese* sp. nov., and emended description of *Methanosarcina baltica. International Journal of Systematic and Evolutionary Microbiology* 55:2531–2538.

Smith, M. C., Bowman, J. P., Scott, F. J., and M. A. Line. 2000. Sublithic bacteria associated with Antarctic quartz stones. *Antarctic Science* 12:2391–2396.

Soo, R., Wood S. A., Grzymski J. J., McDonald, I. R., and S. C. Cary. 2009. Microbial biodiversity of thermophilic communities in hot mineral soils of Tramway Ridge, Mount Erebus, Antarctica. *Environmental Microbiology* 11:715–728.

Venter, J. C., Remington, J., and J. F. Heidelberg. 2004. Environmental genomics and sequencing of the Sargasso Sea. *Science* 304:66–74.

Webster, N. S. and A. P. Negri. 2006. Site-specific variation in Antarctic marine biofilms established on artificial surfaces. *Environmental Microbiology* 8:1177–1190.

Webster, N. S., Negri, A. P., Munro, M. M. H. G., and C. N. Battershill. 2004. Diverse microbial communities inhabit Antarctic sponges. *Environmental Microbiology* 6:288–300.

Woese, C. R. and G. E. Fox. 1977. Phylogenetic structure of the prokaryotic domain: Proposals for the domains Archaea, Bacteria, and Eucarya. *Proceedings of the National Academy of Sciences USA* 74:5088–5090.

Wuchter, C., Abbas, B., Coolen, M. J. L. et al. 2006. Archaeal nitrification in the ocean. *Proceedings of the National Academy of Sciences USA* 103:12317–12322.

Yergeau, E., Kang, S., He, Z., Zhou, J., and G. A. Kowalchuk. 2007. Functional microarray analysis of nitrogen and carbon cycling genes across an Antarctic latitudinal transect. *ISME Journal* 1:163–179.

3 Bacterial Biodiversity of Antarctica: Conventional Polyphasic and rRNA Approaches

Sisinthy Shivaji and G.S.N. Reddy

CONTENTS

3.1 INTRODUCTION

The continent of Antarctica is considered to be one of the most extreme habitats in the world because it is the coldest, the driest (Claridge and Campbell, 1977; Vincent, 1988; Campbell and Claridge, 2000), the windiest, and the iciest of all known habitats with high solar (ultraviolet) radiation at least during the summer season (Smith et al., 1992). Despite the harsh climatic conditions, certain life forms such as mites, ticks, seals, penguins, mosses, lichens, bacteria, yeasts, and algae do exist in Antarctica (Cameron et al., 1970; Vishniac and Mainzer, 1972; Vincent, 1988; Wynn-Williams, 1990; Shivaji, 2005). Among these life forms, the cold-loving (psychrophilic) and cold-tolerant (psychrotrophic) bacteria are predominant and they play a key role in the Antarctic ecosystem, especially with respect to nutrient recycling (Stokes and Redmond, 1966; Delille and Lagarde, 1974; Herbert and Bell, 1974; Kelly et al., 1978; Tanner and Herbert, 1981; Tanner, 1985; Voytek and Ward, 1995; Cavanagh et al., 1996; Cavicchioli and Thomas, 2000; Chessa et al., 2000; Denner et al., 2001). Therefore, studies on the diversity, physiology, and molecular biology of psychrophilic bacteria would provide important inputs on their distribution, survival strategies, and the molecular basis of their adaptation to low temperature. This chapter is essentially focused on the biodiversity of the bacteria of Antarctica.

Erik Ekelöf is regarded as the father of Antarctic microbiology, and he was the first to demonstrate the presence of bacteria, yeasts, and fungi in the soil and air at Snow Hill Island, Antarctic Peninsula (Ekelöf, 1908a,b). These pioneering studies of Ekelöf were confirmed by Pirie (1904, 1912), Gazert (1912), and Tsilinsky (1908), who reported the presence of bacteria in soil and water samples from Antarctica. However, it was McLean (1918a,b) who provided the first evidence that these heterotrophic bacteria from Antarctica not only survive but also multiply at low temperatures (1°C–2°C). Despite this intensive activity for two decades (1901–1919), the impetus for Antarctic microbiology could be traced to the International Geophysical Year (1957–1958) when the emphasis was on enumerating bacteria from various habitats of Antarctica (Flint and Stout, 1960; Straka and Stokes, 1960) and to identify these unique microorganisms (Boyd, 1962; Boyd and Boyd, 1962; Meyer et al., 1962; Margini and Castrelos, 1963, 1965; Pfiser and Burkholder, 1965; Marshall and Ohye, 1966; Friedmann, 1980; Tsyganov, 1970) and ascertaining their distribution in various environmental sites such as coastal waters, sea-ice, soil, glaciers, and lakes. Further, the advent of the culture-independent identification of microorganisms by directly cloning the 16S rRNA gene sequences from environmental samples led to the realization that the bacterial diversity of Antarctica is far greater than that detected by the cultivable approach. The browsing of the NCBI (http://www.ncbi.nlm.nih.gov) database for 16S rRNA gene sequences indicated that about 3236 16S rRNA gene sequences of bacteria from Antarctica have been deposited in the database and a vast majority of the sequences (2563) belong to the noncultivable Gram-negative and Gram-positive bacteria, and the remaining sequences corresponded to the viable isolates so far described from Antarctica (Table 3.1).

3.2 BACTERIAL DIVERSITY OF ANTARCTICA USING THE POLYPHASIC APPROACH

The abundance of cultivable bacteria in Antarctica, as in other continents, is known to vary a great degree depending on the specific habitat and may range from a minimum of < 100 cfu in the case of glacial ice to as many as 0.6×10^{12} cfu g^{-1} in the case of sea-ice (Straka and Stokes, 1960; Delille and Gleizon, 2003). The bacterial number in soil (1.0×10^2 to 10^7 cfu g^{-1}; Straka and Stokes, 1960; Zdanowski and Weglenski, 2001; Shivaji et al., 2004a; Aislabie et al., 2006b), snow (0.2×10^2 to 0.5×10^4 cells mL^{-1} of snow melt; Carpenter et al., 2000), ice (1.0×10^2 to 1.02×10^{12} cells m^{-3}; Straka and Stokes, 1960; Sullivan and Palmisano, 1984), water (10^2 to 10^7 cells mL^{-1}; Takii et al., 1986; Franzmann et al., 1990), and accreted ice (2×10^2 to 3.6×10^4 cfu; Karl et al., 1999; Priscu et al., 1999) varied to a great extent. The variation in the abundance of the bacteria in the different habitats may reflect the water or the nutrient content of the habitat (Archer et al., 1996a; Freckman and Virginia, 1997; Bargagli et al., 1999; Bowman and McCuaig, 2003; Sinclair et al., 2003; Barrett et al., 2006; Howard-Williams et al., 2006; Yergeau et al., 2007a). The abundance and the composition

TABLE 3.1
A List of Bacteria Isolated from Various Habitats from the Antarctic Continent

Name of Species	Habitat	Reference
α-Proteobacteria		
Antarctobacter heliothermusa	Lake water	Labrenz et al. (1998)
Brevundimonas vesicularis	Soil	Wery et al. (2003)
Devosia sp.	Microbial mat	Van Trappen et al. (2002)
Loktanella fryxellensis[a]	Microbial mat	Van Trappen et al. (2004a)
Loktanella salsilacus	Microbial mat	Van Trappen et al. (2004a)
Loktanella vestfoldensis	Microbial mat	Van Trappen et al. (2004a)
Mesorhizobium sp.	Microbial mat	Van Trappen et al. (2002)
Methylobacterium sp.	Lake water	Christner et al. (2001)
Octadecabacter antarcticus	Sea-ice	Gosink and Staley (1995)
Paracoccus sp.	Sea water	Michaud et al. (2004)
Porphyrobacter sp.	Microbial mat	Van Trappen et al. (2002)
Robiginitomaculum antarcticum[a]	Sea water	Lee et al. (2007a)
Roseisalinus antarcticus[a]	Lake water	Labrenz et al. (2005)
Roseobacter denitrificans	Lake water	Soller et al. (2000)
Roseobacter sp.	Sea-ice	Bowman et al. (1997c)
Roseobacter litoralis	Lake water	Soller et al. (2000)
Roseovarius tolerans[a]	Lake water	Labrenz et al. (1999)
Sphingomonas adhaesiva	Sea-ice	Bowman et al. (1997c)
Sphingomonas aerolata	Soil and ice	Busse et al. (2003)
Staleya guttiformis[a]	Lake water	Labrenz et al. (2000)
Sulfitobacter brevis	Lake water	Labrenz et al. (2000)
β-Proteobacteria		
Achromobacter ruhlandii	Quartz stone	Smith et al. (unpublished)
Aquaspirillum sp.	Microbial mat	Van Trappen et al. (2002)
Chromobacterium lividum	Water maritime	Shivaji et al. (2004a)
Hydrogenophaga sp.	Microbial mat	Van Trappen et al. (2002)
Janthinobacterium lividum	Soil	Shivaji et al. (1991)
Massilia timonae	Soil	Wery et al. (2003)
Polaromonas vacuolata[a]	Marine water	Irgens et al. (1996)
Rhodoferax antarcticus	Microbial mat	Madigan et al. (2000)
γ-Proteobacteria		
Acinetobacter johnsonni	Soil	Wery et al. (2003)
Alteromonas stellipolaris	Sea water	Van Trappen et al. (2004e)
Alteromonas haloplanktis	Sea-ice	Gauthier et al. (1995)
Colwellia demingiae	Sea-ice	Bowman et al. (1998a)
Colwellia hornerae	Sea-ice	Bowman et al. (1998a)
Colwellia psychrerythraea	Sea-ice	Bowman et al. (1998a)
Colwellia psychrotropica	Sea-ice	Bowman et al. (1998a)

(continued)

TABLE 3.1 (continued)
A List of Bacteria Isolated from Various Habitats from the Antarctic Continent

Name of Species	Habitat	Reference
Colwellia rossensis	Sea-ice	Bowman et al. (1998a)
Glaciecola pallidula	Sea-ice	Bowman et al. (1998c)
Glaciecola punicea	Sea-ice	Bowman et al. (1998c)
Granulosicoccus antarcticus[ac]	Sea water	Lee et al. (2007c)
Hahella antarctica	Sea water	Lee et al. (2008)
Halomonas alkaliantarctica	Saline lake	Poli et al. (2007)
Halomonas glaciei	Fast ice	Reddy et al. (2003c)
Halomonas subglaciescola	Sea-ice	Dobson and Franzmann unpublished
Halomonas variabilis	Sea-ice	Bowman et al. (1997c)
Idiomarina loihiensis	Sea water	Zeng et al. (2007)
Marinobacter guineae	Marine sediment	Montes et al. (2008)
Marinobacter maritimus	Sea water	Shivaji et al. (2005a)
Marinomonas polaris	Sea water	Gupta et al. (2006)
Marinomonas ushuaiensis	Sea water	Prabagaran et al. (2005)
Marinomonas protea	Water	Mills et al. unpublished
Methylosphaera hansonii	Lake sediment	Bowman et al. (1997e)
Oleispira antarctica[a]	Sea water	Yakimov et al. (2003)
Pseudoalteromonas antarctica	Marine water	Bozal et al. (1997)
Pseudoalteromonas prydzensis	Sea-ice	Bowman (1998)
Pseudoalteromonas tetraodonis	Water	Hagstrom et al. (2000)
Pseudomonas guineae	Soil	Bozal et al. (2007)
Pseudomonas antarctica	Microbial mat	Reddy et al. (2004)
Pseudomonas meridiana	Microbial mat	Reddy et al. (2004)
Pseudomonas proteolytica	Microbial mat	Reddy et al. (2004)
Pseudomonas azotoformans	Water	Hagstrom et al. (2000)
Pseudomonas fluorescens	Soil	Shivaji et al. (1989b)
Pseudomonas putida	Soil	Shivaji et al. (1989b)
Pseudomonas syringae	Soil	Shivaji et al. (1989b)
Psychrobacter salsus	Fast ice	Shivaji et al. (2004b)
Psychrobacter adeliensis	Fast ice	Shivaji et al. (2004b)
Psychrobacter glacialis	Fast ice	Shivaji et al. (2004b)
Psychrobacter vallis	Microbial mat	Shivaji et al. (2005b)
Psychrobacter aquaticus	Microbial mat	Shivaji et al. (2005b)
Psychrobacter glacincola	Sea-ice	Bowman et al. (1997d)
Psychrobacter nivimaris	Organic debris	Heuchert et al. (2004)
Psychrobacter luti	Marine water	Bozal et al. (2003)
Psychrobacter fozii	Marine water	Bozal et al. (2003)
Psychrobacter frigidicola	Ornithogenic soil	Bowman et al. (1996)
Psychrobacter immobilis	Ornithogenic soil	Bowman et al. (1996)
Psychrobacter urativorans	Ornithogenic soil	Bowman et al. (1996)

TABLE 3.1 (continued)
A List of Bacteria Isolated from Various Habitats from the Antarctic Continent

Name of Species	Habitat	Reference
Psychrobacter pacificensis	Sea water	Maruyama et al. (2000)
Psychrobacter proteolyticus	Krill	Denner et al. (2001)
Psychromonas antarctica[a]	Pond sediment	Mountfort et al. (1998)
Saccharospirillum impatiens[a]	Hypersaline lake	Labrenz et al. (2003)
Shewanella colwelliana	Marine water	Coyne et al. (1989)
Shewanella livingstonenis	Marine water	Bozal et al. (2002)
Shewanella gelidimarina	Sea-ice	Bowman et al. (1997a)
Shewanella frigidimarina	Sea-ice	Bowman et al. (1997a)
Stenotrophomonas maltophilia	Soil	Wery et al. (2003)
δ-Proteobacteria		
Myxococcus stipitatus	Soil	Ruckert (1985)
Myxococcus virescens	Soil	Ruckert (1985)
Bacteroidetes		
Aequorivita crocea[a]	Sea water	Bowman and Nichols (2002)
Aequorivita ferruginea	Quartz stone	Bowman and Nichols (2002)
Aequorivita lipolytica	Sea water	Bowman and Nichols (2002)
Aequorivita antarctica	Sea water	Bowman and Nichols (2002)
Aequorivita sublithincola	Quartz stone	Bowman and Nichols (2002)
Algoriphagus antarcticus	Microbial mat	Van Trappen et al. (2004b)
Bizionia algoritergicola	Copepod saline pond	Bowman and Nichols (2005)
Bizionia myxarmorum	Copepod saline pond	Bowman and Nichols (2005)
Bizionia gelidisalsuginis	Coastal fast sea-ice brine	Bowman and Nichols (2005)
Bizionia saleffrena	Coastal fast sea-ice brine	Bowman and Nichols (2005)
Cellulophaga algicola	Algae	Bowman (2000)
Cyclobacterium scophthalmum	Soil	Wery et al. (2003)
Cytophaga sp.	Sea-ice	Bowman et al. (1997c)
Flavobacterium weaverense	Soil	Yi and Chun (2006)
Flavobacterium segetis	Soil	Yi and Chun (2006)
Flavobacterium frigidimaris	Sea water	Nogi et al. (2005)
Flavobacterium fryxellicola	Microbial mat	Van Trappen et al. (2005)
Flavobacterium antarcticum	Soil sample	Yi et al. (2005a)
Flavobacterium gelidilacus	Microbial mat	Van Trappen et al. (2003)
Flavobacterium degerlachei	Microbial mat	Van Trappen et al. (2004c)
Flavobacterium frigoris	Microbial mat	Van Trappen et al. (2004c)
Flavobacterium micromati	Microbial mat	Van Trappen et al. (2004c)
Flavobacterium psychrolimnae	Microbial mat	Van Trappen et al. (2005)
Flavobacterium frigidarium	Marine sediment	Humphry et al. (2001)
Flavobacterium gillisiae	Sea-ice	McCammon and Bowman (2000)

(continued)

TABLE 3.1 (continued)
A List of Bacteria Isolated from Various Habitats from the Antarctic Continent

Name of Species	Habitat	Reference
Flavobacterium tegetincola	Ace Lake	McCammon and Bowman (2000)
Flavobacterium xanthum	Mud pool	McCammon and Bowman (2000)
Flavobacterium hibernum	Fresh water	McCammon et al. (1998)
Gelidibacter algens[a]	Sea-ice	Bowman et al. (1997b)
Gillisia limnaea[a]	Microbial mats	Van Trappen et al. (2004d)
Gelidibacter gilvus	Algal assemblage	Bowman and Nichols (2005)
Gelidibacter salicanalis	Algal assemblage	Bowman and Nichols (2005)
Gillisia illustrilutea	Algal assemblage	Bowman and Nichols (2005)
Gillisia sandarakina	Algal assemblage	Bowman and Nichols (2005)
Gillisia hiemivivida	Algal assemblage	Bowman and Nichols (2005)
Hymenobacter roseosalivarius[a]	Sand stone	Hirsch et al. (1998)
Lacinutrix copepodicola[a]	Copepod species	Bowman and Nichols (2005)
Leeuwenhoekiella aequorea[a]	Sea water	Nedashkovskaya et al. (2005)
Polaribacter glomeratus	Sea water	McGuire et al. (1987)
Polaribacter filamentus	Sea water	Gosink et al. (1998)
Polaribacter irgensii	Sea water	Gosink et al. (1998)
Polaribacter franzmannii	Sea water	Gosink et al. (1998)
Psychroflexus torquis[a]	Sea-ice	Bowman et al. (1998b)
Psychroflexus gondwanensis	Saline water	Dobson et al. (1993); Bowman et al. (1998b)
Psychroserpens burtonensis[a]	Lake water	Bowman et al. (1997b)
Salegentibacter salegens[a]	Hypersaline lake	McCammon and Bowman (2000)
Sejongia marina	Sea water	Lee et al. (2007b)
Sejongia antarctica	Soil	Yi et al. (2005b)
Sejongia jeonii	Soil	Yi et al. (2005b)
Sphingobacterium antarcticum	Soil	Shivaji et al. (1992)
Sphingobacterium sp.	Soil	Wery et al. (2003)
Subsaximicrobium wynnwilliamsii[a]	Quartz stone	Bowman and Nichols (2005)
Subsaximicrobium saxinquilinus	Quartz stone	Bowman and Nichols (2005)
Subsaxibacter broadyi[a]	Quartz stone	Bowman and Nichols (2005)
Ulvibacter antarcticus	Sea water	Choi et al. (2007)
Actinobacteria		
Arthrobacter globiformis	Lake sediment	Shivaji et al. (1989a)
Arthrobacter pascens	Soil sample	Shivaji et al. (1989a)
Arthrobacter agilis	Lake water	Brambilla et al. (2001)
Arthrobacter protophormiae	Penguin rookery	Shivaji et al. (1989a)
Arthrobacter sulfureus	Soil sample	Wery et al. (2003)
Arthrobacter ardleyensis	Sediment	Chen et al. (2005)

TABLE 3.1 (continued)
A List of Bacteria Isolated from Various Habitats from the Antarctic Continent

Name of Species	Habitat	Reference
Arthrobacter gangotriensis	Penguin rookery	Gupta et al. (2004)
Arthrobacter kerguelensis	Sea water	Gupta et al. (2004)
Arthrobacter roseus	Microbial Mat	Reddy et al. (2002a)
Arthrobacter flavus	Microbial Mat	Reddy et al. (2000)
Brachybacterium conglomeratum	Lake water	Christner et al. (2001)
Cryobacterium psychrophilum[a]	Soil sample	Suzuki et al. (1997)
Friedmanniella antarctica	Sandstone	Schumann et al. (1997)
Friedmanniella lacustris	Hypersaline Lake	Lawson et al. (2000)
Janibacter sp.	Seawater	Michaud et al. (2004)
Kocuria polaris	Microbial mat	Reddy et al. (2003d)
Leifsonia rubra	Microbial mat	Reddy et al. (2003b)
Leifsonia aurea	Microbial mat	Reddy et al. (2003b)
Leifsonia antarctica	Ice core	Pindi et al. (2009)
Luteococcus japonicus	Sand stone	Schumann et al. (1997)
Microbacterium sp.	Microbial mat	Van Trappen et al. (2002)
Micrococcus luteus	Lake water	Brambilla et al. (2001)
Micrococcus antarcticus	Soil	Liu et al. (2000)
Micromonospora endolithica	Sandstone rock	Hirsch et al. (2004b)
Micromonospora coerulea	Sandstone rock	Hirsch et al. (2004b)
Modestobacter multiseptatus[a]	Soil	Mevs et al. (2000)
Nesterenkonia lacusekhoensis	Hypersaline Lake	Collins et al. (2002)
Nocardioides aquaticus	Water Lake	Lawson et al. (2000)
Nocardiopsis antarcticus	Ice sheet	Abyzov et al. (1983)
Pseudonocardia antarctica	Moraine sample	Prabahar et al. (2004)
Rhodococcus fascians	Quartz stone	Smith et al. (unpublished)
Rhodoglobus vestalii[a]	Lake water	Sheridan et al. (2003)
Sanguibacter antarcticus	Sea sand	Hong et al. (2008)
Firmicutes		
Alicyclobacillus pohliae[b]	Geothermal soil	Imperio et al. (2008)
Aneurinibacillus terranovensis[b]	Geothermal soil	Allan et al. (2005)
Anoxybacillus amylolyticus[b]	Geothermal soil	Poli et al. (2006)
Bacillus fumarioli	Soil	Logan et al. (2000)
Bacillus shackletonii	Soil	Logan et al. (2004a)
Bacillus luciferensis	Soil	Logan et al. (2002)
Bacillus macquariensis	Soil	Marshall and Ohye (1966)
Bacillus marinus	Sea sediment	Rüger et al. (2000)
Bacillus thuringiensis	Soil	Rüger et al. (2000)
Bacillus thermantarcticus[b]	Geothermal soil	Lama et al. (1996)
Brevibacillus levickii[b]	Geothermal soil	Allan et al. (2005)

(*continued*)

TABLE 3.1 (continued)
A List of Bacteria Isolated from Various Habitats from the Antarctic Continent

Name of Species	Habitat	Reference
Carnobacterium funditum	Anoxic water	Franzmann et al. (1991)
Carnobacterium alterfunditum	Anoxic water	Franzmann et al. (1991)
Clostridium vincentii	Sediment of ice shelf	Mountfort et al. (1997)
Clostridium schirmacherense	Lake sediment	Alam et al. (2006)
Clostridium frigoris	Microbial mat	Spring et al. (2003)
Clostridium lacusfryxellense	Microbial mat	Spring et al. (2003)
Clostridium bowmanii	Microbial mat	Spring et al. (2003)
Clostridium psychrophilum	Microbial mat	Spring et al. (2003)
Clostridium estertheticum subsp. *laramiense*	Microbial mat	Spring et al. (2003)
Clostridium perfringens	Soil	Miwa (1975)
Clostridium bifermentans	Soil	Miwa (1975)
Clostridium sordellii	Soil	Miwa (1975)
Clostridium sporogenes	Soil	Miwa (1975)
Clostridium plagarum	Soil	Miwa (1975)
Clostridium paraperfringens	Soil	Miwa (1975)
Clostridium septicum	Soil	Miwa (1975)
Clostridium tertium	Soil	Miwa (1975)
Clostridium cadaveris	Soil	Miwa (1975)
Clostridium butyricum	Soil	Miwa (1975)
Clostridium felsineum	Soil	Miwa (1975)
Exiguobacterium undae	Pond water	Frühling et al. (2002)
Exiguobacterium antarcticum	Pond water	Frühling et al. (2002)
Jeotgalicoccus pinnipedialis	Elephant seal	Hoyles et al. (2004)
Paenibacillus antarcticus	Lake Sediment	Montes et al. (2004)
Paenibacillus amylolyticus	Soil	Wery et al. (2003)
Paenibacillus cineris	Volcanic soil	Logan et al. (2004b)
Paenibacillus cookie	Volcanic soil	Logan et al. (2004b)
Paenibacillus wynnii	Soil	Rodríguez-Díaz et al. (2005)
Planococcus maitriensis	Microbial mat	Alam et al. (2003)
Planococcus antarcticus	Microbial mat	Reddy et al. (2002b)
Planomicrobium psychrophilum	Microbial mat	Reddy et al. (2002b)
Planomicrobium okeanokoites	Soil sample	Wery et al. (2003)
Planomicrobium mcmeekinii	Sea-ice brine	Junge et al. (1998)
Psychrosinus fermentans[a]	Water	Sattley et al. (2008)
Sporosarcina macmurdoensis	Microbial mat	Reddy et al. (2003a)
Deinococcus/Thermus		
Deinococcus frigens	Soil and rock	Hirsch et al. (2004a)
Deinococcus saxicola	Soil and rock	Hirsch et al. (2004a)
Deinococcus marmoris	Soil and rock	Hirsch et al. (2004a)

TABLE 3.1 (continued)
A List of Bacteria Isolated from Various Habitats from the Antarctic Continent

Name of Species	Habitat	Reference
Spirochaetes		
Anaeroplasma sp.	Soil	Franzmann and Dobson (1992)

Note: Some of the species have previously been undescribed.
[a] Represents novel genera.
[b] Represents thermophilic strains.
[c] Represents a novel family.

of the bacterial populations are also likely to vary depending on the season of the year (Gibson et al., 1990; James et al., 1994; Delille and Rosiers, 1995; Leakey et al., 1996; Delille et al., 1997).

As of now, the various habitats from which bacterial diversity has been studied include soil samples (Flint and Stout, 1960; Boyd and Boyd, 1963a,b; Marshall and Ohye, 1966; Herbert and Bell, 1974; Miwa, 1975; Shivaji et al., 1988, 1989a,b, 1991, 1992, 2004a), pyritic sediments (Barghoorn and Nichols, 1961; Meyer et al., 1962), seawater (Pfiser and Burkholder, 1965; Maugeri et al., 1996; Bruni et al., 1999; Michaud et al., 2004; Prabagaran et al., 2007), sea-ice (Archer et al., 1996b; Gosink and Staley, 1995; Staley and Gosink, 1999; Bowman et al., 1997c), brine channels of sea-ice (Sullivan and Palmisano, 1984), ice cores (Abyzov et al., 2001), frozen lakes (Karl et al., 1999; Priscu et al., 1999), marine plankton (Zeng et al., 2007), penguin rookery (Shivaji et al., 1989b), cryoconite holes (Christner et al., 2003), cyanobacterial mats (Reddy et al., 2000, 2002a,b; 2003a,b,c,d; Van Trappen et al., 2002; Alam et al., 2003; Reddy et al., 2004; Shivaji et al., 2004b; Shivaji, 2005b), hypersaline lakes (Bowman et al., 2000a), and various other habitats (Meyer, 1962; Margini and Castrelos, 1963, 1965; Tsyganov, 1970; Friedmann, 1980).

The bacterial isolates from all the aforementioned habitats were affiliated to the lineages Proteobacteria, Bacteroidetes, Actinobacteria, and Firmicutes, and a few have also been identified to be associated with *Deinococcus/Thermus* and Spirochaetes. Table 3.1 provides a list of the bacterial strains isolated from various Antarctic sources. Some of these cultures attracted more interest because of certain unique features like the ice-nucleating bacterium *Pseudomonas antarctica* IN-74 from Ross Island (Obata et al., 1999) and the thermophilic bacteria *Alicyclobacillus pohliae*, *Aneurinibacillus terranovensis*, *Anoxybacillus amylolyticus*, *Bacillus thermantarcticus*, and *Brevibacillus levickii* from geothermal soils of Mount Melbourne and Mount Rittmann (Lama et al., 1996; Allan et al., 2005; Poli et al., 2006; Imperio et al., 2008) with a growth temperature ranging from 40°C to 65°C. Similarly, Franzmann and Dobson (1992) isolated so far the only known strain of the genus *Anaeroplasma*, a member of the order Spirochaetales.

So far, only one previously undescribed family, Granulosicoccaceae of γ-Proteobacteria, has been described from Antarctica, and it was isolated from seawater. It includes a single species, *Granulosicoccus antarcticus*, of the genus *Granulosicoccus* (Lee et al., 2007c). The number of previously undescribed genera that have been described so far from Antarctica are 26 and include the genera *Loktanella* (Van Trappen et al., 2004a), *Robiginitomaculum* (Lee et al., 2007a), *Roseovarius* (Labrenz et al., 1999), *Roseisalinus* (Labrenz et al., 2005), and *Staleya* (Labrenz et al., 2000) of α-Proteobacteria; *Polaromonas* (Irgens et al., 1996) of β-Proteobacteria, *Granulosicoccus* (Lee et al., 2007c), *Oleispira* (Yakimov et al., 2003), *Psychromonas* (Mountfort et al., 1998), and *Saccharospirillum* (Labrenz et al., 2003) of γ-Proteobacteria; *Aequorivita* (Bowman and Nichols, 2002), *Gelidibacter* (Bowman et al., 1997b), *Gillisia* (Van Trappen et al., 2004d), *Hymenobacter* (Hirsch et al., 1998), *Lacinutrix* (Bowman and Nichols, 2005), *Leeuwenhoekiella* (Nedashkovskaya et al., 2005), *Psychroflexus* (Bowman et al., 1998b), *Psychroserpens* (Bowman et al., 1997b), *Salegentibacter* (McCammon and Bowman, 2000), *Subsaximicrobium* (Bowman and Nichols, 2005), and *Subsaxibacter* (Bowman and Nichols, 2005) of Bacteroidetes, *Antarctobacter* (Labrenz et al., 1998); *Cryobacterium* (Suzuki et al., 1997), *Modestobacter* (Mevs et al., 2000), and *Rhodoglobus* (Sheridan et al., 2003) of Actinobacteria; and *Psychrosinus* (Sattley et al., 2008) of Firmicutes. Out of these 26 genera, 12 (*Roseisalinus, Staleya, Oleispira, Granulosicoccus, Saccharospirillum, Lacinutrix, Psychroserpens, Subsaximicrobium, Subsaxibacter, Antarctobacter, Rhodoglobus,* and *Psychrosinus*) have been uniquely isolated from Antarctica and so far no species affiliated to these genera have been described from other places. It is interesting to note that all the Antarctic isolates of bacteria were psychrotolerant and could be differentiated from other closely related species based on their phenotypic and chemotaxonomic characteristics and also at the 16S rRNA gene sequence level.

3.3 BACTERIAL DIVERSITY OF ANTARCTICA AS STUDIED BY NONCULTIVABLE APPROACHES

The bacterial abundances in Antarctic (Hagson et al,, 1983a,b; Bird and Karl, 1991; Karl et al., 1999) and sub-Antarctic ecosystems (Delille and Bouvy, 1989) do not correlate with heterotrophic activity, implying that the ratio of cultivable bacteria to bacterial abundance was low. Thus, there is an urgent need to study the "viable and noncultivable fractions" of the bacterial assemblage in Antarctic ecosystems using culture-independent methods. Culture-independent approaches based on 16S rRNA gene sequences and DGGE and RFLP fingerprinting were initiated almost a decade ago for analyzing the bacterial diversity of various habitats in Antarctica (Gordon et al., 2000; Bowman et al., 2000b; Brambilla et al., 2001; Bowman and McCuaig, 2003; Shivaji et al., 2004a; Mikucki and Priscu, 2007; Prabagaran et al., 2007; Niederberger et al., 2008). Using the aforementioned approaches, representative examples showing the breadth of the bacterial diversity of various Antarctic habitats are summarized in Table 3.2. The compilation of the bacterial diversity data from the aforementioned studies indicated that irrespective of the habitat and the geographical location, bacterial clones belonging to 214 genera were detected besides unaffiliated

16S rRNA clones or sequences (Table 3.2). The 214 bacterial genera belonged to the classes α-Proteobacteria (21 genera), β-Proteobacteria (22 genera), γ-Proteobacteria (33 genera), δ-Proteobacteria (29 genera), ε-Proteobacteia (3 genera), Bacteroidetes (26 genera), Cyanobacteria (19 genera), Planctomycetes (3 genera), Acidobacteria (4 genera), Firmicutes (17 genera), Actinobacteria (25 genera), Spirochaete (1 genus), *Deinococcus/Thermus* (1 genus), Chloroflexi (3 genera), Gemmatimonadetes (1 genus), Nitrospira (1 genus), Verrucomicrobia (3 genera), and Aquificae (1 genus) besides copious number of sequences belonging to hitherto unidentified groups (Table 3.2). From all the habitats, bacteria belonging to the divisions/classes Proteobacteria, Bacteroidetes, Firmicutes, Actinobacteria, Cyanobacteria, *Deinococcus/Thermus*, and Acidobacteria were dominant. The other lineages such as Planctomycetes, Spirochaete, Chloroflexi, Gemmatimonadetes, Nitrospira, Verrucomicrobia, and Aquificae were reported from only a few habitats such as sediments (Gordon et al., 2000; Bowman et al., 2000b), sediment core (Bowman and McCuaig, 2003), soil (Shivaji et al., 2004a; Smith et al., 2006; Niederberger et al., 2008), glacier melt water (Sjöling and Cowan, 2003), microbial mats, water samples (Pearce, 2003), and dust (Hughes et al., 2004) (Table 3.2).

The occurrence and distribution of bacteria among the earlier studied habitats from Antarctica were found to be variable with respect to the habitat. Some genera, such as *Sphingomonas, Flavobacterium, Arthrobacter, Rubrobacter,* and uncultured members of the divisions *Deinococcus/Thermus* and Acidobacteria (Table 3.2), were present in most of the habitats. Whereas genera like *Arcobacter, Rhodopila,* and *Thiomicrospira* of ε-Proteobacteria, and *Aquifex, Nitrospira,* and *Fibrobacter* appear to be restricted to sediment cores of Mertz glacier and sand stones from McMurdo dry valleys (Bowman and McCuaig, 2003; de la Torre et al., 2003), water samples from South Orkney (Pearce, 2003), glacial water from Bratina Island (Sjöling and Cowan, 2003), soil samples from Schirmacher Oasis (Shivaji et al., 2004a), or a soil sample from Victoria Land (Niederberger et al., 2008) (Table 3.2).

Culture-independent bacterial diversity studies from most of the Antarctic habitats were primarily based on the 16S rRNA gene sequences. However, Karr et al. (2005), investigated the population of sulfate-reducing bacteria in Lake Fryxell, Taylor Valley by both 16S rRNA gene sequence and oligonucleotide primers targeting the *dsrA* gene (dissimilatory sulfite reductase). The sequence information resulted in the detection of a diverse group of sulfate-reducing prokaryotes of the domain *Bacteria* (Table 3.2). In addition, the anoxygenic phototrophic bacterial diversity of Fryxell Lake was also studied by using the *pufM,* a gene coding for photosynthetic pigment-binding protein (Karr et al., 2003). The results revealed 33 unique phylotypes representing alpha and beta subdivisions of Proteobacteria and were distantly related to *Rubrivivax, Acidiphilum, Rhodoferax,* and *Roseateles.*

3.4 INFLUENCE OF HABITAT ON BACTERIAL DIVERSITY

As in other regions of the world, bacterial diversity in Antarctica is also known to be dependent on the specific habitat studied. For instance, Niederberger et al. (2008) while investigating the soil microbial communities from low- and high-productivity habitats in the Luther Vale, Northern Victoria Land, Antarctica, observed

TABLE 3.2

Bacterial Diversity of Antarctica as Determined by Sequencing of 16S rRNA Gene Clones from Environmental Samples[a]

Habitat	Subglacial out flow	Lake sediments	Anoxic meromictic sediments	Sediment core	Water and sediments	Frozen water	Cyanobacterial mats	Mineral soils	Mat sample	Glacial melt water	Crypto-endolithic sand stone	Soil sample	Mat sample	Soil samples	Soil samples	Water sample	Air sample	Melt water	Soil samples	Soil sample	Seawater
Location	Blood Falls, Taylor Glacier	Lake Bonney, McMurdo Dry Valleys	Vestfold Hills	Mertz Glacier	Lake Fryxell, Taylor Valley	Lake Fryxell, Taylor Valley	Bratina Island, McMurdo	McMurdo Dry Valleys	Lake Fryxell, Taylor Valley	Bratina Island, Ross Sea	McMurdo Dry Valleys	Schirmacher Oasis	Lake Fryxell, Taylor Valley	Scott Base	Victoria Land	South Orkney Island	Rothera point	Lake Vostok	Northern Victoria Land	Cape Evans. McMurdo	Ushuaia, Argentina
Reference	Mikucki and Priscu (2007)	Gordon et al. (2000)	Bowman et al. (2000b)	Bowman and McCuaig (2003)	Karr et al. (2005)	Karr et al. (2003)	Jungblut et al. (2005)	Smith et al. (2006)	Taton et al. (2003)	Sjoling and Cowan (2003)	Torre et al. (2003)	Shivaji et al. (2004a)	Brambilla et al. (2001)	Saul et al. (2005)	Aislabie et al. (2006b)	Pearce (2003)	Hughes et al. (2004)	Christner et al. (2001)	Niederberger et al. (2008)	Shravage et al. (2007)	Prabagaran et al. (2007)
Class/Genera																					
I. α-Proteobacteria (21)	4	17		2 10 18		1		22		22	2 19 22	11 15 19	4 5 6 7 11 12 13 15 19	3 19	22		19	6 22	2 8 9 14 19	16 19 22	16 20 21 22

II. β-Proteobacteria (22)	9 11 13	8 17 19		12		17 18 19						1 6 23	4 8 12 14 16 17	21 22 23	23	2 4 5 10 15 23		3 10	3 7 20	11 12 23	
III. γ-Proteobacteria (33)	1 10 14 24 25 29 31		18 30	3 4 6 17 21 22 23 26 29						2 5 8 12	1 34	7 19 24 28 32	12 16 32	24 32	34			24	9 13 27 32	14 25 32	3 11 15 20 24 25 26 33 34
IV. δ-Proteobacteria (24)	4 6 15 17		4 8 9 13 16 18 24	5 13 15 18	7 8 11 12 13 14					12 21			2 10 19 21 23						1 2 3 20 22 23		
V. ε-Proteobacteria (3)				1 3							2										1 4
VI. Bacteroidetes (26)	1 8 20		2 4 6 8 10 12 13 15 18 21	8				4		8	7	10	4 7 8 9 10 17	4 8 14 22	4 7 12 14 19 27 28	8	10 14	22	4 7	3 4 7 8 10 23 26	4 5 7 8 10 11 16 18 24 25 28

VII. Firmicutes (17)			2 8 14										1 2 3 4 7 8 9 10 13 16 17	5 6 15	18			12		11	
VIII. Actinobacteria (25)	4		26					2 4 7 11 12		6	3 4 5 8 15 20 21 22 24 26	4 21 26	1 6	4 10 13 17 18 19 20 21 23	2 4 9 18	25	2 8 17 20 21 26	21	2 14 16 17 19 21	21	
IX. Cyanobacteria (17)		2 6 11 12	3 13 14 15 17				1 7 8 9 10 11	6 9 10 11 17	5 6 8 9 10 11 15		4				21		9		10		
X. Planctomycetes (3)		1 2 3		2							2								2 3		
XI. Acidobacteria (4)		1		1				1			1	1		1	1		1		1 2 3 4		
XII. Spirochaetes (1)			1	1						1											
XIII. *Deinococcus/Thermus* (1)											1			1	1				1	1	
XIV. Chloroflexi (3)		3		2				1			1	1							1		

XV. Gemmatimonadetes (1)												1					1		1		
XVI. Nitrospira (1)												1									
XVII. Verrucomicrobia (3)			3	1				1				1	1						2		
XVIII. Aquificae (1)										1											

Note: **I. α-Proteobacteria:** 1. *Acidiphilum*, 2. *Amaricoccus*, 3. *Brevundimonas*, 4. *Caulobacter*, 5. *Chelatobacter*, 6. *Holospora*, 7. *Hiyphomonas*, 8. *Liberibacter*, 9. *Methylobacterium*, 10. *Olavius*, 11. *Pedomicrobium*, 12. *Paracoccus*, 13. *Rhodobacter*, 14. *Rhodomicrobium*, 15. *Rhodopseudomonas*, 16. *Roseobacter*, 17. *Roseococcus*, 18. *Ruegeria*, 19. *Sphingomonas*, 20. *Staleya*, 21. *Sulfitobacter*, 22. Uncultured; **II. β-Proteobacteria:** 1. *Acidovorax*, 2. *Achromobacter*, 3. *Aquabacterium* , 4. *Aquaspirillum*, 5. *Bordetella*, 6. *Burkholderia*, 7. *Hydrogenophilus*, 8. *Ideonella*, 9. *Janthinobacterium*, 10. *Leptothrix*, 11. *Nitrosomonas*, 12. *Nitrosospira*, 13. *Oxalobacter*, 14. *Polarimonas*, 15. *Polynuclobacter*, 16. *Propionvibrio*, 17. *Rhodoferax*, *18. Roseateles*, 19. *Rubrivivax* , 20. *Taylorella*, 21. *Ultramicrobium*, 22. *Variovorax*, 23. Uncultured; **III. γ-Proteobacteria:** 1. *Acinetobacter*, 2. *Burkholderia*, 3. *Colwellia*, 4. *Coxiella*, 5. *Curacaobacter*, 6. *Ectothiorhodospira*, 7. *Enterobacter*, 8. *Escherichia*, 9. *Frateuria*, 10. *Fulvimonas*, 11. *Glaciecola*, 12. *Legionella*, 13. *Lysobacter*, 14. *Marinobacter*, 15. *Marinomonas*, 16. *Methylocaldum*, 17. *Moritella*, 18. *Myxobacterium*, 19. *Nevskia*, 20. *Oleispira*, 21. *Oceanospirillum*, 22. *Photobacterium*, 23. *Pseudoalteromonas*, 24. *Pseudomonas*, 25. *Psychrobacter*, 26. *Psychromonas*, 27. *Rhodanobacter*, 28. *Serratia*, 29. *Shewanella*, 30. *Syntrophus*, 31. *Thiomicrospira*, 32. *Xanthomonas*, 33. *Vibrio*, 34. Uncultured; **IV. δ-Proteobacteria:** 1. *Bactriovorax* , 2. *Bdellovibrio*, 3. *Chondromyces*, 4. *Desulfobacterium*, 5. *Desulfobulbus*, 6. *Desulfocapsa*, 7. *Desulfococcus*, 8. *Desulfofaba*, 9. *Desulfofrigus*, 10. *Desulfomonile*, 11. *Desulfonema*, 12. *Desulforhopalus*, 13. *Desulfosarcina*, 14. *Desulfovibrio*, 15. *Desulfuromonas*, 16. *Desulfuromusa*, 17. *Geopsychrobacter*, 18. *Myxobacteria*, 19. *Myxococcus*, 20. *Nannocystis*, 21. *Pelobacter*, 22. *Polyangium*, 23. *Stigmatella*, 24. *Syntrophus*. **V. ε-Proteobacteria:** 1. *Arcobacter*, 2. *Rhodopila*, 3. *Thiomicrospira*, 4. Uncultured; **VI. Bacteroidetes:** 1. *Algoriphagus*, 2. *Anaeroflexus*, 3. *Aequorivita*, 4. *Bacteroides*, 5. *Cellulophaga*, 6. *Cyclobacterium*, 7. *Cytophaga*, 8. *Flavobacterium*, 9. *Flectobacillus*, 10. *Flexibacter*, 11. *Formosa*, 12. *Gelidibacter*, 13. *Haliscomenobacter*, 14. *Hymenobacter*, 15. *Lewinella*, 16. *Maribacter*, 17. *Marinilabilia*, 18. *Polaribacter*, 19. *Pontibacter*, 20. *Psychroflexus*, 21. *Psychroserpens*, 22. *Sphingobacterium*, 23. *Subsaxibacter*, 24. *Tenacibaculum*, 25. *Ulvibacter*, 26. *Vitellibacter*, 27. Uncultured; **VII. Firmicutes:** 1. *Acetivibrio*, 2. *Acetobacterium*, 3. *Acetonema*, 4. *Acidaminobacter*, 5. *Bacillus*, 6. *Sporosarcina*, 7. *Caloramator*, 8. *Clostridium*, 9. *Dendrosporobacter*, 10. *Desulfosporosinus*, 11. *Desulfotomaculum*, 12. *Dolosigranulum*, 13. *Eubacterium*, 14. *Fusibacter*, 15. *Paenibacillus*, 16. *Ruminococcus*, 17. *Sporomusa*, 18. Uncultured; **VIII. Actinobacteria:** 1. *Acidimicrobium*, 2. *Actinob acteria*, 3. *Actinomyces*, 4. *Arthrobacter*, 5. *Blastococcus*, 6. *Clavibacter*, 7. *Conexibacter*, 8. *Frankia*, 9. *Friedmanniella*, 10. *Frigoribacterium*, 11. *Geodermatophilus*, 12. *Kribella*, 13. *Microbacterium*, 14. *Microlunatus*, 15. *Microsphaera* , 16. *Microthrix*, 17. *Nocardioides*, 18. *Nesterenkonia*, 19. *Pseudonocardia*, 20. *Rhodococcus*, 21. *Rubrobacter*, 22. *Sporichthya*, 23. Streptomyces, 24. *Terrabacter*, 25. *Tetrasphaera*, 26. Uncultured; **IX. Cyanobacteria:** 1. *Anabaena*, 2. *Chamaesiphon*, 3. *Chlamydomonas*, 4. *Chlorella*, 5. *Hydrocoryne*, 6. *Leptolyngbya*, 7. *Lyngbya*, 8. *Nodularia*, 9. *Nostoc*, 10. *Oscillatoria*, 11. *Phormidium*, 12. *Plectonema*, 13. *Prochlorococcus*, 14. *Schizothrix*, 15. *Skeletonema*, 16. *Synechococcus*, 17. *Trichormus*, 18. Uncultured; **X. Planctomycetes:** 1. *Isosphaera*, 2. *Planctomyces*, 3. *Pirellula;* **XI. Acidobacteria:** 1. *Acidobacterium*, 2. *Fibrobacter*, 3. *Geothrix*, 4. *Holophaga;* **XIV. Chloroflexi:** 1. *Chloroflexus*, 2. *Dehalococcoides*; 3. unclassified **XVII. Verrucomicrobia:** 1. *Verrucomicrobia*, 2. *Opitutus*, 3. *Parachlamydia*.

[a] The number in bold in column indicates the number of genera detected in this class in all the habitats listed in row 1. Bacteria related to a particular genus under a specific class is given a number in columns 2–22 and identified in the note.

distinct differences in community structure. Strains affiliated to *Deinococcus/Thermus* lineage were found exclusively in low-productivity soils, while the genus *Xanthomonas* of class γ-Proteobacteria were found exclusively in nutrient-rich soils. Similarly, Saul et al. (2005) while studying the impacts of hydrocarbon contamination on the diversity of bacterial communities in coastal soil from Ross Island identified a diverse population of bacteria belonging to the divisions Acidobacteria, Bacteroidetes, *Deinococcus/Thermus,* and Firmicutes which occurred almost exclusively in control soils whereas *Pseudomonas, Sphingomonas*, and *Variovorax* of the division Proteobacteria were dominant in contaminated soils. Bacterial diversity in sub-Antarctic seawater, and seawater contaminated with crude oil, was statistically different (*P* value 0.001) (Prabagaran et al., 2007). Clones associated with the genera *Roseobacter, Sulfitobacter, Staleya, Glaciecola, Colwellia, Marinomonas, Cytophaga,* and *Cellulophaga* were common to both the libraries. However, 16S rRNA clones associated with *Psychrobacter, Arcobacter, Formosa algae, Polaribacter, Ulvibacter,* and *Tenacibaculum* were found only in seawater contaminated with hydrocarbons. Further, the percentage of clones of *Roseobacter, Sulfitobacter,* and *Glaceicola* was high in seawater (43%, 90%, and 12%, respectively) compared to seawater contaminated with hydrocarbons (35%, 4%, and 9%, respectively) (Prabagaran et al., 2007). Mikucki and Priscu (2007) reported that in the subglacial outflow (wherein the brine is released from below the glacier) of the Taylor Glacier, McMurdo Dry Valleys, the most abundant community was the γ-Proteobacteria (46%) with a clone closely related to a bacterium *Thiomicrospira arctica*. Bowman and McCuaig (2003) assessed the vertical distribution of bacterial community within a sediment core obtained from Mertz Glacier Polynya (MGP) region and observed that the surface oxic sediments (depth 0–0.4 cm) was dominated by gamma and delta proteobacterial phylotypes whereas at a depth of 20–21 cm, a decline in species richness was accompanied by a decrease in Proteobacteria. Yergeau et al. (2007b) studied bacterial diversity in soils sampled along a >3200 km southern polar transect spanning a gradient of increased climate severity over 27° of latitude. The results indicated a decline in the bacterial diversity with increased latitude. Further, the geographical distance and vegetation cover were also found to significantly influence the bacterial diversity. Among the various habitats studied in Antarctica, Lake Vostok, the largest subglacial lake in Antarctica, is probably the only lake of its kind that has been sampled and the diversity of an ice core obtained from this lake at a depth of 3591–3593 m below sea level, yielded clones belonging to only seven genera (Karl et al., 1999; Priscu et al., 1999; Christner et al., 2001). The seven closest genera were *Holospora* of α-Proteobacteria, *Leptothrix* and *Aquabacterium* of β-Proteobacteria, *Pseudomonas* of γ-Proteobacteria, *Sphingobacterium* of Bacteroidetes, *Dolosigranulum* of Firmicutes, and *Rubrobacter* of Actinobacteria. Christner et al. (2001) suggested that the low bacterial diversity of an ice core from Lake Vostok could be attributed to the pristine isolation of the lake for more than 0.42 million years.

Hughes et al. (2004) were the first to study the airborne bacterial diversity over Rothera point of the Antarctic Peninsula. The results indicated the occurrence of a range of microorganisms including Cyanobacteria, Actinobacteria, and yet to be cultured bacterial phylotypes (Table 3.2). Interestingly, the closest matches for sequences were from Antarctic clones that were already deposited in the database, indicating that the microorganisms have dispersed the Antarctic Peninsula from the continental shelf.

Thus discounting the discrepancies attributed to the reported limitations with respect to DNA isolation, PCR amplification, and cloning techniques (Wintzingerode et al., 1997; Dahllöf et al., 2000; Pontes et al., 2007) associated with 16S rRNA gene sequence based studies, the nonconventional methods used in determining the bacterial diversity from Antarctic habitats resulted in the identification of a large number of unique phylotypes representing previously undescribed taxa.

3.5 CULTIVABLE VERSUS NONCULTIVABLE APPROACH

Bacterial diversity as studied by noncultivable methods has led to the identification of 214 genera (Table 3.2) (Figure 3.1) belonging to the class/divisions such as Proteobacteria, Bacteroidetes, Firmicutes, Acidobacteria, and Actinobacteria lineages. Rare groups such as Spirochaetes, *Deinococcus/Thermus*, Chlorflexi, Gemmatimonadetes, Nitrospira, Verrucomicrobia, and Aquificae have also been detected. In contrast, the conventional methods involving the culturing of bacteria have detected only 96 genera (Table 3.1) (Figure 3.1) belonging mainly to Proteobacteria, Bacteroidetes, Actinobacteria, and Firmicutes.

3.6 BACTERIA FROM ANTARCTICA INVOLVED IN HYDROCARBON DEGRADATION

Increased anthropogenic activities in the polar regions have led to accidental hydrocarbon spills around the scientific bases in Antarctica (Whyte et al., 1999; Aislabie et al., 2004) and in the open sea (Karl, 1992; Simpson et al., 1995) respectively. However, our knowledge and understanding of hydrocarbon degradation in the polar regions by microorganisms is limited. In extreme habitats, such as in the continent of Antarctica, cold and fluctuating temperatures, low nutrient levels, low moisture content, and alkaline pH do not favor efficient biodegradation (Atlas, 1986; Morgan and Watkinson, 1989; Bragg et al., 1994; Atlas and Cerniglia, 1995; Aislabie et al., 1998, 2006a; Margesin and Schinner, 1999; Whyte et al., 1999, 2001; Thomassin-Lacroix et al., 2002; Rike et al., 2005). Generally, biodegradation (Margesin and Schinner, 1999) is delayed and occurs mainly during the austral summer season when soils are thawed and water is available (Atlas, 1986). Therefore, studies have evaluated the benefits of biostimulation by fertilizing the soil with nitrogen and/or phosphorus so as to enhance hydrocarbon mineralization (Braddock et al., 1997; Aislabie et al., 1998; Whyte et al., 1999, 2001) or by addition of fertilizers (INIPOL EAP 22, fish composts etc.) to sub-Antarctic and Antarctic seawater (Delille and Vaillant, 1990; Delille et al., 1998), Antarctic sea-ice (Delille et al., 1997), Antarctic soils (Delille, 2000), and to sub-Antarctic intertidal sediments (Delille and Delille, 2000) and found this approach to be more acceptable. Bioremediation experiments indicated that

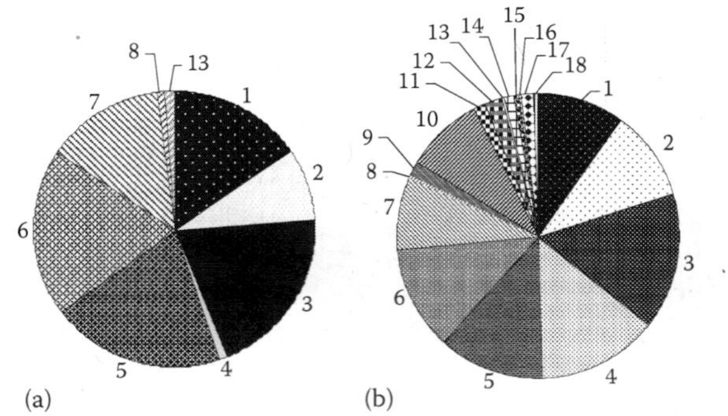

FIGURE 3.1 Bacteria belonging to different classes/divisions in various habitats of Antarctica identified by cultivable (a) and noncultivable methods (b). The numbers 1 to 20 represent the bacteria belonging to the classes/divisions as follows: 1. α-Proteobacteria, 2. β-Proteobacteria, 3. γ-Proteobacteria, 4. δ-Proteobacteria, 5. Bacteroidetes, 6. Actinobacteria, 7. Firmicutes, 8. *Deinococcus/Thermus*, 9. ε-Proteobacteria, 10. Cyanobacteria, 11. Planctomycetes, 12. Acidobacteria, 13. Spirochaetes, 14. Chloroflexi, 15. Gemmatimonadetes, 16. Nitrospira, 17. Verrucomicrobia, and 18. Aquificae.

nutrient and/or water addition to polar soil generally enhance the degradation of hydrocarbon (Snape et al., 2003; Coulon et al., 2004).

Bacteria (Aislabie et al., 2006a) and yeasts (Margesin et al., 2003a,b) mainly bring about the degradation of hydrocarbon in the poles. The bacteria are affiliated to the genera *Rhodococcus, Pseudomonas* (Aislabie et al., 2006a) and the yeasts to the genera *Rhodotorula, Candida,* and *Cryptococcus* (Margesin and Schinner, 1997; Margesin et al., 2003b). Earlier studies had indicated distinct and prominent changes in the bacterial community following biostimulation and/or bioaugmentation by hydrocarbons. However, in these earlier studies the exposure to hydrocarbon was for longer periods ranging from a few weeks to a year (Mohn et al., 2001; Thomassin-Lacroix et al., 2002; Coulon and Delille, 2006). In contrast, in a recent study, bacterial diversity following the addition of water soluble fraction (WSF) of crude oil to seawater showed significant changes though the treatment lasted only for 2 days (Prabagaran et al., 2006). For instance, clones identifying with *Psychrobacter, Arcobacter, Formosa algae, Polaribacter, Ulvibacter,* and *Tenacibaculum* were found only in WSF-added seawater. Increase in their numbers would mean that WSF addition enriched these clones but whether they are involved in biodegradation needs to be demonstrated. It was also observed that clones related to *Psychromonas, Vibrio, Cytophaga,* and *Maribacter,* which were present in seawater, were no longer detectable in seawater with water-soluble fraction (WSF) of crude oil, probably because of toxic effects of the hydrocarbons present (Prabagaran et al., 2006).

Whyte et al. (1996) found that catabolic genes from several aromatic-degrading psychrotolerant strains had homology to those described in mesophilic bacteria and the genes for aromatic catabolism may be chromosomal or plasmid-borne as in *Pseudomonas* sp. strain BI7 (Whyte et al., 1997).

3.7 CONCLUSIONS

Studies on the bacterial diversity of the various habitats of Antarctica have indicated a considerable degree of diversity and have also led to the discovery of many previously undescribed bacterial genera and species. Nevertheless only a miniscule of the total bacterial diversity in Antarctica is known and there is a need to extend studies to habitats unique to Antarctica such as the pyritic sediments, brine channels of sea-ice, ice cores, frozen lakes, cryoconite hole, and cryptoendolithic communities. Further, there is a need to study the population dynamics of psychrophilic bacteria in various habitats of Antarctica with respect to their diversity, abundance, and distribution as a function of environmental parameters such as temperature, salinity, pH, nutrient content, and water activity, so as to identify bacteria which could be explored for various applications such as bioremediation or to identify suitable biomolecules for biotechnological applications. In addition, these cold-loving bacteria could be used as model systems to understand the molecular basis of cold adaptation with respect to their ability to sense temperature and regulate the expression of genes required for survival, growth, and multiplication.

ACKNOWLEDGMENTS

S. Shivaji would like to thank the Department of Biotechnology, the Department of Earth Sciences, and the Council of Scientific and Industrial Research, Government of India, for funding projects related to biodiversity.

REFERENCES

Abyzov, S.S., S.N. Philippova, and V.D. Kuznetsov. 1983. *Nocardiopsis antarcticus*-a new species of actinomyces, isolated from the ice sheet of the central Antarctic glacier. *Izv. Akad. Nauk SSR Ser. Biol.* 4:559–568.

Abyzov, S.S., I.N. Mitskevich, M.N. Poglazova et al. 2001. Microflora in the basal strata at Antarctic ice core above the Vostok lake. *Adv. Space. Res.* 28:701–706.

Aislabie, J., M. McLeod, and R. Fraser. 1998. Potential for biodegradation of hydrocarbons in soil from the Ross Dependency, Antarctica. *Appl. Microbiol. Biotechnol.* 49:210–214.

Aislabie, J., M.R. Balks, J. Foght, and E.J. Waterhouse. 2004. Hydrocarbon spills on Antarctic soils: Effects and management. *Environ. Sci. Technol.* 38:1265–1274.

Aislabie, J., D.J. Saul, and J.M. Foght. 2006a. Bioremediation of hydrocarbon-contaminated polar soils. *Extremophiles* 10:171–179.

Aislabie J.M., K. Chhour, D.J. Saul et al. 2006b. Dominant bacterial groups in soils of Marble Point and Wright Valley, Victoria Land, Antarctica. *Soil Biol. Biochem.* 38:3041–3056.

Alam, S.I., L. Singh, S. Dube, G.S.N. Reddy, and S. Shivaji. 2003. Psychrophilic *Planococcus maitrii* sp. nov. from Antarctica. *Syst. Appl. Microbiol.* 26:505–510.

Alam, S.I., A. Dixit, G.S. Reddy et al. 2006. *Clostridium schirmacherense* sp. nov., an obligately anaerobic, proteolytic, psychrophilic bacterium isolated from lake sediment of Schirmacher Oasis, Antarctica. *Int. J. Syst. Evol. Microbiol.* 56:715–720.

Allan, R.N., L. Lebbe, J. Heyrman, P. De Vos, C.J. Buchanan, and N.A. Logan. 2005. *Brevibacillus levickii* sp. nov. and *Aneurinibacillus terranovensis* sp. nov., two novel thermoacidophiles isolated from geothermal soils of northern Victoria Land, Antarctica. *Int. J. Syst. Evol. Microbiol.* 55:1039–1050.

Archer, S.D., R.J.G. Leakey, I.I.P.H. Burk, and M. Slelgh. 1996a. Microbial dynamics in coastal waters of East Antarctica herbivory by heterotrophic dinoflagellates. *Mar. Ecol. Prog. Ser.* 139:239–255.

Archer, S.D., R.J.G. Leakey, P.H. Burkill, M.A. Sleigh, and C.J. Appleby. 1996b. Microbial ecology of sea ice at a coastal Antarctic site: Community composition, biomass and temporal change. *Mar. Ecol. Progr. Ser.* 135:179–195.

Atlas, R.M. 1986. Fate of petroleum pollutants in Arctic ecosystems. *Water Sci. Technol.* 18:59–67.

Atlas, R.M. and C.E. Cerniglia. 1995. Bioremediation of petroleum pollutants. *Bioscience* 45:332–338.

Bargagli, R., J.C. Sanchez-Hernandez, and F. Monaci. 1999. Baseline concentrations of elements in the Antarctic macrolichen *Umbilicaria decussate. Chemosphere* 38:475–487.

Barghoorn, E.S. and R.L. Nichols. 1961. Sulfate-reducing bacteria and pyritic sediments in Antarctica. *Science* 134:190.

Barrett, J.E., R.A. Virginia, D.H. Wall et al. 2006. Co-variation in soil biodiversity and biogeochemistry in Northern and Southern Victoria Land, Antarctica. *Antarct. Sci.* 18:535–548.

Bird, D.F. and D.M. Karl. 1991. Spatial pattern of glutamate and thymidine assimilation in Bransfield Strait, Antarctica during and following the austral spring bloom. *Deep Sea Res.* 38:1057–1075.

Bowman, J.P. 1998. *Pseudoalteromonas prydzensis* sp. nov., a psychrotrophic, halotolerant bacterium form Antarctic sea ice. *Int. J. Syst. Bacteriol.* 3:1037–1041.

Bowman, J.P. 2000. Description of *Cellulophaga algicola* sp. nov., isolated from the surfaces of Antarctic algae, and reclassification of *Cytophaga uliginosa* (ZoBell and Upham 1944) Reichenbach 1989 as *Cellulophaga uliginosa* comb. nov. *Int. J. Syst. Evol. Microbiol.* 50:1861–1868.

Bowman, J.P. and R.D. McCuaig. 2003. Biodiversity, community structural shifts, and biogeography of prokaryotes within Antarctic continental shelf sediment. *Appl. Environ. Microbiol.* 69:2463–2483.

Bowman, J.P. and D.S. Nichols. 2002. *Aequorivita* gen. nov., a member of the family *Flavobacteriaceae* isolated from terrestrial and marine Antarctic habitats. *Int. J. Syst. Evol. Microbiol.* 52:1533–1541.

Bowman, J.P. and D.S. Nichols. 2005. Novel members of the family *Flavobacteriaceae* from Antarctic maritime habitats including *Subsaximicrobium wynnwilliamsii* gen. nov., sp. nov., *Subsaximicrobium saxinquilinus* sp. nov., *Subsaxibacter broadyi* gen. nov., sp. nov., *Lacinutrix copepodicola* gen. nov., sp. nov., and novel species of the genera *Bizionia*, *Gelidibacter* and *Gillisia. Int. J. Syst. Evol. Microbiol.* 55:1471–1486.

Bowman, J.P., J. Cavanagh, J.J. Austin, and K. Sanderson. 1996. Novel *Psychrobacter* species from Antarctic ornithogenic soils. *Int. J. Syst. Bacteriol.* 46:841–848.

Bowman, J.P., S.A. McCammon, D.S. Nichols et al. 1997a. *Shewanella gelidimarina* sp. nov. and *Shewanella frigidimarina* sp. nov., novel Antarctic species with the ability to produce eicosapentaenoic acid (20:5 omega 3) and grow anaerobically by dissimilatory Fe(III) reduction. *Int. J. Syst. Bacteriol.* 47:1040–1047.

Bowman, J.P., S.A. McCammon, J.L. Brown, P.D. Nichols, and T.A. McMeekin. 1997b. *Psychroserpens burtonensis* gen. nov., sp. nov., and *Gelidibacter algens* gen. nov., sp. nov., psychrophilic bacteria isolated from Antarctic lacustrine and sea ice habitats. *Int. J. Syst. Bacteriol.* 47:670–677.

Bowman, J.P., S.A. McCammon, M.V. Brown, D.S. Nichols, and T.A. McMeekin. 1997c. Diversity and association of psychrophilic bacteria in Antarctic sea ice. *Appl. Environ. Microbiol.* 63:3068–3078.

Bowman, J.P., D.S. Nichols, and T.A. McMeekin. 1997d. *Psychrobacter glacincola* sp. nov., a halotolerant, psychrophilic bacterium isolated from Antarctic sea-ice. *Syst. Appl. Microbiol.* 20:209–215.

Bowman, J.P., S.A. McCammon, and J.H. Skerratt. 1997e. *Methylosphaera hansonii* gen. nov., sp. nov., a psychrophilic, group I methanotroph from Antarctic marine-salinity, meromictic lakes. *Microbiology*, 143:1451–1459.

Bowman, J.P., J.J. Gosink, S.A. McCammon et al. 1998a. *Colwellia demingiae* sp. nov., *Colwellia hornerae* sp. nov., *Colwellia rossensis* and *Colwellia psychrotropica* sp. nov., psychrophilic Antarctic species with the ability to synthesize docosahexaenoic acid (22:6w3). *Int. J. Syst. Bacteriol.* 48:1171–1180.

Bowman, J.P., S.A. McCammon, T. Lewis et al. 1998b. *Psychroflexus torquis* gen. nov., sp. nov., a psychrophilic species from Antarctic sea ice, and reclassification of *Flavobacterium gondwanense* (Dobson et al. 1993) as *Psychroflexus gondwanense* gen. nov., comb. nov. *Microbiology* 144:1601–1609.

Bowman, J.P., S.A. McCammon, J.L. Brown, and T.A. McMeekin. 1998c. *Glaciecola punicea* gen. nov., sp. nov. and *Glaciecola pallidula* gen. nov., sp. nov.: Psychrophilic bacteria from Antarctic sea-ice habitats. *Int. J. Syst. Bacteriol.* 48:1213–1222.

Bowman, J.P., S.A. McCammon, S.M. Rea, and T.A. McMeekin. 2000a. The microbial composition of three limnologically disparate hypersaline Antarctic lakes. *FEMS Microbiol. Lett.* 183:81–88.

Bowman, J.P., S.M. Rea, S.A. McCammon, and T.A. McMeekin. 2000b. Diversity and community structure within anoxic sediment from marine salinity meromictic lakes and a coastal meromictic marine basin, Vestfold Hilds, Eastern Antarctica. *Environ. Microbiol.* 2:227–237.

Boyd, W.L. 1962. Comparison of soil bacteria and their metabolic activities in Arctic and Antarctic regions. *Polar Rec.* 11:319.

Boyd, W.L. and J.W. Boyd. 1962. Presence of *Azotobacter* species in polar regions. *J. Bacteriol.* 85:1121–1123.

Boyd, W.L. and J.W. Boyd. 1963a. Soil microorganisms of the McMurdo Sound area, Antarctica. *Appl. Microbiol.* 11:1121–1123.

Boyd, W.L. and J.W. Boyd. 1963b. Variability of coliform bacteria in Antarctic soil. *J. Bacteriol.* 85:1121–1123.

Bozal, N., E. Tudela, R. Rosselló-Mora, J. Lalucat, and J. Guinea. 1997. *Pseudoalteromonas antarctica* sp. nov., isolated from an Antarctic coastal environment. *Int. J. Syst. Bacteriol.* 47:345–351.

Bozal, N., M.J. Montes, E. Tudela, F. Jiménez, and J. Guinea. 2002. *Shewanella frigidimarina* and *Shewanella livingstonensis* sp. nov. isolated from Antarctic coastal areas. *Int. J. Syst. Evol. Microbiol.* 52:195–205.

Bozal, N., M.J. Montes, E. Tudela, and J. Guinea. 2003. Characterization of several *Psychrobacter* strains isolated from Antarctic environments and description of *Psychrobacter luti* sp. nov. and *Psychrobacter fozii* sp. nov. *Int. J. Syst. Evol. Microbiol.* 53:1093–1100.

Bozal, N., M.J. Montes, and E. Mercadé. 2007. *Pseudomonas guinea* sp. nov., a novel psychrotolerant bacterium from an Antarctic environment. *Int. J. Syst. Evol. Microbiol.* 57:2609–2612.

Braddock, J.F., M.L. Ruth, and P.H. Catterall. 1997. Enhancement and inhibition of microbial activity in hydrocarbon-contaminated arctic soils: Implications for nutrient-amended bioremediation. *Environ. Sci. Technol.* 31:2078–2084.

Bragg, J.R., R.C. Prince, E.J. Harner, and R.M. Atlas. 1994. Effectiveness of bioremediation for the Exxon Valdez oil spill. *Nature* 36:413–418.

Brambilla, E., H. Hippe, A. Hagelstein, B.J. Tindall, and E. Stackebrandt. 2001. 16S rDNA diversity of cultured and uncultured prokaryotes of a mat sample from Lake Fryxell, McMurdo Dry Valleys, Antarctica. *Extremophiles* 5:23–33.

Bruni, V., C. Gugliandolo, T. Maugeri, and A. Allegra. 1999. Psychrotrophic bacteria from a coastal station in the Ross sea (Terra Nova Bay, Antarctica). *New Microbiol.* 22:357–363.

Busse, H.J., E.B. Denner, S. Buczolits, M. Salkinoja-Salonen, A. Bennasar, and P. Kämpfer. 2003. *Sphingomonas aurantiaca* sp. nov., *Sphingomonas aerolata* sp. nov. and *Sphingomonas faeni* sp. nov., air- and dustborne and Antarctic, orange-pigmented, psychrotolerant bacteria, and emended description of the genus *Sphingomonas. Int. J. Syst. Evol. Microbiol.* 53:1253–1260.

Cameron, R.E., J. King, and C.N. David. 1970. In: *Antarctic Ecology*, Vol. 2, M. Holdgate (ed.). Academic Press, New York, pp. 702–716.

Campbell, I.B. and G.G.C Claridge. 2000. In: *Antarctic Ecosystems: Models for Wider Understanding*, W. Davidson, C. Howard-Williams, and P. Broady (eds.). Caxton Press, Christchurch, U.K., pp. 2332–2340.

Carpenter, E.J., S. Lin, and D.G. Capone. 2000. Bacterial activity in South Pole snow. *Appl. Environ. Microbiol.* 66:4514–4517.

Cavanagh, J., J.J. Austin, and K. Sanderson. 1996. Novel *Psychrobacter* species from Antarctic ornithogenic soils. *Int. J. Syst. Bacteriol.* 46:841–848.

Cavicchioli, R. and T. Thomas. 2000. Extremophiles. In: *Encyclopedia of Microbiology*, 2nd Edn., Vol. 2, J. Lederberg (ed.). Academic Press, San Diego, CA, pp. 317–337.

Chen, M., X. Xiao, P. Wang, X. Zeng, and F. Wang. 2005. *Arthrobacter ardleyensis* sp. nov., isolated from Antarctic lake sediment and deep-sea sediment. *Arch. Microbiol.* 183:301–305.

Chessa, J.P., I. Petrescu, M. Bentahir, J. Van Beeumen, and C. Gerday. 2000. Purification, physico-chemical characterization and sequence of a heat labile alkaline metalloprotease isolated from a psychrophilic *Pseudomonas* species. *Biochim. Biophys. Acta* 1479:265–274.

Choi, T.H., H.K. Lee, K. Lee, and J.C. Cho. 2007. *Ulvibacter antarcticus* sp. nov., isolated from Antarctic coastal seawater. *Int. J. Syst. Evol. Microbiol.* 57:2922–2925.

Christner, B.C., E. Mosley-Thompson, L.G. Thompson, and J.N. Reeve. 2001. Isolation of bacteria and 16S rDNAs from Lake Vostok accretion ice. *Environ. Microbiol.* 3:570–577.

Christner, B.C., B.H. Kvitko, and J.N. Reeve. 2003. Molecular identification of bacteria and Eukarya inhabiting an Antarctic cryoconite hole. *Extremophiles* 7:177–183.

Claridge, G.G. and I.B. Campbell. 1977. Salts in Antarctic soils, their distribution and relationship to soil processes. *Soil Sci.* 123:337–384.

Collins, M.D., P.A. Lawson, M. Labrenz, B.J. Tindall, N. Weiss, and P. Hirsch. 2002. *Nesterenkonia lacusekhoensis* sp. nov., isolated from hypersaline Ekho Lake, East Antarctica, and emended description of the genus *Nesterenkonia Int. J. Syst. Evol. Microbiol.* 52:1145–1150.

Coulon, F. and D. Delille. 2006. Influence of substratum on the degradation processes in diesel polluted sub-Antarctic soils (Crozet Archipelago). *Polar Biol.* 29:806–812.

Coulon, F., E. Pelletier, R. St. Louis, L. Gourhant, and D. Delille. 2004. Degradation of petroleum hydrocarbons in two sub-Antarctic soils: Influence of an oleophilic fertilizer. *Environ. Toxicol. Chem.* 23:1893–1901.

Coyne, V.E., C.J. Pillidge, D.D. Sledjeski et al. 1989. Reclassification of *Alteromonas colwelliana* to the genus *Shewanella* by DNA-DNA hybridization, serology and 5S ribosomal RNA sequence data. *Syst. Appl. Microbiol.* 12:275–279.

Dahllöf, I., H. Baillie, and S. Kjelleberg. 2000. rpoB-based microbial community analysis avoids limitations inherent in 16S rRNA gene intraspecies heterogeneity. *Appl. Environ. Microbiol.* 66:3376–3380.

de la Torre, J.R., B.M. Goebel, E.I. Friedmann, and N.R. Pace. 2003. Microbial diversity of cryptoendolithic communities from the McMurdo Dry Valleys, Antarctica. *Appl. Environ. Microbiol.* 69:3858–3867.

Delille, D. 2000. Response of Antarctic soil bacterial assemblages to contamination by diesel fuel and crude oil. *Microb. Ecol.* 40:159–168.

Delille, D. and M. Bouvy. 1989. Bacterial responses to natural organic inputs in a marine subantarctic area. *Hydrobiologia* 182:225–238.

Delille, D. and B. Delille. 2000. Field observations on the variability of crude oil impact on indigenous hydrocarbon-degrading bacteria from sub-Antarctica intertidal sediments. *Marine Environ. Res.* 49:403–417.

Delille, D. and F. Gleizon. 2003. Distribution of enteric bacteria in Antarctic seawater surrounding the Port-aux-Français permanent station (Kerguelen Island). *Mar. Polut. Bull.* 46:1179–1183.

Delille, D. and E. Lagarde. 1974. Contribution a l'etude ecologique des Milieux subantarctiques, V. *Rev. Inst. Pasteur Lyon.* 7:149–165.

Delille, D. and C. Rosiers. 1995. Seasonal changes of Antarctic marine bacterioplankton and sea ice bacterial assemblages. *Polar Biol.* 16:27–34.

Delille, D. and N. Vaillant. 1990. The influence of crude oil on the growth of subantarctic marine bacteria. *Antarct. Sci.* 2:123–127.

Delille, D., A. Bassères, and A. Dessommes. 1997. Seasonal variation of bacteria in sea ice contaminated by diesel fuel and dispersed crude oil. *Microb. Ecol.* 33:97–105.

Delille, D., A. Basseres, and A. Dessommes. 1998. Effectiveness of bioremediation for oil polluted Antarctic seawater. *Polar Biol.* 19:237–241.

Denner, E.B.M., B. Mark, H.J. Busse, M. Turkiewicz, and W. Lubitz. 2001. *Psychrobacter proteolyticus* sp. nov., a psychrotrophic, halotolerant bacterium isolated from the Antarctic krill *Euphausia superba* Dana, excreting a cold-adapted metalloprotease. *Syst. Appl. Microbiol.* 24:44–53.

Dobson, S.J., R.R. Colwell, T.A. McMeekin, and P.D. Franzmann. 1993. Direct sequencing of the polymerase chain reaction-amplified 16S rRNA gene of *Flavobacterium gondwanense* sp. nov. and *Flavobacterium salegens* sp. nov., two new species from a hypersaline Antarctic lake. *Int. J. Syst. Bacteriol.* 43:77–83.

Ekelöf, E. 1908a. Bakteriologische studien wahrend der Schwedishen Sudpolar expedition 1901–1903. In: *Wissenschaftiche Ergebnisse der Schwedischen sudpolar Expedition 1901–1903*, O. Nordenskjold (ed.). Lithogr. Inst. Generalstabs, Stockholm, Sweden.

Ekelöf, E. 1908b. Studien uber den Bakteriengehalt der luft und des Erdbodeus der antarktishen Gegenden, ausgefiihrt wahrend der schwedischen sudpolar expedition 1901–1903. *Z. Hyg. Infekt.* 56:344–370.

Flint, E.A. and J.D. Stout. 1960. Microbiology of some soils from Antarctica. *Nature* 188:767–768.

Franzmann, P.D. and S.J. Dobson. 1992. Cell wall-less, free-living *Spirochaetes* in Antarctica. *FEMS Microbiol. Lett.* 76:289–292.

Franzmann, P.D., P.P. Deprez, A.J. MeGuire, T.A. McMeekin, and H.R. Burton. 1990. The heterotrophic, bacterial microbiota of Burton Lake, Antarctica. *Polar Biol.* 10:261–264.

Franzmann, P.D., P. Hopfl, N. Weiss, and B.J. Tindall. 1991. Psychrotrophic lactic acid producing bacteria from anoxic waters in Ace Lake, Antarctica; *Carnobacterium funditum* sp. nov. and *Carnobacterium alterfunditum* sp. nov. *Arch. Microbiol.* 156:255–262.

Freckman, D.W. and R.A. Virginia. 1997. Low-diversity Antarctic soil nematode communities: Distribution and response to disturbance. *Ecology* 78:363–369.

Friedmann, E.I. 1980. Endolithic microbial life in hot and cold deserts. *Origins Life* 10:223–235.

Frühling, A., P. Schumann, H. Hippe, B. Sträubler, and E. Stackebrandt. 2002. *Exiguobacterium undae* sp. nov. and *Exiguobacterium antarcticum* sp. nov. *Int. J. Syst. Evol. Microbiol.* 52:1171–1176.

Gauthier, G., M. Gauthier, and R. Christen. 1995. Phylogenetic analysis of the genera *Alteromonas, Shewanella* and *Moritella* using genes coding for small-subunit rRNA sequences and division of the genus *Alteromonas* into two genera, *Alteromonas* (emended) and *Pseudoalteromonas* gen. nov., and proposal of twelve new species combinations. *Int. J. Syst. Bacteriol.* 45:755–761.

Gazert, H. 1912. Undersuchungen uber Meeresbakterien und chren Einfluss auf den Stoffwecksel im Meer. *Deutschi Biidpolar Expedition. 1901–1903 Berlin*, 7:268–296.

Gibson, J.A.E., C. Russell, R.C. Garrick, and H.R. Burton. 1990. Seasonal fluctuation of bacterial numbers near the antarctic continent. *Proc. NIPR Symp. Polar Biol.* 3:16–22.

Gordon, D.A., J. Priscu, and S. Giovannoni. 2000. Origin and phylogeny of microbes living in permanent Antarctic Lake Ice. *Microb. Ecol.* 39:197–202.

Gosink, J.J. and J.T. Staley. 1995. Biodiversity of gas vacuolate bacteria from Antarctic sea-ice and water. *Appl. Environ. Microbiol.* 61:3486–3489.

Gosink, J.J., C.R. Woese, and J.T. Staley. 1998. *Polaribacter* gen. nov., with three new species, *P. irgensii* sp. nov., *P. franzmannii* sp. nov. and *P. filamentus* sp. nov., gas vacuolate polar marine bacteria of the *Cytophaga-Flavobacterium-Bacteroides* group and reclassification of *'Flectobacillus glomeratus'* as *Polaribacter glomeratus* comb. nov. *Int. J. Syst. Bacteriol.* 48:223–235.

Gupta, P.K., G.S.N. Reddy, D. Delille, and S. Shivaji. 2004. Psychrophilic *Arthrobacter gangotriensis* sp. nov., and *Arthrobacter kerguelensis* sp. nov. from Antarctica. *Int. J. Syst. Evol. Microbiol.* 54:2375–2378.

Gupta, P., P. Chaturvedi, S. Pradhan, D. Delille, and S. Shivaji. 2006. *Marinomonas polaris* sp. nov., a psychrohalotolerant strain isolated from coastal sea water off the subantarctic Kerguelen islands. *Int. J. Syst. Evol. Microbiol.* 56:361–364.

Hagson, R.B., H.K. Lowrey, D. Shafer, R. Sorocco, and D.H. Pope. 1983a. Microbes in Antarctic waters of the Drake Passage: Vertical patterns of substrate uptake productivity and biomass in January, 1980. *Polar Biol.* 2:179–188.

Hagson, R.B., D. Schafer, T. Ryan, D.H. Pope, and H.K. Lowrey. 1983b. Bacterioplankton in Antarctic ocean waters during late austral winter: Abundances, F.D.C. and estimates of production. *Appl. Environ. Microbiol.* 45:1622–1632.

Hagstrom, A., J. Pinhassi, and U.L. Zweifel. 2000. Biogeographical diversity among marine bacterioplankton. *Aquat. Microb. Ecol.* 21:231–244.

Herbert, R.A. and C.R. Bell. 1974. Nutrient cycling in the Antarctic marine environment. *Br. Antarct. Surv. Bull.* 39:7–11.

Heuchert, A., F.O. Glöckner, R. Amann, and U. Fischer. 2004. *Psychrobacter nivimaris* sp. nov., a heterotrophic bacterium attached to organic particles isolated from the South Atlantic (Antarctica). *Syst. Appl. Microbiol.* 27:399–406.

Hirsch, P., W. Ludwig, C. Hethke, M. Sittig, B. Hoffmann, and C.A. Gallikowski. 1998. *Hymenobacter roseosalivarius* gen. nov., sp. nov. from continental Antartica soils and sandstone: bacteria of the *Cytophaga/Flavobacterium/Bacteroides* line of phylogenetic descent. *Syst. Appl. Microbiol.* 21:374–383.

Hirsch, P., C.A. Gallikowski, J. Siebert et al. 2004a. *Deinococcus frigens* sp. nov., *Deinococcus saxicola* sp. nov., and *Deinococcus marmoris* sp. nov., low temperature and draught-tolerating, UV-resistant bacteria from continental Antarctica. *Syst. Appl. Microbiol.* 27:636–645.

Hirsch, P., U. Mevs, R.M. Kroppenstedt, P. Schumann, and E. Stackebrandt. 2004b. Cryptoendolithic Actinomycetes from Antarctic sandstone rock samples: *Micromonospora endolithica* sp. nov. and two isolates related to *Micromonospora coerulea* Jensen 1932. *Syst. Appl. Microbiol.* 27:166–174.

Hong, S.G., Y.K. Lee, J.H. Yim, J. Chun, and H.K. Lee. 2008. *Sanguibacter antarcticus* sp. nov., isolated from Antarctic sea sand. *Int. J. Syst. Evol. Microbiol.* 58:50–52.

Howard-Williams, C., D. Peterson, W.B. Lyons, R. Cattaneo-Vietti, and S. Gordon. 2006. Measuring ecosystem response in a rapidly changing environment: The Latitudinal Gradient Project. *Antarct. Sci.* 18:465–471.

Hoyles, L., M.D. Collins, G. Foster, E. Falsen, and P. Schumann. 2004. *Jeotgalicoccus pinnipedialis* sp. nov., from a southern elephant seal (Mirounga leonina). *Int. J. Syst. Evol. Microbiol.* 54:745–748.

Hughes, K.A., H.A. McCartney, T.A. Lachlan-Cope, and D.A. Pearce. 2004. A preliminary study of airborne microbial biodiversity over Peninsular Antarctica. *Cell. Mol. Biol.* 50:537–542.

Humphry, D.R., A. George, G.W. Black, and S.P. Cummings. 2001. *Flavobacterium frigidarium* sp. nov., an aerobic, psychrophilic, xylanolytic and laminarinolytic bacterium from Antarctica. *Int. J. Syst. Evol. Microbiol.* 51:1235–1243.

Imperio, T., C. Viti, and L. Marri. 2008. *Alicyclobacillus pohliae* sp. nov., a thermophilic, endospore-forming bacterium isolated from geothermal soil of the north-west slope of Mount Melbourne (Antarctica). *Int. J. Syst. Evol. Microbiol.* 58:221–225.

Irgens, R.L., J.J. Gosink, and J.T. Staley. 1996. *Polaromonas vacuolata* gen. nov., sp. nov., a psychrophilic, marine, gas vacuolate bacterium from Antarctica. *Int. J. Syst. Bacteriol.* 46:822–826.

James, S.R., H.R. Burton, T.A. Mcmeekin, and C.A. Mancuso. 1994. Seasonal abundance of *Halomonas meridiana, Halomonas subglaciescola, Flavobacterium gondwanense* and *Flavobacterium salegens* in four Antarctic lakes. *Antarct. Sci.* 6:325–332.

Jungblut, A.D., I. Hawes, D. Mountfort et al. 2005. Diversity within cyanobacterial mat communities in variable salinity meltwater ponds of McMurdo Ice Shelf, Antarctica. *Environ. Microbiol.* 7:519–529.

Junge, K., J.J. Gosink, H.G. Hoppe, and J.T. Staley. 1998. *Arthrobacter, Brachybacterium* and *Planococcus* isolates identified from antarctic sea ice brine. Description of *Planococcus mcmeekinii*, sp. nov. *Syst. Appl. Microbiol.* 21:306–314.

Karl, D.M. 1992. The grounding of the *Bahia Paraiso*: Microbial ecology of the 1989 Antarctic oil spill. *Microb. Ecol.* 24:77–89.

Karl, D.M., D.F. Bird, K. Björkman, T. Houlihan, R. Shackelford, and L. Tupas. 1999. Microorganisms in the accreted ice of Lake Vostok, Antarctica. *Science* 286:2144–2147.

Karr, E.A., W.M. Sattley, D.O. Jung, M.T. Madigan, and L.A. Achenbach. 2003. Remarkable diversity of phototrophic purple bacteria in a permanently frozen Antarctic lake. *Appl. Environ. Microbiol.* 69:4910–4914.

Karr, E.A., W.M. Sattley, M.R. Rice, D.O. Jung, M.T. Madigan, and L.A. Achenbach. 2005. Diversity and distribution of sulfate-reducing bacteria in permanently frozen Lake Fryxell, McMurdo Dry Valleys, Antarctica. *Appl. Environ. Microbiol.* 71:6353–6359.

Kelly, M.D., S. Lukaschewsky, and C.G. Anderson. 1978. Bacterial flora of Antarctic krill and some of their enzymatic properties. *J. Food Sci.* 43:1196–1197.

Labrenz, M., M.D. Collins, P.A. Lawson, B.J. Tindall, G. Braker, and P. Hirsch. 1998. *Antarctobacter heliothermus* gen. nov., sp. nov., a budding bacterium from hypersaline and heliothermal Ekho Lake. *Int. J. Syst. Bacteriol.* 48:1363–1372.

Labrenz, M., M.D. Collins, P.A. Lawson, B.J. Tindall, P. Schumann, and P. Hirsch. 1999. *Roseovarius tolerans* gen. nov., sp. nov., a budding bacterium with variable bacteriochlorophyll a production from hypersaline Ekho Lake. *Int. J. Syst. Bacteriol.* 49:137–147.

Labrenz, M., B.J. Tindall, P.A. Lawson, M.D. Collins, P. Schumann, and P. Hirsch. 2000. *Staleya guttiformis* gen. nov., sp. nov. and *Sulfitobacter brevis* sp. nov., alpha-3-*Proteobacteria* from hypersaline, heliothermal and meromictic antarctic Ekho Lake. *Int. J. Syst. Evol. Microbiol.* 50:303–313.

Labrenz, M., P.A. Lawson, B.J. Tindall, M.D. Collins, and P. Hirsch. 2003. *Saccharospirillum impatiens* gen. nov., sp. nov., a novel *gamma-Proteobacterium* isolated from hypersaline Ekho Lake (East Antarctica). *Int. J. Syst. Evol. Microbiol.* 53:653–660.

Labrenz, M., P.A. Lawson, B.J. Tindall, M.D. Collins, and P. Hirsch. 2005. *Roseisalinus antarcticus* gen. nov., sp. nov., a novel aerobic bacteriochlorophyll a-producing *alpha-Proteobacterium* isolated from hypersaline Ekho Lake, Antarctica. *Int. J. Syst. Evol. Microbiol.* 55:41–47.

Lama, L., B. Nicolaus, V. Calandrelli, E. Esposito, and A. Gambacorta. 1996. Xylanase produced by *Bacillus thermoantarcticus*, a new thermophilic bacillus. *Enz. Engg. XIII (Ann. NY Acad. Sci.)* 799:285–289.

Lawson, P.A., M.D. Collins, P. Schumann, B.J. Tindall, P. Hirsch, and M. Labrenz. 2000. New LL-diaminopimelic acid-containing actinomycetes from hypersaline, heliothermal and meromictic Antarctic Ekho Lake: *Nocardioides aquaticus* sp. nov. and *Friedmanniella* [correction of Friedmannielly] *lacustris* sp. nov. *Syst. Appl. Microbiol.* 23:219–229.

Leakey, R.J.G., S.D. Archer, and J. Grey. 1996. Microbial dynamics in coastal waters of East Antarctica: Bacterial production and nanoflagellate bacterivory. *Mar. Ecol. Prog. Ser.* 142:3–17.

Lee, K., H.K. Lee, T.H. Choi, and J.C. Cho. 2007a. *Robiginitomaculum antarcticum* gen. nov., sp. nov., a member of the family *Hyphomonadaceae*, from Antarctic seawater. *Int. J. Syst. Evol. Microbiol.* 57:2595–2599.

Lee, K., H.K. Lee, T.H. Choi, and J.C. Cho. 2007b. *Sejongia marina* sp. nov., isolated from Antarctic seawater. *Int. J. Syst. Evol. Microbiol.* 57:2917–2921.

Lee, K., H.K. Lee, T.H. Choi, K.M. Kim, and J.C. Cho. 2007c. *Granulosicoccaceae* fam. nov., to include *Granulosicoccus antarcticus* gen. nov., sp. nov., a non-phototrophic, obligately aerobic chemoheterotroph in the order *Chromatiales*, isolated from Antarctic seawater. *J. Microbiol. Biotechnol.* 17:1483–1490.

Lee, K., H.K. Lee, and J.C. Cho. 2008. *Hahella antarctica* sp. nov., isolated from Antarctic seawater. *Int. J. Syst. Evol. Microbiol.* 58:353–356.

Liu, H., Y. Xu, Y. Ma, and P. Zhou. 2000. Characterization of *Micrococcus antarcticus* sp. nov., a psychrophilic bacterium from Antarctica. *Int. J. Syst. Evol. Microbiol.* 50:715–719.

Logan, N.A., L. Lebbe, B. Hoste et al. 2000. Aerobic endospore-forming bacteria from geothermal environments in northern archipelago, with the proposal of *Bacillus fumarioli* sp. nov. *Int. J. Syst. Evol. Microbiol.* 50:1741–1753.

Logan, N.A., L. Lebbe, A. Verhelst et al. 2002. *Bacillus luciferensis* sp. nov., from volcanic soil on Candlemas Island, South Sandwich archipelago. *Int. J. Syst. Evol. Microbiol.* 52:1985–1989.

Logan, N.A., L. Lebbe, A. Verhelst et al. 2004a. *Bacillus shackletonii* sp. nov., from volcanic soil on Candlemas Island, South Sandwich archipelago. *Int. J. Syst. Evol. Microbiol.* 54:373–376.

Logan, N.A., E. De Clerck, L. Lebbe et al. 2004b. *Paenibacillus cineris* sp. nov. and *Paenibacillus cookii* sp. nov., from Antarctic volcanic soils and a gelatin-processing plant. *Int. J. Syst. Evol. Microbiol.* 54:1071–1076.

Madigan, M.T., D.O. Jung, C.R. Woese, and L.A. Achenbach. 2000. *Rhodoferax antarcticus* sp. nov., a moderately psychrophilic purple nonsulfur bacterium isolated from an Antarctic microbial mat. *Arch. Microbiol.* 173:269–277.

Margesin, R. and F. Schinner. 1997. Bioremediation of diesel-oil contaminated alpine soils at low temperatures. *Appl. Microbiol. Biotechnol.* 47:462–468.

Margesin, R. and F. Schinner. 1999. Biological decontamination of oil spills in cold environments. *J. Chem. Technol. Biotechnol.* 74:381–389.

Margesin, R., D. Labbe, F. Schinner, C.W. Greer, and L.G. Whyte. 2003a. Characterization of hydrocarbon-degrading microbial populations in contaminated and pristine alpine soils. *Appl. Environ. Microbiol.* 69:3085–3092.

Margesin, R., S. Gander, G. Zacke, A.M. Gounot, and F. Schinner, F. 2003b. Hydrocarbon degradation and enzyme activities of cold-adapted bacteria and yeasts. *Extremophiles* 7:451–458.

Margini, R.A. and O.D. Castrelos. 1963. *Examenes bacteriologicos* de aire, rieve y suelo de carbo primaruera y Estacion cientifica Ellsworth. *Inst. Antarct. Argent. Publ.* 76:1–15.

Margini, R.A. and O.D. Castrelos. 1965. Las bacterias del Antarctico y su relacion con la epoca del ano contrib. *Inst. Antarct. Argent.* 141:1–19.

Marshall, B.J. and D.F. Ohye. 1966. *Bacillus macquariensis* n. sp. a psychrotrophic bacterium from sub-antarctic soil. *J. Gen. Microbiol.* 44:41–46.

Maruyama, A., D. Honda, H. Yamamoto, K. Kitamura, and T. Higashihara. 2000. Phylogenetic analysis of psychrophilic bacteria isolated from the Japan Trench, including a description of the deep-sea species *Psychrobacter pacificensis* sp. nov. *Int. J. Syst. Evol. Microbiol.* 50:835–846.

Maugeri, T.L., C. Gugliandolo, and V. Bruni. 1996. Heterotrophic bacteria in the Ross Sea (Terra Nova Bay, Antarctica). *Microbiologica* 19:67–76.

McCammon, S.A. and J.P. Bowman. 2000. Taxonomy of Antarctic *Flavobacterium* species: Description of *Flavobacterium gillisiae* sp. nov., *Flavobacterium tegetincola* sp. nov., and *Flavobacterium xanthum* sp. nov., nom. rev. and reclassification of [*Flavobacterium*] *salegens* as *Salegentibacter salegens* gen. nov., comb. nov. *Int. J. Syst. Evol. Microbiol.* 50:1055–1063.

McCammon, S.A., B.H. Innes, J. Bowman et al. 1998. *Flavobacterium hibernum* sp. nov. a lactose utilizing bacterium from a fresh water Antarctic lake. *Int. J. Syst. Bacteriol.* 48:1405–1412.

McGuire, A.J., P.D. Franzmann, and T.A. McMeekin. 1987. *Flectobacillus glomeratus* sp. nov., a curved, nonmotile, pigmented bacterium isolated from Antarctic marine environments. *Syst. Appl. Microbiol.* 9:265–272.

McLean, A.L. 1918a. Bacteria of ice and snow in Antarctica. *Nature* 102:35–39.

McLean, A.L. 1918b. Bacteriological and other researches Australian Antarctic Expedition 1911–1914. *Scient. Rep.* 7:1–128.

Mevs, U., E. Stackebrandt, P. Schumann, C.A. Gallikowski, and P. Hirsch. 2000. *Modestobacter multiseptatus* gen. nov., sp. nov., a budding actinomycete from soils of the Asgard Range (Transantarctic Mountains). *Int. J. Syst. Evol. Microbiol.* 50:337–346.

Meyer, G.H. 1962. Microbiological populations of Antarctic air, soil, snow and melt pools. *Polar Rec.* 11:317–318.

Meyer, G.H., M.B. Morrow, O. Wyss, T.E. Berg, and J.L. Littlepage. 1962. Antarctica: The microbiology of an Unfrozen Saline Pond. *Science* 138:1103–1104.

Michaud, L., F. Di Cello, M. Brilli, R. Fani, A. Lo Giudice, and V. Bruni. 2004. Biodiversity of cultivable psychrotrophic marine bacteria isolated from Terra Nova Bay (Ross Sea, Antarctica). *FEMS Microbiol. Lett.* 230:63–71.

Mikucki, J.A. and J.C. Priscu. 2007. Bacterial diversity associated with Blood Falls, a subglacial outflow from the Taylor Glacier, Antarctica. *Appl. Environ. Microbiol.* 73:4029–4039.

Miwa, T. 1975. *Clostridia* in soil of the Antarctica. *Jpn. J. Med. Sci. Biol.* 28:201–213.

Mohn, W.W., C.Z. Radziminski, M.C. Fortin, and K.J. Reimer. 2001. On site bioremediation of hydrocarbon-contaminated Arctic tundra soils in inoculated biopiles. *Appl. Microbiol. Biotechnol.* 57:242–247.

Montes, M.J., E. Mercadé, N. Bozal, and J. Guinea. 2004. *Paenibacillus antarcticus* sp. nov., a novel psychrotolerant organism from the Antarctic environment. *Int. J. Syst. Evol. Microbiol.* 54:1521–1526.

Montes, M.J., N. Bozal, and E. Mercadé. 2008. *Marinobacter guineae* sp. nov., a novel moderately halophilic bacterium from an Antarctic environment. *Int. J. Syst. Evol. Microbiol.* 58:1346–1349.

Morgan, R. and R.J. Watkinson. 1989. Hydrocarbon degradation in soils and methods for soil treatment. *CRC Crit. Rev. Biotechnol.* 8:305–333.

Mountfort, D.O., F.A. Rainey, J. Burghardt, H.F. Kaspar, and E. Stackebrandt. 1997. *Clostridium vincentii* sp. nov., a new obligately anaerobic, saccharolytic, psychrophilic bacterium isolated from low-salinity pond sediment of the McMurdo Ice Shelf, Antarctica. *Arch. Microbiol.* 167:54–60.

Mountfort, D.O., F.A. Rainey, J. Burghardt, H.F. Kaspar, and E. Stackebrandt. 1998. *Psychromonas antarcticus* gen. nov., sp. nov., a new aerotolerant anaerobic, halophilic psychrophile isolated from pond sediment of the McMurdo ice shelf, antarctica. *Arch. Microbiol.* 169:231–238.

Nedashkovskaya, O.I., M. Vancanneyt, P. Dawyndt et al. 2005. Reclassification of [*Cytophaga*] *marinoflava* Reichenbach 1989 as *Leeuwenhoekiella marinoflava* gen. nov., comb. nov. and description of *Leeuwenhoekiella aequorea* sp. nov. *Int. J. Syst. Evol. Microbiol.* 55:1033–1038.

Niederberger, T.D., I.R. McDonald, A.L. Hacker et al. 2008. Microbial community composition in soils of Northern Victoria Land, Antarctica. *Environ. Microbiol.* 10:1713–1724.

Nogi, Y., K. Soda, and T. Oikawa. 2005. *Flavobacterium frigidimaris* sp. nov., isolated from Antarctic seawater. *Syst. Appl. Microbiol.* 28:310–315.

Obata, H., N. Muryoi, H. Kawahara, K. Yamade, and J. Nishikawa. 1999. Identification of a novel ice-nucleating bacterium of Antarctic origin and its ice nucleation properties. *Cryobiology* 38:131–139.

Pearce, D.A. 2003. Bacterioplankton community structure in a maritime antarctic oligotrophic lake during a period of holomixis, as determined by denaturing gradient gel electrophoresis (DGGE) and fluorescence *in situ* hybridization (FISH). *Microb. Ecol.* 46:92–105.

Pfiser, R.M. and P.R. Burkholder. 1965. Numerical taxonomy of some bacteria isolated from Antarctic and tropical sea waters. *J. Bacteriol.* 90:863–872.

Pindi, P.K., K.H. Kishore, G.S.N. Reddy, and S. Shivaji. 2009. Description of *Leifsonia kafniensis* sp. nov. and *Leifsonia antarctica* sp. nov. *Int. J. Syst. Evol. Microbiol.* 59: 1348–1352.

Pirie, J.H.H. 1904. First Antarctic voyage of the Scotia. *J. Bacteriol. Scottish Geog. Mag.* 20:129–132.

Pirie, J.H.H. 1912. Notes on Antarctic bacteriology. Report of the Scientific Results of the S.Y. Scotia, Vol. 3, pp. 157–168.

Poli, A., E. Esposito, L. Lama et al. 2006. *Anoxybacillus amylolyticus* sp. nov., a thermophilic amylase producing bacterium isolated from Mount Rittmann (Antarctica). *Syst. Appl. Microbiol.* 29:300–307.

Poli, A., E. Esposito, P. Orlando et al. 2007. *Halomonas alkaliantarctica* sp. nov., isolated from saline lake Russell in Antarctica, an alkalophilic moderately halophilic, exopolysaccharide-producing bacterium. *Syst. Appl. Microbiol.* 30:31–38.

Pontes, D.S., C.I. Lima-Bittencourt, E. Chartone-Souza, and A.M. Amaral Nascimento. 2007. Molecular approaches: Advantages and artifacts in assessing bacterial diversity. *J. Ind. Microbiol. Biotechnol.* 34:463–473.

Prabagaran, S.R., K. Suresh, R. Manorama, D. Delille, and S. Shivaji. 2005. *Marinomonas ushuaiensis* sp. nov., isolated from coastal sea water in Ushuaia, Argentina, sub-Antarctica. *Int. J. Syst. Evol. Microbiol.* 55:309–313.

Prabagaran, S.R., R. Manorama, D. Delille, and S. Shivaji. 2006. Predominance of *Roseobacter, Sulfitobacter, Glaciecola* and *Psychrobacter* in seawater collected off Ushuaia, Argentina, Sub-Antarctica. *FEMS Microbiol. Ecol.* 59:342–355.

Prabagaran, S.R., R. Manorama, D. Delille, and S. Shivaji. 2007. Predominance of *Roseobacter, Sulfitobacter, Glaciecola* and *Psychrobacter* in seawater collected off Ushuaia, Argentina, Sub-Antarctica. *FEMS Microbiol. Ecol.* 59:342–355.

Prabahar, V., S. Dube, G.S.N. Reddy, and S. Shivaji. 2004. *Pseudonocardia antarctica* sp. nov. an *Actinomycetes* from McMurdo Dry Valleys, Antarctica. *Syst. Appl. Microbiol.* 27:66–71.

Priscu, J.C., E.E. Adams, W.B. Lyons et al. 1999. Geomicrobiology of subglacial ice above Lake Vostok, Antarctica. *Science* 286:2141–2144.

Reddy, G.S.N., R.K. Agarwal, G.I. Matsumoto, and S. Shivaji. 2000. *Arthrobacter flavus* sp. nov., a psychrotropic bacterium isolated from a pond in Mc Murdo Dry Valley, Antarctica. *Int. J. Syst. Evol. Microbiol.* 50:1553–1561.

Reddy, G.S.N., J.S.S. Prakash, G.I. Matsumoto, E. Stackebrandt, and S. Shivaji. 2002a. *Arthrobacter roseus* sp. nov., a psychrotropic bacterium isolated from an Antarctic cyanobacterial mat sample. *Int. J. Syst. Evol. Microbiol.* 52:1017–1021.

Reddy, G.S.N., J.S.S. Prakash, M. Vairamani, S. Prabhakar, G.I. Matsumoto, and S. Shivaji. 2002b. *Planococcus antarcticus* and *Planococcus psychrophilus* spp. nov. isolated from cyanobacterial mat samples collected from ponds in Antarctica. *Extremophiles* 6:253–261.

Reddy, G.S.N., G.I. Matsumoto, and S. Shivaji. 2003a. Identification of *Sporosarcina macmurdoensis* sp. nov. from ponds in McMurdo dry valley, Antarctica. *Int. J. Syst. Evol. Microbiol.* 53:1363–1367.

Reddy, G.S.N., J.S.S. Prakash, R. Srinivas, G.I. Matsumoto, and S. Shivaji. 2003b. Psychrophilic *Leifsonia rubra* sp. nov. and *Leifsonia aurea* sp. nov. isolated from a Pond in Antarctica. *Int. J. Syst. Evol. Microbiol.* 53:977–984.

Reddy, G.S.N., P.M.U. Raghavan, N.B. Sarita et al. 2003c. *Halomonas glacies* sp. nov. isolated from fast ice of Adelie Land, Antarctica. *Extremophiles* 7:55–61.

Reddy, G.S.N., J.S.S. Prakash, V. Prabahar, G.I. Matsumoto, E. Stackebrandt, and S. Shivaji. 2003d. *Kocuria polaris* sp. nov., an orange pigmented psychrotrophic bacterium isolated from an Antarctic cyanobacterial mat sample. *Int. J. Syst. Evol. Microbiol.* 53:183–187.

Reddy, G.S.N., G.I. Matsumoto, P. Shuman, E. Stackebrandt, and S. Shivaji. 2004. Psychrophilic *Pseudomonas* from Antarctica: *Pseudomonas antarctica* sp. nov., *Pseudomonas meridianae* sp. nov. and *Pseudomonas proteolytica* sp. nov. *Int. J. Syst. Evol. Microbiol.* 54:713–719.

Rike, A.G., K.B. Haugen, and B. Engene. 2005. *In situ* biodegradation of hydrocarbons in arctic soil at sub-zero temperatures—field monitoring and theoretical simulation of the microbial activation temperature at a Spitsbergen contaminated site. *Cold Reg. Sci. Technol.* 41:189–209.

Rodríguez-Díaz, M., L. Lebbe, B. Rodelas, J. Heyrman, P. De Vos, and N.A. Logan. 2005. *Paenibacillus wynnii* sp. nov., a novel species harbouring the *nifH* gene, isolated from Alexander Island, Antarctica. *Int. J. Syst. Evol. Microbiol.* 55:2093–2099.

Ruckert, G. 1985. *Myxobacteria* from Antarctic soils. *Biol. Fert. Soil.* 1:215–216.

Rüger, H.J., D. Fritze, D., and C. Spröer. 2000. New psychrophilic and psychrotolerant *Bacillus marinus* strains from tropical and polar deep-sea sediments and emended description of the species. *Int. J. Syst. Evol. Microbiol.* 50:1305–1313.

Sattley, W.M., D.O. Jung, and M.T. Madigan. 2008. *Psychrosinus fermentans* gen. nov., sp. nov., a lactate-fermenting bacterium from near-freezing oxycline waters of a meromictic Antarctic lake. *FEMS Microbiol. Lett.* 287:121–127.

Saul, D.J., J.M. Aislabie, C.E. Brown, L. Harris, and J.M. Foght. 2005. Hydrocarbon contamination changes the bacterial diversity of soil from around Scott Base, Antarctica. *FEMS Microbiol. Ecol.* 53:141–155.

Schumann, P., H. Prauser, F.A. Rainey, E. Stackebrandt, and P. Hirsch. 1997. *Friedmanniella antarctica* gen. nov., sp. nov., an LL-diaminopimelic acid-containing actinomycete from Antarctic sandstone. *Int. J. Syst. Bacteriol.* 47:278–283.

Sheridan, P.P., J. Loveland-Curtze, V.I. Miteva, and J.E. Brenchley. 2003. *Rhodoglobus vestalii* gen. nov., sp. nov., a novel psychrophilic organism isolated from an Antarctic Dry Valley lake. *Int. J. Syst. Evol. Microbiol.* 53:985–994.

Shivaji, S. 2005. *Microbial Diversity: Current Perspectives and Potential Applications*, T. Satyanarayana and B.N. Johri (Eds.). I.K. International Pvt. Ltd., New Delhi, India, pp. 3–24.

Shivaji, S., N.S. Rao, L. Saisree, V. Sheth, G.S.N. Reddy, and P.M. Bhargava. 1988. Isolation and identification of *Micrococcus roseus* and *Planococcus* sp. from Schirmacher Oasis, Antarctica. *J. Biosci.* 13:409–414.

Shivaji, S., N.S. Rao, L. Saisree, G.S.N. Reddy, G.S. Kumar, and P.M. Bhargava. 1989a. Isolates of *Arthrobacter* from the soils of Schirmacher Oasis, Antarctica. *Polar Biol.* 10:225–229.

Shivaji, S., N.S. Rao, L. Saisree, V. Sheth, G.S.N. Reddy, and P.M. Bhargava. 1989b. Isolation and identification of *Pseudomonas spp.* from Schirmacher Oasis, Antarctica. *Appl. Env. Microbiol.* 55:767–770.

Shivaji, S., M.K. Ray, G.S. Kumar, G.S.N. Reddy, L. Saisree, and D.D. Wynn-Williams. 1991. Identification of *Janthinobacterium lividum* from the soils of the Island of Scotia Ridge from Antarctic Peninsula. *Polar Biol.* 11:267–272.

Shivaji, S., M.K. Ray, N.S. Rao et al. 1992. *Sphingobacterium antarcticus* sp. nov., a psychrotropic bacterium from the soils of Schirmacher Oasis, Antarctica. *Int. J. Syst. Bacteriol.* 42:102–116.

Shivaji, S., G.S.N. Reddy, R.A. Prasad, R. Kutty, and K. Ravenschlag. 2004a. Bacterial diversity of a soil sample from Schirmacher Oasis, Antarctica. *Cell. Mol. Biol.* 50: 525–536.

Shivaji, S., G.S.N. Reddy, P.M.U. Raghavan, N.B. Sarita, and D. Delille. 2004b. Psychrophilic *Psychrobacter salsus* sp. nov. and *Psychrobacter adeliae* sp. nov. isolated from fast ice from Adelie Land, Antarctica. *Syst. Appl. Microbiol.* 27:628–635.

Shivaji, S., P. Gupta, P. Chaturvedi, K. Suresh, and D. Delille. 2005a. *Marinobacter maritimus* sp. nov., a psychrotolerant strain isolated from sea water off the subantarctic Kerguelen islands. *Int. J. Syst. Evol. Microbiol.* 55:1453–1456.

Shivaji, S., G.S.N. Reddy, K. Suresh et al. 2005b. Two novel *Psychrobacter* species from Antarctica: Description of *Psychrobacter vallis* sp. nov. and *Psychrobacter aquaticus* sp. nov. *Int. J. Syst. Evol. Microbiol.* 55:757–762.

Shravage, B.V., K.M. Dayananda, M.S. Patole, and Y.S. Shouche. 2007. Molecular microbial diversity of a soil sample and detection of ammonia oxidizers from Cape Evans, Mcmurdo Dry Valley, Antarctica. *Microbiol. Res.* 162:15–25.

Simpson, R.D., S.D.A. Smith, and A.R. Pople. 1995. The effects of a spillage of diesel fuel on a rocky shore in the sub-Antarctic region (Macquarie Island). *Mar. Poll. Bull.* 31:4–12.

Sinclair, B.J., C. Jaco Klok, M.B. Scott, J.S. Terblanche, and S.L. Chown. 2003. Diurnal variation in supercooling points of three species of Collembola from Cape Hallett, Antarctica. *J. Insect. Physiol.* 49:1049–1061.

Sjöling, S. and D.A. Cowan. 2003. High 16S rDNA bacterial diversity in glacial meltwater lake sediment, Bratina Island, Antarctica. *Extremophiles* 7:275–282.

Smith, J.J., L.A. Tow, W. Stafford, C. Cary, and D.A. Cowan. 2006. Bacterial diversity in three different Antarctic Cold Desert mineral soils. *Microb. Ecol.* 51:413–421.

Smith, R.C., B.B. Prezelin, K.S. Baker et al. 1992 Ozone depletion: Ultraviolet radiation and phytoplankton biology in Antarctic waters. *Science* 255:952–959.

Snape, I., S. Ferguson, and A. Revill. 2003. Constraints on rates of natural attenuation and *in situ* bioremediation of petroleum spills in Antarctica. In: Assessment and remediation of contaminated sites in Arctic and cold climates (Proceedings), M. Nahir, K. Biggar, G. Cotta (eds.). St. Joseph's Print Group Inc. Edmonton AB, Canada, pp. 257–261.

Soller, R., P. Hirsch, D. Blohm, and M. Labrenz. 2000. Differentiation of newly described Antarctic bacterial isolates related to *Roseobacter* species based on 16S–23S rDNA internal transcribed spacer sequences. *Int. J. Syst. Evol. Microbiol.* 50:909–915.

Spring, S., B. Merkhoffer, N. Weiss, R.M. Kroppenstedt, H. Hippe, and E. Stackebrandt. 2003. Characterization of novel psychrophilic clostridia from an Antarctic microbial mat: Description of *Clostridium frigoris* sp. nov., *Clostridium lacusfryxellense* sp. nov., *Clostridium bowmanii* sp. nov. and *Clostridium psychrophilum* sp. nov. and reclassification of *Clostridium laramiense* as *Clostridium estertheticum* subsp. *laramiense* subsp. nov. *Int. J. Syst. Evol. Microbiol.* 53:1019–1029.

Staley, J.T. and J.J. Gosink. 1999. Poles apart: Biodiversity and biogeography of sea ice bacteria. *Annu. Rev. Microbiol.* 53:189–215.

Stokes, J.L. and M.L. Redmond. 1966. Quantitative ecology of psychrophilic microorganisms. *Appl. Microbiol.* 14:74–78.

Straka, R.P. and L.L. Stokes. 1960. Psychrophilic bacteria from Antarctica. *J. Bacteriol.* 80:622–625.

Sullivan, C.W. and A.C. Palmisano. 1984. Sea ice microbial communities: Distribution, abundance, and diversity of ice bacteria in McMurdo Sound, Antarctica, in 1980. *Appl. Environ. Microbiol.* 47:788–795.

Suzuki, K., J. Sasaki, M. Uramoto, T. Nakase, and K. Komagata. 1997. *Cryobacterium psychrophilum* gen. nov., sp. nov., nom. rev., comb. nov., an obligately psychrophilic actinomycete to accommodate *Curtobacterium psychrophilum* Inoue and Komagata 1976. *Int. J. Syst. Bacteriol.* 47:474–478.

Takii, S., T. Kondal, A. Hiraishi, G.I. Matsumoto, T. Kawano, and T. Torii. 1986. Vertical distribution in and isolation of bacteria from Lake Vanda: an Antarctic lake. *Hydrobiologia* 135:15–21.

Tanner, A.C. 1985. The role of bacteria in the cycling of nutrients within the maritime Antarctic environment. In: *Antarctic Nutrient Cycles and Food Webs*, W.R. Siegfried, P.R., Condy, and R.M. Laws (eds.). Springer Verlag, Heidelberg, Germany, pp. 123–127.

Tanner, A.C. and R.A. Herbert. 1981. Nutrient regenration in Antarctic marine sediments. *Kiel Meeresforsch Sonderh*, 5:390–395.

Taton, A., S. Grubisic, E. Brambilla, R. De Wit, and A. Wilmotte. 2003. Cyanobacterial diversity in natural and artificial microbial mats of Lake Fryxell (McMurdo Dry Valleys, Antarctica): A morphological and molecular approach. *Appl. Environ. Microbiol.* 69:5157–5169.

Thomassin-Lacroix, E.J.M., M. Eriksson, K. J. Reimer, and W. W. Mohn. 2002. Biostimulation and bioaugmentation for on-site treatment of weathered diesel fuel in Arctic soil. *Appl. Microbiol. Biotechnol.* 59:551–556.

Tsilinsky, M. 1908. La flore microbieene dans les regions due pole Sud. *Exp. Antarct. Francais 1903–1905* 3:1–33.

Tsyganov, V.A. 1970. Detection and morphological-cultural characteristics of *Actinomycetes* from the antarctic. *Mikrobiologia* 39:821–826.

Van Trappen, S., J. Mergaert, S. Van Eygen, P. Dawyndt, M.C. Cnockaert, and J. Swings. 2002. Diversity of 746 heterotrophic bacteria isolated from microbial mats from Antarctic lakes. *Syst. Appl. Microbiol.* 25:603–610.

Van Trappen, S., J. Mergaert, and J. Swings. 2003. *Flavobacterium gelidilacus* sp. nov., isolated from microbial mats in Antarctic lakes. *Int. J. Syst. Evol. Microbiol.* 53:1241–1245.

Van Trappen, S., J. Mergaert, and J. Swings. 2004a. *Loktanella salsilacus* gen. nov., sp. nov., *Loktanella fryxellensis* sp. nov. and *Loktanella vestfoldensis* sp. nov., new members of the *Rhodobacter* group, isolated from microbial mats in Antarctic lakes. *Int. J. Syst. Evol. Microbiol.* 54:1263–1269.

Van Trappen, S., I. Vandecandelaere, J. Mergaert, and J. Swings. 2004b. *Algoriphagus antarcticus* sp. nov., a novel psychrophile from microbial mats in Antarctic lakes. *Int. J. Syst. Evol. Microbiol.* 54:1969–1973.

Van Trappen, S., I. Vandecandelaere, J. Mergaert, and J. Swings. 2004c. *Flavobacterium degerlachei* sp. nov., *Flavobacterium frigoris* sp. nov. and *Flavobacterium micromati* sp. nov., novel psychrophilic bacteria isolated from microbial mats in Antarctic lakes. *Int. J. Syst. Evol. Microbiol.* 54:85–92.

Van Trappen, S., I. Vandecandelaere, J. Mergaert, and J. Swings. 2004d. *Gillisia limnaea* gen. nov., sp. nov., a new member of the family *Flavobacteriaceae* isolated from a microbial mat in Lake Fryxell, Antarctica. *Int. J. Syst. Evol. Microbiol.* 54:445–448.

Van Trappen, S., T.L. Tan, J. Yang, J. Mergaert, and J. Swings. 2004e. *Alteromonas stellipolaris* sp. *nov.,* a novel budding, prosthecate bacterium from Antarctic seas, and emended description of the genus *Alteromonas. Int. J. Syst. Evol. Microbiol.* 54:1157–1163.

Van Trappen, S., I. Vandecandelaere, J. Mergaert, and J. Swings. 2005. *Flavobacterium fryxellicola* sp. nov. and *Flavobacterium psychrolimnae* sp. nov., novel psychrophilic bacteria isolated from microbial mats in Antarctic lakes. *Int. J. Syst. Evol. Microbiol.* 55:769–772.

Vincent, C.F. 1988. *Microbial Ecosystems of Antarctica.* Cambridge University Press, Cambridge, U.K., 303p.

Vishniac, V.W. and S.E. Mainzer. 1972. Soil microbiology studied *in situ* in the dry valleys of Antarctica. *Antarct. J. United States* 7:88–89.

Voytek, M.A. and B.B. Ward. 1995. Detection of ammonium-oxidizing bacteria of the beta-subclass of the class *Proteobacteria* in aquatic samples with the PCR. *Appl. Environ. Microbiol.* 61:1444–1450.

Wery, N., U. Gerike, A. Sharman, J.B. Chaudhuri, D.W. Hough, and M.J. Danson. 2003. Use of a packed-column bioreactor for isolation of diverse protease-producing bacteria from antarctic soil. *Appl. Environ. Microbiol.* 69:1457–1464.

Whyte, L.G., L. Bourbonniere, C. Bellerose, and C. W. Greer 1999. Bioremediation assessment of hydrocarbon-contaminated soils from the high Arctic. *Bioremediation J.* 3:69–79.

Whyte, L.G., L. Bourbonniere, and C.W. Greer. 1997. Biodegradation of petroleum hydrocarbons by psychrotropic Pseudomonas strains possessing both alkane (alk) and napthalene (nah) catabolic pathways. *Appl. Environ. Microbiol.* 63:3719–3723.

Whyte, L.G., B. Goalen, J. Harwari, D. Labbe, C.W. Greer, and M. Nahir. 2001. Bioremediation treatability assessment of hydrocarbon contaminated soils from Eureka, Nunavut. *Cold Reg. Sci. Technol.* 32:121–132.

Whyte, L.G., C.W. Greer, and W.E. Inniss. 1996. Assessment of the biodegradation potential of psychrotrophic microorganisms. *Can. J. Microbiol.* 42:99–106.

Wintzingerode, F.V., U.B. Goëbel, and E. Stackebrandt. 1997. Determination of microbial diversity in environmental samples:pitfalls of PCR-based rRNA analysis. *FEMS Microbiol. Rev.* 21:213–229.

Wynn-Williams, D.D. 1990. Ecological aspects of Antarctic microbiology. In: *Advances in Microbial Ecology*, Vol. 11, K.C. Marshall (ed.). Plenum Press, New York, pp. 71–146.

Yakimov, M.M., L. Giuliano, G. Gentile et al. 2003. *Oleispira antarctica* gen. nov., sp. nov., a novel hydrocarbonoclastic marine bacterium isolated from Antarctic coastal sea water. *Int. J. Syst. Evol. Microbiol.* 53:779–785.

Yergeau, E., S. Kang, Z. He, J. Zhou, and G.A. Kowalchuk. 2007a. Functional microarray analysis of nitrogen and carbon cycling genes across an Antarctic latitudinal transect. *ISME J.* 1:163–179.

Yergeau, E., K.K. Newsham, D.A. Pearce, and G.A. Kowalchuk. 2007b. Patterns of bacterial diversity across a range of Antarctic terrestrial habitats. *Environ. Microbiol.* 9:2670–2682.

Yi, H. and J. Chun. 2006. *Flavobacterium weaverense* sp. nov. and *Flavobacterium segetis* sp. nov., novel psychrophiles isolated from the Antarctic. *Int. J. Syst. Evol. Microbiol.* 56:1239–1244.

Yi, H., H.M. Oh, J.H. Lee, S.J. Kim, and J. Chun. 2005a. *Flavobacterium antarcticum* sp. nov., a novel psychrotolerant bacterium isolated from the Antarctic. *Int. J. Syst. Evol. Microbiol.* 55:637–641.

Yi, H., H.I. Yoon, and J. Chun. 2005b. *Sejongia antarctica* gen. nov., sp. nov. and *Sejongia jeonii* sp. nov., isolated from the Antarctic. *Int. J. Syst. Evol. Microbiol.* 55:409–416.

Zdanowski, M.K. and P. Weglenski. 2001. Ecophysiology of soil bacteria in the vicinity of Henryk Arctowski Station, King George Island, Antarctica. *Soil Biol. Biochem.* 33:819–829.

Zeng, Y., W. Liu, H. Li, Y. Yu, and B. Chen. 2007. Effect of restriction endonucleases on assessment of biodiversity of cultivable polar marine planktonic bacteria by amplified ribosomal DNA restriction analysis. *Extremophiles* 11:685–692.

4 L.I.F.E. in Antarctic Lakes

Birgit Sattler and Michael C. Storrie-Lombardi

CONTENTS

4.1 INTRODUCTION

Spread across the Arctic, Antarctic, and Alpine regions of our world, the cryosphere lakes of our planet are most likely an ancient and vital key for the persistence of life on Earth. Strong evidence continues to accumulate indicating that the Earth has been completely covered with ice for 10 million years or more on at least two occasions in a process now known as "Snowball Earth" (Kirschvink et al., 2000). During these periods of massive global glaciation, icy microbial ecosystems would have served as the central reservoir for life, including the photosynthetic primary producers such as the cold-tolerant cyanobacteria (Vincent and Howard-Williams, 2000; Priscu and Christner, 2004). The microbial communities inhabiting this cryosphere would have already adapted and evolved in the annual miniature version of the global Snowball Earth. To survive in one of the most hostile, extreme environments on Earth, to deal with high ultraviolet radiation loads, freeze–thaw cycles, and organic resources dependent over the long term on photosynthetic primary production, these communities would contain the elite specimens of terrestrial evolution (Tranter et al., 2004).

However, the inhabitants of this massive ecosystem are now facing a new set of challenges. Few places on Earth have been affected by climate change more than the Antarctic Peninsula with an increase in annual average temperatures of almost 3°C during the past 50 years (Hansen et al., 1999). The impact of this change is complex. For instance, an increased temperature is accompanied by an increase in humidity and water availability producing a significantly increased growing season. But increases in humidity can alter the ultraviolet-induced inactivation of airborne bacteria (Peccia et al., 2001). Radiation can inactivate airborne bacteria at both moderate

(50%–60%) and high (85%–95%) levels of relative humidity, but inactivation rates are the greatest at moderate levels (Paez-Rubio and Peccia, 2005). In addition, the elevated awareness in the scientific community of the importance of these extreme ecosystems has led to significant increases in human contact with these fragile microbial consortia inhabiting ice ecosystems. The clear price for that contact will be an increase in the risk of forward contamination of the lake cryosphere with human microbiota and the possibility of chemical contamination with the hydrocarbons inherent in human transportation and survival in Antarctica.

In this chapter, we will review the anatomy of Antarctic lakes, and discuss their potential role as a climate change "canary." The complex, fragile structure of the microbial communities of Antarctic lakes reflects the changes experienced by neighboring and distant ecosystems, and can serve as an early warning signal for global climate change or distributed shifts in total hydrocarbon flux. In particular, we will describe recent work to develop laser-induced fluorescence emission (L.I.F.E.) techniques for the nondestructive, remote, and *in situ* detection of microbial life living in the ice and the snow of Earth's cryosphere. To begin our review of the anatomy of Antarctic lakes, we note that the microbial life of these cryosphere lakes inhabits four distinct environmental niches: the air, the annual or perennial ice coverings, the lake water, and the sediments.

4.2 LIFE IN THE CRYOSPHERE AIR

The evolution of microbial communities of isolated Antarctic lakes actually begins in the air. Antarctica is isolated from the other continents of the world by the Southern Ocean and the Antarctic circumpolar current (Smith, 1991; Wynn-Williams, 1991). However, it has been clear for more than six decades that bacteria, algae, and fungi are easily transported through significant distances by the wind (Gislén, 1948). In fact, hundreds of millions of tons of dust containing viable microorganisms, trace metals, and organic materials are transported between continents each year (Choi et al., 1997; Garrison et al., 2003). Bacteria have even been shown to divide on airborne particles (Dimmick et al., 1979a,b) and approximately 20% of the total atmospheric aerosol mass is carbonaceous material (Bauer et al., 2002). Even cloud water can contain high concentrations of organic acids, such as formate and acetate, that bacteria can utilize as energy sources (Herlihy and Mills, 1986; Herlihy et al., 1987), and microorganisms can grow and metabolize even in icy supercooled cloud droplets (Sattler et al., 2001). Whatever its origin may be, biological material present in Antarctic air represents a diverse consortia of life forms including moss spores, pollen, fungal spores, bacteria (including cyanobacteria), algal propagules, viruses, lichen propagules, tardigrade cysts, nematodes, and arthropod fragments (Burckle et al., 1988; Wynn-Williams, 1991; Marshall, 1996a,b).

For fungi, probably the most important dispersal route is air transport. However, airborne distribution of fungi in the Antarctic is constrained by the low levels of airborne particles. Counts of total colony forming units (bacteria and fungi) in low altitude air systems are in the order of 0.5–3 m^{-3} (Cameron et al., 1977; Wynn-Williams, 1991; Marshall, 1996a). Meteorological conditions that would allow for such transport occur regularly, although not necessarily frequently. Nevertheless, more than

900 fungal species have been identified in the Antarctic, and the presence of fungi pathogenic to plants and invertebrates not found in Antarctica suggests dispersion from other regions (Onofri et al., 2000; Tosi et al., 2002). In addition, fungi normally associated with particular hosts have been found in Antarctic ecosystems where their hosts are absent (Göttlich et al., 2003).

Molecular biological studies of airborne bacterial diversity and transport in Antarctica are not as extensive as the work on fungi. However, it does appear that prokaryotic life in the air of Antarctica may be of distant, local, or regional origin. PCR-based investigation of microbial diversity in air samples collected at Rothera Point on Adelaide Island in the Antarctic Peninsula, identified a wide variety of cyanobacteria, actinomycetes, and diatom plasmids (Hughes et al., 2004). The closest 16S rDNA matches for many of the sequences were to organisms already identified in the Antarctic or other cryosphere environments. While the majority of matches were to clones of local origin, wind trajectory calculations indicated that the air had recently traversed the Antarctic Peninsula raising the possibility that a significant portion of the microbiota identified may have been of widely distributed origin. Regardless of origin, the link between the airborne microbial life and the ice of glaciers and Antarctic lakes depends on the fundamental physics of light, heat, and the adhesive characteristics of organic molecules.

4.3 LIFE IN ICE

The icy covers of Earth's polar regions and high mountains have classically been seen as sterile, harsh environments with excessively low temperatures, large pH variations, and dangerous levels of ultraviolet (UV) radiations, too poor in nutrients and liquid water to sustain life (Psenner and Sattler, 1998). However, not only the lakes, but also the ice and the snow covering these lakes, harbor rich, diverse, complex microbial communities. Examples of the extensive ecological networks can be found in the ice of alpine and polar lakes, sea ice, glacier ice, and even the ice of supercooled cloud droplets (Priscu et al., 1998; Psenner and Sattler, 1998; Psenner et al., 1999; Sattler et al., 2001; Priscu and Christner, 2004; Sattler et al., 2004). This icy ecosystem extends across both hemispheres and all continents. It is highly sensitive to environmental changes, making it possible for the health of these cryosphere communities to serve as proxies of the Earth's response to more widespread environmental stressors (Hodson et al., 2008).

The first step in the establishment of microbial life in polar or alpine ice is marked by a change in ice albedo secondary to the adhesion and accumulation of surface dust particles (Warren and Wiscombe, 1980; Cutler and Munro, 1996; Brock et al., 2000). Dust deposition can significantly accelerate glacial melting independent of global warming or cooling trends. The major contributor to the absorption of solar energy by dust is the humic component (Takeuchi et al., 2001; Takeuchi and Li, 2008). A diverse set of microbial species including photosynthetic microorganisms move through the atmosphere of our planet (Morris et al., 2008) either suspended in cloud droplets (Sattler et al., 2001) or attached to dust grains (Figure 4.1) until the assemblage lands on and adheres to the surface of the ice (Psenner, 1999; Porazinska et al., 2004). Dust grains containing high levels of humic material will be more

(a)

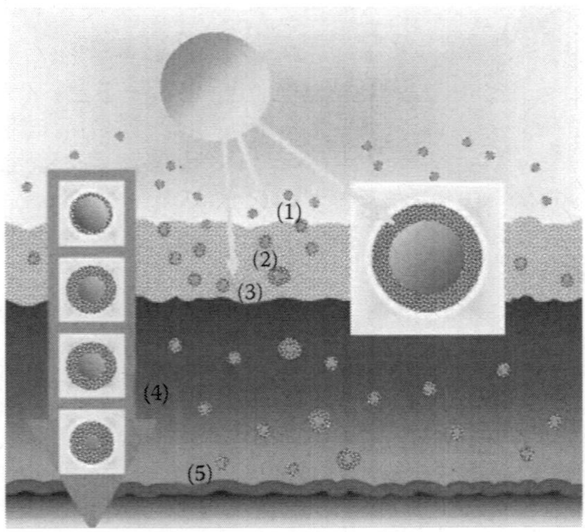

(b)

FIGURE 4.1 (See color insert following page 276.) One of the authors (BS) collecting air samples (a) for 16s RNA analysis of airborne assemblages of microbes on the ice-covered Lake Untersee, Dronning Maud Land, Antarctica. The microbial ecosystem of the lake begins as airborne particles of microbes, humic material, and minerals adhere (b1) to the icy surface. Warmed by the sun the dark mixture of organic and inorganic materials sinks into the ice (b2) and is covered by melt water that soon freezes. Solar energy heats the assemblage sufficiently to produce a thin biofilm on the particle surface comprised of liquid water and multiple microbial species (insert). Photosynthetic cyanobacteria, protected from predators and provided with solar energy for photosynthesis, multiply (b3) until the particle surface is maximally covered (b4). The assemblage passes into the lake water and finally joins the cyanobacteria-dominated mat community (b5) at the lake floor.

likely to adhere to a surface than would a smooth, inorganic mineral grain because of the high coefficient of friction characterizing organic material (Figure 4.1). If the material includes significant quantities of microorganisms, the sticky outer exopolysaccharide coat of many of these microbes will increase the likelihood that the assemblage will adhere to any surface including ice. Once the relatively dark, humic-rich dust attaches to the ice surface, solar energy absorbed by the grains raises the temperature of the mixture of organic and inorganic materials. In contact with mineral or humic surfaces, ice remains unfrozen as a microfilm surrounding the particle as long as it is sufficiently warmed by the sun (Priscu et al., 1998; Price, 2007). The increase in assemblage temperature melts the ice forming a cylindrical depression and makes it possible for gravity to pull the mixture of soil, minerals, and microorganisms into the ice. The descending assemblage is quickly covered by a new icy surface if the atmospheric temperature is sufficient to freeze the surface melt water. The process generates extreme changes in hydrophysical and hydrochemical conditions for the entrapped microorganisms (Tranter et al., 2004). In the clear ice coverings characterizing the Antarctic lakes such as those of the Dry Valleys of Schirmacher Oasis and Lake Untersee in Dronning Maud Land, the process continues as sunlight penetrates deep into the ice (Storrie-Lombardi and Sattler, 2009). Solar energy radiating from the dust melts the surrounding ice, coating both the grains and the living microorganisms with a thin layer of water. The combination of dust, water, organic matter, sunlight, and life now forms a tiny microcosm containing liquid water, energy from the sun, and photosynthetic microbial life capable of performing as primary producers for this miniature world. The assemblage continues to sink into the ice cover until radiant energy from solar heating is balanced by heat loss to the surrounding ice.

Larger macroscopic versions of these accumulations were first called cryoconites (from the Greek kryos, "icy cold;" and konía, "dust") by the Swedish explorer A.E. Nordenskjöld, in 1870 during an exploration of Greenland glaciers (Miteva, 2008). Cryoconite assemblages actually range in scale from the massive surface boulders melting slowly into Antarctic lakes to small accumulations of grains with a few micrometers to a few millimeters in diameter (Figure 4.2). The microbial consortia surviving in these miniature habitats dominated by photosynthetic species, particularly cyanobacteria (Säwström et al., 2002; Anesio et al., 2009), may be the principle contributors to primary productivity and biogeochemical cycling for glaciers throughout the Earth's cryosphere (Tranter et al., 2004; Hodson et al., 2005; Foreman et al., 2007; Hodson et al., 2008).

Phylogenetic analysis of the microbial communities inhabiting lake ice cryospheres reveals some consistent and anticipated patterns. For example, an analysis of the microbial consortium inhabiting the 2.5 m thick perennial ice cover of Lake Paula in Patriot Hills in the West Antarctic revealed that many of the organisms which could be sequenced were spore forming and were most closely related to species within the families of Oxalobacteraceae, Caulobacterales, Sphingomonadales, Microbacteriaceae and Firmicutes (Sattler et al., 2004). Two organisms, *Acinetobacter radioresistans* and *Methylobacterium radiotolerans*, were isolated which are known to be resistant and tolerant, respectively, to radiation. The remainder of the sequences came from species most closely related to cold-tolerant organisms, such as *Cellulomonas* sp., *Sphingomonas* sp. (*paucimobilis*), *Microbacterium brevis*, *Paenibacillus illinoisensis*, *Arthrobacter* sp., *Clavibacter* sp., and *Bacillus megaterium*. The vertical analysis of biomass and carbon production revealed other consistent gradients. Biomass estimated from a consolidated sample (mean cell count $1.35 \times 10^8 L^{-1}$) was 2.17 μg carbon L^{-1}. Bacterial abundance showed high variability

(a)

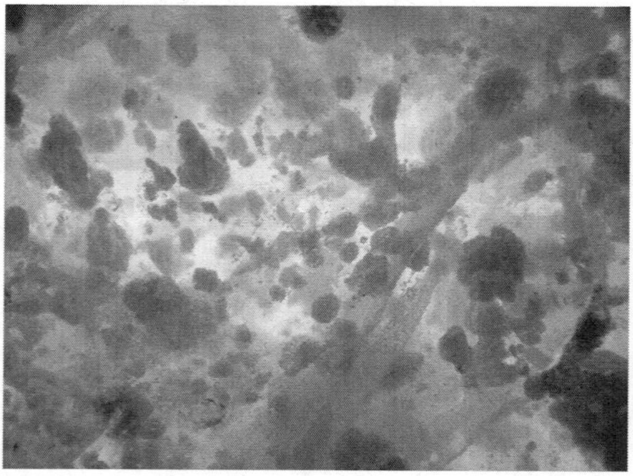

(b)

FIGURE 4.2 (See color insert following page 276.) Cryoconites come in all sizes from accumulations at the base of massive boulders (a) floating on the ice cover of Lake Untersee to millimeter scale particles wrapped in refractive crystals of ice spread across the lake surface (b).

ranging from a minimum of 1.35×10^6 cells L^{-1} in the middle part of the ice core up to a maximum of $3.03 \times 10^8 L^{-1}$ on the top ice surface. The cell production rates measured by ^{3}H-thymidine incorporation varied between a minimum of 16.7 cells $L^{-1} h^{-1}$ in the midportion of the ice, to a maximum of 7.84×10^4 to 10^6 cells $L^{-1} h^{-1}$ at the bottom of the ice at the interface to the liquid phase of the lake. Microbial carbon production was highest in the upper layer at 260 fgC $L^{-1} h^{-1}$, a layer dominated by optically larger cells than were found in the lower layers. Bacterial doubling times show significant variations ranging from 17 days for the most productive lower layer to 2334 days for the minimally productive inner layers of the ice.

4.4 LIFE IN ANTARCTIC LAKE WATER, SEDIMENT, AND MATS

Antarctic lakes exhibit a wide range of physical and chemical conditions, may be freshwater or hypersaline, usually contain simplified truncated food webs, may be perennially or annually ice covered, and may or may not contain bottom-dwelling microbial mat communities (Laybourn-Parry and Pearce, 2007). Many of the lakes are found in the Antarctic oases such as the ice-free dry valleys of the Schirmacher Oasis in Dronning Maud Land or the McMurdo Dry Valleys in Southern Victoria. A second class of lakes, the Epishelf lakes (Smith et al., 2006), can be found between ice shelves and land mass, and appear to be almost unique to the Antarctic continent with only a few examples described in the Arctic (Vincent et al., 2001). In addition, temporary lakes form on glacier surfaces with microbial communities resembling those found in smaller cryoconite holes, and from ground-penetrating radar data we now know that there exists a vast array of subglacial lakes spread across the continent. Cold, oligotrophic, alkaline epiglacial lakes may be one of the most common lake types in Antarctica. 16S rRNA gene T-RFLP data demonstrated considerable heterogeneity between the bacterial communities of eight epiglacial lakes located in the Framnes Mountains 30 km inland from Mawson Station (Twin et al., 2008). The lakes support under-ice algal mats, harboring a variety of metazoan communities. Patterned Lake (pH 10.4, salinity 0.075 pss) photosynthetic microbial community is dominated by cyanobacteria related to *Pseudanabaena* and *Leptolyngbya* spp., while Sonic Lake (pH 10.6, 0.682 pss) is dominated by eustigmatophytes similar to *Nannochloropsis salina* and to uncharacterized freshwater rhodophytes and chrysophytes. Extreme environmental conditions including cold, salinity, minimal levels of photosynthetically active radiation (PAR), and limited availability of nitrogen, phosphorous, and oxygen have all provided a unique set of evolutionary constraints. For the past decade understanding of the phylogenetics and biodiversity of Antarctic lakes has relied on the PCR-amplifed 16S rRNA gene sequencing and the development of clone libraries (Dobson et al., 1993; Bowman et al., 2000a,b; Baker et al., 2003). But now the field determination of microbial community structure and function has become possible through the increasing use of other molecular and physiochemical technologies including fluorescence *in situ* hybridization, spectrophotometry, radioactive isotope incorporation, stable isotope analysis, microautoradiography, whole genome sequencing, and *in situ* L.I.F.E.

In a comparison to culture-dependent and culture-independent techniques to assess Antarctic lake bacterioplankton, six bacterial divisions (including 24 genera) were identified using culture-dependent techniques, while eight bacterial divisions (including 23 genera) were identified using culture-independent techniques (Pearce et al., 2003). Only five genera, *Corynebacterium*, *Cytophaga*, *Flavobacterium*, *Janthinobacterium*, and *Pseudomonas* (representing four bacterial divisions), could be detected using both sets of techniques. Interestingly, pigment production was found within members of each of these genera. This and similar studies emphasize the importance of a comprehensive polyphasic approach in the analysis of lake bacterioplankton communities.

The extreme environmental conditions and geographic isolation of the Antarctic comments led early investigators to speculate that lake microbial communities would exhibit significant biodiversity. Current data indicates that sequence dissimilarity between the rRNA sequences of Antarctic strains and their nearest known relatives is sufficient to imply species divergence from one another much earlier than the establishment of their current Antarctic habitat (Franzmann and Dobson, 1993). 16S rRNA clone library analysis evaluated the biodiversity within the sediments of three hypersaline Antarctic lakes (Bowman et al., 2000a,b). The Vestfold Hills lake system (68°S 78°E) of Eastern Antarctica formed during the Holocene, has been relatively unchanged for 4000 years, and accounts for approximately one-fifth of the world's meromictic lakes. Three hypersaline lakes in the region formed by seawater evaporation are almost perennially ice free. Ekho Lake is a 40 m deep, heliothermal meromictic lake; Organic Lake is a shallow meromictic lake with unusually high levels of dimethylsulfide in its bottom waters; and Deep Lake is a frigid (−14°C to −18°C) 36 m deep monomictic lake with a salinity 10 times that of seawater. The microbial community of Deep Lake was dominated by halophilic Archaea. The sediment communities of the two meromictic hypersaline lakes, Organic Lake and Ekho Lake, were more complex. Microbial phylotypes clustered within the Proteobacteria and Bacteroidetes divisions and with algal chloroplasts. Many phylotypes of these lakes were related to taxa adapted to marine salinity. The Ekho Lake clone library contained several major phylotypes related to the Haloanaerobiales, probably the result of selection pressure from the comparatively high (15°C) *in situ* lake temperature. Diversity indices were determined including indices estimating biodiversity coverage, diversity (Shannon-Weaver index), dominance (Simpson index), evenness, and species richness. The coverage of biodiversity was very high, ranging from 87% to 96%, but the species richness index was 2–20 times lower than was found in the sediments of low to moderate salinity Antarctic lakes (Coolen et al., 2004; Glatz et al., 2006).

Subglacial Lake Vostok, the eighth largest lake of Earth's lakes (area = 14,000 km^2), is covered by 4 km of glacial ice. As the glacier has traversed the lake, lake water has frozen to the bottom of the glacier forming a 200 m accretion ice layer. The Vostok microbial community, while exhibiting some of the lowest recorded concentrations of microbes found on the planet, nevertheless presents as an extremely diverse and complex microbial assemblage. An analysis of seven accretion core sections and two deep glacial cores identified 18 unique bacterial rRNA gene phylotpes. Maximum parsimony analysis indicates that nearest known sequences represent psychrotolerant organisms from *Firmicutes*, *Actinobacteria*, and α-proteobacteria taxa (D'Elia et al., 2008). Microbial diversity in Lake Vostok even includes thermophilic chemoautotrophs. The thermophile bacterium *Hydrogenophilus thermoluteolus* has been identified in the 3561 and 3607 m accretion ice core samples (Lavire et al., 2006). The findings point to the presence of thermophilic chemoautotrophic microorganisms in Lake Vostok accretion ice, presumably living originally in deep faults in the lake bedrock cavity.

Understanding the history of the adaptive responses used in Antarctic lake microbial communities across evolutionary time and predicting likely responses (including species divergence) to current climate change, hydrocarbon contamination, and increased contact with humanity will fall on the shoulders of molecular biology. Portions of the adaptive record of Antarctic microbial communities will be contained in the genomes of the Archaea and bacteria inhabiting Antarctic air, soil, ice, and lakes. Whole genome microbial sequences for Antarctic lake species have only recently appeared (Peck et al., 2005), but the data have already contributed to our understanding of cold adaptive responses for both genome integrity and whole organism function. Complete genomic sequences have now been obtained for *Methanogenium frigidum* and *Methanococcoides burtonii* isolated from the hypolimnion of Ace Lake, Vestfold Hills, Antarctica (Saunders et al., 2003). A comparison of these genomes of cold-adapted methanogens to seven genomes from hyperthermophilic and mesothermophilic methanogens revealed that proteins encoded in the cold-adapted Archaea are characterized by a higher content of noncharged polar amino acids, particularly Gln and Thr, and a lower number of hydrophobic amino acids, particularly Leu. Protein structure models generated using sequence data from the nine genomes indicated that the cold-adapted Archaea showed a strong tendency to use more Gln, Thr, and hydrophobic residues and fewer charged residues in the solvent-accessible area of the proteins. A cold shock domain (CSD) protein (CspA homolog) was identified in *M. frigidum*, two hypothetical proteins with CSD-folds were found in *M. burtonii*, and a unique winged helix DNA-binding domain protein occurs in *M. burtonii*. The findings suggest that these types of nucleic acid binding proteins have a critical role in cold-adapted Archaea, but the exact functions remain undefined. GC content is the major factor influencing tRNA stability in hyperthermophiles, but not in the psychrophile, mesophile or moderate thermophile genomes. These observations imply that any demand for tRNA flexibility in psychrophiles is mediated by other mechanism.

4.5 ANTARCTIC LAKE HYDROCARBON EXPOSURE AND DEGRADATION

Oil spills caused by humans leave irreparable tracks in sensitive ecosystems, such as Antarctic lakes. A helicopter crash in January 2003 deposited a JP5-AN8 mixture aviation diesel fuel on the 5 m thick perennial ice cover of Lake Fryxell in McMurdo Dry Valleys, East Antarctic (Jaraula et al., 2009). Immediate spread of the fuel was enhanced by the presence of meltpools formed in response to aeolian sediment accumulation. But even in the absence of melt pools, slow spreading of the diesel through the porous ice occurred depending on the aqueous solubility of the fuel components. Some ice samples exhibited water-washed diesel, while others contained only relatively soluble low molecular weight aromatic hydrocarbons, such as alkylbenzene and naphthalene homologues. Follow-up analysis a year after the spill identified simple evaporation as a major diesel-weathering variable in both ice and meltpool waters.

Even the scientific exploration of Antarctic Lakes poses considerable contamination risks. For example, a fundamental problem in phylogenetic studies of glacial ice and subglacial lakes is the prevention of both forward and retrograde contaminations of drill sites while obtaining ice cores. At Lake Vostok, after obtaining 3650 m of ice core, it was noted that the core surface was coated with a hard-to-remove film comprised of aliphatic and aromatic hydrocarbons and foranes originating in the drilling fluid (Alekhina et al., 2007). 16S rRNA gene sequencing identified six phylotypes.

The two dominant phylotypes recovered from 3400 to 3600 m were from within the genus *Sphingomonas*, a genus that includes organisms quite capable of PAH degradation. The remaining phylotypes were human- or soil-associated bacteria presumed to be drilling fluid contaminants.

Hydrocarbon bioremediation can certainly be accomplished in cold freshwater lakes. Psychrotrophic phenanthrene-degrading bacteria have been found in sediment samples from Lake Baikal, Russia (Ahn et al., 2005). From 70 phenanthrene-degrading original isolates, the seven demonstrating the highest phenanthrene-degradation rates were identified by 16S rDNA sequencing. The Gram-positive rod-shaped *Rhodococcus erythropolis* exhibited the highest growth and degradation rates removing more than a quarter of the available phenanthrene in 20 days at 15°C. Degradation was less efficient but still significant at 5°C. This psychrotrophic phenanthrene-degrading bacterium has been proposed as a candidate for use in the bioremediation of polycyclic hydrocarbon contamination in low temperature environments.

However, Lake Baikal contains a rich, diverse microbial community with ample opportunity for interspecies horizontal gene transfer of enzyme codes important for bioremediation and rapid response to wide variations in temperature and oxygenation. Under laboratory conditions indigenous Antarctic coastal bacteria can degrade *n*-alkanes and the more recalcitrant PAH, phenanthrene. Bacterial cultures from selective enrichment of sand collected 60 days and 1 year after treatment of sites in an oil spill trial at Airport Beach, Vestfold Hills, East Antarctica, were evaluated for their ability to degrade *n*-alkane and phenanthrene (Cavanagh et al., 1998). Cultures obtained from sites treated with Special Antarctic Blend [SAB] and BP-Visco degraded *n*-alkanes more efficiently than cultures derived from sites contaminated with fish oil [orange roughy]) or inoculated with water from Organic Lake, a site previously shown to contain hydrocarbon-degrading bacteria. Two isolates obtained from samples collected at 60 days efficiently degraded phenanthrene. 1-Hydroxy-naphthoic acid was the major phenanthrene metabolite, but there were detectable levels of salicylic acid, l-naphthol 1,4-naphthaquinone, and phenanthrene 9-10 dihydrodiol.

In specific cold, underwater Antarctic conditions, the biodegradation of some hydrocarbons may be difficult. An *in situ* experiment (Thompson et al., 2006) near the Bailey Peninsula area, Casey Station, East Antarctica, monitored the natural attenuation in the marine sediments of synthetic lubricants including unused and used Mobil lubricants (OW/40) and a biodegradable alternative (OW/20; Fuchs). Clean sediment was contaminated with lubricants and deployed by divers onto the seabed. Alkanoate esters of 1,1,1-tris(hydroxymethyl) propane in the biodegradable and unused lubricants were degraded extensively, but constituted only a minor proportion of the lubricant volume. Potentially toxic additives, such as alkylated naphthalenes and substituted diphenylamines, were fairly resistant to degradation. Since the biodegradable lubricant did not break down under these Antarctic marine conditions it could not be classified as biodegradable.

Critical to our understanding of indigenous bioremediation capability of Antarctic lakes will be our ability to identify the nucleotide codes for useful degradation or metabolic enzymes. Sequencing of these two Archaeal methanogens, *M. frigidum* and *M. burtonii*, will be particularly useful for our exploration of PAH degradation strategies in Antarctica. Both organisms have made significant cold adaptive responses to Antarctic temperatures, yet no biodegradative enzyme pathways have been identified in either genome. If these archea are of relatively ancient origin, such an omission is more than interesting, and makes their genome organization and composition a useful baseline for our understanding of the appearance of biodegradation capabilities in Antarcic Eubacteria and Archaea. However, the even more intriguing possibility exists that they may use strategies we simply do not yet recognize.

One of the most significant genome efforts will be our attempts to understand anaerobic strategies for biodegradation. Both liquid water and ice covers of Antarctic lakes can exhibit local or distributed anoxic regions. The absence of oxygen produces a shift in the metabolic strategies for biodegradations of a variety of hydrocarbon contaminants, the most difficult of which are the PAHs. Originally thought to require the presence of oxygen to initiate the enzymatic attack on PAH rings (Peng et al., 2008), data now exists documenting that the degradation of aromatic compounds including PAHs readily occurs in anaerobic systems. For a comprehensive general review of anaerobic fuel degradation and specifically anaerobic degradation of aromatic compounds see Powell et al. (2006) and Carmona et al. (2009) and references therein, respectively. The ability to degrade aromatic hydrocarbons in anaerobic conditions is of considerable importance since many cryosphere ecosystems that are at risk for contamination by these substances are primarily anaerobic. The stability of the benzene ring has led to its considerable usage in industry, its toxicity, and its persistence at spill sites. The stability poses a significant challenge for anaerobic catabolic or anabolic systems.

Beyond its current environmental significance as a carcinogen, the anaerobic activation of benzene (the 1-ring precursor for all PAH species) is of considerable interest for our understanding of the origin and evolution of microbial life on Earth. The activation in the absence of molecular oxygen of a nonsubstituted aromatic ring, exhibiting the highest C–H bond dissociation energy among all hydrocarbons, is a significant biochemical reaction. The fundamental nature of this reaction forces a single question: just how ancient are the anaerobic PAH degradation pathways? PAH spectral signatures have been documented in meteorites, the dust of the interstellar medium, throughout our local Milky Way galaxy, and neighboring galaxies (for a complete discussion of the ubiquity of PAHs in the local universe see the chapter by Gerakines et al. in this volume). In brief, following the cessation of the heavy bombardment of Earth, the organic resources of the planet would have been restricted to the molecular species provided by planetary infall of interstellar dust, micrometeorites, meteorites, and comets (Chyba et al., 1990; Chyba and McDonald, 1995). Small 1-, 2-, 3-, 4-, and 5-ring PAH moieties would have comprised a significant portion of the organic material available to the microbial community of Archaean Earth. If our sun were, indeed, significantly cooler (Sagan and Mullin, 1872) and the Earth devoid of free oxygen during that time period, then our planet may have experienced repeated episodes of global glaciation significantly earlier than the posited palaeoproterozoic snowball Earth (Kirschvink et al., 2000) and the only survivors may have been cold-adapted anaerobes. Did the anaerobic microbial population of an ancient cryosphere develop PAH degradation pathways during the earliest days of life on this planet? At present little is known about when these pathways appeared. But acquiring and exploring the whole genome sequences of

Antarctica's cryosphere anaerobes for clues to the antiquity of PAH degradation pathways will constitute a significant and exciting challenge for polar science and Astrobiology in the coming decade.

4.6 ENVIRONMENTAL EARLY WARNING SYSTEMS: L.I.F.E.

Current climate warming, ozone depletion, and increased ultraviolet radiation are all altering Antarctic aquatic systems (Quayle et al., 2002). The Antarctic Peninsula, for example, has been subjected to one of the most rapid temperature increases on Earth with mean air temperature rising 2°C across the past 50 years compared to a mean global increase of only 0.6°C ± 0.2°C during the twentieth century. However, the major observable and concerning phenomenon is not the simple global warming trend, but the chaotic, in a mathematical sense, nature of local climate responses across the Antarctic continent. For instance, although the peninsula has experienced excessive warming, other regions such as the environment around the geographic South Pole are cooling. In like fashion, while there is no data to indicate that the shallow southern oceans have experienced significant warming, the mean winter water temperature for Signy Island lakes increased by 0.9°C in 15 years from 1980 to 1995, approximately three times the rise in global mean temperature (Quayle et al., 2002). The rise has been accompanied by a loss of almost half of the ice cover and lake chlorophyll concentrations have doubled (Quayle et al., 2003).

Antarctic lakes are clearly delicate ecosystems capable of rapid, measurable responses to local climatic change (Lyons et al., 2007). However, it is not the individual stressors posed for life in Antarctica that make this continent worthy of intensive study. Cold, desiccation, UV exposure, and organic pollutants are present across the planet. Instead, it is the timely convergence of all of these stressors and the adaptive responses demanded that place the Antarctic continent and its microbial life in a unique position. The response of a living system to these multivariate phenomena can be unexpected. For instance, the investigation of the covariance of increases in mean temperature and increases in UV radiation in the McMurdo Dry Valleys induced a synergistic increase in the production of the UV-screening compound, scytonemin, by the extremophilic cyanobacterium *Chroococcidiopsis* sp. (Dillon et al., 2002), a most serendipitous event for the organism, but not predicted from first principles.

The measurement of Earth's cryosphere biota has classically relied on *in situ*, human-intensive efforts. The techniques survey only a single or a few local systems and samples must be cored from the ice or siphoned from lake water by onsite personnel. The processes generally disrupt the structure of indigenous microbial communities and their interaction with the surrounding matrix. In fact, current techniques employed may even provide their own contamination component, such as the drill oil contamination at Lake Vostok described above (Alekhina et al., 2007).

Global monitoring of Earth's cryosphere will depend on the development of new techniques for *in situ*, airborne and/or orbital surveys. L.I.F.E. from specific biomolecular intracellular and extracellular targets is the most robust technique that does not require sample preparation, sample destruction, or limited consumable resources other than power (Asher, 1993). L.I.F.E. biosignatures may be extracted using either spectra or images, or by combining both using nonlinear artificial neural networks (Storrie-Lombardi et al., 1992; Lahav et al., 1995). High-resolution epifluorescence microscopy with broadband excitation wavelengths at 254 and 375 nm, and laser excitation wavelengths at 532 and 660 nm is the standard equipment for aquatic microbiologists. Laser wavelengths between 220 and 250 nm (deep ultraviolet) probe nucleic acids and amino acids to produce fluorescence at 300–350 nm (Nealson et al., 2002). L.I.F.E. using these wavelengths has been used to detect microbial life 1.3 km below Mauna Kea (Fisk et al., 2003) and has been proposed as a survey tool to search for life on Mars (Storrie-Lombardi et al., 2001). The excitation of metabolites found in all life on Earth (flavin adenine dinucleotide and nicotinamide adenine dinucleotide) using a 375 nm laser evokes fluorescence in the blue and green portions of the visible spectrum (Storrie-Lombardi, 2005) and is being evaluated for use in the European Space Agency's ExoMars mission to search for PAHs in the Mars regolith (Griffiths et al., 2008; Storrie-Lombardi et al., 2008a,b). The excitation between 450 and 650 nm produces fluorescence in microbial photosynthetic pigments (Blinks, 1954) including the phycobiliproteins and the chlorophylls. Phycobiliproteins are found primarily in red algae, cyanobacteria, and cryptomonads including species inhabiting ice-covered Antarctic lakes where they harvest light at energies poorly handled by chlorophyll *a*. The green–brown cyanobacterium *Chroococcidiopsis* sp. and three cyanobacteria (the blue–green *Gloeocapsa* sp., the brown *Gloeocapsa* sp., and the red–orange *Gloeocapsa* sp.) are the most widespread phycobionts of cryptoendolithic lichens in the high-polar regions of Antarctica. All species produce significant fluorescence response to excitation at either 436 nm (Soret band for chlorophyll) or 550 nm (region of maximum absorption by phycobiliproteins).

L.I.F.E. spectral biosignatures, most commonly attributed to the phycobilirubins, have been observed decades from airborne platforms such as NASA's Airborne Oceanographic LIDAR (*Li*ght *D*etection *a*nd *R*anging) since 1979 while surveying Earth's liquid oceans and lakes (Hoge and Swift, 1981). But while the remote detection of L.I.F.E. signatures is relatively common at this time for terrestrial aquatic monitoring and has even been used to map subtle shifts in aquatic biomass composition, no systematic *in situ* or remote sensing attempt has been made to extract such information from the ice of Arctic, Antarctic, or Alpine lakes and glaciers. The absence of such efforts is surprising since, as described above, atmospheric infall material incorporated into the upper ice in the form of cryoconite sediments is relatively rich in organic material (Anesio et al., 2009) and microbial communities dominated by photosynthetic organisms flourish even inside the ice if covered by a microfilm of liquid water (Priscu et al., 1998; Psenner and Sattler, 1998). The detection of such assemblages within the first few centimeters of the ice surface using simple, robust L.I.F.E. techniques was recently accomplished as a part of the Tawani 2008 Antarctic Expedition to Schirmacher Oasis and Lake Untersee, a 2 month campaign in 2008 to lakes in Dronning Maud Land, Antarctica (Storrie-Lombardi and Sattler, 2009). Initial *in situ* survey of lake ice covers in the Schirmacher Oasis and Lake Untersee revealed the massive extent and density of dust infall, entrapment, and the initiation of cryoconite hole evolution (Figure 4.2). The assemblages cover the entire surface of lakes sufficiently large to permit orbital LIDAR imaging, and L.I.F.E. signatures were easily extracted *in situ* with current off-the-shelf laser and imaging technology. The findings make possible the development of *in situ* and orbital monitoring strategies capable of predicting the impact of climate change or contamination on the annual cryosphere contribution to the global carbon budget. L.I.F.E. wavelength choices will permit the monitoring of either simple 3-, 4-, or 5-ring PAH moieties or more complex photosynthetic pigments.

4.7 SUMMARY

The cryosphere lakes of our planet, particularly those in Antarctica, contain microbial communities essential to our understanding of the appearance and survival of life on Earth. Those species sufficiently adapted to survive within the ice itself will most likely carry genomic and phenotypic clues to the strategies that life has employed to withstand the recurrent global glaciations of the Snowball Earth. Of most interest to our understanding of microbial adaptation to current hydrocarbon contamination of pristine cryosphere regions is the possibility that microbial life on Earth had to adapt to primitive hydrocarbons, such as the polycyclic aromatic hydrocarbons during the first few hundred million years after the cessation of the heavy bombardment. In fact, if life arose in the dust clouds of star-forming regions of space, the adaptive process may certainly have predated the appearance of life on a formed terrestrial planet. Monitoring the response of Earth's cryosphere to modern changes in temperature, radiation, humidity, and human contact will require a significant interdisciplinary effort to develop new and more efficient strategies for remote and *in situ* quantifications of life in snow and ice. While the engineering and scientific challenges of developing and deploying a global monitoring network, such as the L.I.F.E. techniques, described here are considerable, the possibility exists that we can draw on two of humanity's most ancient economic and social mandates: our need to protect a unique and fragile ecosystem, and the perennial human desire to explore other worlds for universal signatures of life.

ACKNOWLEDGMENTS

The authors gratefully thank the Tawani Foundation, the Austrian Ministry of Science and Research, and the Kinohi Institute for financial support. B.S. was funded by the Austrian Academy of Sciences, Planetary Studies Foundation, and the Tyrolean Science Fund and wishes to thank Roland Psenner.

REFERENCES

Ahn, T.S., Lee, G.H., and Song, H.G. (2005) Biodegradation of phenanthrene by psychrotrophic bacteria from Lake Baikal. *J. Microbiol. Biotechnol.* 15(5): 1135–1139.

Alekhina, I.A., Marie, D., Petit, J.R., Lukin, V.V., Zubkov, V.M., and Bulat, S.A. (2007) Molecular analysis of bacterial diversity in kerosene-based drilling fluid from the deep ice borehole at Vostok, East Antarctica. *FEMS Microbiol. Ecol.* 59(2): 289–299.

Anesio, A.M., Hodson, A.J., Fritz, A., Psenner, R., and Sattler, B. (2009) High microbial activity on glaciers: Importance to the global carbon cycle. *Global Change Biol.*, doi:10.1111/j.1365–2486.2008.01758.x.

Asher, S.A. (1993) UV resonance Raman spectroscopy for analytical, physical, and biophysical chemistry. 1. *Anal. Chem.* 65(2): A59–66.

Baker, G.C., Tow, L.A., and Cowan, D.A. (2003) PCR based detection of non-indigenous microorganisms in "pristine" environments. *J. Microbiol. Methods* 53: 157–164.

Bauer, H., Kasper-Giebl, A.M.L., Giebl, H., Hitzenberger, R., Zibuschka, F., and Puxbaum, H. (2002) The contribution of bacteria and fungal spores to the organic carbon content of cloud water precipitation and aerosols. *Atmos. Res.* 64: 109–119.

Blinks, L.R. (1954) The photosynthetic function of pigments other than chlorophyll. *Ann. Rev. Plant Physiol.* 5: 93–114.

Bowman, J.P., McCammon, S.A., Rea, S.M., and McMeekin, T.A. (2000a) The microbial composition of three limnologically disparate hypersaline Antarctic lakes. *FEMS Microbiol. Lett.* 183: 81–88.

Bowman, J.P., Rea, S.M., McCammon, S.A., and McMeekin, T.A. (2000b) Diversity and community structure within anoxic sediment from marine salinity meromictic lakes and a coastal meromictic marine basin, Vestfold Hills, eastern Antarctica. *Environ. Microbiol.* 2: 227–237.

Brock, B.W., Willis, I.C., and Sharp, M.J. (2000) Measurement and parameterization of albedo variations at Haut Glacier d'Arolla, Switzerland. *J. Glaciol.* 46(155): 675–688.

Burckle, L.H., Gayley, R.I., Ram, M., and Petit, J.-R. (1988) Diatoms in Antarctic ice cores: Some implications for the glacial history of Antarctica. *Geology* 16: 326–329.

Cameron, R.E., Honour, R.C., and Morelli, F.A. (1977) Environmental impact studies of Antarctic sites. In *Adaptions Within Antarctic Ecosystems*. G.A. Llano (ed.). Washington, DC: Smithsonian Institution, pp. 1157–1158.

Carmona, M., Zamarro, M.T., Blázquez, B., Durante-Rodríguez, G., Juárez, J.F., Valderrama, J.A., Barragán, M.J.L., García, J.L., and Díaz, E. (2009) Anaerobic catabolism of aromatic compounds: A genetic and genomic view. *Microbiol. Mol. Biol. Rev.* 73(1): 71–133.

Cavanagh, J.E., Nichols, P.D., Franzmann, P.D., and McMeekin, T.A. (1998) Hydrocarbon degradation by Antarctic coastal bacteria. *Antarct. Sci.* 10(4): 386–397.

Choi, D.S., Park, Y.K., Oh, S.K., Yoon, H.J., Kim, J.C., Seo, W.J., and Cha, S.H. (1997) Distribution of airborne microorganisms in yellow sands of Korea. *J. Microbiol.* 35: 1–9.

Chyba, C. and McDonald, G.D. (1995) The origin of life in the solar system: Current issues. *Annu. Rev. Earth Planet. Sci.* 23: 215–249.

Chyba, C.F., Thomas, P.J., Brookshaw, L., and Sagan, C. (1990) Cometary delivery of organic molecules to the early Earth. *Science* 249: 366–373.

Coolen, M.J.L., Hopmans, E.C., Rijpstra, W.I.C., Muyzer, G., Schouten, S., Volkman, J.K., and Damsté, J.S.S. (2004) Evolution of the methane cycle in Ace Lake (Antarctica) during the Holocene: Response of methanogens and methanotrophs to environmental change. *Org. Geochem.* 35: 1151–1167.

Cutler, P.M. and Munro, D.S. (1996) Visible and near-infrared reflectivity during the ablation period on Peyto Glacier, Alberta, Canada. *J. Glaciol.* 42(141): 333–340.

D'Elia, T., Veerapaneni, R., and Rogers, S.O. (2008) Isolation of microbes from Lake Vostok accretion ice. *Appl. Environ. Microbiol.* 74(15): 4962–4965.

Dillon, J.G., Tatsumi, C.M., Tandingan, P.G., and Castenholz, R.W. (2002) Effect of environmental factors on the synthesis of scytonemin, a UV-screening pigment, in a cyanobacterium. *Arch. Microbiol.* 177: 322–331.

Dimmick, R.L., Wolochow, H., and Chatigny, M.A. (1979a) Evidence that bacteria can form new cells in airborne particles. *Appl. Environ. Microbiol.* 37: 924–927.

Dimmick, R.L., Wolochow, H., and Chatigny, M.A. (1979b) Evidence for more than one division of bacteria within airborne particles. *Appl. Environ. Microbiol.* 38: 642–643.

Dobson, S.J., Colwell, R.R., McMeekin, T.A., and Franzmann, P.D. (1993) Direct sequencing of the polymerase chain reaction-amplified 16S ribosomal RNA gene of *Flavobacterium salegens* sp. nov. 2 new species from a hypersaline Antarctic lake. *Int. J. Syst. Bacteriol.* 43: 77–83.

Fisk, M.R., Storrie-Lombardi, M.C., Douglas, S., McDonald, G.D., and Popa, R. (2003) Evidence of biological activity in Hawaiian subsurface basalts. *Geochem. Geophys. Geosyst.* 4(1): 1–24.

Foreman, C.M., Sattler, B., Mikucki, J.A., Porazinska, D.L., and Priscu, J.C. (2007) Metabolic activity and diversity of cryoconites in the Taylor Valley, Antarctica. *J. Geophys. Res. G* 112(G4): G04S32.

Franzmann, P.D. and Dobson, S.J. (1993) The phylogeny of bacteria from a modern Antarctic refuge. *Antarct. Sci.* 5: 267–270.

Garrison, V.H., Shinn, E.A., Foreman, W.T., Griffin, D.W., Holmes, C.W., Kellogg, C.A., Majewski, M.S., Richardson, L.L., Ritchie, K.B., and Smith, G.W. (2003) African and Asian dust: From desert soils to coral reefs. *Bioscience* 53: 469–480.

Gislén, T. (1948) Aerial plankton and its conditions of life. *Biol. Rev.* 23: 109–126.

Glatz, R.E., Lepp, P.W., Ward, B.B., and Francis, C.A. (2006) Planktonic microbial community composition across steep physical/chemical gradients in permanently ice-covered Lake Bonney, Antarctica. *Geobiology* 4(1): 53–67.

Göttlich, E., de Hoog, G.S., Genilloud, O., Jones, B.E., and Marinelli, F. (2003) MICROMAT: Culturable fungal diversity in microbial mats of Antarctic lakes. In *Anatarctic Biology in a Global Context*. A.H.L. Huiskes, W.W.C. Gieskes, and J. Rozema, et al. (eds.). Leiden, the Netherlands: Backhuys Publishers, pp. 251–254.

Griffiths, A.D., Coates, A.J., Muller, J.-P., Storrie-Lombardi, M., Jaumann, R., Josset, J.-L., Paar, G., and Barnes, D. (2008) Enhancing the effectiveness of the ExoMars PanCam instrument for astrobiology. *Geophys. Res. Abs.* 10: EGU2008-A-09486.

Hansen, J., Ruedy, R., Glascoe, J., and Sato, M. (1999) GISS analysis of surface temperature change. *J. Geophys. Res.* 104: 30997–31022.

Herlihy, A.T. and Mills, A.L. (1986) The pH regime of sediments underlying acidified waters. *Biogeochemistry* 2(1): 95–99.

Herlihy, A.T., Mills, A.L., Hornberger, G.M., and Bruckner, A.E. (1987) The importance of sediment sulfate reduction to the sulfate budget of an impoundment receiving acid-mine drainage. *Water Resour. Res.* 23(2): 287–292.

Hodson, A.J., Anesio, A.M., Tranter, M., Fountain, A.G., Osborn, M., Priscu, J., Laybourn-Parry, J., and Sattler, B. (2008) Glacial ecosystems. *Ecol. Monogr.* 78: 41–67.

Hodson, A.J., Mumford, P.N., Kohler, J., and Wynn, P.M. (2005) The High Arctic glacial ecosystem: New insights from nutrient budgets. *Biogeochemistry* 72: 67–86.

Hoge, F.E. and Swift, R.N. (1981) Airborne simultaneous spectroscopic detection of laser-induced water Raman backscatter and fluorescence from chlorophyll a and other naturally occurring pigments. *Appl. Opt.* 20(18): 3197–3205.

Hughes, K.A., McCartney, H.A., Lachlan-Cope, T.A., and Pearce, D.A. (2004) A preliminary study of airborne microbial biodiversity over peninsular Antarctica. *Cell Mol. Biol.* 50: 537–542.

Jaraula, C.M.B., Kenig, F., Doran, P.T., Priscu, J.C., and Welch, K.A. (2009) Composition and biodegradation of a synthetic oil spilled on the perennial ice cover of Lake Fryxell, Antarctica. *Environ. Sci. Technol.*, doi: 10.1021/es802655s.

Kirschvink, J.L., Gaidos, E.J., Bertani, L.E., Beukes, N.J., Gutzmer, J., Maepa, L.N., and Steinberger, R.E. (2000) Paleoproterozoic snowball Earth: Extreme climatic and geochemical global change and its biological consequences. *Proc. Natl. Acad. Sci. U.S.A.* 97: 1400–1405.

Lahav, O., Naim, A., Buta, R.J., Corwin, H.G., de Vaucouleurs, G., Dressler, A., Huchra, J.P., van den Bergh, S., Raychaudhury, S., Sodre, L. Jr., and Storrie-Lombardi, M.C. (1995) Galaxies, human eyes and artificial neural networks. *Science* 267: 859–861.

Lavire, C., Normand, P., Alekhina, I., Bulat, S., Prieur, D., Birrien, J.-L., Fournier, P., Hänni, C., and Petit, J.-R. (2006) Presence of *Hydrogenophilus thermoluteolus* DNA in accretion ice in the subglacial Lake Vostok, Antarctica, assessed using rrs, cbb and hox. *Environ. Microbiol.* 8(12): 2106–2114.

Laybourn-Parry, J. and Pearce, D.A. (2007) The biodiversity and ecology of Antarctic lakes: Models for evolution. *Phil. Trans. R. Soc. B* 362: 2273–2289.

Lyons, W.B., Laybourn-Parry, J., and Welch, K.A. (2007) Antarctic lake systems and climate change. In *Trends in Antarctic Terrestrial and Limnetic Systems*. D. Bergstrom, A. Huiskes, and P. Convey (eds.). Heidelberg, Germany, Springer, pp. 273–295.

Marshall, W.A. (1996a) Biological particles over Antarctica. *Nature* 383: 680.

Marshall, W.A. (1996b) Aerial dispersal of lichen soredia in the maritime Antarctic. *New Phytol.* 134: 523–530.

Miteva, V. (2008) Bacteria in snow and glacier Ice. In *Psychrophiles: From Biodiversity to Biotechnology*, vol. 31. R. Margesin, F. Schinner, J.-C. Marx, and C. Gerday (eds.). Berlin, Germany, Springer-Verlag.

Morris, C.E., Sands, D.C., Bardin, M., Jaenicke, R., Vogel, B., Leyronas, C., Ariya, P.A., and Psenner, R. (2008) Microbiology and atmospheric processes: research challenges. *Biogeosci. Dis.* 5: 191–212.

Nealson, K.H., Tsapin, A., and Storrie-Lombardi, M. (2002) Searching for life in the universe: Unconventional methods for an unconventional problem. *Int. Microbiol.* 5: 223–230.

Onofri, S., Fenice, M., Cicalini, A.R., Tosi, S., Magrino, A., Pagano, S., Selbmann, L., Zucconi, L., Vishniac, H.S., Ocampo-Friedmann, R., and Friedmann, F.I. (2000) Ecology and biology of microfungi from Antarctic rocks and soils. *Ital. J. Zool.* 67: 163–167.

Paez-Rubio, T. and Peccia, J. (2005) Estimating solar and non-solar inactivation rates of airborne bacteria. *J. Environ. Eng-ASCE.* 131: 512–517.

Pearce, D.A., van der Gast, C.J., Lawley, B., and Ellis-Evans, J.C. (2003) Bacterioplankton community diversity in a maritime Antarctic lake, determined by culture-dependent and culture-independent techniques. *FEMS Microbiol. Ecol.* 45: 59–70.

Peccia, J., Werth, H., Miller, S., and Hernandez, M. (2001) Effects of relative humidity on the ultraviolet induced inactivation of airborne bacteria. *Aerosol. Sci. Technol.* 35: 728–740.

Peck, L.S., Clark, M.S., Clarke, A., Cockell, C.S., Convey, P., Detrich III, H.W., Fraser, K.P.P., Johnston, I.A., Methe, B.A., Murray, A.E., Karin Römisch, K., and Rogers, A.D. (2005) Genomics: Applications to Antarctic ecosystems. *Polar Biol.* 28: 351–365.

Peng, R.-H., Xiong, A.-S., Xue, Y., Fu, X.-Y., Gao, F., Zhao, W., Tian, Y.-S., and Yao, Q.-H. (2008) Microbial biodegradation of polyaromatic hydrocarbons. *FEMS Microbiol. Rev.* 32: 927–955.

Porazinska, D.L., Fountain, A.G., Nylen, T.H., Tranter, M., Virginia, R.A., and Wall, D.H. (2004) The biodiversity and biogeochemistry of cryoconite holes from McMurdo dry valley glaciers, Antarctica. *Arctic Antarct. Alpine Res.* 36(1): 84–91.

Powell, S., Ferguson, S.H., Snape, I., and Siciliano, S. (2006) Fertilization stimulates anaerobic fuel degradation of Antarctic soils by denitrifying microorganisms. *Environ. Sci. Technol.* 40: 2011–2017.

Price, B.P. (2007) Microbial life in glacial ice and implications for a cold origin of life. *FEMS Microbiol. Ecol.* 59: 217–231.

Priscu, J.C. and Christner, B.C. (2004) Earth's icy biosphere. In *Microbial Diversity and Bioprospecting*. A. Bull (ed.). Washington, DC, ASM Press, pp. 130–145.

Priscu, J.C., Fritsen, C.H., Adams, E.E., Giovannoni, S.J., Paerl, H.W., McKay, C.P., Doran, P.T., Gordon, D.A., Lanoil, B.D., and Pinckney, J.L. (1998) Perennial Antarctic lake ice: An oasis for life in a polar desert. *Science* 280: 2095–2098.

Psenner, R. (1999) Living in a dusty world: Airborne dust as a key factor for alpine lakes. *Water Air Soil Pollut.* 112: 217–227.

Psenner, R. and Sattler, B. (1998) Life at the freezing point. *Science* 280: 2073–2074.

Psenner, R., Sattler, B., Willie, A., Fritsen, C.H., Priscu, J.C., Felip, M., and Catalan, J. (1999) Lake ice microbial communities in Alpine and Antarctic Lakes. In *Cold Adapted Organisms – Ecology, Physiology, Enzymology and Molecular Biology*. R. Margesin and F. Schinner (eds.). Berlin, Germany, Springer-Verlag, pp. 17–31.

Quayle, W.C., Convey, P., Peck, L.S., Ellis-Evans, C.J., Butler, H.G., and Peat, H.J. (2003) Ecological responses of maritime Antarctic lakes to regional climate change. In *Antarctic Peninsula Climate Variability: A Historical and Palaeoenvironmental Perspective*. E. Domack, A. Burnett, A. Leventer, et al. (eds.). Antarctic Research Series, vol. 79, Washington, DC, American Geophysical Union.

Quayle, W.C., Peck, L.S., Peat, H., Ellis-Evans, J.C., and Harrigan, P.R. (2002) Extreme responses to climate change in Antarctic lakes. *Science* 295: 645.

Sagan, C. and Mullin, G. (1872) Earth and Mars: Evolution of atmospheres and surface temperatures. *Science* 177(4043): 52–56.

Sattler, B., Puxbaum, H., and Psenner, R. (2001) Bacterial growth in supercooled cloud droplets. *Geophys. Res. Lett.* 28(2): 239–242.

Sattler, B., Waldhuber, S., Fischer, H., Semmler, H., Sipiera, P., and Psenner, R. (2004) Microbial activity and phylogeny in ice cores retrieved from Lake Paula, a newly detected freshwater lake in Antarctica. *Instruments, Methods, and Missions for Astrobiology VIII.* R.B. Hoover, G.V. Levin, and A.Y. Rozanov (eds.). Denver, CO. *Proc. SPIE* 5555: 170–179.

Saunders, N.F.W., Thomas, T., Curmi, P.M.G., Mattick, J.S., Kuczek, E., Slade, R., Davis, J., Franzmann, P.D., Boone, D., Rusterholtz, K., Feldman, R., Gates, C., Bench, S., Sowers, K., Kadner, K., Aerts, A., Dehal, P., Detter, C., Glavina, T., Lucas, S., Richardson, P., Larimer, F., Hauser, L., Land, M., and Cavicchioli, R. (2003) Mechanisms of thermal adaptation revealed from the genomes of the Antarctic Archaea *Methanogenium frigidum* and *Methanococcoides burtonii. Genome Res.* 13(7): 1580–1588.

Säwström, C., Mumford, P., Marshall, W., Hodson, A., and Laybourn-Parry, J. (2002) The microbial communities and primary productivity of cryoconite holes in an Arctic glacier (Svalbard 79°N). *Polar Biol.* 25: 591–596.

Smith, J.A., Hodgson, D.A., Bentley, M.J., Verleyen, E., Leng, M.J., and Roberts, S.J. (2006) Limnology of two Antarctic epishelf lakes and their potential to record periods of ice shelf loss. *J. Paleolimnol.* 35(2): 373–394.

Smith, R.I.L. (1991) Exotic sporomorpha as indicators of potential immigrant colonists in Antarctica. *Grana* 30: 313–324.

Storrie-Lombardi, M.C. (2005) Post-Bayesian strategies to optimize astrobiology instrument suites: Lessons from Antarctica and the Pilbara. *Astrobiology and Planetary Missions.* R.B. Hoover, G.V. Levin, and A.Y. Rozanov (eds.), *Proc. SPIE*, vol. 5906, San Diego, CA, pp. 288–301.

Storrie-Lombardi, M.C. and Sattler, B. (2009) Laser induced fluorescence emission (L.I.F.E.): Detection of microbial life in the Ice covers of Antarctic lakes. *Astrobiology*. 9: 659–672.

Storrie-Lombardi, M.C., Hug, W.F., McDonald, G.D., Tsapin, A.I., and Nealson, K.H. (2001) Hollow cathode ion lasers for deep ultraviolet Raman spectroscopy and fluorescence imaging. *Rev. Sci. Instrum.* 72(12): 4452–4459.

Storrie-Lombardi, M.C., Lahav, O., Sodre, L., and Storrie-Lombardi, L.J. (1992) Morphological classification of galaxies by artificial neural networks. *Mon. Not. R. Astron. Soc.* 259: 8–12.

Storrie-Lombardi, M.C., Muller, J.-P., Fisk, M.R., Griffiths, A.D., and Coates, A.J. (2008a) Potential for non-destructive astrochemistry using the ExoMars PanCam. *Geophys. Res. Lett.* 35: L12201, doi 10.1029/2008GL034296.

Storrie-Lombardi, M.C., Muller, J.-P., Fisk, M.R., Griffiths, A.D., and Coates, A.J. (2008b) Epifluorescence surveys of extreme environments using PanCam imaging systems: Antarctica and the Mars regolith. *Instruments, Methods, and Missions for Astrobiology XI*. R.B. Hoover, G.V. Levin, and A.Y. Rozanov (eds.). San Diego, CA. *Proc. SPIE* 7097 (25): 1–10.

Takeuchi, N. and Li, Z. (2008) Characteristics of surface dust on Ürümqi glacier No. 1 in the Tien Shan mountains, China. *Arctic Antarct. Alpine Res.* 40(4): 744–750.

Takeuchi, N., Kohshima, S., and Seko, K. (2001) Structure, formation, and darkening process of albedo-reducing material (cryoconite) on a Himalayan glacier: A granular algal mat growing on the glacier. *Arctic Antarct. Alpine Res.* 33(2): 115–122.

Thompson, B.A.W., Davies, N.W., Goldworthy, P.M., Riddle, M.J., Snape, I., and Stark, J.S. (2006) In situ lubricant degradation in Antarctic marine sediments. 1. Short-term changes. *Environ. Toxicol. Chem.* 25(2): 356–366.

Tosi, S., Casado, B., Gerdol, R., and Caretta, G. (2002) Fungi isolated from Antarctic mosses. *Polar Biol.* 25: 262–268.

Tranter, M., Fountain, A., Fritsen, C., Lyons, B., Statham, P., and Welch, K. (2004) Extreme hydrochemical conditions in natural microcosms entombed within Antarctic Ice. *Hydrol. Proc.* 18: 379–387.

Twin, J., Gibson, J.A.E., and Bowman, J.P. (2008) Microbial communities of alkaline epiglacial lakes of the Framnes Mountains, Antarctica. *3rd International Conference on Polar and Alpine Microbiology*, S4-4, Banff, Alberta, Canada.

Vincent, W.F. and Howard-Williams, C. (2000) Life on snowball Earth. *Science* 287: 2421.

Vincent, W.F., Gibson, J.A.E., and Jeffries, M.O. (2001) Ice shelf collapse, climate change and habitat loss in the Canadian high Arctic. *Polar Rec.* 37: 131–136.

Warren, S.G. and Wiscombe, W.J. (1980) A model for the spectral albedo of snow. II. Snow containing atmospheric aerosols. *J. Atoms. Sci.* 37: 2734–2745.

Wynn-Williams, D.D. (1991) Aerobiology and colonization in Antarctica—The BIOTAS programme. *Grana* 30: 380–393.

5 Psychrophilic and Psychrotolerant Microbial Extremophiles in Polar Environments

Richard B. Hoover and Elena V. Pikuta

CONTENTS

5.1 INTRODUCTION

Microbial extremophiles are the dominant life forms of the polar environments. They are able to survive in the extreme polar environments and have developed mechanisms that allow them to cope with a variety of stressors. These include freezing temperatures and repeated freeze-thaw cycles, desiccation, high or low levels of salinity or pH, and lengthy periods of darkness during winter. Polar life forms must also be able to survive exposure to high levels of solar UVB (280–314 nm) radiation due to stratospheric ozone depletion over the Antarctic (McKenzie et al., 2003) and Arctic regions during the summer (Knudsen et al., 2005). The dominant prokaryotes of polar environments are psychrophilic and psychrotolerant cyanobacteria, bacteria and archaea, and the dominant photosynthetic eukaryotes are algae, which are primarily diatoms.

Morita (1975) defined psychrophiles as organisms that have optimal temperature for growth below 15°C and maximal temperature for growth at 20°C. Psychrotolerant organisms are able to grow at low temperatures with much lower rates, and they have optimal growth in the range of mesophilic organisms (20°C–40°C). Psychrotolerant microorganisms are usually the organisms most frequently found in cold environments, perhaps because they may have better nutritional adaptability (Wynn-Williams, 1990) or due to horizontal gene transfer from mesophiles (Aislabie et al., 2004). Many of these psychrotolerant bacteria found in mesophilic environments and may have been transferred to Antarctica.

It has been known since 1887 that some species of bacteria were capable of growth at 0°C or below (Foster, 1887). In 1918, McLean reported the isolation of gram positive cocci and gram negative spore-forming- and non-spore-forming rods from the snow, ice, and frozen algae of Antarctica. Darling and Siple (1941) isolated 178 strains from the snow, ice, soil, and debris of Antarctica. Straka and Stokes (1960) reported psychrophilic bacteria from ice and guano in Antarctica, but due to a difference in terminology the growth characteristics indicate that many of these bacteria were actually psychrotolerant rather than true psychrophiles. However, most of the knowledge about psychrophilic and psychrotolerant microorganisms and their habitats has been gained in the past three decades. Extensive studies of microbial life in the polar environments have been carried out, and new taxa of psychrophiles have been discovered and described (Broady, 1981, 1982; Mountfort et al., 1997; Reddy et al., 2003).

Now, there is a much better understanding of the mechanisms for survival in polar environments. Survival strategies include genetic and acclimation processes, and adaptations (Elster, 1999, 2002; Vincent, 2000). These include mechanisms such as seasonally induced dormancy, production of pigments, exopolysaccharides, and sheaths to protect from UV radiation, and the utilization of specialized enzymes and cryoprotectants to prevent cell lysis during freeze-thaw cycles.

Concerning metabolism, among psychrophiles, the organoheterotrophic and phototrophic types are predominant. The lithoautotrophic type, which is the earliest by the evolutionary scale, and usually is the characteristic for hyperthermophiles, is common only for some species of psychrotolerant methanogenic archaea, bacterial methanotrophs and homoacetogens inhabiting polar peat bogs (Dedysh et al., 2000; Trotsenko and Khmelenina, 2005; Zhang et al., 2008). Reports about lithotrophic, psychrophilic, sulfate-reducing bacteria have also been published (Geittel et al., 2008).

Many of the chemical entities of importance to modern biotechnology are proteins, polysaccharides and other large polymeric molecules, which function at extreme physicochemical parameters. Psychrophilic and psychrotolerant microorganisms and their unique cold shock and cold-acclimation proteins and enzymes (e.g., proteases, lipases, and cellulases) have a host of biotechnology applications (Gounot, 1991). These include cold-water detergents, food additives and flavor modifying agents, biosensors, and environmental bioremediation. The exclusively temperature-sensitive molecules and enzymes of psychrophiles have applications in pharmaceuticals. The cryoprotectors synthesized by psychrophilic and psychrotolerant microorganisms are used in cosmetics, medicine, and agriculture.

A detailed review of the biotechnology applications of psychrophilic and psychrotolerant microorganisms has been presented by Huston (2007). Psychrotolerant microorganisms are of great value in the bioremediation of hydrocarbon-contaminated soil in Antarctica, and they have the ability to maintain activity under the extreme conditions of the polar environment (Bej et al., 2000; Paniker et al., 2002, 2006). Several new strains of psychrotolerant bacteria have been isolated from oil-contaminated soil in Antarctica and studied in detail (Aislabie et al., 2000; Bej et al., 2000). Horizontal gene transfer mechanisms have been suggested to play a role in cold-adapted aromatic degrading bacteria (Aislabie et al., 2004, 2006). Molecular methods help to understand the distribution of microorganisms in polar environments (natural and contaminated ecosystems). Gene mapping has been performed for several ecosystems and the wide microbial diversity in these environments has been demonstrated (Juck et al., 2000).

Investigations of microbial extremophiles in polar environments are important to understand environmental change, global warming, and the newly emerging field of Astrobiology. Using three decades of satellite (Coastal Color Scanner and Sea-Viewing Wide Field-of-View Sensor) and field observations Montes-Hugo et al. (2009) showed that ocean zone biological productivity, estimated by the measurements of surface chlorophyll *a* concentrations along the West Antarctic Peninsula, has declined by 12%. Large scale climate changes could alter the content of microbial communities, especially in polar regions, possibly contributing to the natural process of substitution of true psychrophilic species by psychrotolerant ones, accompanied by gene-transfer and other processes. A key element of Astrobiology is the study of the limits and diversities of microbial life in the most extreme environments on Earth. Since water is crucial for life as we know it, the most important Astrobiology targets are the frozen worlds of the Solar System bodies where water is present either in liquid or solid state. These include comets or water bearing asteroids; the permafrost, craters, or polar ice caps of Mars; and the icy moons of Jupiter (e.g., Europa, Callisto, Io, or Ganymede) or Saturn (e.g., Titan or Enceladus). The polar environments of Earth provide the best terrestrial analogues for these regimes. Consequently, the nature and distribution of microbial extremophiles that inhabit the Earth's polar regions may provide valuable information crucial to the development of instruments and operational techniques needed to search for and to recognize evidence of extant or extinct life elsewhere in the Cosmos.

In this chapter, we review recent information about psychrophilic and psychrotolerant microorganisms isolated from different cold environments. We provide some preliminary data on the isolation, morphology, and physiology of novel anaerobic microbial extremophiles from polar environments (and psychrotolerant strains isolated from penguin guano) in the logic scheme for the biogeochemical cycling of organic matter divided into two major groups: (1) the producers and (2) the decomposers of organic matter. This arrangement does not cover an exact functional meaning for all reviewed microorganisms, since it is difficult to put the metabolic diversity of organisms under one universal definition. For example, some phototrophs can use the photoheterotrophic and, even, organoheterotrophic pathways in metabolism, depending upon conditions. There are species of diatoms and green algae that lack pigments and support their life cycle exclusively by organoheterotrophic metabolism. Furthermore, some anaerobic decomposers, which are lithohetherotrophs or organotrophs, are also able to synthesize organic matter from inorganic compounds; for example, some methanogenic archaea and bacterial homoacetogens are able to use lithoautotrophic metabolic pathways.

5.2 PRODUCERS OF ORGANIC MATTER IN POLAR ENVIRONMENTS

In the polar environments, the dominant primary producers of organic matter are the photosynthetic eukyarotic algae (mainly diatoms and snow algae) and photosynthetic prokaryotes (mainly cyanobacteria). The chemolithoautrophic producers of organic matter in anaerobic communities of polar regions are limited to the methanogenic archaea and the homoacetogenic bacteria that are able to grow on hydrogen and CO_2/CO or just on CO_2 (Simankova et al. 2000).

5.2.1 Eukaryotic Photosynthetic Microorganisms

5.2.1.1 Diatoms

Psychrophilic and psychrotolerant diatoms are the most abundant unicellular eukaryotes of the polar environments. Diatoms have also been extensively studied in the polar oceans, sea ice, high alpine regions and the freshwater lakes, streams, and soils of Antarctica. These photoautotrophic microalgae are one of the most important groups of the phytoplankton of the polar oceans. Diatoms are responsible for most of the primary productivity in the polar oceans and they are the principal component of the entire polar food chain—feeding krill, fish, whales, penguins, and other sea birds. One of the main habitats of polar diatoms is sea ice. Diatoms are so abundant in the polar oceans that they produce vast "blooms" during the polar spring that can best be appreciated from satellite imagery. This is due to the presence of high silica concentrations in the polar oceans coupled with the remarkably successful adaptations of diatoms to the polar environment.

Early investigations of polar marine diatoms were carried out by Castracane (1886) during the 1873–1876 *HMS Challenger* Expedition and by Henri van Heurck during the 1897–1899 voyage of the *S. Y. Belgica* (van Heurck, 1909). Freshwater diatoms from the South Georgia Island were described by Reinsch (1890). The dominant diatoms of the polar oceans include many marine centrics, such as *Chaetoceras criophilum*, *Asteromphalus* spp. *Actinocyclus karstenii*, *A. curvatulus*, *Azpeitia tabularis*, *Coscinodiscus* spp., *Corethron criophilum*, *Thalassiosira antarctica*, *T. pseudonana* as well as pennates such as *Fragilariopsis kerguelensis*. The pack-ice and sea-ice communities are dominated by small pennates, such as *Fragilariopsis curta*, *Fragilariopsis cylindrus*, *Amphiprora* spp., *Navicula* spp., and *Nitzschia* spp. Diatoms have been studied from the Antarctic ice-cores (Burckle et al., 1988). Diatom frustules (and in some cases intact cells) as well as a host of other prokaryotes have been found in some of the very deep ice cores above Lake Vostok (Abyzov et al., 2001, 2006; D'Elia et al., 2008). Freshwater diatoms have been used as indicators of environmental change in the high arctic (Douglas and Smol, 1999) and for monitoring human-induced environmental changes in tundra ecosystems (Smol and Douglas, 1996). While an exhaustive description of the modern work on polar diatoms is beyond the scope of this chapter, the reader is referred to Kellogg and Kellogg (2002) for an updated taxonomy of the nonmarine diatoms from the Antarctic and Subantarctic regions. Reviews of psychrophilic diatoms and their mechanisms for survival in cold environments have more recently been given by Spaulding and McKnight (1999) and Mock and Junge (2007).

5.2.1.2 Snow Algae

Several species of eukaryotic microalgae have adapted to growth at low temperature and thrive in melting snow and can color the snow pink, red, green, or yellow (Hoham, 1975). *Chlamydomonas nivalis* is the most common species of the snow algae. Other algae that grow on snow include species of the genera *Ankistrodesmus*, *Chloromonas*, and *Raphidonema*. *C. nivalis* is a green alga that grows actively as the snow melts during the spring and summer and sunlight penetrates the snow pack. Although the vegetative cells of *C. nivalis* are green, it has bright red spores. Their presence in large numbers produces large expanses of red "watermelon snow." The complex life cycle of the snow algae in which the vegetative cells, sexual stage, and spores alternate have been described by Hoham and Ling (2000). Figure 5.1a shows a snowbank in the Kolyma Lowlands of North Siberia with red patches produced by the spores (Figure 5.1b) of *C. nivalis*. Experiments for the determination of UV tolerance have showed that among the eukaryotic phototrophs the green snow algae *C. nivalis* was resistant to high levels of UV radiation (UV-A, 365 nm, and UV-C, 254 nm) due to the production of flavonoids as an antioxidant and by the accumulation of astaxanthin esterified with fatty acids (Duval et al., 2000; Hoham and Ling, 2000).

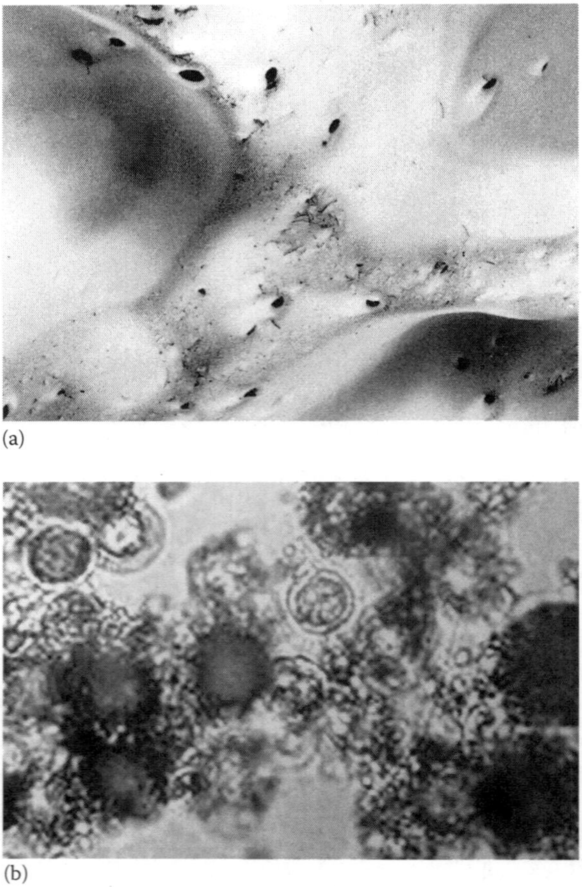

FIGURE 5.1 (See color insert following page 276.) (a) *C. nivalis* on a snowbank in the Kolyma Lowlands, North Siberia. (b) Photomicrograph of red spores of *C. nivalis*.

5.2.1.3 Prokaryotic Photosynthetic Microorganisms

Cyanobacteria are the dominant photosynthetic prokaryotes of the polar regions. While they are rare or absent in the marine environment of the Arctic and the Antarctic, they are found in all the freshwater environments of Antarctica (Seckbach and Oren, 2007). Oscillatorian cyanobacteria are the dominant forms in the benthic mats of polar freshwater ecosystems. The cyanobacteria of the polar regions often contain high levels of canthaxanthin, myxoxanthophyll, and other carotenoids with the ratio carotenoids/chlorophyll *a* being maximal under low temperatures, high light regimes, and moderate UV radiation (Vincent, 2000). All of the species of cyanobacteria that have been isolated from the polar regions are psychrotolerant rather than truly psychrophilic (Nadeau and Castenholz, 2000). And conversely, the polar eukaryotic algae, especially the diatoms and the green algae, that grow at subzero temperatures, are psychrophiles.

5.2.1.4 Bioremediation by Diatoms and Cyanobacteria

Although diatoms and cyanobacteria are the major phototrophic eukaryotic microorganisms in the polar environments, they have not yet found extensive applications in bioremediation. Some species of planktonic and benthic diatoms concentrate heavy metals and radioisotopes by nonspecific bioaccumulation and other processes. (Absil and van Scheppingen, 1996). Adam and Garnier-Laplace (2003) carried out laboratory experiments to assess the uptake of several radioisotopes (^{110}Ag, ^{60}Co, ^{137}Cs, and ^{54}Mn) by the diatom *Cyclotella meneghiana* and the chlorophyte *Scenedesmus obliquus*. Some species of diatoms and oscillatorian cyanobacteria have been shown to be able to partially oxidize hydrocarbons, such as naphthalene and biphenyl, when grown phototrophically (Cerniglia et al., 1979). Phenanthrene is widely distributed in the environment and it is widely used as a model for carcinogenic, polycyclic aromatic hydrocarbons (Bücker et al., 1979). Narro et al. (1992) reported that the marine cyanobacterium *Agmenellum quadruplicatum* PR-6 could oxidize and metabolize the tricyclic aromatic hydrocarbon phenanthrene. Cavanaugh et al. (1998) isolated marine bacteria from the sands of the Vestfold Hills, East Antarctica that were able to degrade phenanthrene. The diatom *Achnanthes minutissima* and the cyanobacterium *Phormidium foveolatum* were capable of degradation of naphthenic hydrocarbons (Antić et al., 2006). The marine diatom *Skeletonema costatum* has been found to have selected detoxification and enzymatic responses to 2,4-dichlorophenol (2,4-DCP). The 2,4-DCP was readily metabolized, but the bioaccumulation and adsorption was negligible (Yang et al., 2002). Cold-adapted planktonic and benthic diatoms and other microalgae may find important applications in bioremediation strategies for the bioaccumulation and the removal of some recalcitrant hydrocarbons, heavy metals and radioisotopes from polar environments.

5.2.2 Psychrophilic and Psychrotolerant Anaerobic Chemolithotrophic Autotrophs

5.2.2.1 Methanogens

Psychrotolerant methanogens from polar peat bogs have been known for a relatively long time. In the last 15 years many true psychrophilic species of methanogenic archaea were reported. The most studied species, *Methanogenium frigidum*, was isolated from an anaerobic water sample at the hypolimnion level in the perennially cold Ace Lake, in Vestfold Hills, Antarctica (Franzmann et al., 1997). It is the first described psychrophilic methanogen with lithoautotrophic metabolism: the best growth was observed on H_2+CO_2. However, the addition of yeast extract significantly stimulated growth. The cells tolerated 10% salinity and were neutrophilic. This species grew optimally at 15°C and did not grow at 18°C–20°C. Another psychrophile, *Methanolobus psychrophilus*, has been described but is not yet validated. It also showed an optimal growth at 15°C (Zhang et al., 2008). This study indicates that psychrophilic methanogenesis in the Zoige wetland is more likely to be methylotrophic, rather than hydrogenotrophic, and plays a significant role in the methane emission of the wetlands. Since methane is a very important greenhouse gas, this research is also relevant to global warming studies.

The psychrotolerant methanogen, *Methanosarcina lacustris*, was isolated from anoxic lake sediments in Switzerland (Simankova et al., 2004). The organism utilized methanol, mono-, di-, trimethylamine and H_2/CO_2 with methane production, and grew at 1°C–35°C with an optimal growth at 25°C. Another psychrotolerant methanogen *Methanogenium marinum* AK-1 grows at 5°C–25°C and was isolated from cold marine sediments at Skan Bay, Alaska (Chong et al., 2002). The cells of this species grow on H_2+CO_2 or formate$+CO_2$ at pH 6.0–6.6 and salinity 0.25–1.25 M.

The isolation of methanogens in enrichment culture from acidic peat bog was reported (Bräuer et al., 2006). Another work (Nozhevnikova et al., 2003) proved the existence of psychrophilic methanogenic communities in sediments of deep lakes with low temperatures (4°C–5°C). It was shown that hydrogen-amended cultures had the highest methane production rates at 6°C. The investigations of methanogens in polar environments are also of utmost interest to Astrobiology. Morozova et al. (2006) carried out experiments that demonstrated the survival of methanogenic archaea from Siberian permafrost under simulated Martian thermal conditions.

The study of biodiversity of methanogens and other Archaea in the permanently frozen Lake Fryxell in Antarctica has also been performed (Karr et al., 2006). The phylogenetic analysis of the 16S rRNA gene sequences obtained showed that at least four clusters of Archaea inhabited Lake Fryxell—three clusters of Euryarchaeota and one cluster of Crenarchaeota. Within the Euryarchaeota, at least two clusters of methanogens were detected. Some phylotypes were closely related to 16S rRNA gene sequences from *Methanosarcina* species. The presence of *Methanosarcina*-like methanogens in Lake Fryxell sediments has been confirmed by the isolation of strain FRX-1, a methylotrophic *Methanosarcina* species growing at 5°C (Singh et al., 2005). The other phylotypes of methanogens in Lake Fryxell sediments clustered with the species of *Methanoculleus*. Unlike *Methanosarcina*, no cultures of *Methanoculleus* species have been obtained from Lake Fryxell; however, a *Methanoculleus*-like 16S rRNA gene was previously detected in a molecular diversity study of the cyanobacterial mats that develop in the peripheral melt waters of this lake (Brambilla et al., 2001; Taton et al., 2003).

5.2.2.2 Acetogens

The first report of a psychrophilic homoacetogenic bacteria was *Acetobacterium carbinolicum* strain HP4, which was isolated by Bak in 1988 (Conrad et al., 1989). As it was shown later that strain HP4 was phylogenetically identified as *A. carbinolicum*, the species that was previously described by Eichler and Schink (1984). This strain had a growth range of 1°C–25°C.

Chronologically, the next three homoacetogenic isolates were described as the new psychrophilic species *Acetobacterium bakii*, *A. paludosum*, and *A. fimetarium* (Kotsyurbenko et al., 1995). Another psychrotolerant homoacetogen *A. tundrae* was isolated from the tundra wetland soil sample collected in polar Ural (Simankova et al., 2000). However, all of these species were able to grow at 30°C or 35°C with an optimal growth at 20°C–30°C, indicating they are psychrotiolerant microorganisms rather than true psychrophiles.

New subspecies *kysingense* of the species *A. carbinolicum* strain SyrA5 was isolated from anaerobic sediments of brackish fjord (Paarup et al., 2005). This isolate was also psychrotolerant and was able to grow lithotrophically on H_2+CO_2 and also on CO.

Two strains of acetogenic bacteria LS1 and LS2 were isolated from superficial sediments of the permanently ice-covered, meromictic Lake Fryxell in McMurdo Dry Valleys, Antarctica (Sattley and Madigan, 2007). It was shown that these isolates represent the first acetogens able to grow at subzero temperatures, but their range of growth −2.5°C to 25°C exceeds the upper limit (20°C) for psychrophiles in accordance with the definition provided by Morita (1975).

The homoacetogenic bacteria are competing with methanogens and sulfate-reducing bacteria for hydrogen and other inorganic and organic substrates within oligotrophic polar environments. It was shown that at low temperatures homoacetogenic bacteria had significantly predominant activity, compared to methanogens and sulfate-reducing bacteria (Simankova et al., 2000; Kotsyrbenko et al., 2001; Nozhevnikova et al., 2001).

Gaidos et al. (2008) studied the microbial diversity of one of Iceland's volcanic subglacial lakes, and they made an unsuccessful attempt to isolate a new psychrophilic acetogenic strain found there, the phylotype of which belongs to the genus *Acetobacterium*.

Here, we are reporting two strains (str. A7AC-96m and str. A7AC-DS7) of truly psychrophilic homoacetogenic bacteria that were isolated from water and sediment samples collected from the 96 m region of the deep anoxic trough of Lake Untersee in Antarctica. Both of these strains have a maximum temperature for growth at 14°C and are able to grow on H_2+CO_2.

5.3 DECOMPOSERS OF ORGANIC MATTER IN POLAR ENVIRONMENTS

All microorganisms that participate in the process of decomposition of organic matter in polar environments are chemo-organotrophs. Some of the best-studied systems of decomposers of organic matter in polar environments are those of the psychrophilic and psychrotolerant microorganisms that inhabit the peat bogs and permafrost soils of Alaska, Canada, North-Eastern Europe, Patagonia, and Siberia. Many species of bacteria and archaea were described and characterized. It is noted that the halo-alkaline lakes on Earth have not previously been investigated for the presence of psychrophiles. All reported studies of the halo-alkaline lakes in Siberia (Russia) and Canada were limited exclusively to mesophilic microorganisms. Perhaps this was because of the lack of specific equipment and techniques required for the isolation of true psychrophiles, or specific conditions for the storage or the transport of samples (extreme sensitivity of psychrophilic microorganisms to warm temperatures).

During the past decade, many new marine psychrophiles, representing several separate new genera and species, have been isolated from cold deep sea samples and described (Bowman et al., 1997; Mountfort et al., 1998; Okamoto et al., 2001; Romanenko et al., 2002). Many of them were called "halophiles" in the original articles, but this is not entirely accurate. A more appropriate term for these microorganisms would be "marine and halotolerant psychrophiles" since their salinity optima occur at 3% NaCl (w/v) and 12%–15% salinity was the maximum limit for the growth of these isolates. True halophiles inhabit brines with salt content as high as 10 times (30% NaCl) the salinity of ordinary seawater. Apart from the green algae *Dunaliella salina*, almost all halophiles are archaea.

The lowest temperature for growth (–2°C) in marine polar ecosystems is recorded for the sulfate-reducing bacteria, which were true psychrophiles and did not grow at room temperature (Knoblauch et al., 1999). The study of the diversity and distribution of sulfate-reducing bacteria in permanently frozen Lake Fryxell in Antarctica had showed that at least four of the six major phylogenetic groups of sulfate-reducing bacteria (*Desulfovibrio*, *Desulfosarcina*, *Desulfotomaculum*, *Desulfobulbus*, *Desulfobacter*, and *Desulfobacterium*) were present in the lake (Karr et al., 2005). All these cultures should be active at 4°C that means they have psychrophilic or psychrotolerant physiology.

For halo-alkaliphilic species the *in situ* observations at temperatures as low as –20°C have been previously reported (Staley and Gosink, 1999). Pure cultures of halophilic psychrophiles with an optimum salinity of 20% have not yet been obtained. *Psychrobacter cryohalolentis* 274-4^T growing at –10°C and 10% salinity was described by Bakermans et al. (2006). This bacterium is not a true psychrophile but rather psychrotolerant and was isolated from the permafrost of the Kolyma Lowlands in North Siberia. The same species was found in the ice-covered Lake Fryxell of the Taylor Valley of Antarctica (Ryan et al., 2007).

In cold peat bogs of the polar regions, psychrophilic and psychrotolerant organoheterotrophic microbial communities developed the tolerance to high concentrations of organic acids, accumulated as a result of the enormous activity of primary decomposers at the aerobic and anaerobic levels. The only explanation for this phenomenon is the predominance of the activity of the primary decomposers over the secondary ones. Perhaps during the process of sharp climatic changes, bacterial communities did not have time to increase the metabolic activity for the secondary decomposers that originally were mesophiles and had to mineralize the organic matter within the community at much higher temperatures. The high concentration of organic acids that decrease the pH to 4–5 could also be explained by the high metabolic acitivity of acetogens. Fungi are responsible for the aerobic decomposition of organic matter in such ecosystems, and it is well known that they are tolerant to moderately low pH. Hughes and Bridge (2009) have reviewed the role of Antarctic fungi and yeasts in the bioremediation of hydrocarbons in Antarctica.

Franzmann et al. (1987) investigated the limnology and the microbiology of Organic Lake, a shallow, hypersaline (130–180 g salt L^{-1}) meromictic lake in the Vestfold Hills of Eastern Antarctica. The anoxic zone of Organic Lake has the highest concentration (up to 300 ng L^{-1}) of dimethyl sulfide (DMS) recorded in a natural aquatic ecosystem. The species diversity of Organic Lake is very low, primarily consisting of halophilic microalgae *Dunaliella* sp. and archaea. They reported the isolation of a strain of the archaea *Halomonas* sp. that was capable of producing DMS from sulfur-bearing amino acids. In addition to species of the *Dunaliella*,

Roberts et al. (1993) reported the detection of the diatom *Chaetoceros* sp. and the choanoflagellate *Acanthaoecopsis unguiculata* and strains of *Halomonas* sp. that produced dimethyldisulfide (DMDS) from methionine and cysteine. They concluded that the reduced sulphur volatiles in the lake were primarily produced by the anaerobic breakdown of algae by bacterial decomposition. Ward and Priscu (1997) reported the detection of denitrifying bacteria in an ice-covered Antarctic Lake. Glatz et al. (2006) investigated the microbial diversity in the water column of permanently ice-covered Lake Bonney, Antarctica. Voytek et al. (1998) used immunofluorescence, *in situ* hybridization, and PCR analysis to determine the abundance of ammonia-oxidizing bacteria in this lake. The denitrification of the hypolimnion and the bacterial productivity in the anoxic zone of Lake Bonney was described by Ward et al. (2003, 2005).

Bowman et al. (2001) obtained 16S rDNA clone libraries and studied the diversity of the bacteria and archaea of the anoxic sediments of meromictic lakes and coastal marine basins. They found little similarity between the phylotypes of the meromictic lake ecosystem with the phylotypes present in marine sediments and reported the detection of a wall-less Antarctic spirochaetes. Milucki and Priscu (2007) investigated the diversity of bacteria in Blood Falls, an iron-rich ancient subglacial marine brine released beneath the Taylor Glacier in Antarctica. The dominant bacterium in the 16S rRNA gene clone library of phylotypes from this unusual polar ecosystem had 99% homology with *Thiomicrospira arctica*, a psychrophilic autotrophic sulfur oxidizer. Their clone library also indicated the presence of phylotypes of the division Bacteroidetes and classes α-proteobacteria, β-proteobacteria, and γ-proteobacteria. The glacier hydrology, lithology and preglacial ecosystem apparently controlled the metabolism and phylogenetic structure of this subglacial ecosystem, with the growth supported by the chemical energy of the reduced sulfur and iron compounds. Perreault et al. (2008) concluded that sulfur compounds were the major energy source for the ecosystem of the perennially cold saline springs at Gypsum Hill and Colour Peak on Axel Heiberg Island in the Canadian High Arctic.

The first moderately acidophilic true psychrophiles were isolated from tundra permafrost soils (Dedysh et al., 2000). All three methanotrophic species of the genus *Methylocella* tolerate acidic pH at 4.5 and grow in the temperature regime of psychrophiles (Dunfield et al., 2003; Dedysh et al., 2004).

Many alkalitolerant psychrophilic or psychrotolerant bacterial species have been described, but none of them were truly alkaliphilic. The study of the unique new ecosystem in Ikka Fjord, Greenland, led to the isolation of the first true alkaliphilic psychrophilic bacterium *Rhodonellum psychrophilum* GCM71^T. This novel strictly aerobic and red pigmented bacterium was isolated from the interior seep water in alkaline ikaite tufa columns (Schmidt et al., 2006a,b). It was found to be very distant from all known species, and it is located on a new separate taxonomic lineage on the genus and species levels. This isolate has an optimal growth at pH 9.2–10.5°C, and 0.6% NaCl. The maximum pH for its growth was determined at 10.7 and the minimum temperature for growth at 0°C. The same group (Schmidt et al., 2007) recently described a new genus and species of marine psychrophilic alkaliphile—*Arsukibacterium ikkense* GCM72^T. These two new "champion extremophiles" with growth at high pH and low temperature (*R. psychrophilum* and *A. ikkense*) are shown in the updated pH/temperature and NaCl/temperature diagrams.

5.4 EXTREMOPHILES WITHIN PHYSICOCHEMICAL MATRIX

The distribution of validly published species of extremophiles was previously presented by us in a matrix form (Pikuta et al., 2007). These diagrams have been updated in accordance with recent discoveries and are presented here as the matrix of pH/temperature (Figure 5.2a), pH/salinity (Figure 5.2b), and temperature/radiation (Figure 5.2c). According to these diagrams, the prokaryotic extremophile that grows at the lowest temperature (–10°C) and 10% salinity currently known is the psychrotolerant mesophilic species *P. cryohalolentis*, which was isolated from the permafrost of the Kolyma Lowlands of North Siberia (Bakermans et al., 2006). *Psychrobacter submarinus* is the most halotolerant true psychrophile growing around 15% salinity (Romanenko et al., 2002). The true halophilic species within psychrophiles remain undiscovered. Methanotrophic psychrophilic species of the genus *Methylocella* tolerate acidic pH at 4.5, but the mesophilic species *Acidiphilum acidophilum* tolerates pH 1.5 at 15°C (the minimum temperature for growth). True psychrophilic acidophiles are unknown or, most probably, do not exist at all. The alkaliphilic psychrophiles, *Rhodonellum psychrophillum* and *A. ikkense*, are the "champions" growing at pH 10.5. We now know that the mesophilies (rather than psychrophilic or psychrotolerant species) have the highest environmental limits for salinity and pH (Figure 5.2a and b).

There are very few known psychrophilic bacteria or archaea that tolerate high levels of gamma radiation (Figure 5.2c). Callegan et al. (2008) described four new species of true psychrophiles of the genus *Deinococcus* isolated from soil of high alpine environments. However, these species were not as resistant to radiation as the type species of this genus *D. radiodurans.*

In terms of either biomass or the number of cells, the majority of the psychrophiles on Earth are the barophilic bacteria and archaea that inhabit the deeper layers of the oceans, which cover almost 75% of the surface of our planet.

We would like to highlight that currently among psychrophilic bacteria only neutrophilic, alkaliphilic, and halotolerant species have been found. The absence of acidophilic psychrophiles remains intriguing. From the cold acidic (pH 1–2) environments the only known acidophiles that have been described are psychrotolerant mesophiles. The true psychrophilic moderate acidophiles, such as species of the genus *Methylocella*, tolerate only pH 4.5–5. This leads to questions: *Does Nature keep a restriction on the combination of the genes for psychrophily and acidophily? If so—Why?* Was it because low-temperature and acidic environments have not coexisted on Earth long enough for evolutionary processes to adjust the genome for supporting life? Or is it because microorganisms capable of inhabiting such environments are extinct or have not yet been discovered? The most likely places to search for such ecosystems are acidic veins in glacial ice (Price, 2000) and beneath glaciers on volcanic peaks, such as Mt. Erebus in Antarctica. Another, intriguing question: *Is it possible to create such combination of genes artificially in the laboratory?* Investigations in the near future may possibly answer these questions.

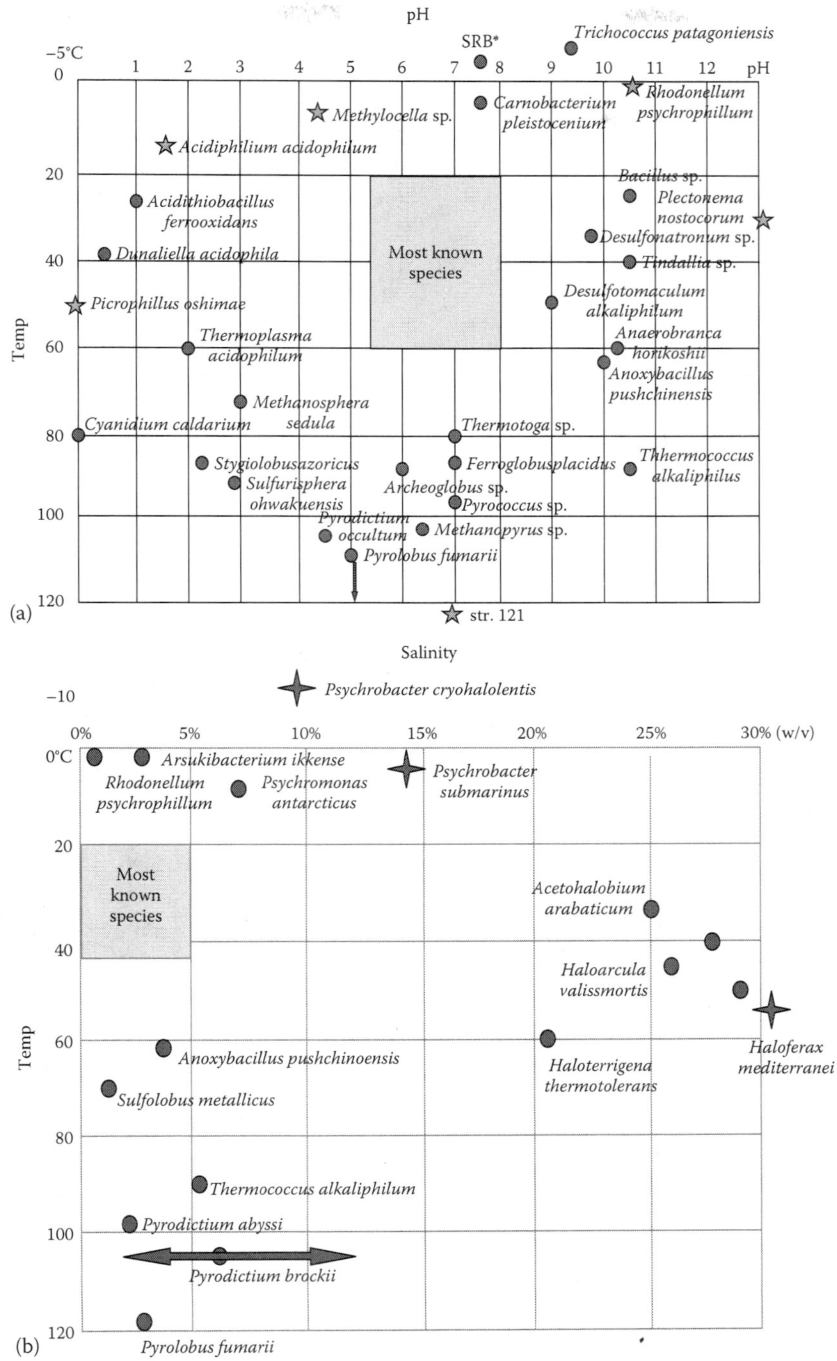

FIGURE 5.2 Diagram of all known validated microorganisms distributed within (a) matrix pH/temperature (* indicates species of five genera of psychrophilic sulfate-reducing bacteria); (b) NaCl—salinity/temperature; and (c) matrix temperature/gamma radiation.

(*continued*)

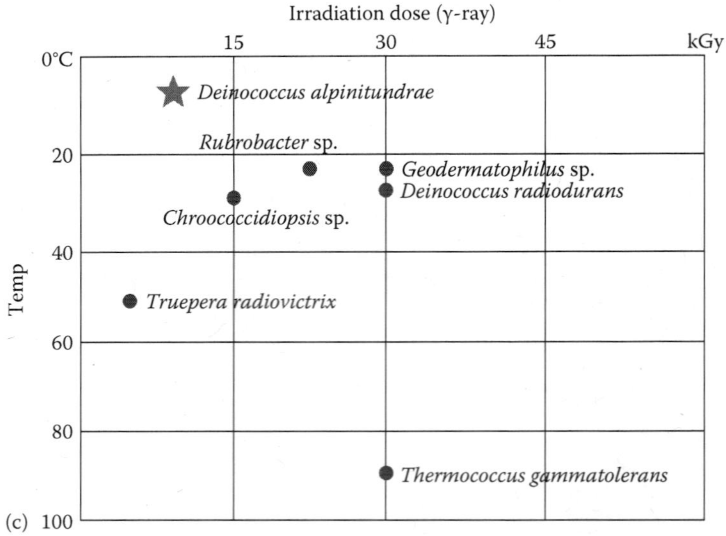

FIGURE 5.2 (continued)

5.5 ANAEROBIC EXTREMOPHILES FROM POLAR EXPEDITIONS

During the past decade, the NASA/MSFC/NSSTC Astrobiology group organized field expeditions to obtain samples of anaerobic extremophiles from several polar and subpolar regions of Earth. These included expeditions to

- Matanuska Glacier and Fox Permafrost Tunnel, Alaska (1998, 1999, 2008)
- Kolyma Lowlands of North Siberia—International Expedition *Beringia* (1999)
- Patriot Hills, Thiel Mountains, and South Pole, Antarctica—*Antarctica 2000 Expedition*
- Patagonia, Chile—*Antarctica 2000 Expedition*
- South Africa; Schirmacher Oasis, Lake Untersee, and Anuchin Glacier, Antarctica—*2008 International Schirmacher Oasis/Lake Untersee, Antarctica Expeditions*

A number of anaerobic bacteria were isolated and characterized from samples obtained during these field expeditions. We review a few interesting microbial extremophiles isolated from these samples.

5.5.1 Novel Psychrotolerant Extremophiles from Expeditions to Alaska

5.5.1.1 Pleistocene Bacterium from Fox Permafrost Tunnel

The study of permafrost samples from Alaska led to the discovery of the psychrotolerant *Carnobacterium pleistocenium* FTR1^T. This unique microorganism was alive after remaining frozen for 32,000 years in the ice of a Pleistocene Thermokarst Pond of the Cold Regions Research and Engineering Laboratory (CRREL) Permafrost Tunnel at Fox, Alaska.

The first study of the geology of the Fox Tunnel was performed by Sellman (1967). Accurate measurement of the ages of the strata in the Fox Tunnel were received by radiocarbon dating of wood from the tunnel using the high-sensitivity liquid scintillation counting method (Long and Pewe, 1996). The paleoecology and the distribution of the taxa of all found fossils of flora and fauna in the Fox Permafrost Tunnel were described in detail by Hamilton et al. (1988). A diagram of the geological cross section is shown in Figure 5.3.

This facultatively anaerobic fermenting bacterium was described as a separate new species since it was different from known species both on genetic and phenotypic levels (Pikuta et al., 2005). Like most other species of the genus *Carnobacterium*, the new strain FTR1^T showed psychrotolerant physiology and demonstrated the ability to grow at 0°C (Figure 5.4a and b). It was neutrophilic (optimum growth at pH 7) and had an alkalitolerant nature (growth observed at a pH 9.5). The optimum temperature for growth was 24°C (maximum 28°C). The NaCl range for growth was 0%–5% (w/v), the optimal concentration of NaCl was 0.5%. The chemo-organotrophic metabolism of this bacterium with ability to utilize sugars and proteolysis products is typical for all known species of this genus. The main metabolic end products were acetate, ethanol, and CO_2. This strain was sensitive to all antibiotics tested, which confirms its wild origin.

All known species of the genus *Carnobacterium* are facultative anaerobes and was reported by Franzmann et al. (1991). They are capable of reducing resazurin in

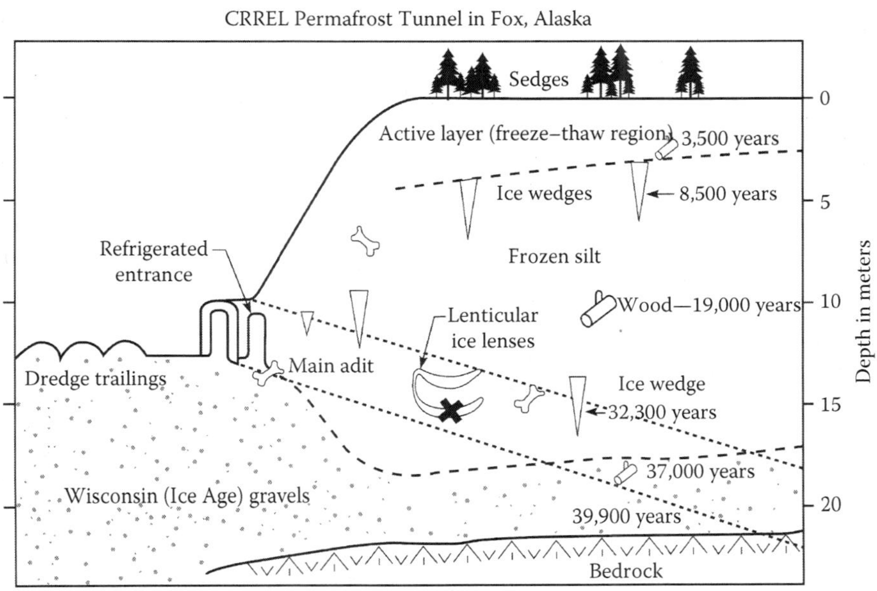

FIGURE 5.3 Cross-section diagram showing the geology of the Fox Permafrost Tunnel with paleontological records. The cross shows the sampling location.

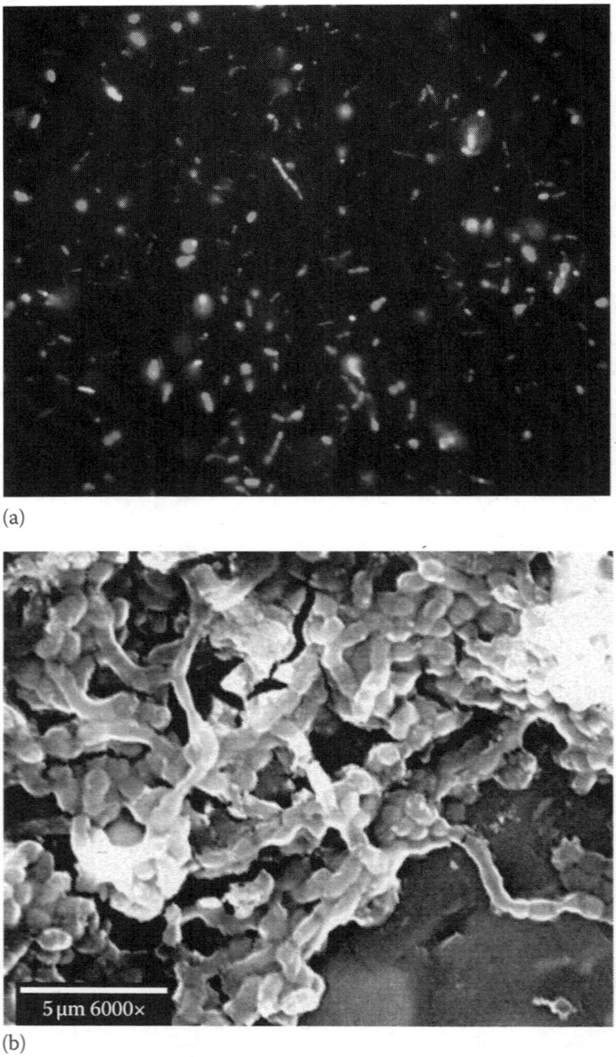

FIGURE 5.4 (See color insert for (a) following page 276.) Images of strain FTR1^T with (a) BacLite live/dead stain (Photo by Prof. A.K. Bej, University of Alabama at Birmingham, Birmingham, AL.); (b) Hitachi S-4000 field emission scanning electron microscope, scale bar is 5 μm. (Photo Courtesy: Gregory A. Jerman and Richard B. Hoover, NASA/Marshall Space Flight Center.)

aerobic media during growth. It is quite possible that in nature these organisms play the role of primary agents responsible for changing the redox potential in ecosystems by the reduction of the oxygen level and as result create conditions that are suitable for the development of obligately anaerobic microorganisms.

This bacterium is of great importance to both astrobiology and paleontology. It is the first validly published species of a living bacterium from the Pleistocene that had been cryopreserved in ancient permafrost. It is also important to Astrobiology and cryobiology as it demonstrates potential relevance to water ice-containing bodies of our Solar System. The presence of viable microbial cells in ancient ice suggests that future space missions may be able to extract ancient cryopreserved and possibly still living microorganisms from the surface crusts of icy moons of Jupiter or Saturn or the permafrost, craters, or polar ice caps of Mars.

5.5.1.2 Novel Acidophile from Chena Hot Springs

The obligately acidophilic strain AGC-2^T was isolated in pure culture (Pikuta and Hoover, 2004) from the Chena Hot Springs in Alaska. The sample was collected from dark-red mud at the bottom of a 0.5 M diameter, 65°C pool near the main spring. This mesophilic, spore-forming strain is able to grow on the medium with high concentrations of Fe^{2+} at pH 1.5–2.0. The taxonomic description of this bacterium is currently in preparation.

Morphology of strain AGC-2: Cell sizes are 0.6×1.2–2.0 μm, shape of cells is straight or slightly curved rods with rounded ends (Figure 5.5). Cells are Gram-positive, motile, endospore-forming with swollen sporangium (only one spore per cell), and the spores have a subterminal location. The cells multiply by binary division and occur singly, in pairs, or in short curved chains.

Physiology of strain AGC-2: Strictly aerobic, obligately acidophilic (no growth at pH 4.0), mesophilic with optimum growth at 29°C–32°C. Mixotroph (using CO_2 and yeast extract for anabolism). No growth on peptone and D-glucose. On agarose medium, it grows only with Fe^{2+} and yeast extract.

Cell-membrane lipids contain only 2% of the ω-alicyclic fatty acids, which are typical for species of the genus *Alicyclobacillus*. The amount of G+C in genomic DNA is 60.7 mol%. Strain AGC-2 was deposited in American Type Culture Collection and Japanese Collection of Microorganisms with corresponding numbers ATCC BAA-757^T (=JCM 12177^T). The GenBank deposition number for this sequence is AF 450135.

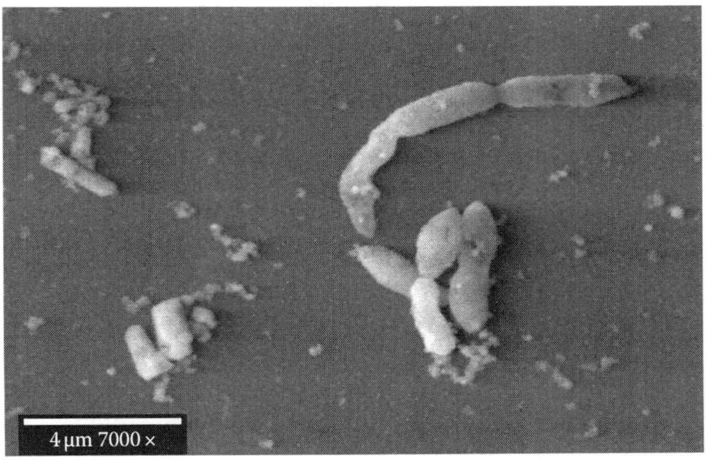

FIGURE 5.5 Cell morphology of strain AGC-2^T. (Photo Courtesy: Gregory A. Jerman, James E. Coston and Richard B. Hoover, NASA/Marshall Space Flight Center.)

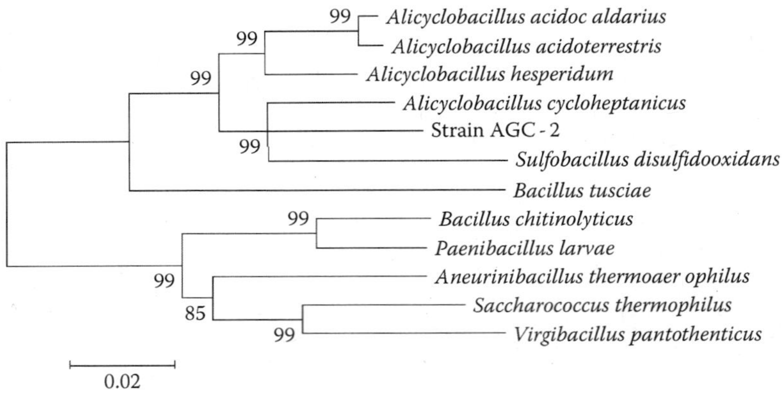

FIGURE 5.6 Phylogenetic tree showing the position of strain AGC-2^T among closest species.

Phylogenetic analysis of strain AGC-2^T: The 16S rRNA gene sequence (Figure 5.6) showed a position distant from all known species of the genus *Alicyclobacillus* and other species.

5.5.2 Novel Extremophiles from *Antarctica 2000* Expedition

5.5.2.1 Psychrotolerant Bacteria from Magellanic Penguin Colony in Southern Patagonia, Chile

Another interesting psychrotolerant sugarlytic bacterium was isolated from a sample of Magellanic Penguin (Figure 5.7) guano collected from the bottom of a tidal pool in Southern Patagonia, Chile. *Trichococcus patagoniensis* PmagG1^T has the ability to grow at −5°C at both aerobic and anaerobic conditions (Pikuta et al., 2006). The cells of strain PmagG1^T grown at −5°C have a tendency to excrete a capsule, composed of a mucopolysaccharide matter, surrounding their cell surface (Figure 5.8). This probably allows this microorganism to keep the water in liquid state for metabolic functions in freezing conditions. Strain PmagG1^T was a catalase-negative facultative anaerobe, with pH range 6.0–10.0, and shows optimum growth at a pH 8.5. It did not require NaCl for growth, showed optimum growth at 0.5% NaCl, and it could tolerate 6.5% salinity. The temperature range for this organism was −5°C to 35°C, and optimum growth was observed at 28°C–30°C. The strain PmagG1^T had heterotrophic metabolism, and could utilize as substrates some sugars and some organic acids (pyruvic and citric). The metabolic end products were lactate, formate, acetate, ethanol (in cultural liquid), and CO_2 in the gas phase. The strain was sensitive to all antibiotics checked, indicating the wild origin of this strain.

During the study it was shown that all species of the genus *Trichococcus* have psychrotolerant physiology with highly stable genomic memory: the sequences of 16S rRNA gene had very close similarity (up to 100%) by the pair-wise distance. This may indicate that the psychrotolerant features of these species evolved to adapt

FIGURE 5.7 Magellanic penguins in Patagonia, Chile.

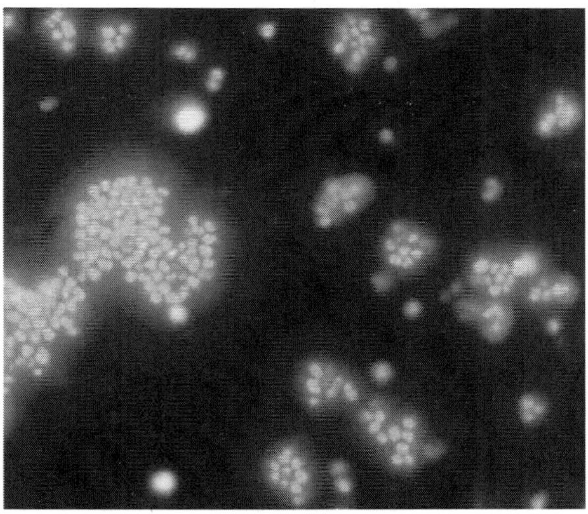

FIGURE 5.8 Image of strain PmagG1^T grown at 3°C with (a). DAPI stained cells under a *Leitz* Diaplan epifluorescent microscope: clumps of cells, single and cells in pairs occur in the culture at the end of exponential growth phase. (Photo by Prof. A.K. Bej, University of Alabama at Birmingham, Birmingham, AL.)

to low temperatures a long time ago and are tightly related to conservative genetic units (chromosomal DNA).

The taxonomical description of the proteolytic, spore-forming, psychrotolerant bacterial strain PPP2^T that represents a separate new genus and species *Proteocatella sphenisci* was completed recently (Pikuta et al., 2009). The cells were motile rods, endospore forming, and gram-positive (Figure 5.9). Strain PPP2^T had the ability to grow exclusively on proteolysis products (peptone, casamino acids, yeast extract), and oxalate. It was unable to grow on separate amino acids nor perform respiration by the Stickland reaction. The main end product of metabolism was acetate; traces

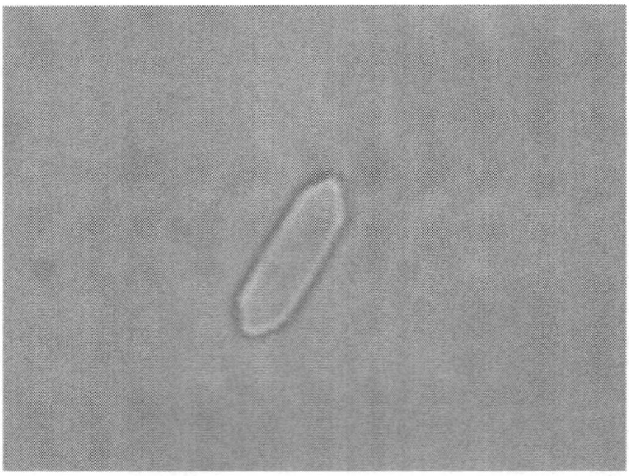

FIGURE 5.9 A typical morphology of the cells of strain PPP2^T grown on sodium oxalate (3 g L^{-1}). It is interesting that although the cells look like hexagonal crystals, they were motile by peritrichous flagella.

of CO_2 and H_2 were gas minor products. Minimum temperature for growth was determined to be 2°C. New isolate was found to belong to Clostridial cluster XI, and it had distant position on the level of separate genus and species.

5.5.2.2 Novel Psychrophilic and Psychrotolerant Anaerobes from Patriot Hills, Antarctica

Strain LL-8 was isolated from the sample collected in the Patriot Hills during the *Antarctica 2000 Expedition*. This strain had features of dimorphic prosthecate type bacteria, and it was difficult to cultivate. It slowly grew during 2–3 months at 3°C on medium with peptone (3 g L^{-1}) and yeast extract (0.1 g L^{-1}). The optical density of the grown enrichment culture never exceeded 0.3 absorption at $\lambda = 510$ nm. The transfer of a small quantity of the substrates (0.02 g L^{-1}) into the medium, which is typical for *Caulobacter*-like bacteria, did not show the growth of the culture.

Strain LL-8 had the unusual morphology of straight rods with the tendency of rectangular cell division forming orthogonal, acute or obtuse junctions as the cells/prosthecae divided (Figure 5.10a and b). Strain LL-8 was spore-forming rods with prosthecae extending up to 20–25 μm. Sometimes star-shape "fury balls" with sizes 100–200 μm could be observed with many straight long stalks or prosthecae (Figure 5.10d). It is interesting that the diameter of the prosthecae varied from 0.2 to 0.8 μm, and sometimes the thicker prosthecae exhibited a shiny reflective material covering the surface. This shiny reflecting cover was not observed in the thin prosthece. Gram-staining of strain PPP2^T cells demonstrated positive reaction. Spores were round with diameter twice that of the cells (Figure 5.10c), the sporangium was not swollen. The morphology of strain LL-8 is unlike that of the known described species of the genera *Caulobacter*, *Asticcacaulis*, *Hyphomicrobium*, *Pedomicrobium*, *Hyphomonas*, or *Thiodendron*. Sometime the dichotomic devision of prosthecae with Y-type bifurcation could be observed during the microscopy of strain LL-8 (Figure 5.10c).

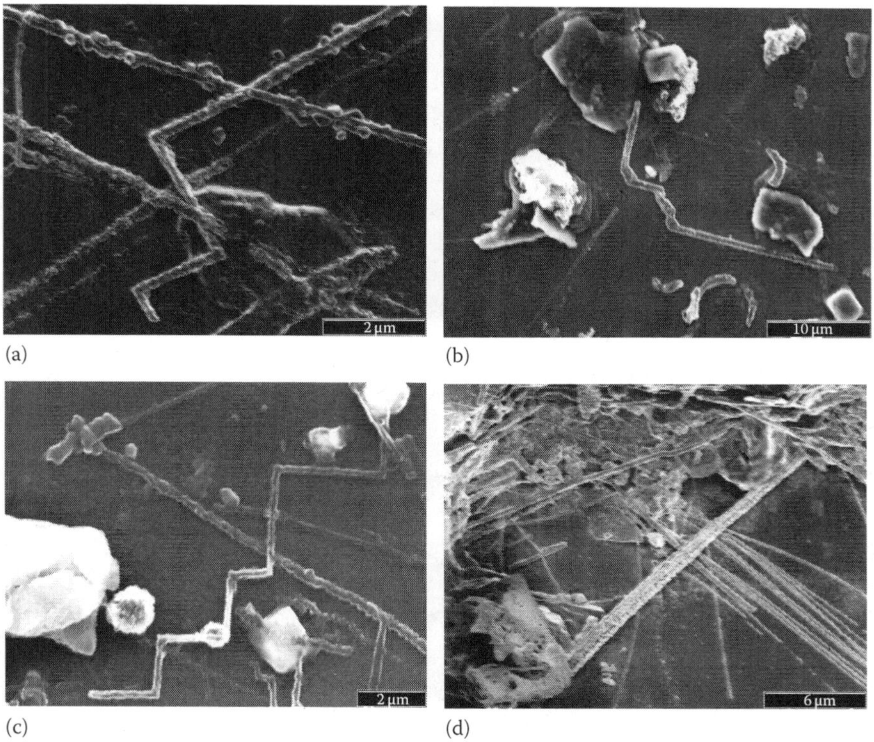

FIGURE 5.10 Electron scan microscopy of strain LL-8: (a) straight and rectangularly bended prosthecae; (b) cells and bended prosthecae; (c) spores and prosthecae, on right lower corner the Y-type bifurcation of the prostheca is shown; (d) the fragment of "fury ball" rays on right lower corner. (Photos Courtesy: Gregory A. Jerman, James E. Coston and Richard B. Hoover, NASA/Marshall Space Flight Center.)

Strain PH-21 was also isolated from the sample collected at the Patriot Hills, Antarctica. The morphology and other phenotypic features of this strain were similar to strain PmagG1^{T}. The only difference was found in the spectrum of substrate utilization, as well as slightly smaller sizes of cells. Phylogenetic analysis of strain PH-21 showed 100% similarity with the sequence of the strain PmagG1^{T}. The DNA–DNA-hybridization analysis of these two strains will confirm the taxonomic position for strain PH-21.

5.5.3 *2008 Tawani International Antarctica Expeditions*

The *2008 Tawani International Schirmacher Oasis/Lake Untersee, Antarctica Expeditions* were conducted as a part of the 2008 International Polar Year. The scientific focus of these two Antarctic expeditions was the investigation of microbial extremophiles and the biodiversity of the lakes and ice of the Schirmacher Oasis and Lake Untersee. The Reconnaissance Expedition was a brief 10 day mission in February, 2008. It was carried out by five members of the international team to the Schirmacher Oasis and Lake Untersee, Antarctica (Hoover et al., 2008). A number of anaerobic microbial cultures were obtained. The samples collected were returned to the NSSTC Astrobiology Laboratory for studies of the anaerobic extremophiles. The samples were distributed to other team members (Asim Bej, Birgit Sattler, and Michael Storrie-Lombardi) for the investigations of aerobic microorganisms and viruses. The validated bacterium capable of growth at –5°C (*T. patagoniensis*) that was isolated from Magellanic Penguin (*Spheniscus magellanicus*) guano collected during the *Antarctica 2000 Expedition* led us to investigate the guano samples from the African Penguin (*Spheniscus demersus*), which also resulted in the detection of interesting novel psychrotolerant anaerobic bacteria.

5.5.3.1 Psychrotolerant Anaerobes from the African Penguin Guano

The African Penguin *S. demersus* is the only species of penguin that breeds around the South African coast (Figure 5.11a). The environmental samples of the penguin guano were collected from a small tidal pool with the assistance of Park Rangers at the Stony Point Nature Reserve, Betty's Bay, South Africa (S 34° 21′25″; E 18° 28′16″).

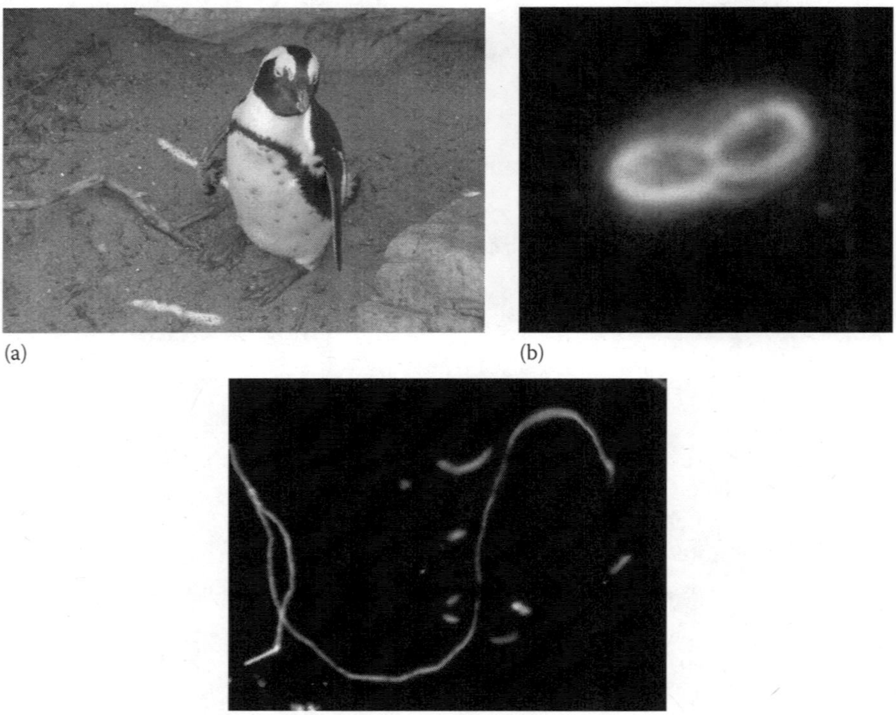

(a) (b) (c)

FIGURE 5.11 (a) African penguins at Stony Point, Betty's Bay, South Africa. (b) Cell morphology of non-spore-forming strain ARHSd-7G; dividing vibrions are shown. Sizes of cells are 0.4×1.2–1.5 μm. Cells are motile due to polar flagella (not shown). Cells are Gram variable, anaerobic and catalase negative. (c) Dark-field microscopy of strain ARHSd-9G shows specific interference coloration effect caused by material in cell walls.

The samples collected had a pH 6.8, salinity 3%, and at the moment of collection the water temperature was 15°C. The sample was immediately examined by visible light, dark field microscopy and then returned directly to the NSSTC Astrobiology Laboratory in Huntsville. Phase Contrast microscopy of the sample at NSSTC showed the presence of diverse motile microbial cells (~10^{13} cells per ml). From these samples, 0.3 ml portions were injected into alkaline (pH 9) anaerobic media with 1% NaCl; an alkaline (pH 9) anaerobic media with 3% NaCl, and marine anaerobic media with pH 7 and 3% NaCl. All tubes were incubated at 3°C for 2 weeks. The enrichment cultures were obtained in Hungate tubes containing medium with pH 7 and 3% NaCl and pH 9 and 1% NaCl. The tube with pH 9 and 3% NaCl did not exhibit any growth.

Strain ARHSd-7G was isolated on marine medium with D-glucose as the substrate. Cells were motile vibrions with sizes 0.6×1.2–1.5 μm (Figure 5.11b). The isolation of colonies was performed by "roll-tubes" method with 3% agar medium at 3°C. The strain ARHSd-7G had an optimum growth at 3% NaCl and did not grow without NaCl or at 0.1% NaCl. A minimal growth was obtained at 0.5% NaCl. The strain ARHSd-7G was determined as highly halotolerant microorganisms, since its maximum growth was at 24% NaCl with a long lag-phase around 3 weeks. No growth was observed at 24.5% NaCl. Growth occurred within the pH range of 5.2 to 10.0, and showed optimum growth at pH 8.9. No growth was observed at pH 5.0 and 10.5.

The range of temperature for the growth of the new isolate was 3°C–40°C, with an optimal growth at 30°C. No growth was observed at 0°C and 45°C. The strain ARHSd-7G was able to use as substrates some sugars and several proteolysis products, but the cultures grown on the proteolysis products were comparatively unstable: the lysis of cells occurred much faster than with sugars. It grew on maltose, D-glucose, D-fructose, sucrose, D-trehalose, D-ribose, D-mannitol, D-cellobiose, peptone, yeast extract, casamino acids, triethylamine, and pyruvate. No growth was observed on D-arabinose, L-arabinose, D-mannose, lactose, starch, trimethylamine, betaine, formate, acetate, lactate, propionate, butyrate, citrate, methanol, ethanol, glycerol, and acetone. The new isolate had strictly fermentative metabolism: any acceptors of electrons (Fe^{3+}, NO_3^{2-}, SO_4^{2-}, SO_3^{2-}, S^0, $S_2O_4^{2-}$) added to the medium did not stimulate growth. The strain ARHSd-7G was unable to reduce sulfur compounds to the H_2S.

Preliminary phylogenetic analysis data showed that strain ARHSd-7G belongs to the genus *Salinivibrio* and had the highest similarity (99.1%) with *S. costicola*. DNA–DNA hybridization analysis will be conducted to finalize the taxonomic position of this new isolate.

Strain ARHSd-9G was obtained from enrichment culture grown on an alkaline medium with pH 9 and 1% NaCl on D-glucose as a substrate. Pure culture was received from 8th dilution on "roll-tubes" with 3% agar medium. The cell morphology as is shown (Figure 5.11c) exhibits elongated straight or slightly curved rods. Cells are motile with peritrichous flagella. The strain is spore forming, and the round spores are located terminally. Sizes of cells are 0.4×3–5 μm. The strain ARHSd-9G is strictly anaerobic, catalase negative, and grows between 3°C and 22°C.

5.5.3.2 Microbial Extremophiles from the Schirmacher Oasis, Antarctica

The Schirmacher Oasis of central Dronning Maud Land, Antarctica is 3 km wide, 20 km long, and contains over 150 lakes. Some of the lakes are perennially ice covered while others show open water in midsummer. The Russian Antarctic Station *Novolazarevskaya* and the Indian Station *Maitri* are located in the Schirmacher Oasis. The 2008 International Tawani Schirmacher Oasis/Lake Untersee Expeditions were designated as official Russian Antarctic Expeditions. The team was very grateful for having been granted permission to utilize the facilities of the *Novolazarevskaya Station*. Numerous samples were obtained from Lakes Glubokoye, Lake Stantionoye, and the nearby Ice Cave. Samples were also collected from Lake Podprudnoye and Lake Priyadarshini. The Indian Antarctic Station *Maitri* is situated on Lake Priyadarshini, which is the largest lake in the Schirmacher Oasis.

5.5.3.2.1 Bacteria from Lake Zub (Lake Priyadarshini)

A living sample of green moss with rhizosphere surrounded by soil was collected from the shore of Lake Zub (also known as Lake Priyadarshini) and returned to NASA/NSSTC in frozen state. This sample gave growth to a rich enrichment culture that led to the isolation of two anaerobic psychrotolerant strains LZ-22G and LZV-3P.

Strain LZ-22G grew exclusively on sugars, on neutral medium with pH 7, and did not require NaCl for growth. The range of NaCl concentration for growth was 0%–3% (w/v) and the optimal growth was at 0% NaCl. The new isolate grew between 3°C and 35°C, no growth was observed at 40°C. After freezing and thawing, the strain was still alive. The pH range for the growth of strain LZ-22G was 4.5–9.0, and the optimal growth was at pH 9.0. Growth was significantly stimulated by yeast extract. This strain is obligately anaerobic and catalase negative. The morphology of the cells had distinguishing characteristics: the cells were comparatively large (1 μm diameter by 10–15 μm length), and the ends of cells were rounded. The cells exhibited gliding motility (Figure 5.12a).

The isolation of colonies of strain LZ-22G was done by serial dilution and "roll-tube" methods. The colonies grown on 3% agar medium had the following characteristics: Light cream color, round shape of concave lens middle of colonies had denser consistency. The colonies were smooth and shiny with diameter 0.5–4 mm; and the older colonies were transparent and surrounded by a shiny circle.

Strain LZ-22G grew on the following compounds: yeast extract, D-glucose, D-fructose, maltose, sucrose, D-trehalose, D-cellobiose, D-ribose, D-cellobiose, lactose, chitin, pectin, and starch. The best growth was observed on lactose. No growth was observed on formate, acetate, pyruvate, lactate, propionate, butyrate, citrate, oxalate, methanol, ethanol, glycerol, acetone, betaine, trimethylamine, triethylamine, peptone, casamino acids, D-mannose, and D-arabinose.

Strain LZV-3P was also isolated from the same sample. Cells of this strain had vibrion shape and also had large sizes of 0.8 μm in diameter and 3–5 μm in length (Figure 5.12b). The cells were motile and spore forming. Strain LZV-3P had proteolytic fermenting metabolism, and grew exclusively on proteolysis products (peptone,

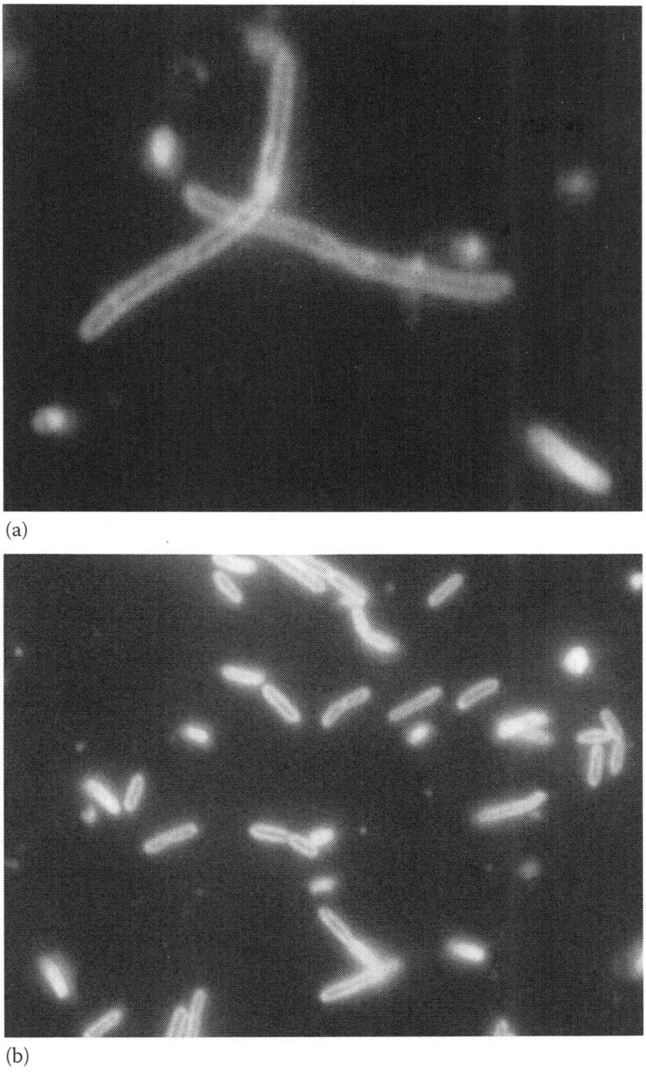

(a)

(b)

FIGURE 5.12 (a) Cell morphology of strain LZ-22G growing on D-glucose at 3°C. Cells are motile due to peretrichous flagella and are spore forming (oval spores are located subterminally); (b) proteolytic vibrion cells of strain LZV-3P from the frozen sample of green moss with soil around rhizosphere collected on the shore of Lake Priyadarshini (*Maitri* station) Antarctica.

yeast extract, casamino acids). Sugars did not support the growth of the new isolate. The culture had a specific strong smell of decomposing proteins (not hydrogen sulfide, but decaying meat). This strain grew between 3°C and 22°C. The cells were catalase negative and strictly anaerobic and the pH for growth was neutral 7.0–7.5. The range of salinity for growth was 0.5%–3% NaCl.

5.5.3.2.2 Bacteria from Ice Sculptures near Lake Podprudnoye

In the vicinity of Lake Podprudnoye (also known as Proglacial Lake 21), the transport vehicle encountered a large field of "ice sculptures" formed by the summer melt and wind erosion processes (Figure 5.13). These "ice sculptures" contained a large number of entrained dark rocks and dust grains. Sunlight penetrates the ice and causes localized melting resulting in the formation of liquid water films around each of these trapped rocks. These water films provide an ideal environment for the growth of photoautotrophic cyanobacteria, bacteria, diatoms, and other algae as well as organotrophic bacteria.

Strain ISLP-3 (Figure 5.14) was isolated from the ice sculpture shown in Figure 5.13. The frozen sample was slowly melted at 3°C in a sterile flask under a pure nitrogen atmosphere. The melted liquid (0.5 ml) was injected into a Hungate tube with anaerobic medium and incubated at 3°C for 2–3 weeks. The cell morphology of strain ISLP-3 had irregular shape of spore-forming vibrions with partially swollen sporangia and pointed ends. Cells occurred in pairs and were slightly angled. Sometimes Y-shape cells were observed. The cells have sizes of 0.5×1.2–$3\,\mu m$. The colonies of strain ISLP-3 grown on 3% agar medium had a circular concave shape with a diameter of 1–3 mm, and the young colonies were smooth and glossy. The aged colonies had a brownish tint, an opaque color, and were not glossy. Strain ISLP-3 was catalase negative and obligately anaerobic microorganism. This strain is psychrotolerant and grows at 0°C–40°C. No growth was observed at 45°C. At 0°C, spore formation was not observed and the cells did not excrete the mucopolysaccharide capsules. Strain ISLP-3 does not require NaCl in the medium and has an optimum of growth at 0.1% NaCl. The range of NaCl for growth is 0%–1% (w/v). The range of pH for growth occurs between 6.2 and 9.8, with an optimum at 8.5. This strain was

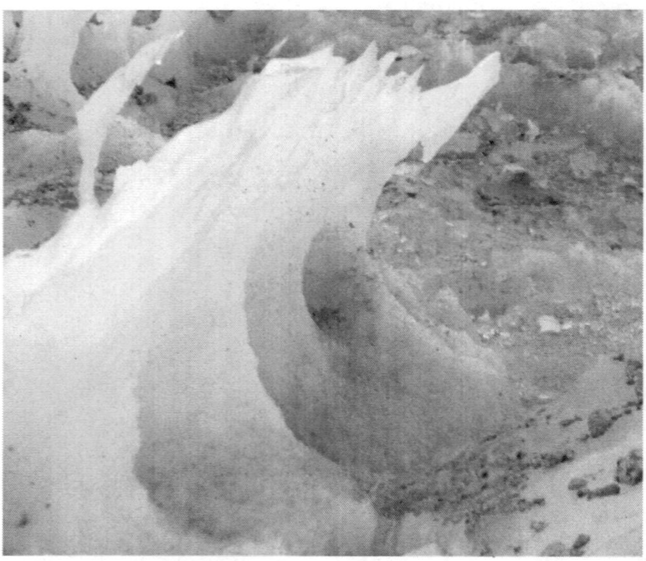

FIGURE 5.13 "Ice sculpture" near Lake Podprudnoye in Schirmacher Oasis, Antarctica.

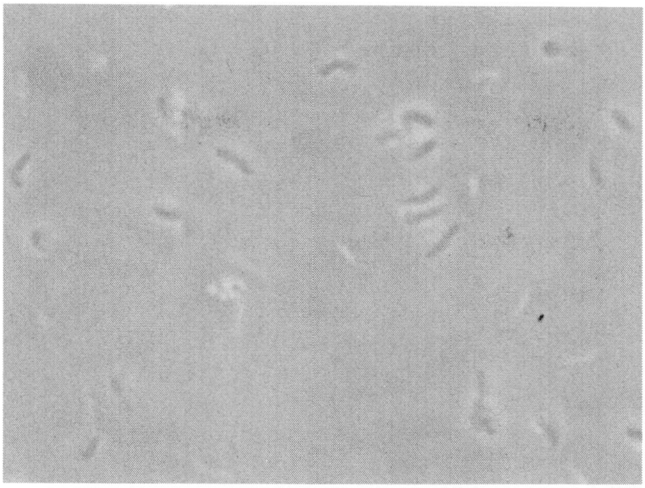

FIGURE 5.14 Cells of strain ISLP-3 have baseball shape and irregular space configurations due to tendency for cells to clump with excreted mucopolysaccharide.

able to grow on following compounds: triethylamine, chitin, *N*-acetylglucosamine, urea, D-glucose, D-arabinose, D-trehalose, maltose, sucrose, D-ribose, D-mannose, lactose, starch, and D-cellobiose. The best growth was observed on D-cellobiose. Growth was absent on following substrates: formate, acetate, pyruvate, lactate, propionate, butyrate, citrate, oxalate, methanol, ethanol, glycerol, D-mannitol, acetone, betaine, trimethylamine, peptone, yeast extract, casamino acids, pectin, and D-fructose.

5.5.4 Microbial Extremophiles from Lake Untersee

Lake Untersee is a perennially ice covered, ultra-oligotrophic, lake in the Otto-von-Gruber-Gebirge (Gruber Mountains) of central Dronning Maud Land. It is 563 m above sea level and has a surface area of 11.4 km^2. It is the largest freshwater lake in east Antarctica. It is the only known highly alkaline lake on Earth that is permanently cold and constantly covered with a glacial ice sheet. It has a low mineralized aquatic system with pH 10, so it must have an unusually unstable buffer system that has not previously been described. This buffer system, as well as the microbial communities that may have adapted to live in it, needs detailed investigation. This lake has extreme physicochemical stratification with dramatic changes in pH, sulfides, oxygen content, and temperature as a function of depth in the water column as well as the highest methane concentration of any natural ecosystem on Earth (Wand et al., 2006). Although the previous physicochemical data suggested that the Lake Untersee methane was biological in origin, the prior researchers had not conducted a detailed study on the microbiology of this cold hyperalkaline system.

While considerable work has been done on extremophiles of the McMurdo Dry Valleys, little is known about microbial ecosystems of the Schirmacher Oasis and Lake Untersee. Studies of the cyanobacteria and algal species diversity of Antarctic lake environments (Singh and Elster, 2007) and streams in the Oasis have been conducted (Gupta et al., 2006), and a novel obligately anaerobic proteolytic bacterium *Clostridium schirmacherense* was isolated from lake sediments (Alam et al., 2006). A survey of microorganisms and microbial communities that may survive in the cold, hyperalkaline environment of Lake Untersee has never been carried out.

Previous expeditions to Lake Untersee have conducted detailed investigations of the water chemistry. Wand et al. (2006) found that the pH was extremely high (mean 11.34 ± 0.12) in the near freezing water (T 0.8°C) immediately beneath a 2 m thick ice-cover. The pH remains high and remarkably uniform between a depth of 3 and 72 m before dropping dramatically from >10 at 72 m to 7.6 at 75 m. These studies suggest the lake contains the two disparate environments characteristic of inland soda lakes: an upper layer that is most likely dominated by cyanobacteria and other oxygen-generating phototrophs and a lower, anoxic sediment and water column producing high levels of methane that probably are the result of a relict community of bacteria and archaea. The perennially ice-covered hyper-alkaliphilic Lake Untersee in central Dronning Maud Land, East Antarctica was the primary target of the Expedition carried out in November–December, 2008. Wand et al. (1997) has previously provided data on the physicochemical stratification in Lake Untersee. Wand and Perlt (1999) described the "floating boulders" that ride across Lake Untersee on the Anuchin Glacier ice sheet. During the 2008 Expedition, Dale Andersen and Chris McKay obtained new GPS data on the current positions of these "floating boulders" for comparison with Wand's data. One of the "floating boulders" is shown in Figure 5.15.

FIGURE 5.15 "Floating Boulder" on the ice sheet that covers Lake Untersee.

5.5.4.1 Psychrophilic and Psychrotolerant Anaerobes from Lake Untersee

Lake Untersee was the source of isolation for several novel psychrophilic and psychrotolerant anaerobes. Two saccharolytic and one proteolytic anaerobic bacterial strains growing at 5°C were isolated and purified. Strain UDS7-G was isolated from a sediment sample taken from the deep anoxic trough. This strain grows on anaerobic medium with pH 7, 0.5% NaCl, and D-glucose. The cell morphology showed straight rods with pointed ends, and the sizes of the cells were 0.7×2–3 μm (see Figure 5.16a). This bacterium is motile and spore forming, and grows at 3°C–22°C.

Another sugarlytic strain UL7-96mG was isolated from 96 m deep water sample of Lake Untersee. It had a similar morphology, and the only difference at the current stage of investigation is its growth rate; strain UL7-96mG grows significantly faster than strain UDS7-G (see Figure 5.16b). Genetic and physiological tests now underway will provide information about the identity of these two strains.

The proteolytic anaerobic bacterium, strain LU-96m7P, was also isolated from 96 m deep water sample of Untersee Lake on a medium with pH 7, 0.5% NaCl, and peptone as a substrate. The morphology of this strain shows rod-shaped, motile, and spore-forming cells with sizes 0.6×3–5 μm. The strain has truly psychrophilic physiology: it grows rapidly at 3°C–5°C and does not grow at 22°C.

Strain A4P-85m was isolated on an anaerobic medium with pH 3.5 and peptone 3 g L^{-1}. The unusual "Zig-Zag" morphology of this strain is similar to strain LL-8 from a Patriot Hills sample obtained during the *Antarctica 2000 Expedition* (Hoover and Pikuta, 2004). The growth rate of this strain is much higher than the LL-8 strain obtained on that expedition (see Figure 5.17).

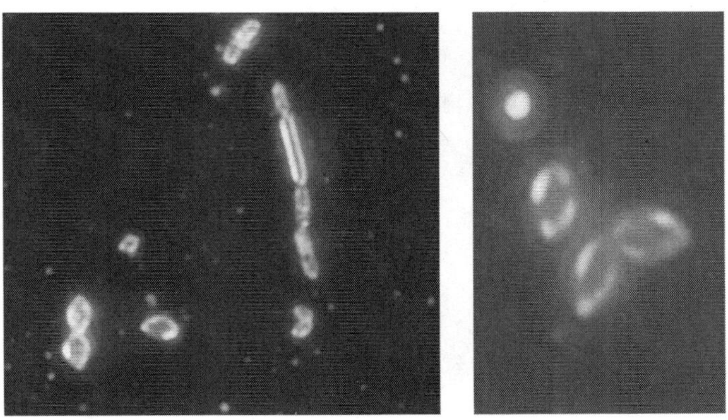

(a) (b)

FIGURE 5.16 (a) Strain UDS7-G and (b) strain UL7-96mG from the Deep Anoxic Trough sediment sample of Lake Untersee.

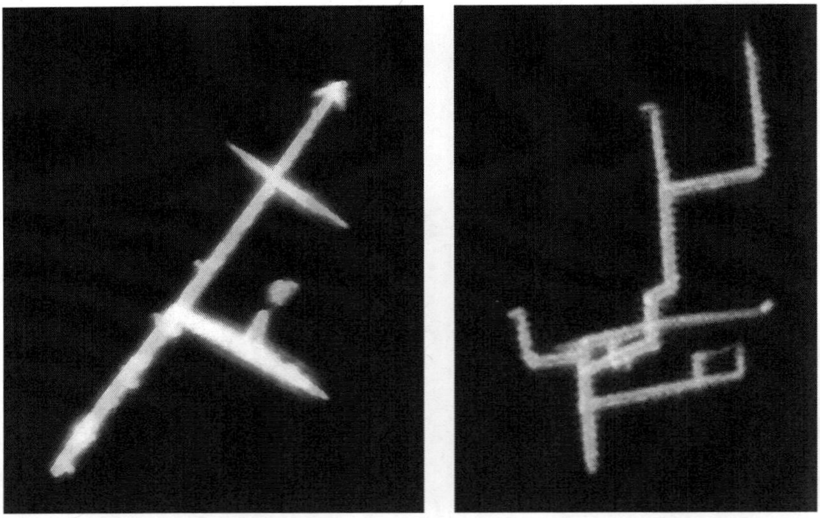

(a) (b)

FIGURE 5.17 Strain A4P-85m was isolated on an anaerobic medium with pH 3.5 and peptone.

Two homoacetogenic strains A7AC-96m and A7AC-DS7 were obtained from Lake Untersee water samples at 96 m depth (Figure 5.18a) and from the deep anoxic trough sediment sample (Figure 5.18b), respectively. Both of these bacteria are motile vibrions that grow at 5°C and pH 7. Cell sizes are $0.6–0.7 \times 2.0–4.0\,\mu m$. Maximum of growth (during 1–2 days) was observed at 15°C. No growth was observed at 22°C. This indicates the truly psychrophilic nature of these strains. Both strains grew lithoautotrophically (during 2 days after inoculation) on $H_2 + CO_2$ as the only source for energy and carbon.

Another proteolytic strain AP7-90 was obtained from a sample of the water at 90 m depth at the Lake Untersee Deep Anoxic Trough on an anaerobic medium at pH 7 with peptone. The morphology of the cells is long thin or curved hair-like rods (Figure 5.19a) (sizes $0.3–0.4 \times 10–15\,\mu m$) with laterally located spores (Figure 5.19b).

5.5.5 Microorganisms *In Situ* in Ice Bubbles

One of the most surprising observations made during the 2008 Tawani Expeditions to the Schirmacher Oasis and Lake Untersee in Antarctica was the great abundance of viable bacteria in the ice bubbles. Spectacular "ice-bubble" systems were observed throughout the upper layers of the ice sheet covering Lake Glubokoye in the Schirmacher Oasis in November, 2008 (Figure 5.20). Similar ice bubble assemblages were also found in the Anuchin Glacier and the ice sheet over Lake Untersee. Even though the ice-bubble samples appeared completely transparent and clear, as soon as the samples were thawed, large numbers of morphologically

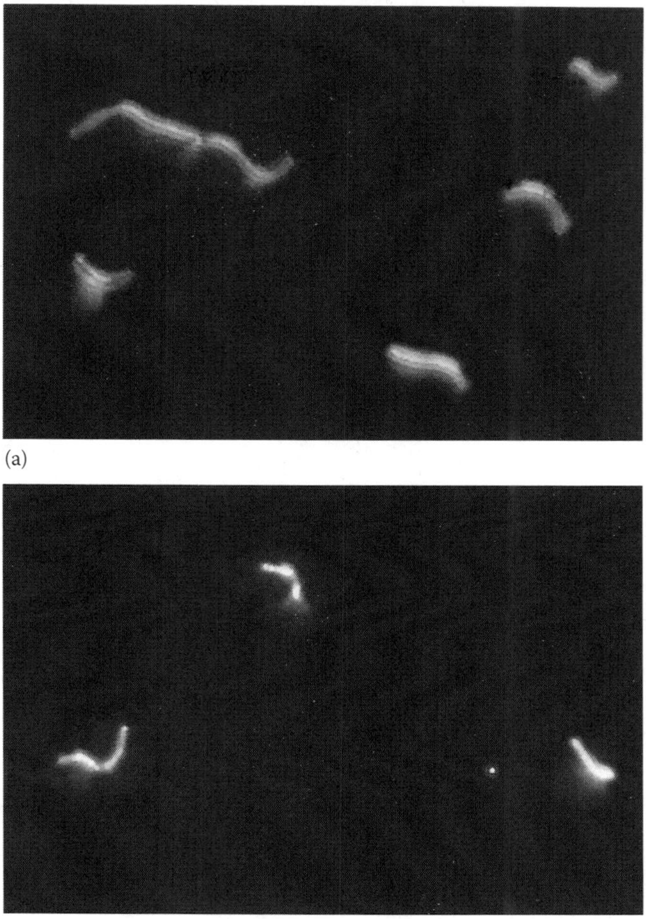

(a)

(b)

FIGURE 5.18 Anaerobic cultures from Lake Untersee, Antarctica: Psychrophilic homoacetogenic bacteria: (a) strain A7AC-96m and (b) strain A7AC-DS7 growing on H_2+CO_2 at 5°C.

diverse, motile bacterial cells were observed. Their motility was documented in the field using high resolution dark field video photomicroscopy methods. We had previously observed motile bacteria in ice bubbles found in the blue ice at the Patriot Hills and the Thiel Mountains during the *Antarctica 2000 Expedition* and during field expeditions to Matanuska Glacier of Alaska. The ice bubbles absorb solar radiation and melt the ice forming thin films or pockets of liquid water trapped within the interior of the ice. Bacteria, cyanobacteria, diatoms, and other microorganisms grow within these bubbles and produce metabolic end products that generate microniches within the ice (Hoover and Gilichinsky, 2001; Bargagli, 2005). Unfortunately, this interesting ecosystem has received very little attention by microbiologists.

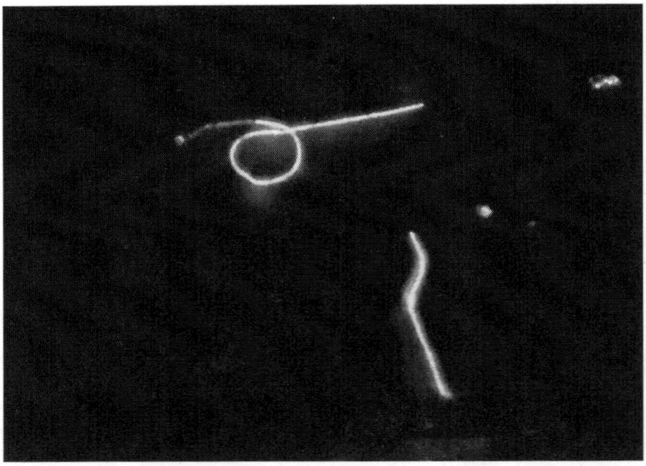

(a)

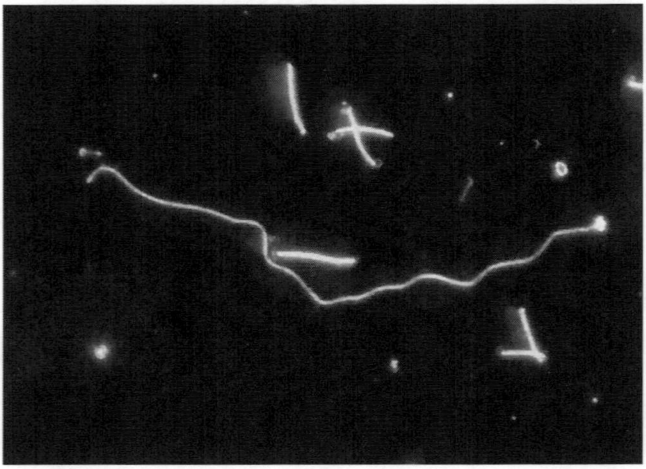

(b)

FIGURE 5.19 Anaerobic cultures from Lake Untersee, Antarctica: Psychrophilic proteolytic strain AP7-90 growing at pH 7 with peptone.

The pioneering investigations of viable and entrapped microorganisms in the deep ice sheet above Lake Vostok, Antarctica were carried out by Sabit S. Abyzov of the Institute of Microbiology of the Russian Academy of Sciences. The research has shown that viable microorganisms can remain alive while trapped deep within the Antarctic Ice Sheet (Abyzov, 1993; Karl et al., 1999; Siegert et al., 2001). The phylogenetic analysis of 16S rRNA gene sequences and bacterial isolates and amplicons obtained from melted ancient (~500 kyr) glacial ice samples and accretion ice above Lake Vostok has yielded six different bacterial lines of α-, β-, and γ-proteobacteria (Christner, 2002). Recent reviews of the studies of the microorganisms of the ice layers above Lake Vostok have been provided (Abyzov et al., 2005, 2006; Bell

FIGURE 5.20 Ice bubbles containing entrained rocks and living bacteria in the ice sheet overlying Lake Glubokoye in Antarctica.

et al., 2005). The microorganisms cultivated from these ancient glacial ice samples were primarily psychrotolerant mesophiles, with the ability to grow at 4°C but the optimal growth occurred above 20°C. However, Bulat et al. (2004) have reported the detection of the signatures of thermophilic microorganisms in the ice above Lake Vostok.

The astonishing ability of microorganisms to remain alive for long periods of time while trapped in deep layers of the Antarctic ice sheet and to grow actively within the ice bubbles is of great significance to astrobiology. Thick ice sheets can provide protection from hard vacuum and the radiation environment of space and create localized conditions suitable for the needs of the specific microorganisms. Consequently, the polar ice caps and permafrost of Mars, the icy cores of comets, and the crustal layers of the icy moons of Jupiter and Saturn may comprise appropriate environments for the growth of psychrophilic and psychrotolerant species such that they are able to inhabit the polar environments, glaciers and ice sheets of planet Earth.

5.6 RELEVANCE OF POLAR MICROBIAL EXTREMOPHILES TO ASTROBIOLOGY

Several of the observations made during these expeditions may have relevance to Astrobiology. During the 2008 Lake Untersee Expedition, Dale Anderson and Chris McKay triggered an "ice geyser" when their drill entered localized pockets of high pressure air in the upper layer of the Anuchin Glacier ice sheet covering the surface of Lake Untersee. This interesting phenomenon may have relevance to understanding the nature of the recently discovered "ice geysers" that erupt from the cracks in "tiger stripe" area of Saturn's moon Enceladus. Ice Bubbles were found throughout the surface of Lake Untersee and Lake Glubokoye in the Schirmacher Oasis. Motile bacteria were observed by high-resolution dark field microscopy at the moment of melting of these "ice bubble" samples. The presence of viable bacteria frozen in the ice of the Anuchin Glacier at Lake Untersee suggests that it may not be necessary to drill through the thick icy crusts to search for life in the seas of the frozen moons of Jupiter and Saturn as viable cells might be cryopreseved in the upper layers of the ice crust.

While it is generally accepted that life on Earth had an endogenous origin, biomarkers in some of the most ancient rocks on Earth suggest that the endogenous origin hypothesis cannot be logically excluded. The study of psychrophiles may provide a means of differentiating these two models. As we previously pointed out (Pikuta et al., 2007), psychrophily among biological forms have to be developed after a significant temperature drop on the Earth's surface. Hence, the phenotypic features of life should be reflected at the genotype level with all adaptation changes encoded during the later period of the Earth's evolution—after the appearance of constant polar ice caps. If this model is correct, then cold-shock proteins and the other cold-adaptive structures and mechanisms for cells should be encoded at the level of plasmids and surface genes, which are more recent and unstable genetic units that are more easily modified. However, if truly psychrophilic microorganisms originated elsewhere in the Cosmos (where the evolution of life experienced different scenarios) and these microorganisms were delivered to Earth by comets or meteorites (Hoover and Pikuta, 2004), then the phenotypes of these microorganisms with psychrophilic features should be tightly connected with the most general life processes and structures and located deeply in conservative and stable genetic units. As a result, such microorganisms would appear on separate phylogenetic lineages of genera, families, and higher taxa levels. While both types of psychrophilic microorganisms may currently coexist in nature, the studies on the isolation of plasmids, expression of genes, etc. may help to differentiate between these two hypotheses regarding the origin of life on Earth.

5.7 SUMMARY

The microbial extremophiles that inhabit the polar regions of our planet are of tremendous significance. The psychrophilic and psychrotolerant microorganisms, which inhabit all of the cold environments on Earth have important applications to Bioremediation, Medicine, Pharmaceuticals, and many other areas of Biotechnology. Until recently, most of the research on polar microorganisms was confined to studies of polar diatoms, yeast, fungi, and cyanobacteria. However, within the past three decades, extensive studies have been conducted to understand the bacteria and archaea that inhabit the Arctic and Antarctic sea-ice, glaciers, ice sheets, permafrost and the cryptoendolithic, cryoconite, and ice-bubble environments. These investigations have resulted in the discovery of many new genera and species of anaerobic and aerobic microbial extremophiles. Exotic enzymes, cold-shock proteins, and pigments produced by some of the extremophiles from polar environments have the potential to be of great benefit to mankind. Knowledge about microbial life in the polar regions is crucial to understanding the limitations and biodiversity of life on Earth and may provide valuable clues to the Origin of Life on Earth. The discovery of viable microorganisms in ancient ice from the Fox Tunnel, Alaska and the deep Vostok Ice has shown that microorganisms can remain alive while cryopreserved in ancient ice. The psychrophilic, lithoautotrophic homoacetogen isolated from the deep anoxic trough of Lake Untersee is an ideal candidate for life that might inhabit comets or the polar caps of Mars. The spontaneous release of gas from within the Anuchin Glacier above Lake Untersee may provide clues to the ice geysers that erupt from the "tiger stripe" regions of Saturn's moon Enceladus. The methane productivity in the lower regimes of Lake Untersee may also provide insights into possible mechanisms for the recently discovered methane releases on Mars. Since most of the other water-bearing bodies of our Solar system are frozen worlds, microbial extremophiles from the polar regions of Earth are of great importance to Astrobiology in understanding where and how to search for evidence of life elsewhere in the Cosmos.

ACKNOWLEDGMENTS

We want to thank James Pritzker and the Tawani Foundation, Valerii Lukin and the Arctic and Antarctic Research Institute/Russian Antarctic Expedition (AARI/RAE), Mirella Kruger of Antarctic Logistics Centre International (ALCI), Marty Kress and the Von Braun Center for Science and Innovation, NASA, NSF, and the entire scientific and logistics team from Russia, Austria, and the United States that worked tirelessly to make possible the 2008 International Schirmacher Oasis/Lake Untersee, Antarctica Expeditions. We also want to acknowledge the contributions to this study by Prof. Gopi Podila of the University of Alabama in Huntsville and his students Melissa Guisler and Alisa Townsend. We are deeply grateful to Gregory A. Jerman and James E. Coston (NASA/MSFC) for Scanning Electron Microscopy support.

REFERENCES

Absil, M. C. P. and Y. van Scheppingen. 1996. Concentrations of selected heavy metals in Benthic diatoms and sediment in the Westerschelde *Estuary. Bull Environ Contam Toxicol* 56: 1008–1015.

Abyzov, S. S. 1993. Microorganisms in ancient ice. In E. I. Friedman (ed.), *Antarctic Microbiology*. Wiley-Liss, Inc., New York, pp. 265–296.

Abyzov, S. S., Duxbury, N. S., Bobin, N. E., Fukuchi, M., Hoover, R. B., Kanda, H., Mitskevich, I. N., Mulyukin, A. L., Naganuma, T., Poglazova, M. N., and M. V. Ivanov. 2006. Super-long anabiosis of ancient microorganisms in ice and terrestrial models for development of methods to search for life on Mars, Europa and other planetary bodies. *Adv Space Res Space Life Sci* 38: 1191–1197.

Abyzov, S. S., Mitskevich, I. N., Poglazova, M. N., Barkov, N. I., Lipenkov, V. Ya., Bobin, N. E., Koudryashov, B. B., Pashkevich, V. M., and M. V. Ivanov. 2001. Microflora in the basal strata at Antarctic ice core above the Vostok lake. *Adv Space Res* 28: 701–706.

Abyzov, S. S., Poglazova, M. N., Mitskevich, I. N., and M. V. Ivanov. 2005. Common features of microorganisms in ancient layers of the Antarctic Ice Sheet. In J. D. Castello and S. O. Rogers (eds.), *Life in Ancient Ice*. Princeton University Press, Princeton, NJ, pp. 240–250.

Adam, C. and J. Garnier-Laplace. 2003. Bioaccumulation of silver-110m, cobalt-60, cesium-137, and manganese-54 by the freshwater algae *Scenedesmus obliquus* and *Cyclotella meneghiana* and by suspended matter collected during a summer bloom event. *Limnol Oceanogr* 48: 2303–2313.

Aislabie, J. M., Balks, M. R., Foght, J. M., and E. J. Waterhouse. 2004. Hydrocarbon spills on Antarctic soils: Effects and management. *Environ Sci Technol* 38: 1265–1274.

Aislabie, J., Foght, J., and D. Saul. 2000. Aromatic hydrocarbon-degrading bacteria from soil near Scott Base, Antarctica. *Polar Biol* 23: 183–188.

Aislabie, J., Saul, D. J., and J. M. Foght. 2006. Bioremediation of hydrocarbon-contaminated polar soils. *Extremophiles* 10: 171–179.

Alam, S. I., Dixit, A., Reddy, G. S. N., Dube, S., Palit, M., Shivaji, S., and L. Singh. 2006. *Clostridium schirmacherense* sp. nov., an obligately anaerobic, proteolytic, psychrophilic bacterium isolated from lake sediment of Schirmacher Oasis, Antarctica. *Int J Syst Evol Microbiol* 56: 715–717.

Antić, M. P., Jovancicevic, B., Vrvić, M. M., and J. Schwarzbauer. 2006. Petroleum pollutant degradation by surface water microorganisms. *Environ Sci Pollut Res* 13: 287–360.

Bakermans, C., Ayala-del-Río, H. L., Ponder, M. A., Vishnivetskaya, Gilichinsky, D., Thomashow, M. F., and J. M. Tiedje. 2006. *Psychrobacter cryohalolentis* sp. nov. and *Psychrobacter arcticus* sp. nov., isolated from Siberian permafrost. *Int J Syst Evol Microbiol* 56: 1285–1291.

Bargagli, R. 2005. Antarctic ecosystems environmental contamination, climate change and human impact. *Ecological Studies vol. 175*. Springer, Heidelberg, Germany, p. 60.

Bell, R., Studinger, M., Tikku, A., and J. D. Castello. 2005. Comparative biological analyses of accretion ice from subglacial Lake Vostok. In J. D. Castello and S. O. Rogers (eds.), *Life in Ancient Ice*. Princeton University Press, Princeton, NJ, pp. 251–267.

Bej, A. K., Saul, D., and Aislabie, J. 2000. Cold-tolerant alkane-degrading *Rhodococcus* species from Antarctica. *Polar Biol* 23: 100–105.

Bowman, J. P., Brown, M. V., and D. S. Nichols. 1997. Biodiversity and ecophysiology of bacteria associated with Antarctic sea ice. *Antarct Sci* 9: 134–142.

Bowman, J. P., Rea, S. M., McCammon, S. A., and T. A. McMeekin. 2001. Diversity and community structure within anoxic sediment from marine salinity meromictic lakes and a coastal meromictic marine basin, Vestfold Hills, Eastern Antarctic. *Environ Microbiol* 2: 227–237.

Brambilla, E., Hippe, H., Hagelstein, A., Tindall, B. J., and E. Stackebrandt. 2001. 16S rDNA diversity of cultured and uncultured prokaryotes of a mat sample from Lake Fryxell, McMurdo Dry Valleys, Antarctica. *Extremophiles* 5: 22–33.

Bräuer, S. L., Cadillo-Quiroz, H., Yashiro, E., Yavitt, J. B., and S. H. Zinder. 2006. Isolation of a novel acidiphilic methanogen from an acidic peat bog. *Nature* 442: 192–194.

Broady, P. A. 1981. The ecology of sublithic terrestrial aglae at the Vestfold Hills, Antarctica. *Br Phycol J* 16: 231–240.

Broady, P. A. 1982. Ecology of non-marine algae at Mawson Rock, Antarctica. *Nova Hedwigia* 36: 209–229.

Bücker, M., Glatt, H. R., Platt, K. L., Avnir, D., Ittah, Y., Blum, J., and F. Oesch. 1979. Mutagenicity of phenanthrene and phenanthrene K-region derivatives. *Mutat Res* 66: 337–348.

Bulat, S. A., Alekhina, I. A., Blot, M., Petit, J. R., Waggenbach, D., Lipenkov, V. Y., Raynaud, D., and V. V. Lukin. 2004. Thermophiles microbe signatures in Lake Vostok, Antarctica. *Eos Trans* 83: B021–A09.

Burckle, L. H., Gayley, R. I., Ram, M., and J. R. Petit, 1988. Diatoms in Antarctic ice cores: Some implications for the glacial history of Antarctica. *Geology* 164: 326–329.

Callegan, R. P., Nobre, M. F., McTernan, P. M., Battista, J. R., Navarro-González, R., McKay, C. P., da Costa, M. S., and F. A. Rainey. 2008. Description of four novel psychrophilic, ionizing radiation-sensitive *Deinococcus* species from alpine environments. *Int J Syst Evol Microbiol* 58: 1252–1258.

Castracane, A. F. A. 1886. Report on the Diatomaceae collected by the *H. M. S. Challenger* during the years 1873–76. In *Reports of the Scientific Voyage of H. M. S. Challenger-Botany* 2, pp. 1–178.

Cavanaugh, J. E., Nichols, P. D., Franzmann, P. D., and T. A. McMeekin. 1998. Hydrocarbon degradation by Antarctic coastal bacteria. *Antarct Sci* 10: 386–397.

Cerniglia, C. E., Gibson, D. T., and C. Van Baalen. 1979. Algal oxidation of aromatic hydrocarbons: Formation of 1-naphthol from naphthalene by *Agmenellum quadruplicatum*, strain PR-6. *Biochem Biophys Res Commun* 88: 50–58.

Chong, S., Liu, Y., Cummins, M., Valentine, D., and D. Boone. 2002. *Methanogenium marinum* sp. nov., a H_2-using methanogen from Skan Bay, Alaska, and kinetics of H_2 utilization. *Antonie van Leeuwenhoek* 81: 263–270.

Christner, B. C. 2002. Detection, Recovery, Isolation and Characterization of Bacteria in Glacial Ice and Lake Vostok Accretion Ice. PhD dissertation. Ohio State University Print, Columbus, OH, pp. 200.

Conrad, R., Bak, F., Seitz, H. J., Thebrath, B., Mayer, H. P., and H. Schulz. 1989. Hydrogen turnover by psychrophilic homoacetogenic and mesophilic methanogenic bacteria in anoxic paddy soil and lake sediment. *FEMS Microbiol Ecol* 62: 285–294.

Darling, C. A. and P. A. Siple. 1941. Bacteria of Antarctica. *J Bacteriol.* 42: 83–98.

Dedysh, S. N., Berestovskaya, Y. Y., Vasylieva, L. V., Belova, S. E., Khmelenina, V. N., Suzina, N. E., Trotsenko, Y. A., Liesack, W., and G. A. Zavarzin. 2004. *Methylocella tundrae* sp. nov., a methanotrophic bacterium from acidic tundra peatlands. *Int J Syst Evol Microbiol* 54: 151–156.

Dedysh, S. N., Liesack, W., Khmelenina, V. N., Suzina, N. E., Trotsenko, Y. A., Semrau, J. D., Bares, A. M., Panikov, N. S., and J. M. Tiedje. 2000. *Methylocella palustris* gen. nov., sp. nov., a new methane-oxidizing acidophilic bacterium from peat bogs, representing a novel subtype of serine-pathway methanotrophs. *Int J Syst Evol Microbiol* 50: 955–969.

D'Elia, T., Veerapaneni, R., and S. O. Rogers. 2008. Isolation of microbes from Lake Vostok Accretion Ice, *Appl Environ Microbiol* 74: 4962–4965.

Douglas, M. S. V. and J. P. Smol. 1999. Freshwater diatoms as indicators of environmental change in the High Arctic. In E. F. Stoermer and J. P. Smol (eds.), *The Diatoms: Applications for the Environmental and Earth Sciences*. Cambridge University Press, Cambridge, U.K., pp. 227–244.

Dunfield, P. F., Khmelenina, V. N., Suzina, N. E., Trotsenko, Y. A., and S. N. Dedysh, 2003. *Methylocella silvestris* sp. nov., a novel methanotroph isolated from an acidic forest cambisol. *Int J Syst Evol Microbiol* 53: 1231–1239.

Duval, B., Shetty, K., and W. H. Thomas. 2000. Phenolic compounds antioxidant properties in the snow alga *Chlamydomonas nivalis* after exposure to UV light. *J Appl Phycol* 11: 559–566.

Eichler, B. and B. Schink. 1984. Oxydation of primary aliphatic alcohols by *Acetobacterium carbinolicum* sp. nov., a homoacetogenic anaerobe. *Arch Microbiol* 140: 147–152.

Elster, J. 1999. Algal versatility in various extreme environments. In J. Seckbach (ed.), *Enigmatic Microorganisms and Life in Extreme Environments*. Kluwer Academic Publishers, Dordrecht, the Netherlands.

Elster, J. 2002. Ecological classification of terrestrial algae communities of polar environment. In L. Beyer and M. Bölter (eds.), *GeoEcology of Terrestrial Oases*, *Ecological Studies* 154. Springer-Verlag, Heidelberg, Germany, pp. 303–326.

Foster, J. 1887. Über eihnigh Eigenschafter leuchtender Bakterien. *Centr Bakteriol Rev Parasitenk* 2: 337–340.

Franzmann, P. D., Deprez, P. P., Burton, H. R., and J. van den Hoff. 1987. Limnology of Organic Lake, Antarctica, a meromictic lake that contains high concentrations of dimethyl sulfide. *Aust J Mar Freshwater Res* 38: 409–417.

Franzmann, P. D., Höpfl, P., Weiss, N., and B. J. Tindall. 1991. Psychrotrophic, lactic acid-producing bacteria from anoxic waters in Ace Lake, Antarctica; *Carnobacterium funditum* sp. nov. and *Carnobacterium alterfunditum* sp. nov. *Arch Microbiol* 156: 255–262.

Franzmann, P. D., Liu, Y., Balkwill, D. L., Aldrich, H. C., Conway de Macario, E., and D. R. Boone. 1997. *Methanogenium frigidum* sp. nov., a psychrophilic, H2-using methanogen from Ace Lake, Antarctica. *Int J Syst Bacteriol* 47: 1068–1072.

Gaidos, E., Marteinsson, V., Thorsteinsson, T., Johanneson, T., Runarsson, A. R., Stefanson, A., Glazer, B., Lanoil, B., Skidmore, M., Han, S., Miller, M., Rusch, A., and W. Foo. 2008. An oligarchic microbial assemblage in the anoxic bottom waters of a volcanic subglacial lake. *ISME J* 1–12. www.nature.com/ismej.

Geittel, A., Mumann, M., Sass, H., Cypionka, H., and M. Konneke. 2008. Identity and abundance of active sulfate-reducing bacteria in deep tidal flat sediments determined by directed cultivation and CARD-FISH analysis. *Environ Microbiol* 10: 2645–2658.

Glatz, R. E., Lepp, P. W., Ward, B. B., and C. A. Francis. 2006. Microbial diversity in the water column of permanently ice-covered Lake Bonney, Antarctica. *Geobiology* 4: 53–67.

Gounot, A.-M. 1991. Bacterial life at low temperature: Physiological aspects and biotechnological implications. *J Appl Bacteriol* 71: 386–397.

Gupta, R. K., Pandey, V. D., Pandey, K. D., and D. Vyas, 2006. Algal species diversity of streams of Schirmacher Oasis, Antarctica. In R. K. Gupta, M. Kumar, and G. S. Paliwal (eds.), *Glimpses of Cyanobacteria*. Daya Publishing House, Delhi, pp. 231–241.

Hamilton, T. D., Craig, J. L., and P. V. Sellman. 1988. The Fox permafrost tunnel: A late quaternary geologic record in central Alaska. *Geol Soc Am Bull* 100: 948–969.

Hoham, R.W. 1975. Optimum temperatures and range for the growth of snow alga. *Arctic Antarct Alpine Res* 7: 13–24.

Hoham, R. W. and H. U. Ling. 2000. Snow algae. The effects of chemical and physical factors on their life cycles and populations. In J. Seckbach (ed.), *Journey to Diverse Microbial Worlds*. Kluwer Academic Publishers, Dordrecht, pp. 131–145.

Hoover, R. B. and D. Gilichinsky. 2001. Significance to Astrobiology of Microorganisms in Permafrost and Ice. In R. Paepe (ed.), *Permafrost Response on Economic Development, Environmental Security and Natural Resource Potential*, NATO-ARW held in Novosibirsk, Siberia, November 12–16, 1998. Kluwer Publishing, New York, pp. 553–580.

Hoover, R. B. and E. V. Pikuta. 2004. Microorganisms on comets, Europa, and the polar ice caps of Mars. *SPIE*, 5163: 191–201.

Hoover, R. B., Pikuta, E. V., Townsend, A., Anthony, J., Guisler, M., McDaniel, J., Bej, A., and M. Storrie-Lombardi. 2008. Microbial extremophiles from the 2008 Schirmacher Oasis Expedition: Preliminary results. *Proc SPIE* 7097: 70970L.

Hughes, K. A. and Bridge, P. 2009. Tolerance of Antarctic soil fungi to hydrocarbons and their potential role in soil bioremediation. In A. K. Bej, J. Aislabie, and R. M. Atlas (eds.), *Polar Microbiology: The Ecology, Biodiversity and Bioremediation Potential of Microorganisms in Extremely Cold Environments*. Taylor & Francis Group, Boca Raton, FL, pp. 279–302.

Huston, A. L. 2007. Biotechnological aspects of cold adapted enzymes. In R. Margesin, F. Schinner, J.-C. Marx, and E. Gerday (eds.), *Psychrophiles: From Biodiversity to Biotechnology*. Springer, Heidelberg, Germany, pp. 347–364.

Juck, D., Charles, T., Whyte, L. G., and C. W. Greer. 2000. Polyphasic microbial community analysis of peteroleum hydrocarbon-contaminated soils from two northern Canadian communities. *FEMS Microbiol Ecol* 33: 241–249.

Karl, D. M., Bird, D. F., Björkman, K., Houlihan, T., Shackelford, R., and L. Tupas. 1999. Microorganisms in the Accreted Ice of Lake Vostok, Antarctica. *Science* 286: 2144–2147.

Karr, E. A., Ng, J. M., Belchik, S. M., Sattley, W. M., Madigan, M. T., and L. A. Achenbach. 2006. Biodiversity of methanogenic and other *Archaea* in the permanently frozen Lake Fryxell, Antarctica. *Appl Environ Microbiol* 72: 1663–1666.

Karr, E. A., Sattley, W. M., Rice, M. R., Jung, D. O., Madigan, M. T., and L. Achenbach. 2005. Diversity and distribution of sulphate-reducing bacteria in permanently frozen Lake Fryxell, McMurdo Dry Valleys, Antarctica. *Appl Environ Microbiol* 71: 6353–6359.

Kellogg, T. B. and D. E. Kellogg, 2002. Non-marine diatoms from the Antarctic and Subantarctic regions, distribution and updated taxonomy. In A. Witkowski (ed.), *Diatom Monographs*. Ganter Verlag, Ruggell.

Knoblauch, C., Sahm, K., and B. B. Jørgensen. 1999. Psychrophilic sulfate-reducing bacteria isolated from permanently cold Arctic marine sediments: Description of *Desulfofrigus oceanense* gen. nov., sp. nov., *Desulfofrigus fragile* sp. nov., *Desulfofaba gelida* gen. nov., sp. nov., *Desulfotalea psychrophila* gen. nov., sp. nov. and *Desulfotalea arctica* sp. nov. *Int J Syst Bacteriol* 49: 1631–1643.

Knudsen, B. M., Jønch-Sørensen, H., Eriksen, P., Johnsen, B. J., and G. E. Bodeker. 2005. UV radiation below an Arctic vortex with severe ozone depletion. *Atmos Chem Phys* 5: 2981–2987.

Kotsyurbenko, O. R., Glagolev, M. V., Nozhevnikova, A. N., and R. Conrad. 2001. Competition between homoacetogenic and methanogenic archaea for hydrogen at low temperature. *FEMS Microbiol Ecol* 38: 153–159.

Kotsyurbenko, O. R., Simankova, M. V., Nozhevnikova, A. N., Zhilina, T. N., Bolotina, N. P., Lysenko, A. M., and G. A. Osipov. 1995. New species of psychrophilic acetogens: *Acetobacterium bakii* sp. nov., *A. paludosum* sp. nov., *A. fimetarium* sp. nov. *Arch Microbiol* 163: 29–34.

Long, A. and T. L. Pewe. 1996, Radiocarbon dating by high sensitivity liquid scintillation counting of wood from the Fox Permafrost Tunnel near Fairbanks, Alaska. *Permafrost Periglacial Process* 7: 281–285.

McClean, A. L. 1918. Bacteria of the Ice and Snow in Antarctica. *Nature* 102: 35–39.

McKenzie, R. L., Bjorn, L. O., Bais, A., and M. Iayis. 2003. Changes in biologically active ultraviolet radiation reaching the Earth's Surface. *Photochem Photobiol Sci* 2: 5–15.

Milucki, J. A. and J. C. Priscu, 2007. Bacterial diversity associated with blood falls, a subglacial outflow from the Taylor Glacier, Antarctica. *Appl Environ Microbiol* 73: 4029–4039. http://www.pubmedcentral.nih.gov/articlerender.fcgi?artid=1932727

Mock, T. and K. Junge, 2007. Psychrophilic diatoms: Mechanisms for survival in freeze-thaw cycles. In J. Seckbach (ed.), *Algae and Cyanobacteria in Extreme Environments*. Springer, Dordrecht, the Netherlands, pp. 345–364.

Montes-Hugo, M., Doney, S. C., Ducklow, H. W., Fraser, W., Martinson, D., Stammerjohn, S. E., and O. Schofield. 2009. Recent changes in phytoplankton communities associated with rapid regional climate changes along the Western Antarctic Peninsula. *Science* 323: 1470–1473.

Morita, R. Y. 1975. Psychrophilic bacteria. *Bacteriol Rev* 39: 144–167.

Morozova, D., Möhlmann, D., and D. Wagner. 2006. Survival of methanogenic archaea from Siberian permafrost under simulated Martian thermal conditions. *Origins Life Evol Biospheres* 37: 189–200.

Mountfort, D. O., Rainey, F. A., Burghardt, J., Kaspar, H. F., and E. Stackebrandt. 1997. *Clostridium vincentii* sp. nov., a new obligately anaerobic, saccharolytic, psychrophilic bacterium isolated from low-salinity pond sediment of the McMurdo Ice Shelf, Antarctica. *Arch Microbiol* 167: 54–60.

Mountfort, D. O., Rainey, F. A., Burghardt, J., Kaspar, H. F., and E. Stackebrandt, 1998. *Psychromonas antarcticus* gen. nov., sp. nov., a new aerotolerant anaerobic, halophilic psychrophile isolated from pond sediment of the McMurdo Ice Shelf, Antarctica. *Arch Microbiol* 169: 231–238.

Nadeau, T.-L. and R. W. Castenholz. 2000. Characterization of psychrophilic oscillatorians (Cyanobacteria) from Antarctic Meltwater Ponds. *J Phycol* 36: 914–923.

Narro, M. L., Cerniglia, C. E., Van Baalen, C., and D. T. Gibson. 1992. Metabolism of Pnenanthrene by the Marine Cyanobacterium *Agmenellum quadruplicatum* PR-6. *Appl Environ Biol* 58: 1351–1359.

Nozhevnikova, A. N., Simankova, M. V., Parshina, S. N., and O.R. Kotsyurbenko. 2001. Temperature characteristics of methanogenic archaea and acetogenic bacteria isolated from cold environments. *Water Sci Technol* 44: 41–48.

Nozhevnikova, A. N., Zepp, K., Vazquez, F., Zehnder, A. J. B., and C. Holliger. 2003. Evidence for the existence of psychrophilic methanogenic communities in anoxic sediments of deep Lakes. *Appl Environ Microbiol* 69: 1832–1835.

Okamoto, T., Fujioka, K., and T. Naganuma, 2001. Phylogenetic similarity of aerobic gram-negative halophilic bacteria from a deep-sea hydrothermal mound and Antarctic habitats. *Polar Biosci* 14: 1–9.

Paarup, M., Friedrich, M. W., Tindall, B. J., and K. Finster. 2005. Characterization of the psychrotolerant acetogen strain SyrA5 and the emended description of the species *Acetobacterium carbinolicum. Antonie van Leeuvenhoek* 89: 55–69.

Paniker, G., Aislabie, J., and A. K. Bej. 2006. Analysis of aggregative behavior of *Pseudomonas* sp. 30-3 isolated from Antarctic Soil. *Soil Biol Biochem* 38: 3152–3157.

Paniker, G., Aislabie, J., Saul, D., and A. K. Bej. 2002. Cold tolerance of *Pseudomonas* sp. 30-3 from oil-contaminated soil, Antarctica. *Polar Biol* 25: 5–11.

Perreault, N. N., Greer, C. W., Andersen, D. T., Tille, S., Lacrampe-Couloume, G., Lollar, B. S., and L. G. Whyte. 2008. Heterotrophic and autotrophic microbial populations in cold perennial springs of the high Arctic. *Appl Environ Microbiol* 74: 6898–6907.

Pikuta, E. V. and R. B. Hoover. 2004. Astrobiological significance of chemo-lithoautotrophic acidophiles. *Proc SPIE* 5163: 179–190.

Pikuta, E. V., Hoover, R. B., Bej, A. K., Marsic, D., Whitman, W. B., Krader, P. E., and J. Tang. 2006. *Trichococcus patagoniensis* sp. nov., a facultative anaerobe that grows at –5°C, isolated from penguin guano in Chilean Patagonia, *Int J Syst Evol Microbiol* 56: 2055–2062.

Pikuta, E. V., Hoover, R. B., Marsic, D., Whitman, W. B., Lupa, B., Krader, P. E., and J. Tang. 2009. *Proteocatella sphenisci* gen. nov., sp. nov., a novel psychrotolerant, spore-forming anaerobe isolated from Magellanic penguin guano in Patagonia, Chile. *Int J Syst Evol Microbiol* 59: 2302–2307.

Pikuta, E. V., Hoover, R. B., and J. Tang. 2007. Microbial extremophiles at the limits of life, *Crit Rev Microbiol* 33: 183–209.

Pikuta, E. V., Marsic, D., Bej, A. K., Tang, J., Krader, P., and R. B. Hoover. 2005. *Carnobacterium pleistocenium* sp. nov., a novel psychrotolerant, facultative anaerobe isolated from permafrost of the Fox Tunnel in Alaska. *Int J Syst Evol Microbiol* 55: 473–479.

Price, R. B. 2000. A habitat for psychrophiles in deep Antarctic ice. *PNAS* 37: 1247–1251.

Reddy, G. S. N., Prakash, J. S. S., Srinivas, R., Matsumoto, G. I., and S. Shivaji. 2003. *Leifsonia rubra* sp. nov. and *Leifsonia aurea* sp. nov., psychrophiles from a pond in Antarctica. *Int J Syst Evol Microbiol* 53: 977–984.

Reinsch, P. F. 1890. Die Süsswasseralgenflora von Süd-Georgien. In G. Neumeyer (ed.), *Die Deutschen Expeditionen und ihre Ergebnisse*, vols. 1882–01883, pp. 329–365.

Roberts, N. J., Burton, H. R., and G. A. Pitson. 1993. Volatile organic compounds from Organic Lake, an Antarctic hypersaline, meromictic lake. *Antarct Sci* 5: 361–366.

Romanenko, L. A., Schumann, P., Rohde, M., Lysenko, A. M., Mikhailov, V., and E. Stackebrandt. 2002. *Psychrobacter submarinus* sp. nov. and *Psychrobacter marincola* sp. nov., psychrophilic halophiles from marine environments. *Int J Syst Evol Microbiol* 52: 1291–1297.

Ryan, G., Priscu, J., and C. Takacs-Veshbach. 2007. Extremophilic bacteria from Antarctic Lake Fryxell: Evidence for phenotypic convergence. Montana State University, Bozeman, MT. http://mcm-dvlakesmo.montana.edu/images/Data/posters/ASLO_RYAN_poster.pdf.

Sattley, W. M. and M. T. Madigan. 2007. Cold-active acetogenic bacteria from surficial sediments of perennially ice-covered Lake Fryxell, Antarctica. *FEMS Microbiol Lett* 272: 48–54.

Schmidt, M., Priemé, A., and P. Stougaard. 2006a. *Rhodonellum psychrophillum* gen. nov., sp. nov., a novel psychrophilic and alkaliphilic bacterium of the phylum *Bacteroidetes* isolated from Greenland. *Int J Syst Evol Microbiol* 56: 2887–2892.

Schmidt, M., Priemé, A., and P. Stougaard. 2006b. Bacterial diversity in permanently cold and alkaline ikaite columns from Greenland. *Extremophiles* 10: 551–562.

Schmidt, M., Priemé, A., and P. Stougaard. 2007. *Arsukibacterium ikkense* gen. nov., sp. nov., a novel alkaliphilic, enzyme-producing γ-Proteobacterium isolated from a cold and alkaline environment in Greenland. *Syst Appl Microbiol* 30: 197–201.

Seckbach, J. and A. Oren. 2007. Oxygenic photosynthetic microorganisms in extreme environments: Possibilities and limitations. In J. Seckbach (ed.), *Algae and Cyanobacteria in Extreme Environments*. Springer, Dordrecht, the Netherlands, pp. 5–25.

Sellman, P. V. 1967. Geology of the USA CRREL Permafrost Tunnel, Fairbanks, Alaska. U.S. Army CRREL Technical Report 199. Hanover, NH, pp. 1–22.

Siegert, M. J., Ellis-Evans, J. C., Tranter, M., Mayer, C., Petit J., Salamatin, A., and J. C. Priscu. 2001. Physical, chemical, and biological processes in Lake Vostok and other Antarctic subglacial lakes. *Nature* 414: 603–609.

Simankova, M. V., Kotsyurbenko, O. R., Stackebrandt, E., Kostrikina, N. A., Lysenko, A. M., Osipov, G. A., and A. N. Nozhevnikova. 2000. *Acetobacterium tundrae* sp. nov., a new psychrophilic acetogenic bacterium from tundra soil. *Arch Microbiol* 174: 440–447.

Simankova, M. V., Parshina, S. N., Tourova, T. P., Kolganova, T. V., Zehnder, A. J. B., and A. N. Nozhevnikova. 2004. *Methanosarcina lacustris* sp. nov., a new psychrotolerant methanogenic Archaeon from Anoxic Lake Sediments. *Syst Appl Microbiol* 24: 362–367.

Singh, S. M. and J. Elster. 2007. Cyanobacteria in Antarctic Lake Environments. In J. Seckbach (ed.), *Algae and Cyanobacteria in Extreme Environments*. Springer, Dordrecht, the Netherlands, pp. 305–320.

Singh, N., Kendall, M. M., Liu, Y., and D. R. Boone. 2005. Isolation and characterization of methylotrophic methanogens from anoxic marine sediments in Skan Bay, Alaska, description of *Methanococcoides alaskense* sp. nov., and emendation of *Methanosarcina baltica*. *Int J Syst Evol Microbiol* 55: 2531–2538.

Smol, J. P. and M. S. V. Douglas, 1996. Long term environmental monitoring in Arctic Lakes and ponds using diatoms and other biological indicators. *Geosci Can* 23, 225–230.

Spaulding S. A. and D. M. McKnight. 1999. Diatoms as indicators of environmental change in Antarctic freshwaters. In J. P. Smol and E. F. Stoermer (eds.), *The Diatoms: Applications for the Environmental and Earth Sciences*. Cambridge University Press, Cambridge, U.K., pp. 249–263.

Staley, J. T. and J. J. Gosink. 1999. Poles apart: Biodiversity and biogeography of sea ice bacteria. *Ann Rev Microbiol* 53: 189–215.

Straka, R. P. and J. L. Stokes 1960. Psychrophilic Bacteria from Antarctica. *J Bacteriol* 80: 622–625.

Taton, A., Grubisic, S., Brambilla, E., De Wit, R., and A. Wilmotte. 2003. Cyanobacterial diversity in natural and artificial microbial mats of Lake Fryxell (McMurdo Dry Valleys, Antarctica): A morphological and molecular approach. *Appl Environ Microbiol* 69: 5157–5159.

Trotsenko, Y. A. and V. N. Khmelenina. 2005. Aerobic methanotrophic bacteria of cold ecosystems. *FEMS Microbiol Ecol* 53: 15–26.

Van Heurck, H. 1909. Diatomées. In *Expédition Antarctic Belge, Résultats du Voyage du S. Y. Belgica en 1897–1899. Botanique*, 6: 1–126. Antwerp: Buschmamn.

Vincent, W. F. 2000. Cyanobacterial dominance in the polar regions. In B.A. Whitton and M. Potts (eds.), *Ecology of Cyanobacteria: Their Diversity in Time and Space*. Kluwer Academic Publishers, Dordrecht, the Netherlands, pp. 321–340.

Voytek, M. A., Ward, B. B., and J. C. Priscu. 1998. The abundance of ammonia-oxidizing bacteria in Lake Bonney, Antarctica determined by immunofluorescence, PCR and in situ hybridization. Antarctic Research Series. *The McMurdo Dry Valleys*, pp. 217–228.

Wand, U. and J. Perlt. 1999. Glacial boulders "floating" on the ice cover of Lake Untersee, East Antarctica. *Antarct Sci* 11: 256–260.

Wand, U., Samarkin, V. A., Nitzsche, H.-M., and H.-W. Hubberten. 2006. Biogeochemistry of methane in the permanently ice-covered Lake Untersee, central Dronning Maud Land, East Antarctica. *Limnol Oceanogr* 51: 1180–1194.

Wand, U., Schwarz, G., Bruggemann, E., and K. Brauer. 1997. Evidence for physical and chemical stratification in Lake Untersee (central Dronning Maud Land, East Antarctica). *Antarct Sci* 9: 43–45.

Ward, B. B. and J. C. Priscu. 1997. Detection and characterization of denitrifying bacteria in an ice-covered Antarctic Lake. *Hydrobiologia* 347: 57–68.

Ward, B. B., Granger, J., Maldonado, M. T., Casciotti, K. L., Harris, S., and M. L. Wells. 2005. Denitrification in the hypolimnion of permanently ice-covered Lake Bonney, Antarctica. *Aquat Microb Ecol* 52: 197–205.

Ward, B. B., Granger, J., Maldonado, M. T., and M. L. Wells. 2003. What limits bacterial production in the suboxic region of permanently ice-covered Lake Bonney, Antarctica. *Aquat Microb Ecol* 31: 33–47.

Wynn-Williams, D. W. 1990. Ecological aspects of Antarctic microbiology. *Adv Microb Ecol* 11: 71–146.

Yang, S., Wu, R. S., and R. Y. C. Kong. 2002. Biodegradation and enzymatic responses in the marine diatom *Skeletonema costatum* upon exposure to 2,4-dichlorophenol. *Aquat Toxicol* 59: 191–200.

Zhang, G., Jiang, N., Liu, X., and X. Dong. 2008. Methanogenesis from methanol at low temperatures by a novel psychrophilic methanogen "*Methanolobus psychrophilus*" sp. nov., prevalent in Zoige wetland of the Tibetan Plateau. *Appl Environ Microbiol* 74: 6114–6120.

6 Cold Adaptation in Antarctic Biodegradative Microorganisms

Asim K. Bej and Nazia Mojib

CONTENTS

6.1 INTRODUCTION

The icy continent of the Southern hemisphere on this planet consists most part frozen water and ~0.32% ice-free Dry Valleys (Ugolini and Bockheim 2008). The explorative nature of humans had encouraged early explorers to sail to this continent simply by steam-powered ships and wind. Since then there has been a steady increase in the human presence on this continent, as the means of transportation by sea and air improved and the capability of carrying larger quantity of supplies increased. Historically Australia, France, Argentina, Chile, New Zealand, United Kingdom, and Norway have territorial claims; however, currently over 30 countries have signed the Antarctic treaty and maintain research stations supporting over 4000 research and logistics personnel during polar summer months and 1000 support personnel during polar winter period. Although the United States has the largest operation, they never claimed any territory on the Antarctic continent. Besides the scientific activities, the International Association of Antarctic Tour Operators (http://www.iaato.org) reported that in 2007–2008 season over 33,000 tourists landed on this continent either by sea or by air. According to their estimate, the tourism in Antarctica has increased almost three times since 2000–2001. To maintain the scientific stations supporting the effective science missions and a steady increase in the tourism, it is necessary to transport and store adequate volume of fossil fuels (jet fuels, diesel fuel, kerosene, and other petroleum hydrocarbon-based fuels) to this continent. Moreover transportation within the continent requires the use of either land vehicles or fixed- or rotary-winged airplane. Storage and the use of the fuel for transportation and maintaining the permanent and temporary research stations and tourism would inevitably lead to small or large quantities of fuel spills. One may argue that the Antarctic continent is stretched about 14.4 million km^2 (5.4 million sq mi) and relatively a very small fraction of this vast ice-covered continent has been explored and traversed by humans. Moreover, the Antarctic treaty of environmental protection mandates that all visitors to the continent avoid oil spill or at least minimize such events. Although these arguments have some merit, the steady increase in the establishment of additional research stations, their extended activities at remote locations, and more importantly an unprecedented surge of tourism during the last decade have led to more events of oil spills and contamination of the pristine environment affecting the ecosystems and food web starting from the microorganisms to mammals, who may not have previously experienced the toxic effects of the petroleum hydrocarbon and other fossil fuels. For example, in 1989, an Argentina supply ship with tourists on board spilled over 158,000 gallons of fuel near the U.S. Palmer station after running into an underwater reef. Although this event did not cause major impact to the natural ecosystems in Antarctic continent, a serious concern has been raised that such accidents could affect the regional food web including the birds, mammals, crustacean and other invertebrates, macroalgae, lichens, and numerous other microscopic organisms. A containment and cleanup policy for small or large oil spills due to the research activities, maintenance of the scientific stations, and tourism is an obvious issue that has been discussed (Aislabie et al. 2004). Several biodegradative microbial species indigenous to Antarctica have been isolated, identified, and shown to possess biodegradative function (Aislabie et al. 2006, Leys et al. 2005). Therefore, the use of the natural bioremediation process inherent to these biodegradative microorganisms to the cleanup of the spilt oil has been proposed as a method for cleaning up such oil spills. Unlike a mesophilic environment, petroleum hydrocarbons behave differently in a cold temperature environment and when trapped on ice. Unfortunately, Antarctica's relatively low microbial biomass and the extreme conditions of polar winter months without sunlight and freezing conditions would predict relatively slow or ineffective biodegradative processes. Another obvious issue is the microbial adaptation to evolve rapidly and maintain optimum biodegradative activity in Antarctic cold and dry environmental conditions when confronted with novel hydrocarbon contaminants. Although nonbiodegradative microbial adaptation in cold temperature environment in Antarctic continent has been described, such adaptive mechanisms are yet to be clearly elucidated in biodegradative species isolated from this continent. So far only a few biodegradative microorganisms have been subjected to the study of the cold adaptation (Ayub et al. 2009, Baraniecki et al. 2002, Kawamoto et al. 2009, Panicker et al. 2002). The biodegradative Antarctic bacteria possess enzymes, which are active at low temperature. They help these organisms biodegrade fuels or petroleum hydrocarbons and obtain nutrient from surroundings; continue the vital processes of cellular function or life, i.e., the replication of DNA, the synthesis of RNA from DNA, and the synthesis of proteins; and maintain an optimum cell membrane fluidity. All of these capabilities are required for these microbial communities to remain viable and metabolically active at low temperatures. To sustain life in persistent cold environment, these microorganisms produce specific sets of proteins, which are regulated both at transcriptional and translational levels that may impart metabolic differences from their mesophilic relatives. However many of the microbial species isolated from the oil-contaminated soils in Antarctica are also found in mesophilic environments, and some of them have helped to advance the study of the mechanisms of the cold adaptation. At present it is useful to assume that the cold adaptive profile of Antarctic biodegradative microorganisms could be similar to the mesophilic microorganisms. The possible roles of cryoprotectants that are spilt near scientific bases in protecting the bacterial cells have been elaborated by Aislabie and Foght (Chapter 9). This chapter will discuss the physiological and genetic mechanisms of cold adaptations in Antarctic microorganisms and their mesophilic relatives that may well apply to the Antarctic biodegradative microorganisms. Four classes of proteins for coping with the cold temperatures have been described: (1) Csps or cold-shock proteins that are expressed immediately after downshift in temperature; (2) Caps or cold acclimation proteins that are expressed during prolonged growth at cold temperatures (acclimation); (3) AFP or antifreeze proteins that are expressed to avoid freezing at subzero temperatures; and (4) IBP or ice-binding proteins whose exact function is still unknown but may act as recrystallization inhibitors to protect membranes in the frozen state. Moreover, the maintenance of membrane fluidity in cold adaptive bacteria; role of pigments; and extracellular polymeric substance (EPS) secreted by Antarctic bacteria in cold adaptation will also be discussed.

6.2 COLD-SHOCK PROTEINS

Cold-shock proteins (CSPs) are small (~7 kDa), nucleic acid-binding proteins that are involved in various cellular processes, mainly transcription and protein folding. The functions of CSP have been studied extensively in mesophilic bacteria (e.g., *Escherichia coli, Bacillus subtilis*), but fewer studies been reported for microorganisms living in cold environment. Nevertheless, the functional analysis of CSP in mesophilic bacteria will help in the understanding of the role of CSP in cold-tolerant bacteria since a number of these mesophilic bacterial genera have been isolated from Antarctica. A well studied and one of the first described examples is the c*spA* family of genes in *E. coli*, which consists of nine homologues (*cspB* to *cspI*) to the major cold-shock protein CspA. CspA, CspB, CspG, and CspI are cold inducible, and CspC and CspE are expressed at both high (37°C) and low temperatures, while CspD is induced during stationary phase and upon nutrition starvation (Bae et al. 1999, Yamanaka and Inouye 1997, Yamanaka et al. 1994). CspA is expressed during times of temperature decrease from 37°C to 10°C. During the cold-shock response, CspA accounts for 10% or more of the total cellular protein. In *B. subtilis*, three proteins homologous to *E. coli* CspA (CspB, CspC, and CspD) are expressed upon temperature downshift (Ermolenko and Makhatadze 2002). In general, c*spA* from *E. coli* and other Gram-negative bacteria share an approximately 43% homology with the eukaryotic Y-box transcription factors and contain two RNA binding motifs, RNP-1 and RNP-2 (Wolffe et al. 1992). The CspA functions as RNA chaperone by minimizing the formation of secondary structures on mRNA, and allowing the efficient translation of proteins necessary for adaptation to cold temperatures and remain viable (Bae et al. 2000). The *cspA* promoter is constitutively expressed at 37°C, though its activity is increased following cold shock (Fang et al. 1997). Therefore, the regulation of *cspA* at cold temperatures occurs at the posttranscriptional level by stabilizing the *cspA* transcript. The deletion analysis of genetic components such as the AT-rich UP element, downstream box (DB), the cold-shock domain (CS), and the 5′-untranslated sequence (5′-UTR) helped to evaluate the role of upstream DNA segments necessary for the expression of CspA in *E. coli* cultures grown at 37°C, but stable expression at colder temperatures (Ross et al. 1998). CspA is a 70 amino acid residue protein with a molecular weight of 7.4 kDa. The protein consists of five antiparallel beta-sheets.

One of the earliest studies on the cold-shock proteins was done in Antarctic psychrotolerant Gram-positive *Arthrobacter protophormiae* and Gram-negative *Pseudomonas fluorescens* bacteria (Ray et al. 1994). The northern blot hybridization study showed a homologue of *cspA* of *E. coli* in these bacteria. The transcript size of the *cspA* homologues in these two Antarctic psychrotolerant bacteria was similar to *E. coli cspA*. These two *cspA* homologues were found to be expressed constitutively at low levels at 4°C and 22°C. In *P. fluorescens*, the *cspA* mRNA was found to be cold inducible, i.e., its expression was increased after a temperature downshift (22°C–4°C).

In 1992, there were reports of the occurrence of major cold-shock protein homologues in Arctic soil organisms. Thirteen common CSPs were found in all Arctic *Rhizobium*, and the 11.1 kDa protein was a major protein expressed after temperature downshift (Cloutier et al. 1992). The cold-adapted arctic rhizobia produced more CSPs under freezing conditions (–10°C) than temperate rhizobia. The expression of CSPs is also been reported in an Arctic psychrophiles *Aquaspirillum arcticum*. Fourteen cold-shock proteins were identified in this bacterium using O-Farrell two-dimensional gel electrophoresis and computing scanning laser densitometry following exposure to 10°C–0°C, 5°C–0°C, and 10°C–5°C (Roberts and Inniss 1992). However, the role of these CSPs in cold adaptation has not been determined. A family of dimeric major cold-shock protein homologs was identified and purified from the psychrotolerant *B. cereus* WSBC 10201 (Mayr et al. 1996). CspA of *B. cereus,* a small polypeptide of 7.5 kDa, was the most abundant of these cold-induced proteins. In addition, four small proteins similar in size to CspA were detected on O'Farrell two-dimensional protein gels. Further analysis with immunoblotting using *B. cereus* anti-CspA antibodies indicated that the five proteins mentioned above along with an additional sixth protein not visible on silver-stained two-dimensional gels are members of a *B. cereus* CSP family. Two low-molecular-mass proteins, designated C7.0 and C8.0 are highly expressed at 4°C–10°C in psychrotolerant bacterium, *P. fragi*. Immunoblot analysis exhibited that these proteins were similar to CspA of *E. coli* (Hebraud et al. 1994). A cold-shock protein CS7.4 similar to *E. coli* CspA and

A9 (CS7.4-like protein) was recognized in psychrotolerant bacterium *A. globiformis* SI55 (Berger et al. 1996). Cold-shock proteins were also identified in *Psychrobacter articus* 273-4 from 20,000 to 40,000 year old Siberian permafrost core (Zheng et al. 2006). Psychrotrophic and pathogenic *Yersinia* species carry two almost identical major cold-shock protein coding regions (*cspA1* and *cspA2*) located approximately 300 bp apart from one another (Neuhaus et al. 2000). Four small cold-induced, cold-shock bacterial proteins were detected from the Antarctic bacterium *Streptomyces* sp. AA8321 (Kim et al. 2007).The purified cold-shock protein $CspA_{St}$ from *Streptomyces* sp. AA8321 exists as a homodimer under physiological conditions and binds to single-stranded, but not double-stranded, DNA. Cold-shock proteins were also reported in psychrotolerant *Acinetobacter* sp. (Barbaro et al. 2002). The complete genome sequencing of marine Antarctic bacterium *Pseudoalteromonas haloplanktis* TAC125 revealed that it consists of nine paralogs of *cspA* gene (Médigue et al. 2005). A cold-shock protein A (CspA) involved in cold adaptation at near freezing temperature was identified in an Antarctic psychrophilic bacterium *Colwellia* sp. NJ341 with the help of two-dimensional gel electrophoresis (2-D) and matrix-assisted laser desorption/ionization mass spectrometry (MALDI-TOF MS) analysis (Wang et al. 2006). CSPs are also found in low-temperature-adapted Archaea from Antarctica. DEAD-box RNA helicase was identified in *Methanococcoides burtonii* (Antarctic methanogen) (Goodchild et al. 2004b, Lim et al. 2000, Saunders et al. 2003). Cold-shock proteins were also identified in *Methanogenium frigidum,* which was also isolated from Antarctica (Franzmann et al. 1997, Giaquinto et al. 2007).

6.3 ROLE OF COLD-SHOCK PROTEINS

The exact role of Csps in Antarctic bacteria is yet to be described. However, since the Csps share significant similarity with the Csps from mesophilic bacteria, the mechanism of action of this protein in *E. coli* or *B. subtilis* could be useful in elucidating the role of Csps in bacteria from cold environment. It can be predicted that in Antarctic biodegradative bacteria in which CspA expression has been identified, possibly it binds to the mRNA therefore preventing any secondary structures from forming, thus keeping mRNA unfolded permitting the continued expression of essential proteins. As a result, CspA in these bacteria has an effect on both the transcription and regulation of mRNA, thereby maintaining essential physiological functions at low temperatures in the Antarctic environment. Recent studies have been conducted to elucidate the specific functions of cold-shock protein A from *Streptomyces* sp. AA8321 ($CspA_{St}$). The gene encoding $cspA_{St}$ was cloned and expressed in *E. coli*. The overexpression of $CspA_{St}$ in *E. coli* severely impaired the ability of the host cells to form colonies, and the cells developed an elongated morphology. The incorporation of a deoxynucleoside analogue, 5-bromo-2′-deoxyuridine, into newly synthesized DNA was also drastically diminished in $CspA_{St}$-overexpressing cells. This result suggests that $CspA_{St}$ acts as DNA replication inhibitor during cold adaptation (Kim et al. 2007). Although how this helps Antarctic microorganisms in coping with the cold temperature environments has not been elucidated, it is possible that the transient expression of CspA plays an important role in the regulation of DNA replication, which is closely associated with the cell growth and division that is in concert with the reduced metabolism in the oligotrophic Antarctic environment. The cessation of cell division results in the elongation of the cells, thereby increasing the volume and the surface area for an efficient nutrient uptake, membrane transport of solutes, and respiration in a low nutrient environment such as Antarctica until more favorable conditions prevail. In contrast, it has been found that various C group antibiotics (chloramphenicol, tetracycline, erythromycin, spiramycin, and fusidic acid) mimic cold shock at high temperatures in *E. coli*, suggesting that ribosome could be the primary sensor of cold-shock response (Van Bogelen and Neidhardt 1990). However, the regulation of induction of CS7.4 expression by low levels of chloramphenicol was found to be at the transcriptional level (Jiang et al. 1993). Based on above findings and induction of expression of A9 (CS7.4-like protein) at 25°C by chloramphenicol and the existence of CS7.4-like proteins in thermophilic bacilli (Schröder et al. 1993), a more general role of CS7.4-like proteins as a chaperone rather than their sole implication in cold adaptation processes has been suggested by Berger et al. (1996). Though much study has not been done on the genetics and physiology of Archaeal responses to cold temperatures, the structure and function study of archaeal Csps from *M. frigidum* provides some information about the role of bacterial Csps in the cold environment. The complementation studies showed that archaeal Csp and CSD protein folding in Euryarchaeota (*M. frigidum* and *M. burtonii*) and Crenarchaeota (uncultured marine picoplankton) functions effectively to rescue a cold-sensitive growth defect in *E. coli* (Giaquinto et al. 2007). This illustrates that there is a high degree of functional similarity of archaeal Csp and CSD folding to their bacterial counterparts and may have evolutionary significance. The folded Csp and CSD fold proteins have been predicted to have a higher content of solvent-exposed basic residues, which are located on the nucleic acid binding surfaces, similar to their arrangement in *E. coli* CspA, indicating a role in the interaction of the protein with the nucleic acid. There is an electrostatic interaction between positive charges and the negatively charged nucleic acid leading to attraction, and the aromatic patches bind and melt nucleic acid secondary structure to facilitate transcription and translation at low temperatures (Phadtare et al. 2004). The purified Csp from *M. frigidum* is a single-domain protein that folds by a reversible two-state mechanism and exhibits a low conformational stability typical of cold-adapted proteins. Moreover, Csps proteins are encoded within a region of archaeal genomes that was previously described as a superoperon involved in RNA and protein processing, and predicted to contain RNA-binding fold, which further supports their role in RNA processing in *M. burtonii* (Goodchild et al. 2004a,b). A summary of the known CSPs isolated from microorganisms from cold environment has been elaborated in Table 6.1.

6.4 COLD ACCLIMATION PROTEINS

The cold acclimation proteins (Caps) seem to form the major component of cold-tolerant Antarctic bacteria. The Caps are expressed constitutively at low levels during prolonged or continuous growth at low temperatures. So far, the Caps have been identified only in cold-adapted bacteria isolated from cold environments. Twenty Caps have been identified in psychrotrophic bacterium, *P. fragi* (Hebraud et al. 1994),

Table 6.1
Description of the Known Cold-Shock, Cold Acclimation, Antifreeze, and Ice-Binding Proteins Isolated from Microorganisms from Cold Environments

Organism	Type/Origin/Source	CSP/CAP/AFP/IBP	Mechanism	Reference
A. protophormiae	Gram-positive Antarctic psychrotroph	CspA homologs	Cold adaptation; specific mechanism not described	Ray et al. (1994)
P. fluorescens	Gram-negative Antarctic psychrotrophs	CspA homologs	Cold adaptation; specific mechanism not described	Ray et al. (1994)
B. cereus WSBC 10201	Psychrotroph isolated from pasteurized milk	CspA of *Bacillus cereus* Other cold stress proteins	Cold adaptation; specific mechanism not described	Mayr et al. (1996)
P. fragi	Psychrotroph from minced beef	C7.0 C8.0 20 Caps	Cold adaptation; specific mechanism not described	Hebraud et al. (1994)
A. globiformis SI55	Psychrotroph	A9 (CS7.4-like protein) 18 Caps	Cold adaptation; specific mechanism not described	Berger et al. (1996)
Yersinia enterocolitica	Psychrotolerant pathogen	CspA1/A2	Cold adaptation; specific mechanism not described	Neuhaus et al. (2000)
M. frigidum	Antarctic Archaeon	Csp	Cold adaptation; specific mechanism not described	Giaquinto et al. (2007)
Colwellia sp. NJ341	Psychrotrophic Antarctic sea ice bacterium	CspA	Cold adaptation; specific mechanism not described	Wang et al. (2006)
Arctic Rhizobium	Arctic psychrotroph	Csps M.W-136.2–11.1	Cold adaptation; specific mechanism not described	Cloutier et al. (1992)
P. 273-4	Siberian permafrost soil psychrotroph	Two major Csp homologs	Cold adaptation; specific mechanism not described	Zheng et al. (2006)
Pseudoalteromonas haloplanktis TAC125	Antarctic marine psychrophile	Nine paralogs of *cspA*,	Cold adaptation; specific mechanism not described	Médigue et al. (2005)

(continued)

Table 6.1 (continued)
Description of the Known Cold-Shock, Cold Acclimation, Antifreeze, and Ice-Binding Proteins Isolated from Microorganisms from Cold Environments

Organism	Type/Origin/Source	CSP/CAP/AFP/IBP	Mechanism	Reference
Methanococcoides burtonii	Antarctic Archaeon	DEAD-box RNA Helicase	Helps in removing cold-stabilized secondary structures in mRNA	Lim et al. (2000)
Streptomyces sp. AA8321	Antarctic psychrotroph	$CspA_{St}$	Inhibition of DNA replication during cold adaptation	Kim et al. (2000)
A. arcticum	Arctic psychrophile	14 Csps and 8 caps	Cold adaptation	Roberts et al. (1992)
187 bacteria from α- and γ-Proteobacteria genera	Psychrotrophs and psychrophiles from Vestfold Hills and Larsemann Hills of Eastern Antarctica	AFP	Antifreeze activity	Gilbert et al. (2004)
P. putida GR12-2(Arctic Rhizobia)	Psychrotroph from soil sample from the Canadian High Arctic	AfpA (lipoglycoprotein)	Antifreeze and ice nucleation activities	Muryoi et al. (2004)
P. fluorescens KUAF-68	Antarctic marine psychrotroph	INP and AFP (uncharacterized)	Ice-nucleating protein and antifreeze protein activities	Kawahara et al. (2004)
Ma. protea	Antarctic bacterium	AFP(uncharacterized)	Antifreeze activity	United States Patent 6,887,984
Moraxella	Psychrotroph from Ross Island, Antarctica.	Antifreeze lipoprotein (AFLP)	Antifreeze activity	Yamashita et al. (2002)
Ma. primoryensis	Antarctic bacterium	AFP	Hyperactive Ca^{2+}-dependent antifreeze activity	Gilbert et al. (2005)
Flavobacterium xanthum IAM 12026	Antarctic bacterium	AFP(uncharacterized)	Antifreeze activity and recrystallization inhibition activity	Kawahara et al. (2007)
Colwellia, strain SLW05	Antarctic sea ice bacterium	IBP	Ice-binding activity and ice recrystallization inhibition activity	Raymond et al. (2007)
Bacterium 3519-10	Antarctic bacterium from Vostok ice core	IBP	Ice-binding activity	Raymond et al. (2008)

and 18 caps have been identified in psychrotrophic bacterium *A. globiformis* SI55 (Berger et al. 1996). Cold acclimation protein, CapB, was also identified and cloned from *Pseudomonas* sp. 30-3 (Panicker et al. 2002). The Caps were also identified in *A. arcticum* an Arctic psychrophiles along with Csps (Roberts and Inniss 1992). So far, only four *caps* have been sequenced and characterized: *capA* and *capB* from *P. fragi* (Michel et al. 1997), *capA* from *A. globiformis* (Berger et al. 1996, 1997), and the *capB* from *Pseudomonas* strain 30-3 (Panicker et al. 2002).

6.5 ROLE OF COLD ACCLIMATION PROTEINS

The Caps share 60%–70% homology with CspA amino acid residues and possess the conserved RNP-1 and RNP-2 ssDNA or RNA-binding motifs (Figure 6.1). This observation suggests that *cspA* and *capB* genes may have similar functions in bacterial cultures following an exposure to cold temperatures as described in the previous section. However, the regulatory sequences including the promoter, 5′-UTR, and UP element are dissimilar between these two genes (A. Bej, unpublished). The CspA expression in *E. coli* and other mesophilic bacteria upon cold shock seems transient, whereas in bacteria inhabiting perennially cold environments including Antarctica the sustained expression of CspA or an equivalent protein with the similar RNA destabilization function seems to be necessary so that the expression of essential proteins remain in place for active growth and metabolism. A summary of the known Caps isolated from microorganisms from cold environment has been elaborated in Table 6.1.

6.6 ANTIFREEZE AND ICE-BINDING PROTEINS

Antifreeze proteins (AFPs) either prevent ice nucleation or inhibit water molecules by attaching to tiny ice crystals so that they cannot form bigger crystals, which, once formed, may be detrimental to the living organisms during freezing conditions. The production of AFPs is one of the most important strategies to survive under subzero temperatures by the cold-tolerant Antarctic fishes, insects, and plants from freezing environments. Similarly, antifreeze proteins are also produced by many bacteria and function as one of their major survival strategies. A decade after the first report of the presence of the antifreeze functions in a group of bacteria (Duman and Olsen 1993), a series of reports are now available describing the genetic regulation and the protein structure and function of bacteria from cold temperature environments including Antarctica, and these bacteria exhibit antifreeze activity. An antifreeze lipoprotein (AFLP) was identified in *Moraxella* sp. from Ross Island of Antarctica (Yamashita et al. 2002). Later, 187 bacteria from α- and γ-Proteobacteria genera from Vestfold Hills and Larsemann Hills of Eastern Antarctica were found to possess the antifreeze activity (Gilbert et al. 2004). The first characterized gene encoding antifreeze protein is *afpA*, which was cloned from *P. putida* GR12–2 (Arctic Rhizobia) from soil sample of Canadian High Arctic (Muryoi et al. 2004). An uncharacterized AFP showing antifreeze protein was found in Antarctic *Marinomonas protea* (Berry et al. 2005). Ice nucleation protein and uncharacterized AFP has been identified in *P. fluorescens* KUAF-68, which was isolated from Antarctica (Kawahara et al. 2004).

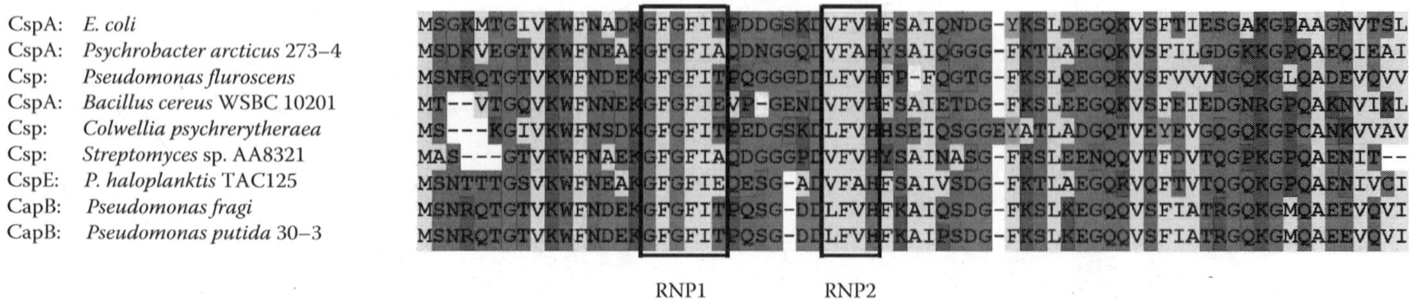

FIGURE 6.1 (See color insert following page 276.) Multiple alignment of cold-shock proteins (Csps) and cold acclimation protein (Caps) from Antarctica with CspA of *E. coli* with the help of CLUSTAL X (http://www.clustal.org/). The RNP1/RNP2 motifs shown in square boxes are conserved domains in Csps and Caps isolated from various bacterial species.

A hyperactive, Ca^{2+} dependent antifreeze protein was identified in an Antarctic bacterium, *Ma. primoryensis*. (Gilbert et al. 2005). A novel, intracellular antifreeze protein (uncharacterized) was found in an Antarctic bacterium, *Flavobacterium xanthum* (Kawahara et al. 2007). Another class of protein, which binds to ice is called ice-binding protein (IBP) and functions as an inhibitor of ice recrystallization. IBP was first identified and characterized in *Colwellia* strain SLW05, an Antarctic sea ice bacterium (Raymond et al. 2007). Ice-binding protein was also identified in bacterium 3519-10 from Vostok ice core (Raymond et al. 2008). A summary of the known AFPs and IBPS isolated from microorganisms from cold environment has been elaborated in Table 6.1. Interestingly, almost all the reported AFP or IBP have been identified in bacteria isolated from Antarctica. This indicates that these specific sets of proteins play a major role in cold adaptation at subzero temperatures. From the evolution point of view, there is no sequence similarity in AFPs from different organisms (Chattopadhyay 2007) An Antarctic fish AFP appears to have evolved from a pancreatic trypsinogen-like protease (Cheng et al. 1999). Some plant AFPs are homologs of plant pathogenesis proteins (Ökke and Barbaros 2003). It appears that a series of very different proteins have independently evolved to perform a common function (i.e., ice binding) to form AFPs, even though they have no sequence similarity (Figure 6.2). Due to the absence

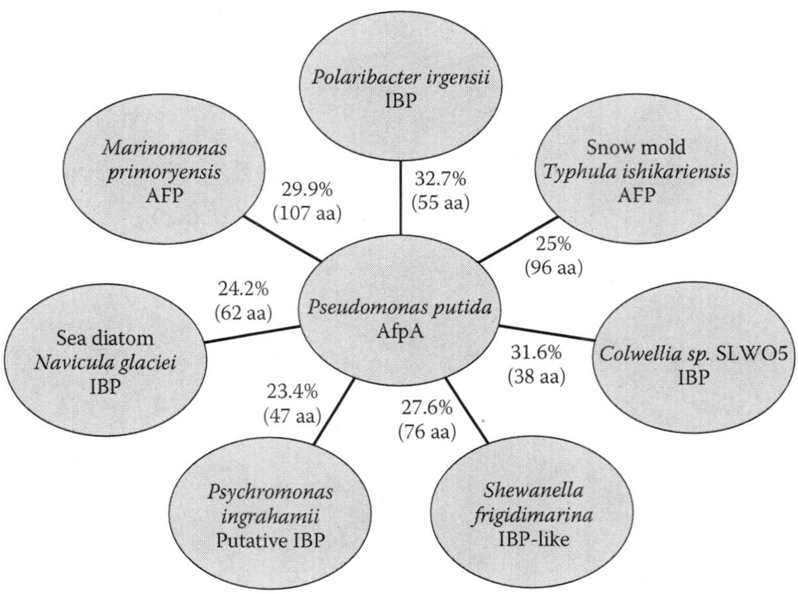

FIGURE 6.2 Schematic diagram showing the percent identity of ice-binding protein (IBP) and antifreeze protein (AFP) from different bacteria, snow mold, and sea diatom with antifreeze protein A (AfpA) from *P. putida*. The diagram depicts that the AFP and the IBP are very diverse proteins, which evolved acquiring similar functions with time but are of different structures. The percent identity was calculated by aligning the amino acid sequences of the proteins (NCBI) using LALIGN software (http://www.ch.embnet.org/software/LALIGN_form.html). The number indicates % identity with number of amino acid overlap in parentheses.

of a consensus ice-binding domain, PCR-based studies on these proteins became unsuccessful. It seems that reverse genetics is the preferable approach for the identification of gene sequence of the AFPs.

6.7 ROLE OF AFP AND IBP

Thermal hysteresis (TH) is a measure of the antifreeze activity. Thermal hysteresis is the difference between melting point and freezing point. When solutes are dissolved in water, it increases the boiling point and lowers the freezing point of water. In Antarctica, the seawater consists of sodium chloride and other solutes, which lower the freezing point of water from 0°C to −1.9°C. Therefore in order to survive in such freezing conditions, the organisms living in water must depress the freezing point of either their blood (in case of fish), or their surrounding water (in case of bacterium) below −1.9°C to avoid freeze damage. It has been reported that the solutes present in the blood of fish cannot lower the freezing point by more than 1.4°C. It is the presence of these AFPs, which causes further lowering of freezing point to 0.5°C or more (DeVries Wohlschlag 1969). Depending on the antifreeze activity and the concentration of AFP, a single ice crystal with a well-defined shape and size can stay in its supercooled solution down to a certain degree below 0°C, while the melting point of the ice crystal could not be significantly changed from 0°C. The temperature difference for such a single crystal to grow and shrink with a rate more than $0.2\,\mu s^{-1}$ is defined as the thermal hysteresis (Figure 6.3). The mechanism of such process is due to the adsorption of AFP molecules on an ice surface inducing a dense AFP-water layer, which can significantly decrease the mole fraction of the interfacial water and thus, lower the temperature for a seed ice crystal to grow in a supercooled AFP solution (Mao and Ba 2006a). In other words, water's chemical potential in

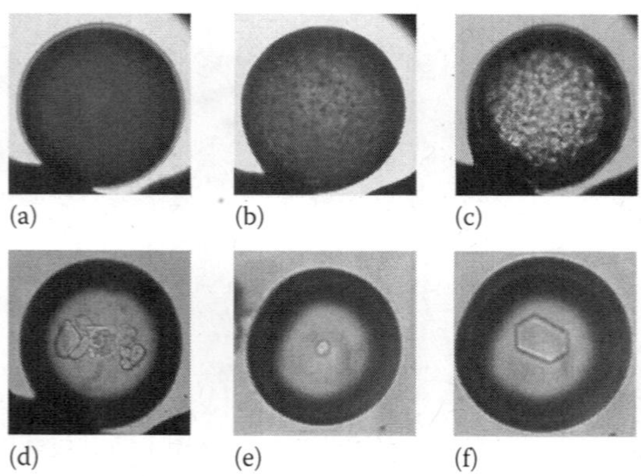

FIGURE 6.3 (See color insert following page 276.) A sequence of cryomicroscopic images (a–f) in a Clifton nanoliter osmometer monitoring the growth of a single ice crystal when temperature from −40°C is elevated to just below the freezing point. The effect of the antifreeze function of the crude extract of the total cellular protein from a biodegradative bacterium *P.* sp. 30/3 isolated from oil-contaminated soil from Antarctica is shown by the growth along the *c*-axis of a stable hexagonal ice crystal.

the interfacial AFP-water layer is lowered due to the significantly increased AFP concentration compared with that in the bulk solution, which results in a freezing point depression for water in the interfacial AFP-water layer. The temperature has to be decreased below the freezing point of water in the interfacial AFP-water layer for the AFP-surrounded ice crystal to grow. The AFP molecules can be overgrown inside the ice crystal, once the temperature is decreased below the freezing point and the ice surface advancing rate is higher than the AFP's desorption and diffusion rate. Thus, in terms of thermodynamics, the unusual phenomenon of AFP's thermal hysteresis and also the nearly unchanged melting point for the ice crystal due to the AFP's ice-surface adsorption was explained by colligative effect mechanism (Mao and Ba 2006b).

6.8 MAINTENANCE OF THE MEMBRANE FLUIDITY

At low temperatures, the cell membrane fluidity decreases because the lipids found in the cell membrane become more rigid, which interfere with the function of the membrane and membrane-bound components like pumps and channels (Rodriguez-Vargas et al. 2007). In Antarctic environment, it becomes crucial for the bacteria to maintain the fluidity of the membrane to survive. The adjustment of cell membrane fluidity as a function of temperature change is known as homeoviscous adaptation (Sinensky 1974). This occurs to maintain constant membrane fluidity. It has been reported that membrane desaturase enzymes, fatty acid, and pigments produced in Antarctic bacteria contribute to homeoviscous adaptation (Chattopadhyay 2006). The desaturase enzymes are induced at low temperatures, which catalyze the conversion of saturated fatty acids into unsaturated fatty acids thereby increasing the fluidity of the cell membrane. The increased synthesis of unsaturated fatty acids at lower temperature was observed in Antarctic soil bacterium *Micrococcus roseus* and *Sphingobacterium antarcticus* (Chattopadhyay and Jagannadham, 2001). During cold adaptation, the lipid profile of the cell membrane of the bacteria changes. Instead of long-chain fatty acids, straight-chain fatty acids and iso-fatty acids, there is an increased synthesis of short-chain fatty acids, branched-chain fatty acids, and anteiso-fatty acids, respectively, all of which render the membrane more fluid (Suutari and Laakso, 1994). Using DNA microarray analysis, it has been observed that there is a upregulation of many genes at transcription level, which code for the degradation of branched-chain amino acids when *B. subtilis* was exposed to low temperature (Kaan et al. 2002). *In vitro* study on the increased amount of hydroxyl fatty acids in the lipopolysaccharides from an Antarctic bacterium *P. syringae* incubated at low temperatures with a concomitant increase in the fluidity suggests the role of hydroxyl fatty acids in outer membrane fluid state (Kumar et al. 2002). *Listeria monocytogenes*, a food borne pathogen has the ability to adjust to a wide range of environmental stresses. When *L. monocytogenes* is grown at low temperatures, it modifies its plasma membrane to maintain its fluidity with an increase in the amount of odd-numbered fatty acids and branched-chain fatty acids, a switch from iso to anteiso branching patterns, and, later, a shortening of the length of the branched-chain fatty acids. The membrane fluidity of *L. monocytogenes* cold-sensitive mutant (*cld*1) was measured to be less than the membranes of the other strains when grown at 30°C, but when grown at 15°C it was able to adjust its membrane to approach the rigidity of the other strains (Najjar et al. 2007). It was found that the time required for various *Listeria monocytogenes* strains to adjust their membrane fluidity were 130–180 s, which suggests the multistage adjustment of the cell membrane to temperatures.

The majority of Antarctic bacterial species are pigmented implying a possible role of these pigments in cold adaptation. Most of the pigments are found to be associated with the membrane, which indicates their role in the maintenance of membrane fluidity (Chattopadhyay 2006). It was found that the major carotenoid pigments bind to vesicles, which are made of both synthetic and natural lipids, and make them rigid. A polar carotenoid is reported to decrease the membrane fluidity (Subczynski et al. 1992). An increased amount of polar carotenoids and a decreased amount of nonpolar carotenoids were observed in Antarctic strains *Micrococcus roseus* and *Sphingobacterium antarcticus* when grown at low temperature (Chattopadhyay and Jagannadham 2001). Based on this observation, Chattopadhyay and Jagannadham (2001) suggested that the synthesis of membrane-rigidifying polar carotenoids was increased to counterbalance the effects of increased synthesis of membrane-fluidizing fatty acids in the Antarctic bacteria. In another study, it has been reported that a psychrotolerant Antarctic seawater bacterium, *Shewanella livingstonensis* Ac10 produces eicosapentaenoic acid (EPA) as part of the membrane phospholipids when exposed to cold temperatures (Kawamoto et al. 2009). The EPA constitutes about 5% of the total fatty acids when cultures are grown at 4°C. The genetic components for the EPA have been identified (*orf2*, *orf5*, *orf6*, *orf7*, and *orf8*), and the deletion of any of these genes resulted in a significant growth retardation filamentous cell morphology. The cessation of growth was restored by supplementing with the EPA. This suggests that at 4°C the absence of EPA causes the defects in cell division. Also culture with it has been shown that a number of membrane proteins are affected due to the absence of EPA resulting in the overall organization of the cell membrane that is necessary for an effective cell division process. Although, it has been established that polyunsaturated fatty acids are necessary to increase the fluidity of the hydrophobic segment of the membrane, EPA contributes to the membrane organization and cell division at cold temperatures.

Recently, the role of polyhydroxyalkanoates (PHAs) in the adaptation to cold temperatures has been elaborated in an Antarctic *Pseudomonas* sp. 14-3 isolate (Ayub et al. 2009). The genes *phaRBAC* have been identified and a mutation on the *phaC* had caused the depletion in the accumulation of the PHAs in the cell resulting in the sensitivity to cultures exposed to cold and freezing temperatures. It has been concluded that the PHAs are necessary for maintaining a steady membrane redox state (NADH/NAD ratio and NADPH content) by activating the oxidative stress system in this bacterium at cold temperatures.

6.9 ROLE OF EXTRAPOLYMERIC SUBSTANCES

The synthesis of extrapolymeric substances (EPSs) in Antarctic microorganisms has been envisaged to be an essential strategy to maintain an ice-free microenvironment in which nutrient acquisition, motility, respiration, and other essential cellular metabolic functions are sustained (Shivaji et al. 2004, 2007). Such EPS microenvironment harbors living and physiologically "active" microbial

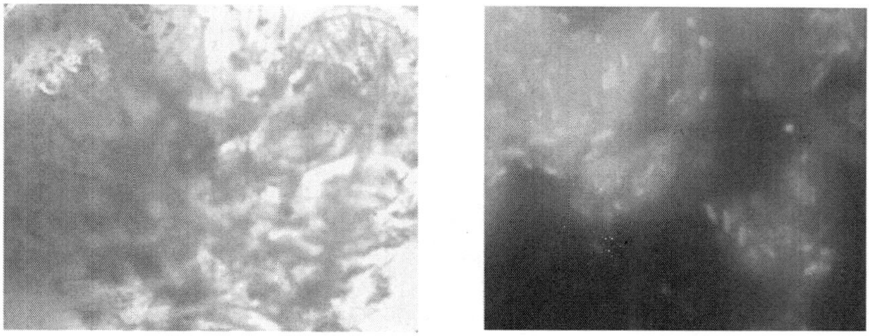

(a) (b)

FIGURE 6.4 (See color insert following page 276.) Extracellular polymeric substance (EPS) produced by *Janthinobacterium* sp. strain Ant 5-2 isolated from a Proglacial lake P9 located at the Schirmacher Oasis of East Antarctic Dronning Maud Land: (a) "slimy" EPS secreted by *Janthinobacterium* sp. Ant 5-2 in a stationary culture after 7 days of incubation at 15°C; (b) microscopic examination of the bacillus *Janthinobacterium* sp. Ant 5-2 aggregated flocs of cells stained with Live/Dead *Bac*Light fluorescent staining kit (Molecular Probes) embedded in thick EPS moiety.

consortium; its study is often useful to better understand the microbial interactions and biological processes necessary to sustain life by adapting to the Antarctic cold and subzero environments (Figure 6.4). The synthesis of EPS by Cyanobacteria has been reported on Antarctic giant rocks in Antarctica (De los Rios et al. 2007). Also, EPS has been linked to the formation of cyanobacterial mats (Klock et al. 2007), and the removal of heavy metals from the surrounding waters (Micheletti et al. 2008, Paperi et al. 2006). EPS synthesis has also been reported in other heterotrophic Antarctic bacterial communities (Shivaji et al. 2007, Nichols et al. 2005a,b, Mancuso Nichols et al. 2004). Most of these EPSs are acidic heteropolysaccharides with carbohydrate and non-carbohydrate constituents. However the homopolysaccharide EPSs constituting of alginate, levan, and cellulose have also been reported (Leigh and Coplin, 1992). The qualitative nature of the EPS in microbial consortium is normally determined by the enzyme-linked lectinosorbent assay (ELLA) using enzyme linked or fluor-conjugated lectins (Klock et al. 2007, Leriche et al. 2000, Panicker et al. 2006). The synthesis of EPS often leads to the formation of biofilms conforming a microenvironment of microbial consortium primarily in Antarctic lakes has been suggested to function to protect heterotrophic and autotrophic microorganisms from freeze damage, dehydration and maintaining basal levels of metabolic activities and categorized as to be one of the methods of survival mechanism in sustained cold and subzero temperature environments. A list of bacterial EPSs cited in World Wide Web and their nature, biochemistry, and role have been listed by Wackett (2009).

6.10 CONCLUDING REMARKS

It is apparent that the bacterial cell physiology, metabolism, and the stability of membrane components undergo adjustments to cope with the temperature fluctuations during the diurnal freeze–thaw cycles during polar summer and persistent freezing conditions during winter months. These adjustments have been investigated in combinations or individually in many of the Antarctic microorganisms. The genetic regulations for the adaptation in cold temperatures have been elucidated in only a few Antarctic biodegradative microorganisms. It is obvious from the reports that in order to maintain the cellular functions at cold temperatures, mRNA stability and functionality is necessary. At cold temperatures, mRNA tends to form secondary structures by complementary base pairing at various segments. Failure to accomplish this could hinder the translation of the necessary proteins for an optimum cellular metabolism necessary to survive, maintain growth, and division of the cells. Studies have implicated a role for the CSPs in destabilizing the secondary structures of the mRNA and other RNA enabling cells to engage in an effective expression of the necessary proteins and enzymes for continued growth and division. Interestingly, the regulatory nucleotide sequences located on the upstream of the ORF of the *capB* (one of the CAPs) have been found to be significantly different from the gene *cspA* that codes for the major cold-shock protein in most masophilic bacteria. The *capB* regulatory sequence in all Antarctic biodegradative *Pseudomonas* isolates seem to be conserved (A. Bej, unpublished). This result suggests that either the strains consisting of the *capB* have been selected to persist and function in Antarctic environment or been evolved to better cope with the continuous cold temperatures. Whether the *cspA* has been evolved to destabilize RNA in microorganisms inhabiting at cold temperatures or had a different physiological role in the cell, and been recruited to support survival in cold temperature is still unknown.

Besides the cellular function, growth, and division at cold temperatures, the other scenario for the microorganisms to remain viable in subzero freezing temperatures and remain viable during repeated freeze–thaw cycles seems crucial. The strategies to cope in such conditions could be more profound. There are two situations one would assume in this condition: (a) the cessation of cellular metabolism and remain in "dormant" state especially during the polar winter until the favorable conditions prevail during the polar summer months; and (b) continue to perform cellular activities but at a much slower rate (basal level). In addition, it is important to protect the cell membrane from damage and cytoplasm to undergo dehydration during the formation of ice crystals. The primary focus to better understand the adaptive mechanisms in microorganisms inhabiting such environments has been to look for the genetic systems that allow the expression of proteins such as AFP, IBP, and membrane desaturases at subzero temperatures. The AFP and the IBP have justifiably been expressed to protect the cells from damage due to freezing and dehydration during the repeated diurnal freeze–thaw cycles that occur or during the polar winter when icy condition persists. Besides the adaptive mechanism of being able to maintain the cellular function, one of the most important components that must adjust and remain structurally intact and functionally active is the cell membrane. The membrane integrity is maintained by converting the saturated fatty acids into the unsaturated fatty acids using desaturase enzymes. Alternatively, it has been proposed that bacteria in cold environment form a microenvironment within EPS in which nutrient uptake, genetic exchange and chemical signaling among organisms, competition, and phage–host interactions continue, but perhaps at a much slower rate. Microbial survival strategies in cold and subzero temperature environments seem to be a complex mixture of various cellular functions some of which could simply be inherent to the cells and help them to be selected to function or further evolved to become successful in polar environment. The study of cold temperature adaptation in biodegradative microorganisms in polar environment is still in its infancy, and the full extent of the cold-adaptive mechanisms common to the extended microbial community of this unique cryosphere has yet to be unraveled.

ACKNOWLEDGMENTS

We thank Jackie Aislabie for the Antarctic biodegradative strains, and Michael Storrie-Lombardi for reviewing this chapter and giving helpful comments. Some of the studies in the authors' lab cited in this chapter are part of the International Scientific Expedition Tawani 2008/NASA 2008 Antarctic Expedition, which was supported by Col James Pritzker and the Tawani Foundation, Chicago, IL. Logistics support was provided by the Arctic and Antarctic Research Institute/Russian Antarctic Expedition (AARI/RAE) (Valerii Lukin); Antarctic Logistics Centre International (ALCI) (Mirella Kruger); Von Braun Center for Science and Innovation, NASA; Rasik Ravindra (Director) of the National Center for Antarctic Ocean Research (NCAOR), Maitri Station, India, and Novolazarevskaya Station, Russia. We also thank Ed Tracy of Tawani Foundation, Marty Kress of NSSTC/NASA, and Robert Fischer of Biology, UAB for the support.

REFERENCES

Aislabie, J.M., M.R. Balks, J.M. Foght, and E.J. Waterhouse. 2004. Hydrocarbon spills on Antarctic soils: Effects and management. *Environmental Science and Technology* 38(5):1265–1274.

Aislabie, J., D.J. Saul, and J.M. Foght. 2006. Bioremediation of hydrocarbon contaminated polar soils. *Extremophiles* 10(3):171–179.

Ayub, N.D., P.M. Tribelli, and N.I. López. 2009. Polyhydroxyalkanoates are essential for maintenance of redox state in the Antarctic bacterium *Pseudomonas* sp. 14-3 during low temperature adaptation. *Extremophiles* 13(1):59–66.

Bae, W., S. Phadtare, K. Severinov, and M. Inouye. 1999. Characterization of *Escherichia coli* cspE, whose product negatively regulates transcription of cspA, the gene for the major cold shock protein. *Molecular Microbiology* 31:1429–1441.

Bae, W., B. Xia, M. Inouye, and K. Severinov. 2000. *Escherichia coli* CspA-family RNA chaperones are transcription antiterminators. *Proceedings of the National Academy of Sciences* USA 97:7784–7789.

Baraniecki, C.A., J. Aislabie, and J.M. Foght. 2002. Characterization of *Sphingomonas* sp. Ant 17, an aromatic hydrocarbon-degrading bacterium isolated from Antarctic soil. *Microbial Ecology* 43(1):44–54.

Barbaro, S.E., J.T. Trevors, and W.E. Inniss. 2002. Effect of different carbon sources and cold shock on protein synthesis by a psychrotrophic *Acinetobacter* sp. *Canadian Journal of Microbiology* 48(3):239–244.

Berger, F., N. Morellet, F. Menu, and P. Potier. 1996. Cold shock and cold acclimation proteins in the psychrotropic bacterium *Arthrobacter globiformis* SI55. *Journal of Bacteriology* 178:2999–3007.

Berger, F., P. Normand, and P. Potier. 1997. capA, a cspA-like gene that encodes a cold acclimation protein in the psychrotrophic bacterium *Arthrobacter globiformis* SI55. *Journal of Bacteriology* 179:5670–5676.

Berry, M. J., A. Griffitts, P. J. Hill, J. Laybourne-Parry, and S. V., Mills. 2005. Processes and organisms for the production of anti-freeze proteins. United States Patent 6,887,984 B2.

Chattopadhyay, M.K. 2006. Mechanism of bacterial adaptation to low temperature. *Journal of Biosciences* 31(1):157–165.

Chattopadhyay, M.K. 2007. Antifreeze protein of bacteria. *Resonance* 12(12):25–30.

Chattopadhyay, M.K. and M.V. Jagannadham. 2001. Maintenance of membrane fluidity in Antarctic bacteria. *Polar Biology* 24:386–388.

Cheng, C.-H.C. and C. Liangbiao. 1999. Evolution of an antifreeze glycoprotein. *Nature* 401:443–444.

Cloutier, J., D. Prévost, P. Nadeau, and H. Antoun. 1992. Heat and cold shock protein synthesis in arctic and temperate strains of rhizobia. *Applied and Environmental Microbiology* 58(9):2846–2853.

De los Ríos, A., M. Grube, L. Sancho, and C. Ascaso. 2007. Ultrastructural and genetic characteristics of endolithic cyanobacterial biofilms colonizing Antarctic granite rocks. *FEMS Microbiology Ecology* 59:386–395.

DeVries, A.L. and D.E. Wohlschlag. 1969. Freezing resistance in some Antarctic fishes. *Science* 163(3871):1073–1075.

Duman, J.G. and T.M. Olsen. 1993. Thermal hysteresis protein activity in bacteria, fungi, and phylogenetically diverse plants. *Cryobiology* 30:322–328.

Ermolenko, D.N. and G.I. Makhatadze. 2002. Bacterial cold-shock proteins. *Review. Cell Life Sciences* 59:1902–1913.

Fang, L., W. Jiang, W. Bae, and M. Inouye. 1997. Promoter-independent cold-shock induction of cspA and its derepression at 37°C by mRNA stabilization. *Molecular Microbiology* 23:355–364.

Franzmann, P.D., Y. Liu, D.L. Balkwill, H.C. Aldrich, E.C. de Macario, and D.R. Boone. 1997. *Methanogenium frigidum* sp. nov., a psychrophilic, H2-using methanogen from AceLake, Antarctica. *International Journal of Systematic Bacteriology* 47(4): 1068–1072.

Giaquinto, L., P.M. Curmi, K.S. Siddiqui, A. Poljak, E. DeLong, S. DasSarma, and R. Cavicchioli. 2007. Structure and function of cold shock proteins in archaea. *Journal of Bacteriology* 189(15):5738–5748.

Gilbert, J.A., P.J. Hill, C.E.R. Dodd, and J.L. Parry. 2004. Demonstration of antifreeze protein activity in Antarctic lake bacteria. *Microbiology* 150:171–180.

Gilbert, J.A., P.L. Davies, and J.L. Parry. 2005. A hyperactive, Ca^{2+}-dependent antifreeze protein in an Antarctic bacterium. *FEMS Microbiology Letters* 245(1):67–72.

Goodchild, A., M. Raftery, N. F. Saunders, M. Guilhaus, and R. Cavicchioli. 2004a. Biology of the cold adapted archaeon *Methanococcoides burtonii* determined by proteomics using liquid chromatography-tandem mass spectrometry. *Journal of Protein Research* 3:1164–1176.

Goodchild, A., N.F. Saunders, H. Ertan, M. Raftery, M. Guilhaus, P.M. Curmi, and R. Cavicchioli. 2004b. A proteomic determination of cold adaptation in the Antarctic archaeon, *Methanococcoides burtonii. Molecular Microbiology* 53(1):309–321.

Hebraud, M., E. Dubois, P. Potier, and J. Labadie. 1994. Effects of temperatures on the protein levels in a psychrotropic bacterium, *Pseudomonas fragi. Journal of Bacteriology* 176:4017–4024.

Jiang, W., P. Jones, and M. Inouye. 1993. Chloramphenicol induces the transcription of the major cold shock gene of *Escherichia coli*: cspA. *Journal of Bacteriology* 175:5824–5828.

Kaan, T., G. Homuth, U. Mader, J. Bandow, and T. Schweder. 2002. Genome-wide transcriptional profiling of the *Bacillus subtilis* cold-shock response; *Microbiology* 148:3441–3455.

Kawahara, H., N. Yoko, O. Kazuhiro, N.Muryoi, N. Jiro, and O. Hitoshi. 2004. Production of two types of ice crystal-controlling proteins in Antarctic bacterium. *Journal of Bioscience and Bioengineering* 98(3):220–223.

Kawahara, H., I. Yoshiko, H. Sakura, N. Muryoi, S. Mika, H. Michinori, O. Hironori, and O. Hitoshi. 2007. A novel, intracellular antifreeze protein in an Antarctic bacterium, *Flavobacterium xanthum. Cryo-Letters* 28:39–49.

Kawamoto, J., T. Kurihara, K. Yamamoto, M. Nagayasu, Y. Tani, H. Mihara, M. Hosokawa, T. Baba, S.B. Sato, and N. Esaki. 2009. Eicosapentaenoic acid plays a beneficial role in membrane organization and cell division of a cold-adapted bacterium, *Shewanella livingstonensis* Ac10. *Journal of Bacteriology* 191(2):632–40.

Kim, M.-J. Y.K. Lee, H.K. Lee, and H. Im. 2007. Characterization of cold-shock protein A of Antarctic *Streptomyces* sp. AA8321. *The Protein Journal* 26(1):51–59.

Klock, J-H., A. Wieland, R. Seifert, and W. Michaelis. 2007. Extracellular polymeric substances (EPS) from cyanobacterial mats: Characterization and isolation method optimization. *Mars Biology* 152:1077–1085.

Kumar, G.S., M.V. Jagannadham, and M.K. Ray. 2002. Low-temperature induced changes in composition and fluidity of lipopolysaccharides in the Antarctic psychrotrophic bacterium *Pseudomonas syringae. Journal of Bacteriology* 184:6746–6749.

Leigh, J.A. and D.L. Coplin. 1992. Exopolysaccharides in plant–bacterial interactions. *Annual Review of Microbiology* 46:307–346.

Leriche, V., P. Sibille, and B. Carpentier. 2000. Use of an enzyme-linked lectinsorbent assay to monitor the shift in polysaccharide composition in bacterial biofilms. *Applied and Environmental Microbiology* 66:1851–1856.

Leys, N.M., A. Ryngaert, L. Bastiaens, E.M. Top, W. Verstraete, and D. Springael. 2005. Culture independent detection of Sphingomonas sp. EPA 505 related strains in soils contaminated with polycyclic aromatic hydrocarbons (PAHs). *Microbes Ecology* 49(3):443–450.

Lim, J., T. Thomas, and R. Cavicchioli. 2000. Low temperature regulated DEAD-box RNA helicase from the Antarctic archaeon, *Methanococcoides burtonii. Journal of Molecular Biology*. 297(3):553–567.

Mancuso Nichols, C.A., S. Garon, J.P. Bowman, G. Raguénès, and J. Guézennec. 2004. Production of *exopolysaccharides by Antarctic marine bacterial isolates. Journal of Applied Microbiology* 96:1057–1066.

Mao, Y. and Y. Ba. 2006a. Insight into the binding of antifreeze proteins to ice surfaces via ^{13}C spin lattice relaxation solid-state NMR. *Biophysical Journal* 91(3):1059–1068.

Mao, Y. and Y. Ba. 2006b. Ice-surface adsorption enhanced colligative effect of antifreeze proteins in ice growth inhibition. *The Journal of Chemical Physics* 125(9):091102.

Mayr, B., T. Kaplan, S. Lechner, and S. Scherer. 1996. Identification and purification of a family of dimeric major cold Shock protein homologs from the psychrotrophic *Bacillus cereus* WSBC 10201. *Journal of Bacteriology* 178:2916–2925.

Médigue, C., K. Evelyne, P. Géraldine, B. Valérie, A. Bernsel, P.N. Bertin, F. Cheung, S. Cruveiller, S. D'Amico, A. Duilio, G. Fang, G. Feller, C. Ho, S. Mangenot, G. Marino, J. Nilsson, E. Parrilli, E.P.C. Rocha, Z. Rouy, A. Sekowska, M. L. Tutino, D. Vallenet, G.V. Heijne, and A. Danchin. 2005. Coping with cold: The genome of the versatile marine Antarctica bacterium *Pseudoalteromonas haloplanktis* TAC125. *Genome Research* 15:1325–1335.

Michel, V., I. Lehoux, G. Depret, V. Anglade, J. Labadie, and M. Hebraud. 1997. The cold shock response of the psychrotrophic bacterium *Pseudomonas fragi* involves four low-molecular-mass nucleic acid-binding proteins. *Journal of Bacteriology* 179:7331–7342.

Micheletti, E., G. Colica, C. Viti, P. Tamagnini, and R. De Philippis. 2008. Selectivity in the heavy metal removal by exopolysaccharide-producing cyanobacteria. *Journal of Applied Microbiology* 105(1):88–94.

Muryoi, N., M. Sato, S. Kaneko, H. Kawahara, H. Obata, M.W. Yaish, M. Griffith, and B.R. Glick. 2004. Cloning and expression of afpA, a gene encoding an antifreeze protein from the arctic plant growth-promoting rhizobacterium *Pseudomonas putida* GR12–2. *Journal of Bacteriology* 186(17):5661–5671.

Najjar, M.B., M. Chikindas, and T.J. Montville. 2007. Changes in *Listeria monocytogenes* membrane fluidity in response to temperature stress. *Applied and Environmental Microbiology* 73(20):6429–6435.

Neuhaus, K., S. Rappsch, K.P. Francis, and S. Scherer. 2000. Restart of exponential growth of cold-shocked *Yersinia enterocolitica* occurs after down-regulation of cspA1/A2 mRNA. *Journal of Bacteriology* 182:3285–3288.

Nichols, C.M., S.G. Lardière, J.P. Bowman, P.D. Nichols, J. Gibson, and J. Guézennec. 2005a. Chemical characterization of exopolysaccharides from Antarctic marine bacteria. *Microbial Ecology* 49:578–589.

Nichols, C.M., J.P. Bowman, and J. Guezennec. 2005b. Effects of incubation temperature on growth and production of exopolysaccharides by an Antarctic sea ice bacterium grown in batch culture. *Applied and Environmental Microbiology* 71:3519–3523.

Ökke, A. and N. Barbaros. 2003. Antifreeze proteins in higher plants. *Phytochemistry* 64(7):1187–1196.

Panicker, G., J. Aislabie, D. Saul, and A.K. Bej. 2002. Cold tolerance of *Pseudomonas* sp. 30-3 isolated from oil-contaminated soil, Antarctica. *Polar Biology* 25(1):5–11.

Panicker, G., J. Aislabie, and A.K. Bej. 2006. Analysis of aggregative behavior of *Pseudomonas* sp. 30-3 isolated from Antarctic soil. *Soil Biology and Biochemistry* 38:3152–3157.

Paperi, R., E. Micheletii, and R. De Philipis. 2006. Optimization of copper sorbing-desorbing cycles with confined cultures of the exopolysaccharide-producing cyanobacterium *Cyanospira capsulate. Journal of Applied Microbiology* 101:1351–1356.

Phadtare, S., M. Inouye, and K. Severinov. 2004. The mechanism of nucleic acid melting by a CspA family protein. *Journal of Molecular Biology* 337:147–155.

Ray, M.K., T. Sitaramamma, S. Ghandhi, and S. Shivaji. 1994. Occurrence and expression of cspA, a cold shock gene, in Antarctic psychrotrophic bacteria. *FEMS Microbiology Letters* 116(1):55–60.

Raymond, J.A., C. Fristen, and K. Shen. 2007. An ice-binding protein from an Antarctic sea ice bacterium. *FEMS Microbiology Ecology* 61(2):214–221.

Raymond, J.A., B.C. Christner, and S.C. Schuster. 2008. A bacterial ice-binding protein from the Vostok ice core. *Extremophiles* 12(5):713–717.

Roberts, M.E. and W.E. Inniss. 1992. The synthesis of cold shock proteins and cold acclimation proteins in the psychrophilic bacterium *Aquaspirillum arcticum. Current Microbiology* 25(5):275–278.

Rodriguez-Vargas, S., A. Sánchez-Garcia, J.M. Martinez-Rivas, J.A. Prieto, and F. Randez-Gil. 2007. Fluidization of membrane lipids enhances the tolerance of *Saccharomyces cerevisiae* to freezing and salt stress. *Applied and Environmental Microbiology* 73:110–116.

Ross, W., S.E. Aiyar, J. Salomon, and R.L. Gourse. 1998. *Escherichia coli* promoters with UP elements of different strengths: Modular structure of bacterial promoters. *Journal of Bacteriology* 180:5375–5383.

Saunders, N.F., T. Thomas, P.M. Curmi, J.S. Mattick, E. Kuczek, R. Slade, J. Davis, P.D. Franzmann, D. Boone, K. Rusterholtz, R. Feldman, C. Gates, S. Bench, K. Sowers, K. Kadner, A. Aerts, P. Dehal, C. Detter, T. Glavina, S. Lucas, P. Richardson, F. Larimer, L. Hauser, M. Land, and R. Cavicchioli. 2003. Mechanisms of thermal adaptation revealed from the genomes of the Antarctic Archaea *Methanogenium frigidum* and *Methanococcoides burtonii. Genome Research* 13(7):1580–1588.

Schröder, K., P. Zuber, G. Willimsky, B. Wagner, and M.A. Mahariel. 1993. Mapping of the Bacillus subtilis cspB gene and cloning of its homologs in thermophilic, mesophilic and psychrophilic bacilli. *Gene* 136:277–280.

Shivaji, S., G.S. Reddy, R.P. Aduri, R. Kutty, and K. Ravenschlag. 2004. Bacterial diversity of a soil sample from Schirmacher Oasis, Antarctica. *Cell and Molecular Biology* (Noisy-le-grand) 50:525–536.

Shivaji, S., M.D. Kiran, S. Chintalapati. 2007. Perception and transduction of low temperature in bacteria. In *Physiology and Biochemistry of Extremophiles*, C. Gerday and N. Glansdorff (eds.), ASM Press, Washington, D.C.

Sinensky, M. 1974. Homeoviscous adaptation—a homeostatic process that regulates the viscosity of membrane lipids in *Escherichia coli. Proceedings of the National Academy of Sciences USA* 71:522–525.

Subczynski, W.K., E. Markowska, W.I. Gruszecki, and J. Sielewiesiuk. 1992. Effect of polar carotenoids on dimyristoylphosphatidylcholine membranes: A spin-label study. *Biochimica Biophysica Acta* 1105:97–108.

Suutari, M. and S. Laakso. 1994. Microbial fatty acids and thermal adaptation. *Critical Reviews in Microbiology*. 20:285–328.

Ugolini, F.C. and J.G. Bockheim. 2008. Antarctic soil and soil formation in a changing environment. *Geoderma* 144:1–8.

Van Bogelen, R.A. and F.C. Neidhardt. 1990. Ribosomes as sensors of heat and cold shock in *Escherichia coli. Proceedings of the National Academy of Sciences USA* 87:5589–5593.

Wackett, L.P. 2009. Microbial exopolysaccharides: An annotated selection of World Wide Web sites relevant to the topics in Environmental microbiology. *Environmental Microbiology* 11:729–730.

Wang, Q.F., J.L. Miao, Y.H. Hou, Y. Ding, and G.Y. Li. 2006. Expression of CspA and GST by an Antarctic psychrophilic bacterium Colwellia sp. NJ341 at near-freezing temperature. *World Journal of Microbiology and Biotechnology* 22(4):311–316.

Wolffe, A.P., S. Tafuri, M. Ranjan, and M. Familari. 1992. The Y-box factors: A family of nucleic acid binding proteins conserved from *Escherichia coli* to man. *The New Biologist* 4:290–298.

Yamanaka, K. and M. Inouye. 1997. Growth-phase-dependent expression of cspD, encoding a member of the CspA family in *Escherichia coli. Journal of Bacteriology* 179:5126–5130.

Yamanaka, K., T. Mitani, T. Ogura, H. Niki, and S. Hiraga. 1994. Cloning, sequencing, and characterization of multicopy suppressors of a mukB mutation in *Escherichia coli. Molecular Microbiology* 13:301–312.

Yamashita, Y., N. Nakamura, K. Omiya, J. Nishikawa, H. Kawahara, and H. Obata. 2002. Identification of an antifreeze lipoprotein from *Moraxella* sp. of Antarctic origin. *Bioscience, Biotechnology, and Biochemistry* 66(2):239–247.

Zheng S., M.A. Ponder, J.Y.J. Shih, J.M. Tiedje, M.F. Thomashow, and D.M. Lubman, 2006. A proteomic analysis of *Psychrobacter articus* 273-4 adaptation to low temperature and salinity using a 2-D liquid mapping approach. *Electrophoresis* 28(3):467–488.

7 Possible Role of Bacteriophage-Mediated Horizontal Gene Transfer on Microbial Adaptation to Environmental Stressors in Polar Ecosystems

Michael C. Storrie-Lombardi and Shannon J. Williamson

CONTENTS

7.1 INTRODUCTION

During the twentieth and twenty-first centuries, human exploration of remote polar environments has most often been powered by fossil fuels. When the British Antarctic Expedition led by Ernest Shackleton sailed out of London's East India Dock aboard the *Nimrod* on July 30, 1907, the cargo manifest included an experimental vehicle for Antarctic transportation: a brand new Arrol-Johnston motor car (Sipiera 2002). The car would prove capable of traveling a maximum of only 8 miles at 3–15 miles h^{-1} across the smooth ice of Cape Royds at the foot of Mt. Erebus. Although the harsh freezing conditions and unpredictable terrain common to most of the continent rendered the car unusable, the era of hydrocarbon-powered exploration had reached Antarctica.

The use of vehicles and tools requiring hydrocarbon fuels for power and/or lubrication in the extreme cold of both polar regions makes environmental contamination from accidental spillage an ever-present risk (Aislabie et al. 2004). In fact, repeated exposure of the fragile environment to the incomplete combustion products of these fuels, including a wide variety of polycyclic aromatic hydrocarbons (PAHs), would seem to make contamination a statistical certainty. Paradoxiacally, these hydrocarbon products can either serve as a new source of carbon and energy for indigenous microorganisms, or behave as lethal toxins or mutagens. The natural bioremediation potential of the polar microbial community will be a function of at least four variables:

1. The presence of functional genes coding for enzymes capable of degrading an offending contaminant and/or using it as a fuel source.
2. The diversity of the biodegradation enzymes encoded in the metagenome of the multiple species comprising the extended polar microbial community.
3. The ability of the microbial consortium to transfer the genes that code for these enzymes across closely or distantly related species.
4. The availability of indigenous molecular biological mechanisms to modify current enzyme codes to metabolize future novel contaminants.

The polar ecosystems are geographically well defined and would likely have initially supported a relatively small number of microbial species. The combination of limited metabolic resources, brief growing seasons, and tight ecological niches provides a framework that should reward, across evolutionary timescales efficient sharing of information between microbes, produce a gradual increase in microbial diversity, and leave a rich genomic record of inter-organismal transfer of genetic information. In this chapter, we will discuss the most likely method of information transfer between microbes in extreme environments, bacteriophage-mediated horizontal gene transfer, and review current evidence for such information transfer in polar bioremediation.

7.2 HORIZONTAL GENE TRANSFER

The classical transmission of genetic information occurs vertically from parent to offspring within a specific species. Horizontal gene transfer (HGT), also known as lateral gene transfer (LGT), refers to the passage of genetic information from one organism to another that may be closely or only distantly related (Johansen and Ustaheim 1968, Jain et al. 1999, Ochman et al. 2000, Koonin et al. 2001). HGT, along with associated mutations and genome reorganization, is now known to be a ubiquitous, central survival mechanism capable of producing rapid adaptive responses in eukaryotes, prokaryotes, and viruses (Zeph et al. 1988, Medigue et al. 1991, Davison 1999, Choi and Kim 2007). The impact of HGT first became apparent during the genomic analyses of clinically significant pathogenic organisms (Saunders et al. 2001, Schmidt 2001). Novel epidemics, the appearance of new pathogenic strains in nonpathogenic bacteria, and the rapid spread of drug resistance have all been attributed to HGT (Dowson et al. 1989, 1990, Jiang and Paul 1998, Jain et al. 1999, Koonin et al. 2001).

7.2.1 Mechanisms

Prokaryotes have a wide range of mechanisms available for the exchange of genetic information including processes such as conjugation, transformation, and transduction (Mindlin et al. 2002). However, for organisms inhabiting extreme environments, the first two possibilities, conjugation and transformation, are not particularly efficient methods for information transmission. Conjugation requires direct cell-to-cell contact between a donor bacterium harboring fertility plasmids and recipient organisms expressing the appropriate pilus receptor. Multiple examples of information transfer by conjugation have been observed in laboratory settings, but are more difficult to document in the field (Dahlberg et al. 1998, Marcinek et al. 1998, Hausner and Wuertz 1999). The second mechanism, transformation, involves the uptake of free, unprotected genomic material from the local environment. For environments exhibiting extremes in pH, temperature, desiccation, and/or ultraviolet (UV) or heavy particle ionizing radiation, the survival of naked DNA or RNA polymers would be significantly compromised (Pinkhart and Storrie-Lombardi 2007, Storrie-Lombardi and Pinkhart 2007).

7.2.2 Phage-Mediated Transduction

In extreme environments, the third possibility, transduction, is of considerable interest. The process of replicating, modifying, storing, and accessing information encoded in either a DNA or RNA format can be readily facilitated by bacteriophages. Bacteriophages (phages) are viruses that specifically infect prokaryotes (Bacteria and Archaea). Phages are capable of mediating two forms of transduction: generalized and specialized (Snyder and Champness 1997). Both lytic and temperate phages (i.e., have the ability to integrate into the genetic material of their host) can transfer DNA between host cells via generalized transduction. Generalized transduction occurs when the phage accidentally packages a piece of the host's DNA during replication.

Temperate phages are solely responsible for gene transfer via specialized transduction. Temperate phages have the ability to establish a "silent infection" with a host cell upon integration into the host's DNA (Ackermann and DuBow 1987). This type of virus–host interaction is referred to as lysogeny, the phage genome is termed a prophage upon integration, and the host cell is termed a lysogen. The prophage genome is passively propagated along with the host genome across multiple bacterial generations. The excision of the prophage genome, followed by active replication through the lytic cycle can be triggered either chemically under controlled laboratory conditions (most efficiently with Mitomycin C) or in response to a suite of environmental variables that are poorly understood and are likely environment-specific (Williamson et al. 2002, Williamson and Paul 2006). Specialized transduction is initiated during an error in the excision process of the prophage genome. Portions of the host genome that flank the phage attachment site can become packaged along with the phage's DNA into replicating viral particles and subsequently transferred to a new host cell.

Once the transduced DNA is deposited in a new microbial host cell, homologous recombination can occur between the donor and recipient genes if there is a sufficient degree of similarity between the two. This introduction of new DNA can be likened to a genomic search engine capable of producing multiple new phenotypes ready to be evaluated for evolutionary fitness in changing environments. Such a search mechanism is a fundamental requirement for adaptive evolutionary processes. The transfer of novel genes from one microbial species to another can also occur if the packaged DNA is a broad host range plasmid or contains a transposon that has the ability to "jump" onto the DNA of the recipient microbe (Snyder and Champness 1997). The introduction of novel genes can result in a swift alteration in phenotype, providing the recipient host cell with new metabolic capabilities. Given the constraints of pH, salinity, and temperature in polar lakes, soils and endolithic environments, phage-mediated transduction seems the most likely avenue for rapid genetic information transfer between disparate microbial species and the subsequent appearance of adaptive diversity. The resilient protein capsids of phages shield nucleotide polymers from degradation, therefore transduced DNA will persist for longer periods of time in the environment than naked DNA.

7.2.3 Frequency

With the advent of whole genome sequencing, the frequency and impact of HGT on the evolution of bacterial genomes is now becoming clear. An analysis of the complete sequence of *Escherichia coli* strain MG1655 revealed that 755 of 4288 open reading frames (ORFs) (547.8 kb) had been introduced into the genome through a series of 234 HGT events, many of which appeared to be phage-mediated, since *E. coli* diverged from the *Salmonella* linage approximately 100 Myr ago (Lawrence and Ochman 1998). Although most of the acquired genes have since been deleted or inactivated, persistent functional sequences (accounting for approximately 18% of the current genome) code for properties that have permitted *E. coli* to exploit otherwise hostile environments. While the adaptive power of HGT was initially demonstrated by the rapid evolution of bacterial pathogenesis, it now seems clear that this is only a small example of how phages contribute to the adaptive evolution of bacterial genomes (Villarreal 1999, 2005). The generation of microbial diversity through the expansion of gene repertoires coupled with the reorganization of host chromosomes apparently occurs so frequently that the effects must be considered when investigating possible evolutionary histories back to the last common ancestor (Mirkin et al. 2003, Nakagawa et al. 2003, Van Sluys et al. 2003). Even though determining the exact frequency of HGT is difficult to accomplish, estimates of phage-mediated gene transfer were produced for the Tampa Bay estuarine environment at ~10^{14} transduction events per year (Jiang and Paul 1998). Globally, HGT may occur in the world's oceans at the rate of 2×10^{16} times s^{-1} (Bushman 2002, Canchaya et al. 2003). Whole bacterial genome analyses indicate that approximately 16 kb of genomic information is transferred every million years (Lawrence and Ochman 1998).

With the increasing availability of complete microbial genome sequences, evidence has accumulated for the occurrence of HGT not only at the strain and species level, but also between more distantly related organisms (Ochman et al. 2000). Of considerable human and environmental concern is the ability for bacteriophage transfer of antibiotic resistance between distantly related species. For example, highly mobile toxin-carrying *Staphylococcus aureus* pathogenenicity islands, containing superantigen genes, and other mobile elements are transferred to *Listeria monocytogenes* at the same high frequencies as they transfer within *Staphylococcus aureus* species (Chen and Novik 2009). Multiple staphylococcal phages transduced *L. monocytogenes* but could not form plaques. An experiment modeling phage therapy for bovine mastitis demonstrated pathogenicity island transfer between *S. aureus* and *L. monocytogenes* in raw milk. The findings imply that phages may participate in a far more expansive network of genetic information exchange among bacteria of different species than originally assumed. Consequently, HGT has been proposed as a major factor in the spread of critical protective phenotypic adaptations in the natural environment (Bogdanova et al. 1998, Davison 1999). A case in point, examination of mercury resistance determinants in environmental bacterial populations has revealed highly homologous (and recombinant) mercury resistance (*mer*) operons and transposons in bacteria of diverse taxonomic groups existing in geographically distant parts of the Earth (Mindlin et al. 2002). These observations reinforce the notion that HGT is a dynamic process that can strongly influence the adaptation of microorganisms to challenging environmental conditions.

7.2.4 Transfer between Phylogenetically Distant Organisms

The relative importance of HGT to prokaryotic evolution and, in particular, the degree to which distantly related organisms participate in this information transfer has been a matter of controversy since the discovery of the phenomenon (Koonin and Wolf 2008). Clearly, HGT occurs between closely related species and is easily seen in pathogenicity islands, regions approximately 100 kb in length containing multiple prophages and typically located near tRNA genes suggesting the insertion of these islands into a genome can be mediated by a bacteriophage. Moreover, as outlined later, HGT probably plays a pivotal role in the evolution of the photosynthetic gene clusters of closely related cyanobacteria. Unfortunately, detecting HGT across great evolutionary distances is compromised by the rapid amelioration or loss of distinguishing sequence information as the newly incorporated nucleotide and codon usage frequencies shift to host frequencies.

Nevertheless, while the issue of HGT across significant evolutionary distance remains contentious, evidence continues to accumulate that HGT occurs between much more disparate organisms than was once thought (Wolfe and Li 2003). The first clear evidence of massive HGT between Archaea and Bacteria was the determination of significant increases in archaeal protein homologues in the genomes of the hyperthermophilic bacteria *Aquifex aeolicus* (Aravind et al. 1998) and *Thermotoga maritima* (Nelson et al. 1999). *T. maritima*, a non-spore-forming, rod-shaped bacterium belonging to the order Thermotogales, was originally isolated from geothermal heated marine sediment at Vulcano, Italy, exhibiting a growth temperature optimum of 80°C. The whole genome analysis of this eubacterium reveals significant evidence implicating HGT mechanisms in the transfer of multiple genes from archaeal sources. Small subunit ribosomal RNA (SSUrRNA) phylogenetic analysis places *T. maritima* as one of the deepest rooted and most slowly evolving lineages of the eubacteria. The 451 Archaea-like genes in *T. maritima* are not uniformly distributed across the commonly accepted biological roles. The majority of housekeeping genes involved in transcription, translation, DNA replication, and cell division are most similar to eubacteria orthologues. In contrast, 49% of transporters, 60% of electron transport proteins, and 42% of conserved hypothetical proteins are most similar to archaeal genes; 81 of the archaea-like genes are clustered in 15 discrete regions of the chromosome. The genes and gene order conservation in seven of these regions, have only been described in the genomes of thermophilic Archaea. Two of the clusters are associated with Archaea 30 bp repeat elements, lending support to the possibility that these repeat elements may be involved in gene transfer.

HGT between Archaea and Eubacteria is not confined to hyperthermophilic systems. *Salinibacter ruber*, a large, rod hyperhalophilic bacterium isolated from saltern crystallizers (~37% NaCl) exhibits physiological characteristics common to the haloarchaea including high intracellular potassium levels and excessive levels of carotenoids in its membranes (Mongodin et al. 2005). The *S. ruber* pigment is chemically related to carotenoids of *Rhodothermus marinus* than the haloarchaea bacterioruberins. Like the haloarchaea, *S. ruber* is an aerobic heterotroph. The genome highlights the major questions arising in the identification of HGT events. The shared phenotypes may have arisen via physiological convergence, molecular convergence, or via HGT. Phylogenetic analysis places *S. ruber* in the Bacteroides/Chlorobi group with *R. marinus* as its nearest neighbor. *coxB2* and *coxA2* genes appear to have been imported from haloarchaea. Genes coding for proteins critical for survival as a hyperhalophile including three *trkA* homologues (K+uptake/efflux) and one *trkH* (cationic amino acid transporter) are closely related to haloarchaeal homologues. The discovery of the proton pump bacteriorhodopsin in the haloarchaea *Halobacterium salinarum* has been seen as a unique archaea evolutionary contribution. Rhodopsin photobiology has now been found in other groups of prokaryotes, but as occurs in *Halobacterium* sp. NRC-1 and *Haloarcula marismortui*, *S. ruber* exhibits a total of four rhodopsin genes, a feat unmatched by any other characterized bacterium. The functionally important residues in *S. ruber* rhodopsins match those of the haloarchaeal proteins to which the rhodopsins are homologous, and the two SR-transducer homologues both appear to be sensory rhodopsin photoreceptor transducers, like their haloarchaeal *Htrl* homologues. Unless the last universal common ancestor had a complete set of rhodopsins, the presence of several identifiable classes of this protein in *S. ruber* and haloarchaea is best explained by multiple HGT events.

7.3 INVESTIGATING HGT IN POLAR REGIONS: THE PROBLEM OF MULTIPLE STRESSORS

An investigation of the impact of HGT on the response of polar microbial communities to environmental stressors is in its infancy. A central problem compounding the analysis of genomic information from these regions is the simultaneous occurrence of multiple environmental stressors, particularly in Antarctica. The primary stressors requiring adaptive responses from indigenous organisms are

1. Significant increases in UV radiation as a result of ozone depletion.
2. Exposure to environmental toxins introduced by human exploration.
3. In the coming decade, the transformation of freeze-thaw cycles will be instigated by global warming. In the case of ice-covered glacial melt lakes, rising ambient temperatures will result in the exposure of previously isolated microbial communities to invasion by new competitors for limited resources.

The interactions between UV photonic and hydrocarbon stressors pose significant challenges both to the indigenous microbial communities and to *in situ* attempts to monitor and de-convolve the impacts of each factor (Pelletier et al. 2006). It was recently shown that damage from UV-B radiation to marine microorganisms can be completely masked by the even stronger deleterious effects of soluble petroleum hydrocarbons (Sargian et al. 2005). Changes in phytoplankton pigment content and *in vivo* fluorescence, as well as alterations in cell abundance and morphology of phyto- and bacterioplankton populations from Ushuaia Bay (Southern Argentina) have been demonstrated when the community was exposed to either increased UV-B irradiation or the water soluble fraction of a crude oil contaminant (Sargian et al. 2007). Simultaneous exposure to the two environmental stressors produced a synergistic impact. Unfortunately, nothing is known about the frequency of HGT in these specific communities.

7.3.1 Bacteriophages in Polar Regions

Studies conducted in the Arctic polar region have documented the presence of cultivable phage–host systems capable of instigating HGT events. Three phage–host systems from Arctic sea ice and melt pond samples were isolated northwest of Svalbard (Borriss et al. 2003). On the basis of 16S rRNA sequences, the three psychrophilic hosts appear most similar to *Shewanella frigidimarina, Flavobacterium hiberman,* and *Colwellia psychrerythraea.* Their phages are host-specific and all belong to the tailed, double-stranded DNA phage families Siphoviridae and Myoviridae. Both the hosts and their associated phages exhibit significant adaptation to cold temperatures, yet the specific genes responsible for these adaptive capabilities are unknown.

Appreciable *in situ* phage abundances have been observed in polar inland waters ranging from 2.0×10^7 L^{-1} (Beaver Lake, MacRobertson Land, Antarctica) to 1.0×10^{11} L^{-1} (saline lakes, Vestfold Hills, Antarctica) (Sawstrom et al. 2008), indicating that phages are important components of microbial communities in these environments. In general, there is a higher abundance of detectable virus-like particles (VLP) in Antarctic inland waters than those of the Arctic (Sawstrom et al. 2008). VLPs have also been detected in cryoconite sediments originating from an Arctic glacier as well as ice cores from the ice sheet covering Lake Vostok in Antarctica (Sawstrom et al. 2008).

Viral production rates measured for Antarctic inland waters vary from 2.0×10^6 to 2.0×10^9 VLP L^{-1} h^{-1}, depending on the method used to estimate production (Sawstrom et al. 2008). Variability in the production of viruses can also be attributed to the trophic status of the lakes, with ultra-oligotrophic lakes (Crooked Lake and Lake Druzhby) exhibiting the lowest rates in viral production compared to more productive saline lakes (Ace Lake, Pendant Lake, and Lake Hoare) (Sawstrom et al. 2008). Viruses inhabiting these ecosystems strongly influence nutrient cycling and play a crucial role in structuring polar food webs. The latter is especially true for Antarctic lakes that are characterized by condensed food webs primarily constructed of microbes (Sawstrom et al. 2008). The viral-mediated mortality of microbial hosts results in the release of carbon-rich lysis products. Heterotrophic microbes can rapidly assimilate dissolved organic carbon (DOC), resulting in a stimulation of bacterial production and respiration of fixed carbon to carbon dioxide. Although there is no information regarding how polar viral populations respond to naturally occurring or anthropogenic ecosystem stressors, it is likely that perturbations to these environments will influence viral community structure, function, the nature of virus–host interactions and the rate of viral-mediated transduction.

Bacteriophage methods for dealing with life in the extreme cold remain obscure. The mesophilic phage λ has been shown to significantly shift toward lysogeny at the expense of lytic activity when exposed to low temperatures (Obuchowski et al. 1997). The phage essentially serves as a temperature sensor and shuts down lytic pathways at 20°C. The controlling process appears to be the phage CII activity as a transcriptional activator. While this is an interesting response for a mesophilic phage, other processes must be operative for psychrophilic or psychrotolerant phage. Presumably, these strategies will include choices in protein residues and slight alterations in protein tertiary structure similar to changes documented for their bacterial hosts.

7.3.2 Microbial Adaptation to Polar Environmental Stressors

In the polar inland waters of the Arctic and Antarctica, the primary drivers for adaptation during global warming are increases in evaporative water loss, salinity, alkalinity, and surface UV radiation. During the annual springtime ozone depletion over Antarctica, surface UV radiation levels are estimated to be ~130% higher than the values recorded during the 1970s (Perin and Lean 2004). Current models predict significant reduction in primary productivity in Antarctic waters at depths of 0–20 m as a result of the increased UV flux (Smith et al. 1992).

Solar UV radiation modifies genetic information in multiple ways. Ribonucleic acid (RNA) and deoxyribonucleic acid (DNA) can be damaged directly by the absorption of high-energy UV photons or by contact with energetic superoxide free radicals formed by the UV activation of intracellular water molecules. Mutations or lethal damage can occur by assault on either the nucleotide bases or the polymer phosphate backbone. In either case, genomic alterations can significantly impact the formation of critical metabolic and structural proteins (Buma et al. 2001a).

The primary productivity for many of these aquatic ecosystems is determined by the health of a complex community of photosynthetic phytoplankton (Neale 2001). These consortia of autotrophic microbes inhabit the top few meters of the water column and serve as the first line of defense for the remainder of the ecosystem's microbial community during periods of increased UV radiation (Buma et al. 2001b). The inhibition of phytoplankton productivity by UV radiation for Antarctic marine phytoplankton (Smith et al. 1992, Helbling et al. 1996, Neale et al. 1998) most likely results in photosystem II damage (Melis et al. 1992, Schofield et al. 1995, Bouchard et al. 2005), and the impairment of the carbon-fixing enzyme, ribulose-1,5-biphosphate carboxylase-oxygenase (RuBisCO) (Bischof et al. 2000, Xue et al. 2005).

The high cellular abundance and diversity of photosynthetic, nitrogen-fixing cyanobacteria in both Arctic and Antarctic lakes may be indicative of multiple HGT events since the original colonization of these ecological niches (Vincent et al. 2000a,b). These cyanobacteria exhibit a wide range of protective strategies against UV radiation damage including mechanical escape (gliding) and enhanced production of multiple UV-absorbing pigments (Quesada and Vincent 1997). Recently, the *psbA* and *psbD* genes encoding for important photosystem II proteins were identified in multiple cyanophage genomes (Mann et al. 2003, Lindell et al. 2004, Millard et al. 2004, Sullivan et al. 2005) and marine metagenomes (DeLong et al. 2006, Bench et al. 2007, Sharon et al. 2007, Williamson et al. 2008). The analysis of G+C content and codon usage in *psbA* genes of environmental viruses implicate HGT exchange and reshuffling of *psbA* genes between *Synechococcus* and *Prochlorococcus* via phage intermediates (Zeidner et al. 2005, Sullivan et al. 2006). These important findings, along with evidence of additional cellular metabolic genes in environmental viral genomes suggest that phages behave as genetic reservoirs capable of altering the evolution and adaptation of microbial hosts during times of environmental change.

7.3.3 HGT in Antarctica

A complete genomic analysis of the Antarctic biome is still in its early years. Nevertheless, reports of HGT in Antarctic prokaryotes have begun to appear. *Methanococcoides burtonii* is a cold-adapted methanogen isolated from Ace Lake, Antarctica known to use membrane lipid unsaturation as a cold-adaptive mechanism (Goodchild et al., 2004). During the bioinformatic evaluation of 528 proteins that are expressed in the organism when grown at 4°C, 135 proteins were annotated only as unique and conserved, but were suspected of playing important roles during low temperature growth. Subsequent analysis (Saunders et al. 2005) identified gene 150 as a functional *cas* gene with clustered regularly interspaced short palindromic repeats (CRISPR). This family of repetitive DNA sequences and the associated *cas* genes are thought to be involved with phage and plasmid immunity. CRISPRs are subject to HGT and have been detected in several species of Archaea and Bacteria. However, on the basis of gene order and sequence similarity, the *M. burtonii* CRISPR locus is more similar to that found in several pathogenic species of Bacteria. The similarities make the locus a strong candidate for evidence of archaeal–bacterial horizontal gene transfer. The nature of the transfer mechanism is currently unknown.

Phylogenetic screening of 3200 clones from a metagenomic library of Antarctic mesopelagic picoplankton produced two bacterial clones belonging to the Deltaproteobacteria designated DeepAnt-1F12 and DeepAnt-32C6 (Moreira et al. 2006). Many of the genes encoded were informational and involved in replication, transcription, and translation. Nevertheless, several instances of HGT were detected, including an apparent transposase. DeepAnt-1F12 contained one putative gene encoding a long cysteine-rich protein, probably membrane-bound and Ca^{2+} -binding, with only eukaryotic homologues. DeepAnt-32C6 carried some predicted genes involved in metabolic pathways that suggested the organism may be anaerobic and able to ferment and degrade complex compounds extracellularly. Several conserved proteins gave good phylogenetic signals identifying them as very distant from the Deltaproteobacteria. These putative HGT regions exhibited both low-G+C and high-G+C levels. DeepAnt-32C6 ORF 3 encodes an adenosylmethionine-8-amino-7-oxononanoate aminotransferase closely related to high-G+C Gram-positive homologues. DeepAnt-1F12 ORF 8 encodes a conserved transposase domain similar to transposases seen frequently in Gammaproteobacteria, and in one example from the alphaproteobacterium, *Sphingobium herbicidovorans*, and one from the betaproteobacterium, *Delftia acidovorans*. The transposase is also found in one high-G+C Gram-positive species, *Arthrobacter aurescens*, where it exhibits an almost identical sequence, suggesting very recent acquisition from the Gammaproteobacteria. However, for DeepAnt-1F12, the gene is truncated and very divergent, indicating a more ancient acquisition.

Finally, evidence of HGT has also begun to appear in prokaryotes capable of metabolizing potential hydrocarbon contaminants. Twenty-one Antarctic marine bacteria able to utilize diesel fuel as the sole carbon and energy source were isolated from the Ross Sea, analyzed by 16S rDNA restriction analysis using the enzyme *Alu*I, and assigned to two different bacterial species in the genera *Rhodococcus* and *Alcaligenes* (Pini et al. 2007). A 550-bp DNA fragment coding for the inner region of an alkane mono-oxygenase was PCR-amplified from each strain. The phylogenetic analysis of the putative *AlkB* protein coded for by the fragment of these genes suggested HGT of the gene on the basis of G+C content variation.

7.4 EVIDENCE FOR HGT IN BIOREMEDIATION

Horizontal gene transfer appears to play a significant role in the spread of metabolic pathways capable of facilitating the degradation of a variety of environmental contaminants. Even though it is feasible for viruses to mediate the transfer of enzymes with remediation potential, the majority of well-known HGT examples are plasmid related. For example, in a coal tar-contaminated field site in Glen Falls,

New York, bacterial isolates were found to contain self-transmissible plasmids homologous to *pDTG1*, a naphthalene-catabolic plasmid found in *Pseudomonas putida* NCIB 9816-4 (Hohnstock et al. 2000). In a clear example of HGT, genes for enzymes catalyzing naphthalene catabolism were transferred from donor microorganisms to co-occurring microbes that were previously unable to metabolize this compound.

Hexachlorocyclohexane (HCH) is a cyclic, saturated hydrocarbon prepared commercially by the chlorination of benzene in the presence of UV. The extensive and indiscriminant use of HCH for the past five decades against a variety of agricultural pests and in antimalarial health campaigns has produced a profound environmental contamination problem. Bacteria (from the family Sphingomonadaceae) that are capable of degrading HCH have been isolated from geographically diverse HCH-contaminated soils (Lal et al. 2006). These geographically distant isolates contain nearly identical *lin* genes responsible for HCH degradation, all associated with plasmids and the insertion sequence element IS6100; both of which appear to play fundamental roles in this example of widespread HGT.

Synthetic haloalkanes (alkanes such as methane or ethane linked to a halogen such as chlorine or fluorine) are potential carcinogens widely used in a variety of industrial and agricultural products worldwide including flame retardants, solvents, refrigerants, and propellants. One particular family of haloalkanes, the chlorofluorocarbons, has been a focus of considerable investigation for their role in environmental contamination and atmospheric ozone depletion. Microorganisms do not normally express metabolic enzymes capable of efficiently catabolizing these synthetic molecules. The result is the long-term survival of these synthetic hydrocarbons in the environment. It is of considerable interest for our understanding of information exchange and rapid evolutionary adaptation that microbial communities repeatedly exposed to synthetic haloalkanes can devise degradation pathways for these xenobiotic molecules, exploit them as growth substrates, and appear capable of sharing these strategies among a diverse group of microbes. Three phylogenetically distinct bacteria *Rhodococcus rhodochrous* NCIMB13064, *Pseudomonas pavonaceae* 170, and *Mycobacterium* sp. strain GP1 share a highly conserved haloalkane dehalogenase gene (*dhaA*) (Poelarends et al. 2000a). *R. rhodochrous* NCIMB1306 can use the haloalkane 1-chlorobutane as its sole carbon and energy source. The critical step in 1-chlorobutane catabolism is the cleavage of the carbon–halogen bond, a process catalyzed by the inducible hydrolytic haloalkane dehalogenase, *dhaA*, to produce *n*-butanol. The *dhaA* gene for NCIMB13064 resides within a 1-chlorobutane catabolic gene cluster. This cluster of genes also encodes a putative invertase (*invA*), a regulatory protein (*dhaR*), an alcohol dehydrogenase (*adhA*), and an aldehyde dehydrogenase (*aldA*). The alcohol and aldehyde dehydrogenases may catalyze an oxidative conversion of *n*-butanol to *n*-butyric acid, a readily accessible growth substrate for a variety of bacteria. *Pseudomonas pavonaceae* 170, and *Mycobacterium* sp. strain GP1 contain a putative integrase gene next to the conserved *dhaA* segment. The proximity of this integrase to coding regions for catabolic pathways capable of degrading molecules only recently synthesized by humanity suggests that HGT is responsible for the inclusion of these DNA segments in the genomes of these phylogenetically diverse microorganisms.

Along these lines, 16S rRNA genes of the haloalkane-degrading bacterium *Rhodococcus* sp. strain NCIMB13064 and five Gram-positive haloalkane-utilizing bacteria isolated from contaminated sites in Europe, Japan, and the United States have been shown to exhibit less than 1% sequence divergence (Poelarends et al. 2000b). All sequences indicated that the strains belonged to the genus *Rhodococcus* and all the strains shared a completely conserved haloalkane dehalogenase gene (*dhaA*), suggesting these genes derived from a recent common ancestor. The genomes of all strains also contain a conserved region of 12.5 kb that includes the haloalkane-degradative gene cluster previously identified in strain NCIMB13064. Each haloalkane degrader carried the *dhaA* gene region on both the chromosome and on a plasmid suggesting that an ancestral plasmid was transferred between these *Rhodococcus* strains and has since undergone a series of insertions or deletions. These data suggest that the haloalkane dehalogenase catabolic gene cluster of these Gram-positive haloalkane-utilizing bacteria has been distributed to geographically disparate groups of bacteria across the globe on very recent evolutionary timescales.

An epoxyalkane:coenzyme M (CoM) transferase (EaCoMT) enzyme, recently found to be active in aerobic vinyl chloride (VC) and ethene assimilation pathways, was investigated in 10 different mycobacteria isolated on VC or ethene from diverse environmental samples (Coleman and Spain 2003). The *Mycobacterium* EaCoMT genes formed a distinct cluster that appeared to be closely related to the EaCoMT gene of *Rhodococcus* strain B-276. Notable discrepancies between the EaCoMT and 16S rRNA phylogenetic trees coupled with the fact that isolates from geographically distant locations exhibit almost identical EaCoMT genes suggest that EaCoMT has been horizontally transferred among the *Mycobacterium* strains. These observations suggest that the CoM-mediated pathway of epoxide metabolism may be universal in alkene-assimilating mycobacteria because of widespread plasmid-mediated HGT.

One example of potential phage-mediated HGT in bioremediation is illustrated by a species of *Geobacter metallireducens* isolated from contaminated subsurface sediments with elevated concentrations of uranium and nitrate generated during the production of weapons were collected from the Field Research Center at the Y-12 complex within the Oak Ridge National Laboratory reservation at Oak Ridge, Tennessee. Genomic analysis revealed a 300 kb genomic island harboring enzymes capable of degrading phenol, *p*-crestol, 4-hydroxybenzaldehyde, 4-hydroxybenzoate, benzyl alcohol, benzaldehyde, and benzoate (Butler et al. 2007). All four regions of aromatics metabolism genes are flanked on at least one side by tRNA genes. These genes are often associated with genomic islands and are often target sites for phage attachment and integration. The 3′ end of the large aromatics island is flanked by a tRNA-Gly gene, and a pseudo-tRNA appears 6.6 kb upstream. Three genes with homology to phage integrases, Gmet_2271, Gmet_2278, and Gmet_2279, are encoded nearby. There is a tRNA-Gly near the *bamB* homologue (Gmet_1802) and a putative phage integrase (Gmet_1783). A region containing the *oah* homologue has a tRNA-Ala at the 5′ end, and the toluene region is flanked on both sides by tRNA genes: a tRNA-Met at the 5′ end and a tRNA-Val at the 3′ end. This region of the genome exhibits many of the characteristics associated with genomic islands: the genes are species-specific and they are encoded in discrete units flanked by genes with atypical nucleotide composition as well as by tRNA genes and phage-like integrases. The regional signs of genetic mobility include identical transposons and repetitive sequences. The overabundance of phage-like genes in the island suggests that the transfer of these enzymes may have been the result of a transduction event. Furthermore, the aromatic enzymes are flanked on at least one side by tRNA genes, sights that are often targets of temperate phage attachment. The lack of detectable differences in the nucleotide composition of the island and the surrounding genome and the lack of syntenic organization with other aromatic degraders suggest that the transfer event did not occur recently.

7.5 HGT ACTIVITY IN POLAR BIOREMEDIATION

The impact of HGT has been previously documented in a variety of psychrophilic marine waters, freshwater lakes, and soil environments, but our understanding of the extent of HGT in polar bioremediation is in its infancy (Aislabie et al. 2004). As discussed earlier, 21 Antarctic marine bacteria isolated from Terra Nova Bay were capable of utilizing diesel fuel as their sole carbon and energy source (Pini et al. 2007). 16S rDNA restriction analysis and random amplified polymorphic DNA (RAPD) analysis suggested minimal diversity within the microbial community with all organisms belonging to one of two bacterial species in the genera *Rhodococcus* and *Alcaligenes*. The DNA fragment coding for the inner region of alkane monooxygenase was identified as encoding for a putative *AlkB* protein that was most likely acquired via HGT. The microbial isolates utilized two different strategies to acquire hydrocarbons. *Rhodococcus* isolates utilized biosufactants and *Alcaligenes* isolates deal with hydrocarbon energy sources through multiple membrane modifications.

Marinobacter is a widely distributed bacterium indigenous to coastal marine environments in Europe, Asia, North America, Africa, and Antarctica. Strain NCE312 can utilize naphthalene as a sole carbon source and has been shown to degrade 2-methylnaphthalene and 1-methylnaphthalene. The comparative sequence analysis of the *Marinobacter* dioxygenase gene with those from *Pseudomonas* and *Burkholderia* strains revealed a high level of sequence similarity suggesting horizontal transfer between these bacterial lineages (Hedlund et al. 2001).

Twenty-two PAH-degrading bacterial strains using only naphthalene or phenanthrene as a sole carbon source have been isolated from hydrocarbon-contaminated Antarctic soils (Ma et al., 2006), but no degrader species could be isolated from an unpolluted control site. Phylogenetic analysis placed all isolates within the genus *Pseudomonas* except for one identified as most likely belonging to the genus *Rahnella*. Some of the isolates were closely related to known cold-tolerant species, but others clearly represent novel, previously unseen strains. Unlike mesophilic species, psychrophilic isolates were capable of the degradation of naphthalene at 4°C. The maximum naphthalene degradation efficiency for both mesophilies and psychrophiles occurred at 30°C. Psychrophilic *Pseudomonas* strain LCY exhibited naphthalene degradation enzymatic activity level at 30°C of 12 35.19 nmol min^{-1} mg^{-1} compared to 22.41 nmol min^{-1} mg^{-1} for the mesophilic control. A comprehensive amplification of the *ndo* gene encoding naphthalene dioxygenase indicated that all of the isolates possessed the potential to degrade this toxic substance. The dioxygenase sequences from the isolates were not only highly similar to each other at the nucleotide level (97%), they were also remarkably similar to mesophilic *ndo* gene sequences. Significant differences were noted in the phylogeny derived from the analysis of 16S rRNA genes and *ndo* genes for the isolates suggesting that the dioxygenase genes were horizontally transferred. Considering that the *ndo* genes encoded by the isolates are located on large self-transmissible plasmids, there appears to be strong evidence for HGT among these isolates and suggests that certain *Pseudomonas* species that are capable of playing an important role in PAH biodegradation in Antarctic soils.

Similarly, the analysis of 135 psychrotolerant microorganisms isolated from multiple Canadian ecosystems revealed strains that are capable of mineralizing toluene, naphthalene, dodecane, and hexadecane (Whyte et al. 1996). The aromatic degradation genes of the pyschrotolerant strains were either plasmid or chromosomally located and demonstrated homology to genes previously identified in mesophilic bacteria. These studies imply that cold-adapted microbes may have acquired genes that are integral to the degradation of toxic substances from organisms inhabiting environments that are more temperate. However, the relative paucity of Antarctic HGT geographically diverse epidemiological studies of both mesophilic psychrotolerant and psychrophilic biodegradative strains make it impossible to assess the relative importance of HGT and the physical transport of microbial species from South America, Africa, Australia, and New Zealand in the appearance of biodegradative metabolism on the continent.

7.6 SUMMARY

Microbial life has been continuously present on Earth since it appeared relatively rapidly after the cessation of planetary late heavy bombardment: approximately 3.8 billion years ago (Mojzsis et al. 1996). Since then, the rate of evolutionary change has on multiple occasions, accelerated dramatically to produce extraordinary adaptive responses to environmental challenges (Eldredge 1971, Eldredge and Gould 1972, Gould and Eldredge 1977, 1993). In fact, the evolution of microbial diversity on Earth has been attributed by one group to recurrent cycles of global warming and cooling that form, maintain, and then significantly alter the local environments of microbial communities (Zavarzin 1993, 1999). Such cyclical changes in temperature, total solar radiation, and shifts in solar photon energy distributions including the relative abundances of far and near UV light appear to have been a fundamental part of the geo-biological history of Earth.

In some cases, the driving force for change has been the introduction of a particular "contaminant" of biological origin. Certainly, the production of oxygen by photosynthetic cyanobacteria would have functioned as a lethal toxin to many of the anaerobic microbes dominating archean Earth. To other microbial communities capable of producing and then sharing oxygen-utilizing metabolic enzymes, the transition from a reduced to an oxidized planet would have presented significant evolutionary advantages.

Rapid shifts in climate or the appearance of potentially lethal contaminants demand that microbial communities adapt relatively quickly to changes in the availability of structural materials, the abundance or quality of energy sources, and the competitive advantages of coresident microorganisms. The significant variables may include pH, temperature, physical maceration, and hydration/desiccation. On geological or classical evolutionary timescales, the species response times required for such phenotypic adaptation to environmental change is extraordinarily brief, sometimes on the order of tens to hundreds of years. This implies that the extreme phenotypic diversity we see in the surviving eukaryotic and prokaryotic organisms (Ma et al. 2004) has most likely not been the result of the slow accumulation of individual point mutations, but a much more efficient and sophisticated series of events. Evidence continues to accumulate that rapid evolutionary adaptation could have been repeatedly facilitated by strategies that are more efficient; ideally demonstrated by bacteriophage-mediated horizontal gene transfer (Ochman et al., 2000, Hendrix 2002).

While the rapid spread of information by HGT is efficient and dramatic, our understanding of the exact statistics of this process is profoundly deficient. To polar microbial communities, many of the challenges posed by contamination mimic the threat presented to common pathogens by antibiotics. Does HGT occupy a central position in rapid microbial adaptation to the pollution of fragile ecosystems similar to its role in human epidemics and microbial drug resistance? A series of additional questions can be posed. How exactly do viruses and their hosts interact during times of environmental change? Is the selection and excision of a series of microbial host genes a completely random process (during generalized transduction) or is there some kind of selective mechanism that we have yet to discover? What environmental factors trigger the activation of quiescent prophage genomes? In a rapidly changing, isolated environment, how often do lytic and temperate phages and other viruses facilitate HGT among disparate prokaryotic and eukaryotic taxa? How profoundly have these rapid genomic and phenotypic adaptations, mediated by HGT, influenced our record of Darwinian evolution and the development of diversity in viral, prokaryotic, and eukaryotic genomes? Lastly, how do we improve our abilities to detect and monitor HGT information transfer in a community of diverse organisms?

The latter question is of particular importance if we wish to harness the adaptive power of HGT for targeted bioremediation efforts. Clearly, natural microbial communities exposed to toxic compounds can adapt to new environmental demands by mutation or through the acquisition of the appropriate metabolic or biodegradative genes. In other cases, when a novel material defies the solutions available to indigenous microbes, genetically engineered microorganisms have been considered (van der Meer 2006). Current treaties, particularly those covering Antarctica, specifically prohibit the introduction of novel species and require careful monitoring of attempts to modify the current polar ecosystems. The most logical line of investigation for the next decade is to intensify our exploration of pristine polar habitats to identify the metabolic pathways available to indigenous microbes that are capable of accomplishing anticipated bioremediation tasks. The metagenomic exploration (i.e., examining the collective genetic contents of indigenous microbial communities) of polar microbes will help to shed light on the metabolic potential of these organisms and we suspect that increasing efforts will be made to produce metagenomic datasets from select, previously unexplored Arctic and Antarctic ecological niches. The in-depth characterization of polar viral communities using both cultivation-dependent and independent approaches (e.g., metagenomics) coupled with improvements to current bioinformatic algorithms to identify HGT events in metagenomic data will facilitate our understanding of how bacteriophage in these environments mediate the transfer of information between organisms.

Finally, the development of novel, *in situ*, airborne, and satellite optical techniques may be of assistance in the rapid detection of hydrocarbon-contaminated sites and may facilitate the monitoring of bioremediation efforts. Techniques currently being tested to detect both life in the ice on Earth and PAH in-fall material in the Mars regolith may have direct application to polar bioremeditaion strategies. Instrumentation development work currently underway focuses on *in situ* and remote imaging of laser induced fluorescence emission (L.I.F.E.) following 365 nm laser excitation of 3-, 4-, and 5-ring PAH structures presumed present beneath the Mars regolith during the European Space Agency's ExoMars mission (Storrie-Lombardi et al. 2008, Muller et al. 2009) and inducing fluorescence in photosynthetic cyanobacterial pigments in Antarctic cryoconite assemblages resident in transparent Antarctic lake and glacier ice (Storrie-Lombardi et al. 2009). The ability to systematically probe individual biomolecular targets makes it possible to monitor both the health of the indigenous dominant photosynthetic microbial life and the appearance of exogenous PAH contaminants in the ice. The development of these techniques will require extensive interdisciplinary cooperation between microbiology, earth sciences, optics, and materials engineering. However, such cooperation has been a cornerstone of polar exploration for many decades and will continue to bring novel techniques to bear on threats to these fragile, bell-weather environments.

ACKNOWLEDGMENTS

The authors thank the Kinohi Institute, the J. Craig Venter Institute, J. N. Pritzker, and the Tawani Foundation for their ongoing support and encouragement for this project.

REFERENCES

Ackermann, H. W. and M. S. DuBow. 1987. *Viruses of Prokaryotes: General Properties of Bacteriophages*. Boca Raton, FL: CRC Press, Inc.

Aislabie, J. M., M. R. Balks, J. M. Foght, and E. J. Waterhouse. 2004. Hydrocarbon spills on Antarctic soils: Effects and management. *Environ Sci Technol* 38 (5):1265–1274.

Aravind, L., R. L. Tatusov, Y. I. Wolf, D. R. Walker, and E. V. Koonin. 1998. Evidence for massive gene exchange between archaeal and bacterial hyperthermophiles, *Trends Genet.*, 14:442–444.

Bench, S. R., T. E. Hanson, K. E. Williamson, D. Gosh, M. Radosovich, K. Wang, and K. E. Wommack. 2007. Metagenomic characterization of Chesapeake Bay virioplankton. *Appl Environ Microbiol* 73 (23):7629–7641.

Bischof, K., D. Hanelt, and C. Wiencke. 2000. Effects of ultraviolet radiation on photosynthesis and related enzyme reactions of marine macroalgae. *Planta* 211 (4):555–562.

Bogdanova, E. S., I. A. Bass, L. S. Minakhin, M. A. Petrova, S. Z. Mindlin, A. A. Volodin, E. S. Kalyaeva et al. 1998. Horizontal spread of mer operons among gram-positive bacteria in natural environments. *Microbiology* 144 (Pt 3):609–620.

Borriss, M., E. Helmke, R. Hanschke, and T. Schweder. 2003. Isolation and characterization of marine psychrophilic phage-host systems from Arctic sea ice. *Extremophiles* 7 (5):377–384.

Bouchard, J. N., S. Roy, G. Ferreyral, D. A. Campbell, and A. Curtosi. 2005. Ultraviolet-B effects on photosystem II efficiency of natuaral phytoplankton communities from Antarctica. *Polar Biol* 28 (8):607–618.

Buma, A. G., E. W. Helbling, M. K. de Boer, and V. E. Villafane. 2001a. Patterns of DNA damage and photoinhibition in temperate South-Atlantic picophytoplankton exposed to solar ultraviolet radiation. *J Photochem Photobiol B* 62 (1–2):9–18.

Buma, A. G. J., M. K. de Boer, and P. Boelen. 2001b. Depth distributions of DNA damage in Antarctic marine phyto- and bacterioplankton exposed to summertime UV radiation. *J Phycol* 37 (2):200–208.

Bushman, F. 2002. *Lateral DNA Transfer: Mechanisms and Consequences*. Cold Spring Harbor, NY: Cold Spring Harbor Laboratory Press.

Butler, J. E., Q. He, K. P. Nevin, Z. He, J. Zhou, and D. R. Lovley. 2007. Genomic and microarray analysis of aromatics degradation in *Geobacter metallireducens* and comparison to a *Geobacter* isolate from a contaminated field site. *BMC Genom* 8:180.

Canchaya, C., G. Fournous, S. Chibani-Chennoufi, M. L. Dillmann, and H. Brussow. 2003. Phage as agents of lateral gene transfer. *Curr Opin Microbiol* 6 (4):417–424.

Chen, J. and Novick, R. P. 2009. Phage-mediated intergeneric transfer of toxin genes, *Science*, 323 (5910):139–141.

Choi, I. G. and S. H. Kim. 2007. Global extent of horizontal gene transfer. *Proc Natl Acad Sci U S A* 104 (11):4489–4494.

Coleman, N. V. and J. C. Spain. 2003. Distribution of the coenzyme M pathway of epoxide metabolism among ethene- and vinyl chloride-degrading *Mycobacterium* strains. *Appl Environ Microbiol* 69 (10):6041–6046.

Dahlberg, C., M. Bergström, and M. Hermansson. 1998. *In situ* detection of high levels of horizontal plasmid transfer in marine bacterial communities. *Appl Environ Microbiol* 64 (7):2670–2675.

Davison, J. 1999. Genetic exchange between bacteria in the environment. *Plasmid* 42 (2):73–91.

DeLong, E. F., C. M. Preston, T. Mincer, V. Rich, S. J. Hallam, N. U. Frigaard, A. Martinez et al. 2006. Community genomics among stratified microbial assemblages in the ocean's interior. *Science* 311 (5760):496–503.

Dowson, C. G., A. Hutchison, J. A. Brannigan, R. C. George, D. Hansman, J. Linares, A. Tomasz, J. M. Smith, and B. G. Spratt. 1989. Horizontal transfer of penicillin-binding protein genes in penicillin-resistant clinical isolates of *Streptococcus pneumoniae. Proc Natl Acad Sci U S A* 86 (22):8842–8846.

Dowson, C. G., A. Hutchison, N. Woodford, A. P. Johnson, R. C. George, and B. G. Spratt. 1990. Penicillin-resistant viridans streptococci have obtained altered penicillin-binding protein genes from penicillin-resistant strains of *Streptococcus pneumoniae. Proc Natl Acad Sci U S A* 87 (15):5858–5862.

Eldredge, N. 1971. The allopatric model and phylogeny in Paleozoic invertebrates. *Evolution* 25 (1):156–167.

Eldredge, N. and S. J. Gould. 1972. Punctuated equilibria: An alternative to phyletic gradualism. In *Models in Paleobiology*, T. J. M. Schopf (ed.). San Francisco, CA: Freeman Cooper and Co.

Goodchild, A., M. Raftery, N. F. W. Saunders, M. Guilhaus, and Cavicchioli, R. 2004. Cold adaptation in the Antarctic Archaeon Mehtanococcoides burtonii involves membrane lipid unsaturation. *J Proteome Res* 3:1164–1176.

Gould, S. J. and N. Eldredge. 1977. Punctuated equilibrium: The tempo and mode of evolution reconsidered. *Paleobiology* 3 (2):115–151.

Gould, S. J. and N. Eldredge. 1993. Punctuated equilibrium comes of age. *Nature* 366 (6452):223–227.

Hausner, M. and S. Wuertz. 1999. High rates of conjugation in bacterial biofilms as determined by quantitative in situ analysis. *Appl Environ Microbiol* 65 (8):3710–3713.

Hedlund, B. P., A. D. Geiselbrecht, and J. T. Staley. 2001. *Marinobacter* strain NCE312 has a *Pseudomonas*-like naphthalene dioxygenase. *FEMS Microbiol Lett* 201 (1):47–51.

Helbling, E. W., B. E. Chalker, W. C. Dunlap, O. HolmHansen, and V. E. Villafane. 1996. Photoacclimation of Antarctic marine diatoms to solar ultraviolet radiation. *Journal of Experimental Marine Biology and Ecology* 204 (1–2):85–101.

Hendrix, R. W. 2002. Bacteriophages: evolution of the majority. *Theor Popul Biol* 61 (4):471–480.

Hohnstock, A. M., K. G. Stuart-Keil, E. E. Kull, and E. L. Madsen. 2000. Naphthalene and donor cell density influence field conjugation of naphthalene catabolism plasmids. *Appl Environ Microbiol* 66 (7):3088–3092.

Jain, R., M. C. Rivera, and J. A. Lake. 1999. Horizontal gene transfer among genomes: The complexity hypothesis. *Proc Natl Acad Sci U S A* 96 (7):3801–3806.

Jiang, S. C. and J. H. Paul. 1998. Gene transfer by transduction in the marine environment. *Appl Environ Microbiol* 64 (8):2780.

Johansen, I. and A. Ustaheim. 1968. Enhancement of the viability of irradiated bacteria by chromosome transfer. *Radiat Res* 36 (3):610–621.

Koonin, E. V. and Wolf, Y. I. 2008. Genomics of Bacteria and Archaea: The emerging dynamic view of the prokaryotic world. *Nucl Acid Res* 36 (21):6688–6719.

Koonin, E. V., K. S. Makarova, and L. Aravind. 2001. Horizontal gene transfer in prokaryotes: Quantification and classification. *Annu Rev Microbiol* 55:709–742.

Lal, R., C. Dogra, S. Malhotra, P. Sharma, and R. Pal. 2006. Diversity, distribution and divergence of lin genes in hexachlorocyclohexane-degrading sphingomonads. *Trends Biotech* 24 (3):121–130.

Lawrence, J. G. and H. Ochman. 1998. Molecular archaeology of the *Escherichia coli* genome. *Proc Natl Acad Sci U S A* 95 (16):9413–9417.

Lindell, D., M. B. Sullivan, Z. I. Johnson, A. C. Tolonen, F. Rohwer, and S. W. Chisholm. 2004. Transfer of photosynthesis genes to and from Prochlorococcus viruses. *Proc Natl Acad Sci U S A* 101 (30):11013–11018.

Ma, Y., W. Zhang, Y. Xue, P. Zhou, A. Ventosa, and W. D. Grant. 2004. Bacterial diversity of the Inner Mongolian Baer Soda Lake as revealed by 16S rRNA gene sequence analyses. *Extremophiles* 8 (1):45–51.

Ma, Y., L. Wang, and Z. Shao. 2006. *Pseudomonas*, the dominant polycyclic aromatic hydrocarbon-degrading bacteria isolated from Antarctic soils and the role of large plasmids in horizontal gene transfer. *Environ Microbiol* 8 (3):455–465.

Mann, N. H., A. Cook, A. Millard, S. Bailey, and M. Clokie. 2003. Marine ecosystems: Bacterial photosynthesis genes in a virus. *Nature* 424 (6950):741.

Marcinek, H., R. Wirth, A. Muscholl-Silberhorn, and M. Gauer. 1998. *Enterococcus faecalis* gene transfer under natural conditions in municipal sewage water treatment plants. *Appl Environ Microbiol* 64 (2):626–632.

Medigue, C., T. Rouxel, P. Vigier, A. Henaut, and A. Danchin. 1991. Evidence for horizontal gene transfer in *Escherichia coli* speciation. *J Mol Biol* 222 (4):851–856.

Melis, A., J. A. Nemson, and M. A. Harrison. 1992. Damage to functional components and partial degradation of photosystem II reaction center proteins upon chlorplast exposure to ultraviolet-B radiation. *Biochim Biophys Acta* 1100:312–320.

Millard, A., M. R. J. Clokie, D. A. Shub, and N. H. Mann. 2004. Genetic organization of the psbAD region in phages infecting marine *Synechococcus* strains. *Proc Natl Acad Sci USA* 101 (30):11007–11012.

Mindlin, S. Z., I. A. Bass, E. S. Bogdanova, M. Gorlenko Zh, E. S. Kaliaeva, M. A. Petrova, and V. G. Nikiforov. 2002. [Horizontal transfer of mercury resistance genes in natural bacterial populations]. *Mol Biol (Mosk)* 36 (2):216–27.

Mirkin, B. G., T. I. Fenner, M. Y. Galperin, and E. V. Koonin. 2003. Algorithms for computing parsimonious evolutionary scenarios for genome evolution, the last universal common ancestor and dominance of horizontal gene transfer in the evolution of prokaryotes. *BMC Evol Biol* 3:2.

Mojzsis, S. J., G. Arrhenius, K. D. McKeegan, T. M. Harrison, A. P. Nutman, and C. R. Friend. 1996. Evidence for life on Earth before 3,800 million years ago. *Nature* 384 (6604):55–59.

Mongodin, E. F., K. E. Nelson, S. Daugherty, R. T. DeBoy, J. Wister, H. Khouri, J. Weidman et al. 2005. The genome of Salinibacter ruber: Convergence and gene exchange among hyperhalophilic Bacteria and Archaea, *Proc Natl Acad Sci U S A* 102 (50):18147–18152.

Moreira, D., F. Rodriguez-Valera, and P. Lopez-Garcia. 2006. Metagenomic analysis of mesopelagic Antarctic plankton reveals a novel deltaproteobacterial group 1, *Microbiol-SGM*, 152 (2):505–517.

Muller, J.-P., M. C. Storrie-Lombardi, M. R. Fisk, A. Coates, A. Griffiths, K. Rees, D. Walton, C. Theobald, C. Brockley-Blatt, and A. Rousseau. 2009. WALI—Wide Angle Laser Imaging enhancement to ExoMars PanCam: A system for organics and life detection. *Geophys Res Abs* 11, EGU2009-0.

Nakagawa, I., K. Kurokawa, A. Yamashita, M. Nakata, Y. Tomiyasu, N. Okahashi, S. Kawabata et al. 2003. Genome sequence of an M3 strain of *Streptococcus pyogenes* reveals a large-scale genomic rearrangement in invasive strains and new insights into phage evolution. *Genome Res* 13 (6A):1042–1055.

Neale, P. J. 2001. Modeling the effects of ultraviolet radiation on ȩstuarine phytoplankton production: Impact of variations in exposure and sensitivity to inhibition. *J Photochem Photobiol B* 62 (1–2):1–8.

Neale, P. J., R. F. Davis, and J. J. Cullen. 1998. Interactive effects of ozone depletion and vertical mixing on photosynthesis of Antarctic phytoplankton. *Nature* 392 (6676):585–589.

Nelson, K. E., R. A. Clayton, S. R. Gill, M. L. Gwinn, R. J. Dodson, D. H. Haft, E. K. Hickey et al. 1999. Evidence for lateral gene transfer between Archaea and Bacteria from genome sequence of *Thermotoga maritima*, *Nature*, 399 (6734):323–329.

Obuchowski, M., Y. Shotland, S. Koby, H. Giladi, M. Gabig, G. Wegrzyn, and A. B. Oppenheim. 1997. Stability of CII is a key element in the cold stress response of bacteriophage lambda infection, *J Bact* 179 (19):5987–5991.

Ochman, H., J. G. Lawrence, and E. A. Groisman. 2000. Lateral gene transfer and the nature of bacterial innovation. *Nature* 405 (6784):299–304.

Pelletier, E., P. Sargian, J. Payet, and S. Demers. 2006. Ecotoxicological effects of combined UVB and organic contaminants in coastal waters: A review. *Photochem Photobiol* 82 (4):981–993.

Perin, S., and D. R. S. Lean. 2004. The effects of ultraviolet-B radiation on freshwater ecosystems of the Arctic: Influence from stratospheric ozone depletion and climate change. *Environ Rev* 12:1–70.

Pini, F., C. Grossi, S. Nereo, L. Michaud, A. Lo Giudice, V. Bruni, F. Baldi, and R. Fani. 2007. Molecular and physiological characterization of psychrotrophic hydrocarbon-degrading bacteria isolated from Terra Nova Bay (Antarctica). *Eur J Soil Biol* 43 (5–6):368–379.

Pinkhart, H. C. and M. C. Storrie-Lombardi. 2007. Diversity, evolution and horizontal gene transfer (HGT) in soda lakes. In *Instruments, Methods and Missions for Astrobiology X*. San Diego, CA: SPIE.

Poelarends, G. J., L. A. Kulakov, M. J. Larkin, J. E. van Hylckama Vlieg, and D. B. Janssen. 2000a. Roles of horizontal gene transfer and gene integration in evolution of 1,3-dichloropropene- and 1,2-dibromoethane-degradative pathways. *J Bacteriol* 182 (8):2191–2199.

Poelarends, G. J., M. Zandstra, T. Bosma, L. A. Kulakov, M. J. Larkin, J. R. Marchesi, A. J. Weightman, and D. B. Janssen. 2000b. Haloalkane-utilizing *Rhodococcus* strains isolated from geographically distinct locations possess a highly conserved gene cluster encoding haloalkane catabolism. *J Bacteriol* 182 (10):2725–2731.

Quesada, A. and W. F. Vincent. 1997. Strategies of adaptation by Antarctic cyanobacteria to ultraviolet radiation. *Eur J Phycol* 32 (4):335–342.

Sargian, P., B. Mostajir, K. Chatila, G. A. Ferreyra, E. Pelletier, and S. Demers. 2005. Non-synergistic effects of water-soluble crude oil and enhanced ultraviolet-B radiation on a natural plankton assemblage. *Marine Ecol-Prog Ser* 294:63–77.

Sargian, P., S. Mas, E. Pelletier, and S. Demers. 2007. Multiple stressors on an Antarctic microplankton assemblage: Water soluble crude oil and enhanced UVBR level at Ushuaia (Argentina). *Polar Biol* 30 (7):829–841.

Saunders, J. R., H. Allison, C. E. James, A. J. McCarthy, and R. Sharp. 2001. Phage-mediated transfer of virulence genes. *J Chem Technol Biotechnol* 76 (7):662–666.

Saunders, N. F. W., A. Goodchild, M. Raftery, M. Guilhaus, P. M. G. Curmi, and R. Cavicchioli. 2005. Predicted roles for hypothetical proteins in the low-temperature expressed proteome of the Antarctic Archaeon Methanococcoides burtonii. *J Proteome Res* 4:464–472.

Sawstrom, C., J. Lisle, A. M. Anesio, J. C. Priscu, and J. Laybourn-Parry. 2008. Bacteriophage in polar inland waters. *Extremophiles* 12 (2):167–175.

Schmidt, H. 2001. Shiga-toxin-converting bacteriophages. *Res Microbiol* 152 (8):687–695.

Schofield, O., B. M. A. Kroon, and B. B. Prezelin. 1995. Impact of ultraviolet-B radiation on photosystem-Ii activity and its relationship to the inhibition of carbon fixation rates for Antarctic ice algae communities. *J Phycol* 31 (5):703–715.

Sharon, I., S. Tzahor, S. Williamson, M. Shmoish, D. Man-Aharonovich, D. B. Rusch, S. Yooseph et al. 2007. Viral photosynthetic reaction center genes and transcripts in the marine environment. *Isme J* 1 (6):492–501.

Sipiera, P. 2002. *Ernest Shackleton: A Life of Antarctic Exploration*. Dubuque, IA: Kendall Hunt Publishing Company.

Smith, R. C., B. B. Prezelin, K. S. Baker, R. R. Bidigare, N. P. Boucher, T. Coley, D. Karentz et al. 1992. Ozone depletion: Ultraviolet radiation and phytoplankton biology in antarctic waters. *Science* 255 (5047):952–959.

Snyder, L. and W. Champness. 1997. *Molecular Genetics of Bacteria*. Washington, D.C.: ASM Press.

Storrie-Lombardi, M. C. and H. C. Pinkhart. 2007. Co-evolution of cyanophage and cyanobacteria in Antarctic lakes: Adaptive responses to high UV flux and global warming. *SPIE* 66941E:1–13.

Storrie-Lombardi, M. C., J.-P. Muller, M. R. Fisk, A. D. Griffiths, and A. J. Coates. 2008. Potential for non-destructive astrochemistry using the ExoMars PanCam. *Geophys Res Lett* 35, L12201.

Storrie-Lombardi, M. C., B. Sattler, J.-P. Muller, M. R. Fisk, C. Cousins, and L. Dartnell. 2009. Laser induced fluorescence imaging: Searching for organics from the Dry Valleys of Queen Maud Land Antarctica to the regolith and ices of Mars, *Geophys Res Abs* 11: EGU2009-0.

Sullivan, M. B., M. L. Coleman, P. Weigele, F. Rohwer, and S. W. Chisholm. 2005. Three Prochlorococcus cyanophage genomes: signature features and ecological interpretations. *PLoS Biol* 3 (5):e144.

Sullivan, M. B., D. Lindell, J. A. Lee, L. R. Thompson, J. P. Bielawski, and S. W. Chisholm. 2006. Prevalence and evolution of core photosystem II genes in marine cyanobacterial viruses and their hosts. *PLoS Biol* 4 (8):e234.

van der Meer, J. R. 2006. Environmental pollution promotes selection of microbial degradation pathways. *Frontiers Ecol Environ* 4 (1):35–42.

Van Sluys, M. A., M. C. de Oliveira, C. B. Monteiro-Vitorello, C. Y. Miyaki, L. R. Furlan, L. E. A. Camargo, A. C. R. da Silva et al. 2003. Comparative analyses of the complete genome sequences of Pierce's disease and citrus variegated chlorosis strains of *Xylella fastidiosa. J Bacteriol* 185 (3):1018–1026.

Villarreal, L. P. 1999. DNA virus contribution to host evolution. In *Origin and Evolution of Viruses*, E. Domingo, R. G. Webster, and J. J. Holland (eds.). San Diego: Academic Press.

Villarreal, L. P. 2005. *Viruses and the Evolution of Life*. Washington D.C.: ASM Press.

Vincent, W. F., J. P. Bowman, L. M. Rankin, and T. A. McMeekin. 2000a. Phylogenetic diversity of picocyanobacteria in Arctic and Antarctic ecosystems. In *Microbial Biosystems: New Frontiers*, C. R. Bell, M. J. Brylinsky, and P. Johnson-Green (eds.). Kentville, Nova Scotia, Atlantic Canada Society for Microbial Ecology.

Vincent, W. F., J. A. E. Gibson, R. Pienitz, V. Villeneuve, P. A. Broady, P. B. Hamilton, and C. Howard-Williams. 2000b. Ice shelf microbial ecosystems in the high Arctic and implications for life on snowball earth. *Naturwissenschaften* 87 (3):137–141.

Whyte, L. G., C. W. Greer, and W. E. Inniss. 1996. Assessment of the biodegradation potential of psychrotrophic microorganisms. *Can J Microbiol* 42 (2):99–106.

Williamson, S. J. and J. H. Paul. 2006. Environmental factors that influence the transition from lysogenic to lytic existence in the phi HSIC/*Listonella pelagia* marine phage-host system. *Microbiol Ecol* 52 (2):217–225.

Williamson, S. J., L. A. Houchin, L. McDaniel, and J. H. Paul. 2002. Seasonal variation in lysogeny as depicted by prophage induction in Tampa Bay, Florida. *Appl Environ Microbiol* 68 (9):4307–4314.

Williamson, S. J., D. B. Rusch, S. Yooseph, A. L. Halpern, K. B. Heidelberg, J. I. Glass, C. Andrews-Pfannkoch. et al. 2008. The Sorcerer II Global Ocean Sampling Expedition: Metagenomic characterization of viruses within aquatic microbial samples. *PLoS ONE* 3 (1):e1456.

Wolfe, K. H. and W. H. Li. 2003. Molecular evolution meets the genomics revolution, *Nat Genet*, 33 (Suppl S):255–265.

Xue, L., Y. Zhang, T. Zhang, L. An, and X. Wang. 2005. Effects of enhanced ultraviolet-B radiation on algae and cyanobacteria. *Crit Rev Microbiol* 31 (2):79–89.

Zavarzin, G. A. 1993. Epicontinental soda lakes as probable relict biotopes of terrestrial biota formation. *Mikrobiologiya* 62:473–479.

Zavarzin. 1999. The alkaliphilic microbial community and its functional diversity. *Mikrobiologiya* 68 (5):503–521.

Zeidner, G., J. P. Bielawski, M. Shmoish, D. J. Scanlan, G. Sabehi, and O. Beja. 2005. Potential photosynthesis gene recombination between *Prochlorococcus* and *Synechococcus* via viral intermediates. *Environ Microbiol* 7 (10):1505–1513.

Zeph, L. R., M. A. Onaga, and G. Stotzky. 1988. Transduction of *Escherichia coli* by bacteriophage P1 in soil. *Appl Environ Microbiol* 54 (7):1731–1737.

8 Sources of Organic Matter for the Archean Cryosphere

Perry A. Gerakines and
Michael C. Storrie-Lombardi

CONTENTS

8.1 INTRODUCTION

8.1.1 Goals of This Chapter

The goals of this chapter are to offer a brief overview of the origin and evolution of organic molecules in astrophysical environments, and further, to provide a broader context to the task of understanding the origins of the materials that made up the original biosphere, particularly the cryosphere, on Earth. Much of the chemical reservoir for prebiotic and biotic chemistry in the early biosphere was likely delivered from space by comet, asteroid, and meteorite impacts. The compositions of comets and meteorites include complex organics, even molecular species as complex as amino acids and polycyclic aromatic hydrocarbons (PAHs). These materials and their precursors originate in the cold, vast interstellar clouds of gas and dust from which the Sun and other solar-system bodies were formed around 5 billion years ago. Organic molecules observed in these interstellar systems include complex hydrocarbon chains and branched structures, simple sugars, and PAHs. Finally, we will discuss the observation that if complex hydrocarbons including PAHs were present from the time life appeared on Archean

Earth, then it is not unexpected that we find microorganisms efficiently degrading petroleum hydrocarbons and even using them as a metabolic resource in remote polar ecosystems.

8.1.2 The Astrophysical Context

The core set of elements comprising organic matter on Earth—hydrogen, carbon, oxygen, and nitrogen—are four of the five most abundant elements in the observable universe (Lederberg and Cowie 1958). Together, these four elements also account for more than 75% of the total baryonic mass of the universe. Ultraviolet (UV) radiation energy gradients and multiple mechanical protection mechanisms against UV damage inherent in the structure of dust clouds of active star-forming regions provide a rich environment to create increasingly complex organic molecules (Allamandola et al. 1987, Ehrenfreund and Sephton 2006, Ehrenfreund et al. 2006). Laboratory simulations predict that the formation of aromatic organic molecules should be a fundamental feature of the creation of protoplanetary disks (Bernstein et al. 1999) and may play a central role in the appearance of life in the Universe.

The Sun was formed from an overly dense region within a giant interstellar cloud of gas and dust that collapsed under its own gravity. Around 5 billion years ago, the matter orbiting the collapsing proto-Sun formed a disk, within which the planets, comets, and asteroids condensed. Much of the chemical reservoir from which life emerged on Earth is thought to have been delivered from space by way of cometary impacts and collisions with other small planetary bodies during an era of heavy bombardment that took place between 4.5 and 4.0 billion years ago, when the leftover fragments of planet formation were abundant in the inner solar system. Although this topic is the subject of some scrutiny (e.g., Campins et al. 2004, Drake and Campins 2005), comets and asteroids from the heavy bombardment era are widely regarded as a significant source for a fraction (perhaps as much as 50%) of the total volume of water contained within Earth's oceans. These bodies are also known to contain large organic molecules that might have also been delivered in this process, bringing to the young Earth the necessary building blocks for prebiotic chemistry out of which life emerged in a relative short period of time (the oldest evidence for life on Earth is about 3.6–3.8 billion years old—only 200–400 million years after the end of the heavy bombardment era). Indeed, if our Sun were significantly cooler during that period, then life itself may have first appeared during a time of global glaciation and the likelihood of a cold origin of life takes on a new significance (Price 2007).

When considering ancient life on Earth, or when speculating as to the origins of life on Earth, it is therefore important to study this connection between the chemical makeup of Earth's biosphere and the chemistry of astrophysical environments, such as may be found inside comets, asteroids, or giant interstellar clouds. In the remaining sections of this chapter, the chemistry of interstellar clouds, comets, and meteorites will be discussed in the context of following the chemical evolution of organic materials relevant to biological systems on Earth.

8.2 INTERSTELLAR ORGANIC MATTER

Giant clouds of gas and micron-sized dust particles are present in the expanses of space between the stars, or interstellar medium (ISM). The birthplaces of new stars and planetary systems, they are typically hundreds of light-years away (1 light-year is the distance traveled by a beam of light in 1 year, equal to approximately 9.5×10^{12} km). They are composed primarily of hydrogen (90% by number), helium (9%), and heavier atoms (1%), and account for approximately 10% of the total mass of the Milky Way galaxy. Most of the heavier atoms in these clouds are contained within the dust grains. On average, the density of the ISM is only about 1 atom cm^{-3}, but most of the mass of an interstellar cloud is contained in several small cores where the particle density may exceed $10^8\,cm^{-3}$. While the gases in more diffuse regions are almost purely atomic in nature, matter within the dense cores is shielded from the harsh radiation field of interstellar space, and stable molecules may form without being photodissociated. The molecular gas in dense cores is dominated by molecular hydrogen (H_2), where the next most abundant molecule is carbon monoxide (CO), with an abundance on the order of 1 part in 10^4 relative to H_2 (e.g., Whittet 2002).

Due to the relatively high density and the light-absorbing properties of the dust particles, dense cores are opaque to visible radiation and their temperatures fall below 10 K. At such low temperatures, all atoms heavier than hydrogen readily accrete onto the dust particles, where they may interact by surface chemistry to produce molecules that otherwise could not form in the low-density gas, such as H_2, H_2O, or CH_3OH. Some molecules are thermally desorbed from the dust grain by the released binding energy, while others remain to form a solid mantle of condensed volatiles. The exposure of the icy dust grain mantles to ionizing radiation, such as UV, from nearby stars or by cosmic rays may also drive reactive chemical pathways to complex molecular species.

The compositions of the icy mantles are probed by spectroscopy of the characteristic fundamental vibrational modes in the IR portion of the electromagnetic spectrum with wavelengths from about 2 to 25 μm. There are spatial variations in icy grain mantle chemistry, but the general composition in all observed dense interstellar clouds is dominated by H_2O. Other important mantle constituents (10%–30% by number relative to H_2O) include carbon monoxide (CO), carbon dioxide (CO_2), and methanol (CH_3OH). Minor species (1%–10% relative to H_2O) in dense cloud cores may also include methane (CH_4) and ammonia (NH_3). In regions of active star formation, where energetic UV photons, x-rays, and elevated temperatures may influence the ice mantle chemistry, IR absorptions due to cyanate anions (OCN^-), carbonyl sulfide (OCS), formaldehyde (H_2CO), formic acid (HCOOH), and complex organics have also been observed (e.g., Whittet 2002, Gibb et al. 2004). The complete infrared spectrum of the heavily obscured protostar W33A is displayed in Figure 8.1.

In low-density environments, interstellar dust experiences higher temperatures (about 100 K) and radiation fields due to UV and visible light from stars. Due to the relatively high temperatures, most of the icy mantles have been totally evaporated, with the occasional exception of H_2O in certain lines of sight. In these regions

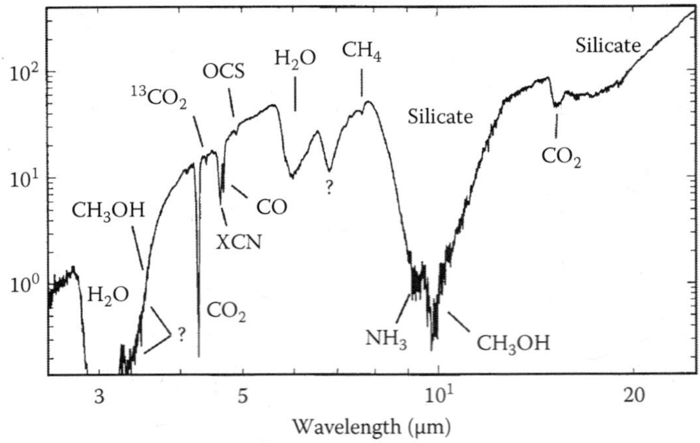

FIGURE 8.1 The infrared spectrum of the heavily obscured protostar W33A. Molecular absorption features are labeled according to the identified or possible carrier species. (From Gibb, E.L. et al., *Astrophys. J.*, 536, 347, 2000. With permission.)

of the ISM, the observed infrared spectra also reveal the presence of refractory organic materials by way of the characteristic absorption of the C–H aromatic and aliphatic bonds at wavelengths near 3.3 and 3.4 μm, respectively, as well as the C–C deformation modes near 6.2 μm (Sandford et al. 1991, Pendleton et al. 1994, Schutte et al. 1998, Pendleton and Allamandola 2002). Since these features are characteristic of all large hydrocarbons, knowledge of the precise composition of this organic material is not possible, and laboratory studies have demonstrated the resemblance of these interstellar spectral features to those obtained from common hydrocarbon-bearing substances (e.g., Pendleton and Allamandola 2002). The 3.4 μm infrared spectrum of the diffuse ISM is shown in Figure 8.2 in comparison

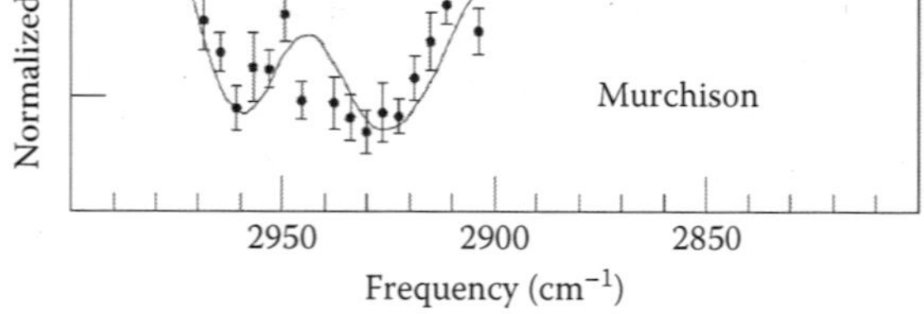

FIGURE 8.2 Comparison of the CH stretch of the 600°C organic extract sublimate from the Murchison meteorite (line) to the diffuse interstellar medium 3.4 μm absorption feature (circles with error bars). (From Pendleton, Y.J. and Allamandola, L.J., *Astrophys. J. Suppl. Ser.*, 138, 75, 2002. With permission.)

with the organic material obtained from the Murchison meteorite. In general, this organic refractory material may be regarded as "kerogen-like," containing highly branched, largely saturated and unsaturated hydrocarbon species, as well as various aromatics. PAHs, which we discuss in more detail in Section 8.5, have also been found to be widely present in the gas phase of interstellar clouds (e.g., Tielens et al. 1999, Tielens 2008).

8.3 ORGANIC MATERIALS IN COMETS

Comets are arguably the most pristine objects in the solar system. Preserved in cold storage at very low temperature (10–20 K) at the outermost reaches of the Sun's influence (as far as 1 light-year away), many of these objects likely contain the unmodified ices and dust from the protoplanetary nebula out of which the Sun and the planets were first formed. Trillions of comets, for a total estimated mass of about 1.9 Earth masses (Weissman 1983), probably exist in the Oort Cloud (after noted Dutch astronomer Jan Oort, who first proposed the possibility of its existence; see Oort 1950). Another grouping of comets exists at distances that span from just beyond the orbit of Neptune to the inner edge of the Oort cloud, in a flattened structure known as the Kuiper Belt, or the Edgeworth–Kuiper Belt (after noted astronomers Gerard Kuiper and Kenneth Edgeworth; see Kuiper 1951 and a comprehensive review of these topics by Luu and Jewitt 2002). Comets in the more distant Oort cloud are thought to have been formed within the inner solar system and subsequently ejected by gravitational interactions with the gas giant planet Jupiter (e.g., Weissman 1983). Comets in the Kuiper Belt are thought to be more pristine, having formed in more or less in their current locations, with some orbital modifications by gravitational interactions with the planets Uranus and Neptune (e.g., Luu and Jewitt 2002).

Despite their vast distances from the Sun, an occasional orbital instability through comet–comet collisions or by the close passage of a nearby star will send one of these objects into the inner solar system for a close study. As they are heated, material from within the nucleus of the comet sublimes into space, forming a coma and tail structure(s). Observations of the solar-induced fluorescence in cometary comae reveal a variety of molecular species (see, e.g., Mumma et al. 2003). The overall composition of comets resembles, qualitatively, the composition of the ISM, and many trace species have also been detected. A listing of major components of cometary comae is given in Table 8.1, which can be directly compared to the major components of interstellar ices given in Table 8.2. The major components of comets include ethane (C_2H_6), acetylene (C_2H_2), HCN, CO, CO_2, and H_2O (Mumma et al. 2003). In comparing the distribution of the molecular species present in interstellar ices, comets, and the Murchison meteorite (Table 8.3) to the microbial life on Earth in the modern era (Table 8.4), the most impressive shift, of course has been from small precursor molecules to complex polymers (proteins, nucleic acid chains, lipids). However, the fundamental components of life today are still hydrogen, carbon, nitrogen, and oxygen present in the organic infall material and atmosphere of early Earth's cryosphere.

TABLE 8.1
Selected Molecules Observed in Comets

Molecule	Abundance Relative to H_2O (%)
H_2O	100
CO	2–20
CO_2	2–6
CH_4	0.6
C_2H_6	0.3
C_2H_2 (acetylene)	0.1
H_2CO	0.04–5
CH_3OH	1–7
HCOOH	0.1
HNCO (isocyanic acid)	0.07
NH_3	0.5
HCN	0.1–0.2
CS_2	0.1
OCS	0.4

Source: Whittet, D.C.B., *Dust in the Galactic Environment*, 2nd edn, Institute of Physics Publishing, Philadelphia, PA, 2002.

8.4 ORGANIC MATERIALS IN METEORITES

About 5% of all meteorites on Earth comprise the class known as carbonaceous chondrites. These meteorites have bulk elemental abundances very similar to those of the Sun, indicating that they have remained essentially unaltered since their formation in the solar nebula (Anders and Grevesse 1989, Burnett et al. 1989). Meteorites can therefore yield information about processes that occurred during planetesimal formation.

Carbonaceous chondrites contain a significant amount of carbon, although it represents only about 5% of the meteorite's total mass (Sears 1978). The primary significance of this carbon lies in its organization into a variety of free organic molecules, including amino acids (Cronin and Moore 1971), sugars (Cooper et al. 2001), nucleic acids (Stoks and Schwartz 1981), fatty acids, and light aromatic compounds, as well as a macromolecular material containing PAH structures (Hayatsu et al. 1977). It is spectroscopically very similar to the organic materials observed in the IR spectrum of the diffuse ISM (see Figure 8.2). The nature of the organic content of these meteorites spans the full range of the most basic materials involved in living systems on Earth and may provide a link between life on Earth and the prebiotic chemistry in the early solar system. Moreover, since PAHs are a significant

TABLE 8.2
Molecules Observed in Icy Grain Mantles in the Dense Interstellar Medium and in Star-Forming Regions

Molecule	Abundance Relative to H_2O (%)
H_2O	100
CO	25–60
CO_2	20–35
CH_3OH	3–30
NH_3	2–15
OCN^-	0–6
H_2CO	0–6
HCOOH	0–6
CH_4	0–2
OCS	0–0.5

Source: Crovisier, J. and Encrenaz, T., *Comet Science*, Cambridge University Press, New York, 2000.

TABLE 8.3
Organic Analysis of the Murchison Meteorite

Class of Material	Concentration (ppm)
Carboxylic acids	>300
Amides	55–70
Amino acids	60
Aliphatic hydrocarbons	>35
Dicarboxylic acids	>30
Aromatic hydrocarbons	15–28
Aldehydes and ketones	27
Hydroxycarboxylic acids	15
Amines	8
Purines and pyrimidines	1.3

Source: Data from Cronin, J.R. et al., *Meteorites and the Early Solar System*, University of Arizona Press, Tucson, AZ, 1988, pp. 819–857.

TABLE 8.4
Organic Composition of a Modern Microbial Cell

Class of Material	Percent of Total Cell Weight
Water	70
Inorganic ions	1
Sugars	1
Amino acids	0.4
Nucleotides	0.4
Fatty acids	1
Other small molecules	0.2
Proteins	15.4
RNA	4.7
DNA	0.9
Polysaccharides	5.0

Sources: Adapted from Alberts, B. et al., *Molecular Biology of the Cell*, 3rd edn, Garland Publishing, Inc., New York, 1994; Vaccari, D.A. et al., *Environmental Biology for Engineers and Scientists*, John Wiley & Sons, Inc., Hoboken, NJ, 2005.

repository of carbon in the ISM (Allamandola et al. 1985), they represent a direct link from the pre-solar nebula to interstellar materials.

The connection between the organics in carbonaceous chondrites and the prebiotic chemistry of the solar nebula is the carbon content of these meteorites and, in particular, the possible association between the free organic material and the macromolecular material. This link remains poorly understood, but there is some indirect evidence that they may be related (e.g., Sephton et al. 1998).

The fact that this macromolecular, kerogen-like material has an interstellar origin is evidenced by enrichments in the relative amounts of the isotope deuterium (D) relative to hydrogen in these materials. The results of Kerridge et al. (1987) imply a D enrichment of up to 1800‰ (parts per mil) relative to terrestrial standards for samples of this material from Murchison, suggesting that this material is at least partially interstellar in origin. Yang and Epstein (1983) have measured the D/H ratios of organic matter in a variety of meteorites, including non-carbonaceous chondrites, and in all instances have found D enrichments, relative to terrestrial standards, that are comparable to those observed by Penzias (1980) and Wannier (1980) for HCN and HCO^+ in interstellar clouds in our galaxy.

Recent measurements published by Dworkin et al. (2001) and reviewed by Deamer et al. (2002) show that the water-soluble organic material extracted from a carbonaceous chondrite meteorite (Murchison) spontaneously forms micron-sized, membranous, vesicle structures when exposed to liquid water. These vesicles seem to be composed of amphiphilic molecules that also fluoresce when exposed to UV radiation. This result is interesting from the point of view of astrobiology, since it provides a possible answer to the question of how the first proto-cells were formed in a prebiotic environment (Deamer et al. 2002).

8.5 PAHs AND DELIVERY TO THE EARLY EARTH

Astrophysical surveys have repeatedly documented the presence of broad infrared spectral emission features at 3.3, 6.2, 7.7, 8.6, 11.3, and 12.7 μm closely matching the spectral features of PAH molecules (van Dishoeck 2004). These molecular signatures have been found in the ISM of nearby star-forming regions (Ehrenfreund and Charnley 2000), around protoplanetary disks (Visser et al. 2007), throughout our galaxy (Flagey et al. 2006), in neighboring galaxies (Sajina et al. 2007), and have now been found in Titan's atmosphere (Coates et al. 2007). The IR spectra of three nearby active galactic nuclei are shown in Figure 8.3.

Estimations of PAH abundances in our solar system from the examination of carbonaceous chondrite meteorites indicate that the most common PAH species are 2-, 3-, and 4-ring structures such as naphthalene, phenanthrene, anthracene, pyrene, and chrysene (Zolotov and Shock 2001). The Murchison meteorite exhibits PAH concentrations of 15–38 ppm with phenanthrene, fluoranthene, pyrene, chrysene, perylene, benzoperylene, and coronene being the most abundant species (Sephton et al. 2004). Electron delocalization over their carbon skeleton makes PAHs remarkably stable. Indeed, these robust molecules may be the most abundant free organic molecules in space (Ehrenfreund and Charnley 2000).

Calculations indicate that comets, meteorites (including micrometeorites), and dust from our local solar system deliver significant quantities of organic materials daily to the surfaces of the terrestrial planets Mercury, Venus, Earth, and Mars (Chyba and Sagan 1992, Flynn 1996). However, the infall rates were significantly greater during the period of heavy bombardment 3.5–4.5 Gy ago. Extending work by Anders (1989), Chyba and colleagues estimated that organic infall rates attributable to interplanetary dust particles (IDPs) 4.5 Gy ago were 10^8 to 10^{10} kg/year and the cometary organic infall rate was on the order of 10^6 to 10^7 kg/year (Chyba et al. 1990). The current terrestrial oceanic biomass is about 3×10^{12} kg, and the total Earth biomass is about 6×10^{14} kg. IDPs are about 10% organic material, primarily kerogen. While these particles are difficult to analyze because of their small size, their Raman spectra are remarkably similar to the interstellar infrared emission features attributed to vibrational transitions in free PAHs (Allamandola et al. 1987), and measurements with a two-step laser mass spectrometer has identified a variety of PAH species (Clemett et al. 1993, Chyba and McDonald 1995).

The deposition of massive quantities of PAHs to an early Earth is of significance for two reasons. First, in attempts to understand the origin of life either here or in the ISM, it is important to note that the simple act of irradiating naphthalene in ice in the presence of methane is sufficient to produce 13 of the 20 amino acids critical to biological function on this planet (Chen et al. 2008). Second, in an early Earth

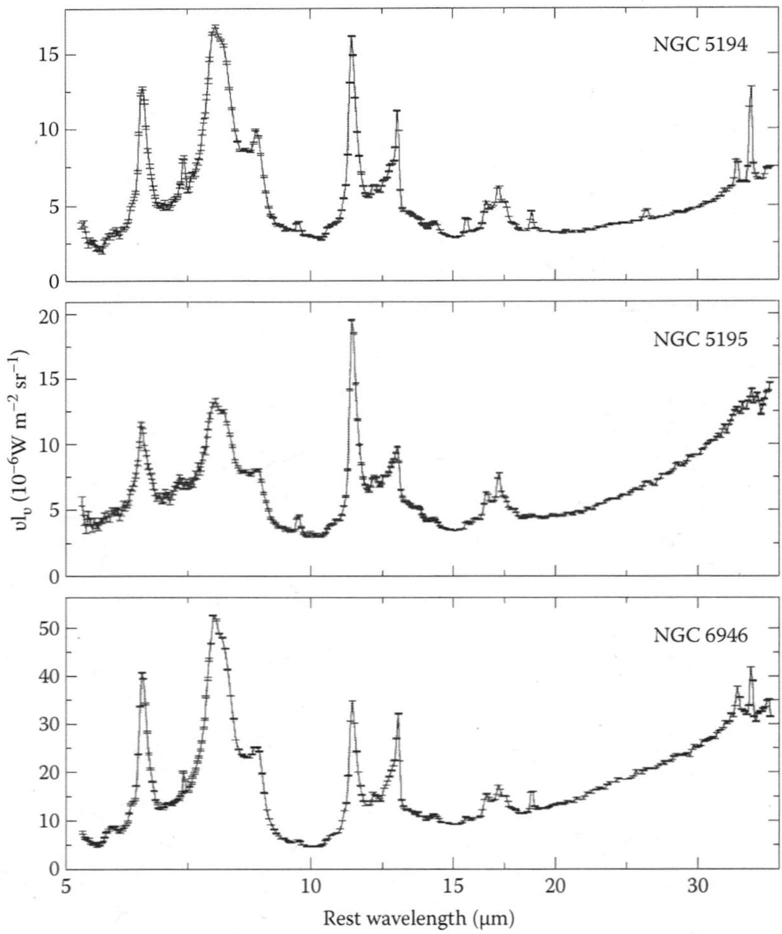

FIGURE 8.3 Infrared spectra of the nuclei of three nearby galaxies, displaying the UIR emission features due to PAHs. (From Tielens, A.G.G.M., *Annu. Rev. Astron. Astrophys.*, 46, 289, 2008. With permission.)

relatively devoid of biological activity, particularly in the absence of significant photosynthetic primary productivity, infall or abiotically generated PAHs would serve as a precious source of organic carbon for nascent prokaryote communities. Such an environment governed by limited organic carbon resources is quite similar to the substrate challenge facing microbial life in PAH-contaminated Arctic and Antarctic ecosystems. The early appearance and evolution of PAH degradation pathways in the microbial life of Archean Earth would have constituted a significant selective advantage for the fortunate species. Increasing the availability of full genomic information on multiple PAH-tolerant and PAH-metabolizing species inhabiting similar carbon-depleted environments may eventually lead us to a new appreciation for the ancient nature of these pathways. If PAHs formed early in the evolution of the universe and arrived on Earth as one of the primary organic carbon sources, it is unlikely that early microbial life would not adapt to both the challenge and the opportunity. (For a review of the impact of such an early interaction between life and PAHs on current efforts to facilitate bioremediation, see Chapter 4.)

8.6 SUMMARY

In this chapter, a brief outline of the pathways of organic chemistry has been presented, beginning in the ISM (where stars and planetary systems are born), to comets (the most pristine objects in the solar system), and finally to meteorites (whose organic molecules may have been the seeds from which prebiotic chemistry emerged on Earth). The low-temperature, low-density environment of interstellar space produces a variety of organic molecules, including PAHs. These molecules become incorporated into the building blocks of comets and planets, where they may be further modified by the Sun. Finally, within the rocks of asteroids and meteorites, these complex organic compounds may find their way to the surface of a planet like Earth—supplying an inventory of prebiotic materials including complex hydrocarbons out of which a biosphere may emerge. If complex hydrocarbons including PAHs were present from the time that life first appeared on Earth, then it is not surprising that we find microorganisms adapt at degrading petroleum hydrocarbons and even using them as a core metabolic resource in remote polar ecosystems.

REFERENCES

Alberts, B., Bray, D., Lewis, J., Raff, M., Roberts, K., and Watson, J. D. (1994). *Molecular Biology of the Cell*, 3rd Edition. New York: Garland Publishing, Inc.

Allamandola, L. J., Tielens, A. G. G. M., and Barker, J. R. (1985). Polycyclic aromatic hydrocarbons and the unidentified infrared emission bands—Auto exhaust along the Milky Way. *Astrophysical Journal* 290, L25–L28.

Allamandola, L. J., Sandford, S. A., and Wopenka, B. (1987). Interstellar polycyclic aromatic hydrocarbons and carbon in Interplanetary dust particles and meteorites. *Science* 237, 56–59.

Anders, E. (1989). Pre-biotic organic matter from comets and asteroids. *Nature* 342, 255–257.

Anders, E. and Grevesse, N. (1989). Abundances of the elements—Meteoritic and solar. *Geochimica et Cosmochimica Acta* 53, 197–214.

Bernstein, M. P., Sandford, S. A., Allamandola, L. J., Gillette, J. S., Clemett, S. J., and Zare, R. N. (1999). UV irradiation of polycyclic aromatic hydrocarbons in ices: Production of alcohols, quinones, and ethers. *Science* 283, 1135–1138.

Burnett, D. S., Woolum, D. S., Benjamin, T. M., Rogers, P. S. Z., Duffy, C. J., and Maggiore, C. (1989). A test of the smoothness of the elemental abundances of carbonaceous chondrites. *Geochimica et Cosmochimica Acta* 53, 471–481.

Campins, H., Swindle, T. D., and Kring, D. A. (2004). Evaluating comets as a source of Earth's water. *Cellular Origin, Life in Extreme Habitats and Astrobiology* 6, 571–591.

Chen, Y.-J., Nuevo, M., Yih, T.-S., Ip, W.-H., Fung, H.-S., Cheng, C.-Y., Tsai, H.-R., and Wu, C.-Y.R. (2008). Amino acids produced from the ultraviolet/extreme-ultraviolet irradiation of naphthalene in a $H_2O + NH_3$ ice mixture. *Monthly Notices of the Royal Astronomical Society* 384, 605–610.

Chyba, C. and McDonald, G. D. (1995). The origin of life in the solar system: Current issues. *Annual Reviews of Earth and Planetary Sciences*. 23, 215–249.

Chyba, C. and Sagan, C. (1992). Endogenous production, exogenous delivery and impact shock synthesis of organic molecules: An inventory for the origins of life. *Nature* 355, 125–132.

Chyba, C. F., Thomas, P. J., Brookshaw, L., and Sagan, C. (1990). Cometary delivery of organic molecules to the early Earth. *Science* 249, 366–373.

Clemett, S. J., Maechling, C. R., Zare, R. N., Swan, P. D., and Walker, R. M. (1993). Identification of complex aromatic molecules in individual interplanetary dust particles. *Science* 262, 721–725.

Coates, A. J., Crary, F. J., Lewis, G. R., Young, D. T., Waite, J. H., and Sittler Jr., E. C. (2007). Discovery of heavy negative ions in Titan's ionosphere. *Geophysical Research Letters* 34, L22103.

Cooper, G., Kimmich, N., Belisle, W., Sarinana, J., Brabham, K., and Garrel, L. (2001). Carbonaceous meteorites as a source of sugar-related organic compounds for the early Earth. *Nature* 414, 879–883.

Cronin, J. R. and Moore, C. B. (1971). Amino acid analyses of the Murchison, Murray, and Allende carbonaceous chondrites. *Science* 172, 1327–1329.

Cronin, J. R., Pizzarello, S., and Cruikshank, D. P. (1988). Organic matter in carbonaceous chondrites, planetary satellites, asteroids and comets. In *Meteorites and the Early Solar System*. Tucson, AZ: University of Arizona Press, pp. 819–857.

Crovisier, J. and Encrenaz, T. (2000). *Comet Science*. New York: Cambridge University Press.

Deamer, D., Dworkin, J. P., Sandford, S. A., Bernstein, M. P., and Allmandola, L. J. (2002). The first cell membranes. *Astrobiology* 2, 371–381.

Drake, M. J. and Campins, H. (2005). Origin of water on the terrestial planets. *Proceedings of the International Astronomical Union* 1, 381–394.

Dworkin, J. P., Deamer, D. W., Sandford, S. A., and Allamandola, L. J. (2001). Self-assembling amphiphilic molecules: Synthesis in simulated interstellar/precometary ices. *Proceedings of the National Academies of Science* 98, 815–819.

Ehrenfreund, P. and Charnley, S. B. (2000). Organic molecules in the interstellar medium, comets, and meteorites: A voyage from dark clouds to the early earth. *Annual Reviews of Astronomy and Astrophysics* 38, 427–483.

Ehrenfreund, P. and Sephton, M. A. (2006). Carbon molecules in space: From astrochemistry to astrobiology. *Faraday Discussions* 133, 277–288.

Ehrenfreund, P., Rasmussen, S., Cleaves, J. and Chen, L. (2006). Experimentally tracing the key steps in the origin of life: The aromatic world. *Astrobiology* 6, 490–520.

Flagey, N., Boulanger, F., Verstraete1, L., Miville Deschênes, M. A., Noriega Crespo, A., and Reach, W. T. (2006). Spitzer/IRAC and ISOCAM/CVF insights on the origin of the near to mid-IR Galactic diffuse emission. *Astronomy and Astrophysics* 453, 969–978.

Flynn, G. J. (1996). The delivery of organic matter from asteroids and comets to the early surface of Mars. *Earth Moon Planets* 72, 469–474.

Gibb, E. L., Whittet, D. C. B., Schutte, W. A., Boogert, A. C. A., Chiar, J. E., Ehrenfreund, P., Gerakines, P. A., Keane, J. V., Tielens, A. G. G. M., van Dishoeck, E. F., and Kerkhof, O. (2000). An inventory of interstellar ices toward the embedded protostar W33A. *Astrophysical Journal* 536, 347–356.

Gibb, E. L., Whittet, D. C. B., Boogert, A. C. A., and Tielens, A. G. G. M. (2004). Interstellar ice: The Infrared Space Observatory legacy. *Astrophysical Journal Supplement Series* 151, 35–73.

Hayatsu, R., Matsuoka, S., Anders, E., Scott, R. G., and Studier, M. H. (1977). Origin of organic matter in the early solar system. VII—The organic polymer in carbonaceous chondrites. *Geochimica et Cosmochimica Acta* 41, 1325–1339.

Kerridge, J. F., Chang, S., and Shipp, R. (1987). Isotopic characterisation of kerogen-like material in the Murchison carbonaceous chondrite. *Geochimica et Cosmochimica Acta* 51, 2527–2540.

Kuiper, G. P. (1951). On the origin of the solar system. In *Astrophysics*, J. A. Hynek (ed.). New York: McGraw-Hill, pp. 357–424.

Lederberg, J. and Cowie, D. B. (1958). Moondust. *Science* 127, 1473–1475.

Luu, J. X. and Jewitt, D. C. (2002). Kuiper Belt Objects: Relics frôm the accretion disk of the Sun. *Annual Reviews of Astronomy and Astrophysics* 40, 63–101.

Mumma, M. J., DiSanti, M. A., Dello Russo, N., Magee-Sauer, K., Gibb, E., and Novak, R., (2003). Remote infrared observations of parent volatiles in comets: A window on the early solar system. *Advances in Space Research* 31 (12), 2563–2575.

Oort, J. H. (1950). The structure of the cloud of comets surrounding the Solar System and a hypothesis concerning its origin. *Bulletin of the Astronomical Institutes of the Netherlands* 11, 91–110.

Pendleton, Y. J. and Allamandola, L. J. (2002). The organic refractory material in the diffuse interstellar medium: mid-infrared spectroscopic constraints. *Astrophysical Journal Supplement Series* 138, 75–98.

Pendleton, Y. J., Sandford, S. A., Allamandola, L. J., Tielens, A. G. G. M., and Sellgren, K. (1994). Near-infrared absorption spectroscopy of interstellar hydrocarbon grains. *Astrophysical Journal* 437, 683–696.

Penzias, A. A. (1980). Nuclear processing and isotopes in the Galaxy. *Science* 208, 663–669.

Price, B. P. (2007) Microbial life in glacial ice and implications for a cold origin of life. *FEMS Microbiology Ecology* 59: 217–231.

Sajina, A., Yan, L., Armus, L., Choi, P., Fadda, D., Helou, G., and Spoon, H. (2007). Spitzer mid-infrared spectroscopy of infrared luminous galaxies at z similar to 2. II. Diagnostics. *Astrophysical Journal* 664, 713–737.

Sandford, S. A., Allamandola, L. J., Tielens, A. G. G. M., Sellgren, K., Tapia, M., and Pendleton, Y. J. (1991). The interstellar C-H stretching band near 3.4 microns—Constraints on the composition of organic material in the diffuse interstellar medium. *Astrophysical Journal* 371, 607–620.

Schutte, W. A., van der Hucht, K. A., Whittet, D. C. B., Boogert, A. C. A., Tielens, A. G. G. M., Morris, P. W., Greenberg, J. M., Williams, P. M., van Dishoeck, E. F., Chiar, J. E., and de Graauw, Th. (1998). ISO-SWS observations of infrared absorption bands of the diffuse interstellar medium: The 6.2 μm feature of aromatic compounds. *Astronomy and Astrophysics* 337, 261–274.

Sears, J. H. (1978). *The Nature and Origin of Meteorites*. New York: Oxford University Press.

Sephton, M. A., Pillinger, C. T., and Gilmour, I. (1998). Small-scale hydrous pyrolysis of macromolecular material in meteorites. *Planetary and Space Science* 47, 181–187.

Sephton, M. A., Love, G. D., Watson, J. S., Verchovsky, A. B., Wright, I. P., Snape, C. E., and Gilmour, I. (2004). Hydropyrolysis of insoluble carbonaceous matter in the Murchison meteorite: New insights into its macromolecular structure. *Geochimica et Cosmochimica Acta* 68, 1385–1393.

Stoks, P. G. and Schwartz, A. W. (1981). Nitrogen-heterocyclic compounds in meteorites—significance and mechanisms of formation. *Geochimica et Cosmochimica Acta* 45, 563–569.

Tielens, A. G. G. M (2008). Interstellar polycyclic aromatic hydrocarbon molecules. *Annual Reviews of Astronomy and Astrophysics* 46, 289–337.

Tielens, A. G. G. M., Hony, S., van Kerckhoven, C., and Peeters, E. (1999). Interstellar and circumstellar PAHs. In *The Universe as Seen by ISO*, P. Cox and M. F. Kessler (eds.). ESA-SP 427, the Netherlands, p. 579.

Vaccari, D. A., Strom, P. F., and Alleman, J. E. (2005). *Environmental Biology for Engineers and Scientists*. Hoboken, NJ: John Wiley & Sons, Inc.

van Dishoeck, E. F. (2004). ISO spectroscopy of gas and dust: From molecular clouds to protoplanetary disks. *Annual Reviews of Astronomy and Astrophysics* 42, 119–167.

Visser, R., Geers, V. C., Dullemond, C. P., Augereau, J. C., Pontoppidan, K. M., and van Dishoeck, E. F. (2007). PAH chemistry and IR emission from circumstellar disks. *Astronomy and Astrophysics* 466, 229–241.

Wannier, P. G. (1980). Nuclear abundances and evolution of the interstellar medium. *Annual Reviews of Astronomy and Astrophysics* 18, 399–437.

Weissman, P. R. (1983). The mass of the Oort cloud. *Astronomy and Astrophysics* 118, 90–94.

Whittet, D. C. B. (2002). *Dust in the Galactic Environment*, 2nd Edition. Philadelphia, PA: Institute of Physics Publishing.

Yang, J. and Epstein, S. (1983). Interstellar organic matter in meteorites. *Geochimica et Cosmochimica Acta* 47, 2199–2216.

Zolotov, M. Y. and Shock, E. L. (2001). Stability of condensed hydrocarbons in the Solar Nebula. *Icarus* 150, 323–337.

9 Response of Polar Soil Bacterial Communities to Fuel Spills

Jackie Aislabie and Julia M. Foght

CONTENTS

9.1 INTRODUCTION

Human activities in polar regions (continental Antarctica and regions lying north of the tree line in continuous permafrost in the Arctic) can result in hydrocarbon contamination of terrestrial ecosystems. Crude oil is extracted in the Arctic, and petroleum hydrocarbons are required for power generation, heating, and the operation of vehicles, aircraft, and ships. Accidental spills of crude oil and past disposal practices of refined products, petroleum contamination of polar soils has occurred, especially around sites exploiting and distributing the northern oil reserves in the Arctic (Poland et al. 2003) and settlements including current and former scientific bases and field camps in the Antarctic (Aislabie et al. 2004).

Polar regions are remote, hence remediation at or near the contaminated site is desirable. For this reason, bioremediation is considered to be an attractive treatment option. As for temperate soils, the success of bioremediation is dependent on biodegradative microbes, the contaminant(s), and environmental conditions. In polar environments, microbes adapted to *in situ* conditions are most likely to be the appropriate organisms for bioremediation. In this chapter, we review polar soils, the sources and fate and types of hydrocarbons that accumulate in the soils, and the response of the bacterial community to fuel spills.

9.2 CHARACTERISTICS OF POLAR SOILS

Most of the world's polar soils (4.5×10^6 km^2) occur in the Arctic, encompassing northern regions of Canada, Russia, Alaska, and all of Greenland and Svalbard. While the continent of Antarctica is large, most of it is covered by ice caps, and less than 0.3% is ice free (0.05×10^6 km^2). The ice-free areas of Antarctica, which include soils, occur in many widely scattered small areas, the largest being the McMurdo Dry Valleys (Tarnocai and Campbell 2002).

The climate of the polar regions is characterized by short cold summers, with the active layer above the permafrost typically thawing for 1–2 months every year, and extremely cold winters. Mean daily air temperatures above 0°C occur only during the warmest part of the summer, and the temperatures fluctuate. In summer, the surface temperature of snow-free soils may range from below 0 to +20°C in a single day (Balks et al. 2002). The annual precipitation in polar regions is generally low, with most occurring as snow. In Antarctica, the annual precipitation averages 50 mm, but effective precipitation is low, as most snowfall blows away or sublimes. Hence, Antarctic soils are typically dry. Arctic soils, in contrast, generally have higher moisture contents, especially near the permafrost. Arctic soils may also contain ice as ice crystals, vein ice, ice wedges, or massive ground ice several meters thick (Tarnocai and Campbell 2002).

Permafrost is continuous in both the Arctic and the Antarctic, and patterned ground is a feature of both areas. The permafrost of Antarctica is commonly ice cemented in coastal regions and on younger land surfaces, but dry frozen in areas of high aridity. The depth of the active layer depends on soil climate. In Arctic soils, the active layer is 25–200 cm below the surface of the soil, whereas in Antarctic soils the depth of the active layer ranges from <5 cm to >70 cm (Tarnocai and Campbell 2002).

Polar soils are generally coarse texture, low in nutrients, and poorly buffered. Soil pH varies depending on the chemistry of the parent material. In the Antarctic, most soils are unvegetated, and soil pH ranges from weakly acidic in inland soils to highly alkaline in coastal regions. The Arctic vegetation is a nearly continuous cover of tundra shrub in the south, grading to a sparse cover of dwarf shrubs, herbs, mosses, and lichens in the North (Tarnocai and Campbell 2002). As with Antarctic soils, those in coastal Arctic environments may be alkaline (Whyte et al. 1999a).

9.3 HYDROCARBON SPILLS ON POLAR SOILS

Hydrocarbons are introduced into polar environments through natural and anthropogenic sources; however, most hydrocarbons on land are derived from human activity. In the Arctic, crude oil spills from ruptured pipelines are the largest source of hydrocarbon contamination, followed by shoreline spills from tankers and resupply vessels (Poland et al. 2003). The Komi crude oil spill of 1994 in Russia, resulting from the rupture of a pipeline, saw thousands of tons of crude oil contaminating the Siberian tundra and the Kolva River. Diesel fuels are the next most common spills. Incidents are typically caused by infrastructure failure, human error during fuel transfer, sabotage, or natural hazards. In contrast to the Arctic, hydrocarbon contaminants in Antarctic soils are largely derived from refined petroleum products (Aislabie et al. 2004). McMurdo Station, the largest scientific research station in Antarctica, has a storage capacity of approximately 34 million L of fuel, most of which is JP-8, a special mix of light petroleum distillate aviation fuel. At other Antarctic locations, jet fuel and diesel variants used include Special Antarctic Blend and Diesel Fuel Arctic (DFA). Hydrocarbon products used in lesser quantities include gasoline, and lubricating and engine oils.

9.3.1 Fate of Fuel Spills on Polar Soils

Hydrocarbons spilled on polar soils undergo naturally occurring processes that reduce the mass of the contaminants. Physical processes can cause the contaminants to disperse and become diluted and/or volatilize, whereas chemical and biological processes can transform the contaminants to other compounds causing them to precipitate or sorb to the soil. These mechanisms typically occur at all sites to varying degrees, depending on the type and concentration of the fuel spills, soil characteristics, and climatic conditions.

The movement of petroleum through polar soils is influenced by the formation and presence of ice. Light fuels such as jet fuel and gasoline readily volatilize from polar surface soil (Green et al. 1992), but being of low viscosity, they are also mobile and thus migrate down through the unfrozen active layer (Gore et al. 1999). In comparison, heavier fuels, such as lubricating and engine oils, are less volatile and more viscous, and do not appear to migrate far from the point of deposition (Gore et al. 1999). When the active layer is thawed, the downward movement of hydrocarbons may be limited by the presence of an ice-saturated lens or layer that often occurs at the top of the permafrost (Chuvilin et al. 2001). Although the ice-cemented layer is a seemingly impermeable barrier, hydrocarbons do move through this layer most likely through air voids in frozen soil (Biggar et al. 1998, McCarthy et al. 2004). In winter, oil spreads mainly on the surface of snow cover and/or frozen soils. A winter spill may therefore cover a larger surface area than a summer spill. In summer, the spill may penetrate the active layer and then spread laterally when it reaches the ice-cemented layer (Chuvilin et al. 2001).

Freeze–thaw processes may also influence hydrocarbon movement in soils. It has been shown in sandy soil that hydrocarbons moved ahead of the freezing front, implying that when soils are cooled from the surface down through the active layer, hydrocarbons may migrate to the permafrost interface. Unlike sandy soil, oil can accumulate in clay soils in the frozen zone under certain temperature conditions (Chuvilin et al. 2001). Dissolved and particle-associated hydrocarbons in soils can be mobilized with soil during the thaw and contribute to the contamination of surface and groundwater (Kennicutt et al. 1992).

As discussed later in this chapter, elevated numbers of hydrocarbon degraders and culturable bacteria in polar soils following an oil spill indicate that hydrocarbons spilled on polar soils can serve as substrates for microbial growth.

9.3.2 Petroleum-Derived Contaminants in Soil

Hydrocarbons spilled on polar soils contain a mix of aliphatic and aromatic compounds. For example, Prudhoe Bay Crude oil, produced in Alaska, contains 78% saturates and 18% aromatics, whereas Norman Wells and Bent Horn crude oils from Canada's Northwest Territories contain 85% and 95% saturates and 11% and 5% aromatics, respectively; Oseberg oil from the Norwegian sector of the North Sea has a lower proportion of saturates (65%) and correspondingly higher aromatics (25%); and Bunker C heavy fuel oil often has almost equivalent proportions of saturates and aromatics, in addition to very high asphaltenes and polar fractions (Wang et al. 1998). The distribution of molecular weight classes and the proportion of alkyl homologues within the aromatic fractions vary considerably with the source and the degree of the abiotic weathering of the oil in the environment. These factors affect biodegradation potential (Wang et al. 1998). In contrast, petroleum products used as fuels have a narrower, more defined composition. For example, JP8 jet fuel has about 80% *n*-alkanes in the range of C_6–C_{18}, with a maximum at C_{12}, and 18% aromatics with <0.5% PAHs having three or more rings (Ritchie 2003).

The total petroleum hydrocarbon (TPH) analysis of soil samples from the sites of known contamination sources on land (fuel storage tanks, fuelling stations, machine shops) confirms the presence of hydrocarbon contamination in polar soils (Aislabie et al. 2004, McCarthy et al. 2004). Although alkanes may predominate, the contaminants of greatest concern are polycyclic aromatic hydrocarbons (PAHs) due to their potential toxicity and environmental persistence. PAH contamination is source dependent. PAHs entering the Arctic environment with crude oil spills could include naphthalene, phenanthrene, fluorene and chrysene, and their alkyl derivatives (Wang et al. 1998). Other sources of PAHs in the Arctic, at the landscape scale, derive from atmospheric deposition from fossil fuel combustion and natural sources (such as oil seeps, peats, and bitumen outcrops) (Macdonald et al. 2000). In Antarctica, methylnaphthalenes account for the majority of the PAH contaminants (Aislabie et al. 2004, Kim et al. 2006). This reflects the prevalence of naphthalene and methylnaphthalene, and low concentrations of larger molecular weight PAHs in JP-8 jet fuel, the major fuel used and spilled (Ritchie 2003). At some spill sites, residual hydrocarbons were detected predominantly as an unresolved complex mixture (UCM). This is indicative of a number of sources including lubricating oils, motor oils, or severely biodegraded or weathered oils (Frysinger et al. 2003).

Other classes of compounds present in fuel oils used in polar regions that could contaminate soils include anti-icing agents such as ethylene glycol, monomethyl ether, or diethylene glycol monomethyl ether. Tetraethyl lead was formerly also added to gasoline, and military grade fuels used in Antarctica by the U.S. scientific bases that contain antistatic and anticorrosive additives, and antioxidants.

9.4 RESPONSE OF SOIL MICROBIAL COMMUNITIES TO FUEL SPILLS

Despite the prevailing environmental conditions in polar soils that include cold and fluctuating temperatures, low water-holding capacities, and low levels of nutrients, diverse bacterial communities are widespread in polar soils and the spillage of fuel typically results in an increase in culturable hydrocarbon-degrading bacteria and heterotrophs. Hydrocarbon degradative activity in soil indicates that the hydrocarbon degraders are active *in situ*, conditions permitting. The observed persistence of hydrocarbons in soils over decades, however, indicates that *in situ* biodegradation rates must be very low. The effects of hydrocarbon contamination on the *in situ* bacterial communities are not yet well documented.

9.4.1 Ubiquity of Hydrocarbon-Degrading Microbes

Cold-adapted hydrocarbon-degrading microorganisms are widely distributed in polar soils. They have been reported, for example, in soils from Alaska (Sexstone and Atlas 1977), Svalbard (Rike et al. 2001) and Antarctica (Delille 2000, Aislabie et al. 2001). The response of the soil microbial population to crude oil or fuel spillage depends on soil type and depth as well as the composition of the contaminant. The spillage of hydrocarbons on polar soils can result in the enrichment of hydrocarbon-degrading microbes so that they become a significant proportion of the total culturable microbiota. Whereas the number of hydrocarbon degraders is often low or below detection limits in pristine polar soils, hydrocarbon degraders have been cultivated at levels $> 10^5$ g^{-1} from contaminated polar soils. Similarly, the number of culturable heterotrophic microbes can increase following oil spillage but only by 1–2 orders of magnitude in hydrocarbon-contaminated soils compared with pristine soils. The detection of elevated numbers of hydrocarbon degraders in contaminated subsurface soils suggests increases in microbial populations parallel the downward migration of oil (Sexstone and Atlas 1977, Aislabie et al. 2001, Rike et al. 2001). While most studies have focused on hydrocarbon degraders in the thawed active layer of the soil, a significant number of hydrocarbon degraders have been detected at a depth of approximately 2.0 m in frozen permafrost soil where oil was present (Rike et al. 2001). The ratio of hydrocarbon degraders to culturable heterotrophs ranges from 1 to <0.2, depending on the time elapsed since oil spillage (Delille 2000, Aislabie et al. 2001). Following the application of diesel or crude oil to Antarctic soil, the number of hydrocarbon degraders and heterotrophs was stimulated in the first few months after application and then declined (Delille 2000). Hydrocarbon degraders persist in polar soils and have been detected in contaminated soils more than 30 years after the oil was spilled (Aislabie et al. 2001). These results indicate that hydrocarbon contaminants in polar soils can serve as substrates for microbial growth and can result in a sustained and enhanced number of culturable microbes, and an increased proportion of hydrocarbon-degrading microbes. An exception was a site at Scott Base, Antarctica where gasoline had been stored (Aislabie et al. 1998). Low numbers of culturable heterotrophs and no detectable hydrocarbon degraders at the site were attributed to co-contamination with organic lead. Along with hydrocarbons, the site had organic lead levels 25-fold higher than that of nearby pristine sites. The other co-contaminants in the fuels could also influence hydrocarbon degradation in polar soils. Anti-icing compounds, for example ethylene glycols, have antimicrobial activity (Ritchie 2003). It is possible that co-contaminants with antimicrobial activity limit the establishment and subsequent activity of hydrocarbon degraders in polar soils and therefore contribute to the observed persistence of hydrocarbons.

9.4.2 Hydrocarbon Biodegradative Activity

Hydrocarbon biodegradation studies in laboratory microcosms demonstrate that hydrocarbon degraders can be active in polar soils when environmental conditions permit. The microbial degradation of hydrocarbons has been detected in polar soils at low (≤10°C) and even subzero temperatures (Aislabie et al. 1998, Whyte et al. 1999a, 2001, Mohn and Stewart 2000, Eriksson et al. 2001, Børessen et al. 2003, 2007, Ferguson et al. 2003, Børessen and Rike 2007), indicating that biodegradation can occur in cold and frozen soils. In laboratory studies, it has been demonstrated that although hydrocarbon mineralization occurs in soils at low temperatures, the rate and perhaps the extent of degradation are higher at elevated temperatures. ^{14}C-hexadecane and phenanthrene were mineralized in soil at –5°C, although at a considerably reduced rate than at 5°C (Børessen et al. 2007). Similarly, hexadecane was mineralized at a faster rate and to a greater extent in Arctic soil incubated at 23°C than at 5°C (Whyte et al. 1999a). In contrast, Mohn and Stewart (2000) reported that a change in incubation temperature from 7°C to 22°C did not affect the extent of mineralization of ^{14}C-dodecane, but did increase the rate of mineralization and decrease the lag time.

In addition to low temperatures, polar soils are subjected to short-term fluctuating temperatures and freeze–thaw cycles (Balks et al. 2002). Freeze–thaw cycles, however, may not inhibit hydrocarbon degradation and in some circumstances may enhance activity (Eriksson et al. 2001, Børessen et al. 2007), possibly through the release of nutrients that are otherwise limiting. Eriksson et al. (2001) reported the stimulation of hydrocarbon biodegradation by alternating 24 h periods at +7°C and –5°C. However, in soil exposed to 4-, 8- or 16 day freeze–thaw cycles between +5°C and –5°C, the extent of hexadecane mineralization was stimulated in unfertilized soil but not fertilized soil, and phenanthrene mineralization was reduced when compared to soil microcosms exposed to constant temperature regardless of nutrient availability (Børessen et al. 2007). This indicates that nutrients released from cells lysed by freeze–thaw cycle in unfertilized soils may have been sequestered by surviving alkane degraders that were more tolerant of freeze–thaw cycle than those that degrade aromatic hydrocarbons.

Because coarse-textured soils, typical of those that occur in polar regions, have poor water-holding capacity, it can be difficult to maintain optimum soil moisture levels for hydrocarbon degradation. Low moisture content may limit hydrocarbon degradation, whereas high moisture content may lead to reduced oxygen availability for the aerobic degradation of hydrocarbons. Optimum water contents for hydrocarbon degradation will depend on soil type. For a sandy Arctic soil, hexadecane mineralization was reduced in soils with a moisture content of 20% compared with soils with 10%, 12%, or 15% moisture contents (Børessen and Rike 2007). In soils where moisture levels are elevated and oxygen is limiting, hydrocarbon degradation may proceed under nitrate-reducing conditions (Eriksson et al. 2003, Powell et al. 2006a), given sufficient electron acceptor concentrations. PAHs, including naphthalene, 2-methylnaphthalene, fluorene, and phenanthrene, were degraded in Arctic soils incubated in nitrate-reducing conditions at 7°C (Eriksson et al. 2003). Similarly, toluene and hexadecane were degraded in fertilized oil-contaminated Antarctic soil under denitrifying conditions (Powell et al. 2006a).

As with temperate soils, the amendment of polar soils with nitrogen and/or phosphorus can lead to the enhancement of hydrocarbon mineralization (Braddock et al. 1997, Aislabie et al. 1998, Mohn and Stewart 2000, Ferguson et al. 2003, Børessen and Rike 2007, Børessen et al. 2007). The mineralization of ^{14}C-alkanes (dodecane or hexadecane) or ^{14}C-aromatics (naphthalene or phenanthrene) was enhanced in polar soil following the addition of nitrogen as nitrate or ammonium (Aislabie et al. 1998, Mohn and Stewart 2000, Ferguson et al. 2003, Børessen and Rike 2007, Børessen et al. 2007) or commercial preparations such as 20:20:20 fertilizer (Braddock et al. 1997, Whyte et al. 1999a, 2001), or Inipol EAP22 (Mohn and Stewart 2000). As polar soils often have low water-holding capacities, it can be difficult to maintain optimal nutrient concentrations, and hence the use of slow release fertilizers such as cod bone meal has been tested (Walworth et al. 2003). Although nitrogen is considered to be the major limiting nutrient, maximal hydrocarbon biodegradation occurs with the supplementation of both N and P (Braddock et al. 1997, Mohn and Stewart 2000).

9.4.3 Hydrocarbon-Degrading Microbes Isolated from Polar Soils

Hydrocarbon-degrading bacteria have been readily isolated from contaminated polar soils. The bacteria are commonly psychrotolerant rather than psychrophilic, growing at low temperatures (<10°C) but having an optimum growth temperature >15°C. As in temperate soils, hydrocarbon contamination results in the enrichment of heterotrophs including alkane- and/or aromatic-degrading bacteria that may persist at elevated levels for decades after the initial spill (Aislabie et al. 2004). Hydrocarbon-degrading bacteria isolated from polar soils have been assigned to the bacterial phyla Actinobacteria or Proteobacteria (Table 9.1). The cultivation of hydrocarbon-degrading microbes from cold soils typically results in the isolation of bacteria; however, fungi may play a role in hydrocarbon biodegradation. Fungi have been isolated from contaminated polar soils (Kerry 1990, Aislabie et al. 2001), and a recent report suggests that some fungi may use dodecane as a carbon source (Hughes et al. 2007).

9.4.3.1 Hydrocarbon-Degrading Actinobacteria

Alkane degraders belonging to the genus *Rhodococcus* are frequently isolated from polar soils (Table 9.1). *Rhodococcus* spp. strains 7/1, 5/1, and 5/14 isolated from Antarctic soil grew on a range of alkanes from hexane (C_6) through at least eicosane (C_{20}) and the isoprenoid compound pristane (2,6,10,14-tetramethyl-pentadecane) (Bej et al. 2000). The phylogenetic analysis of 16S rRNA genes from alkane-degrading

TABLE 9.1
Hydrocarbon-Degrading Bacteria from Polar Soils[a] and Their Hydrocarbon Substrates for Growth

Bacterial Strains	Hydrocarbon Growth Substrate(s)	Reference
	Actinobacteria	
Rhodococcus 5/1, 5/14, and 7/1	JP8 jet fuel, C_6–C_{20} *n*-alkanes, pristane	Bej et al. (2000)
Rhodococcus Rho10	Jet A1, jet fuel, dodecane	Thomassin-Lacroix et al. (2001)
Rhodococcus 43/02	JP5 jet fuel, dodecane, hexadecane, pristane	Saul et al. (2005)
Rhodococcus 3/1 and *3/2*	Hexadecane	Aislabie et al. (2008)
Gordonia 3/3, 3/4, 5/2, and 5/4	Hexadecane	Aislabie et al. (2008)
Unclassified *Corynebacterineae* 5/3	Hexadecane	Aislabie et al. (2008)
	Proteobacteria	
Pseudomonas BI7 and BI8	C_5-C_{12} *n*-alkanes, toluene, naphthalene	Whyte et al. (1997)
Pseudomonas Cam-1 and Sag-50G	Biphenyl	Master and Mohn (1998)
Pseudomonas DhA-91	Jet A-1 jet fuel, octane, dodecane	Yu et al. (2000)
Pseudomonas IpA-92 and IpA-93	Toluene and/or benzene	Yu et al. (2000)
Pseudomonas Ant 5	JP8 jet fuel, naphthalene, 2-methylnaphthalene	Aislabie et al. (2000)
Pseudomonas Ant 9	JP8 jet fuel, *p*-xylene, 1,2,4-trimethylbenzene naphthalene, 1- and 2-methyl naphthalene	Aislabie et al. (2000)
Pseudomonas 7/22	JP8 jet fuel, toluene, *m*- and *p*-xylene, 1,2,4-trimethyl benzene	Aislabie et al. (2000)
Pseudomonas Ps 8	Jet A-1 fuel, hexadecane, pristane	Thomassin-Lacroix et al. (2001)
Pseudomonas PK4	Pyrene, dodecane, hexadecane	Eriksson et al. (2002)
Pseudomonas 5B	JP8 jet fuel, hexane	Eckford et al. (2002)
Pseudomonas 5A	JP-8, benzene, toluene, *m*-xylene	Eckford et al. (2002)
Pseudomonas K319	Pyrene	Eriksson et al. (2002)
Sphingomonas Ant 20	JP8 jet fuel, 1-methyl naphthalene, phenanthrene	Aislabie et al. (2000)
Sphingomonas 43/03	Phenanthrene	Saul et al. (2005)
Sphingobium ("Sphingomonas") DhA-95	Jet A-1 jet fuel, dodecane, pristane	Yu et al. (2000)
Sphingobium ("Sphingomonas") 44/02	Phenanthrene	Saul et al. (2005)

TABLE 9.1 (continued)
Hydrocarbon-Degrading Bacteria from Polar Soils[a] and Their Hydrocarbon Substrates for Growth

Bacterial Strains	Hydrocarbon Growth Substrate(s)	Reference
Sphingobium (*"Sphingomonas"*) Ant 17	JP8 jet fuel, *m*-xylene, 1- and 2-methylnaphthalene, dimethylnaphthalenes, 2-ethylnaphthalene, fluorene, phenanthrene	Aislabie et al. (2000)
Alkanindiges CH71	Hexadecane	Aislabie et al. (2009)

Note: All of the bacteria listed have been identified to genus level by 16S rRNA gene sequence analysis.
[a] Updated from Aislabie et al. (2006).

bacterial isolates indicates that *Rhodococcus* spp. from polar soils group with *R. erythropolis* or *R. fascians* and other *Rhodococcus* isolates from cold soils such as *Rhodococcus* Q15 (Aislabie et al. 2006). Alkane-degrading Actinobacteria isolated from ornithogenic Antarctic soil, including isolates identified as *Rhodococcus*, *Gordonia*, or an unclassified Corynebacterineae, mineralized hexadecane when grown with either ammonium nitrate or uric acid as sole nitrogen source (Aislabie et al. 2008). Although *Rhodococcus* isolates are reported to degrade aromatic compounds, the isolates from polar soils appear to degrade alkanes exclusively (Table 9.1). To aid alkane degradation, some Actinobacteria produce biosurfactants with emulsifying properties that assist direct contact with alkanes (Whyte et al. 1999b, Vasileva-Tonkeva and Gesheva 2005).

9.4.3.2 Hydrocarbon-Degrading Proteobacteria

Hydrocarbon-degrading Proteobacteria isolated from polar soils frequently belong to the genera *Pseudomonas, Sphingomonas* or *Sphingobium* (including some isolates formerly classified as *Sphingomonas*) (Table 9.1). They may utilize aromatics and/or alkanes as growth substrates. *Pseudomonas* sp. strains BI7 and BI8, isolated from petroleum-contaminated Arctic soil, for example, utilized C_5 to C_{12} *n*-alkanes and the aromatic compounds naphthalene and toluene (Whyte et al. 1997). In contrast, *Sphingobium* sp. Ant 17 did not degrade alkanes but did degrade numerous compounds in the aromatic fraction of crude oil, jet fuel, and diesel fuel (Baraniecki et al. 2002), and utilized many aromatic compounds for growth, including *m*-xylene, naphthalene and its methyl derivatives, and fluorene and phenanthrene.

Cold-tolerant hydrocarbon-degrading *Pseudomonas* strains, including the Arctic isolate BI7 and the Antarctic strain Ant 9, cluster together by 16S rRNA gene analysis with *Pseudomonas syringae* (Aislabie et al. 2006), whereas others cluster with *P. fluorescens* or *P. stutzeri*. *Sphingobium* isolates Ant 17, 44/02, and DhA-95 cluster with *Sphingobium chlorophenolica* and are related to aromatic-degrading *Sphingobium* spp. from globally distributed sources (Aislabie et al. 2006). *Sphingomonas* sp. Ant 20 has been assigned to the species *S. aerolata*.

Some Proteobacteria such as *Pseudomonas* spp. 5A and 5B fix nitrogen in addition to utilizing mono-aromatic hydrocarbons or hexane, respectively (Eckford et al. 2002), whereas others degrade hexadecane under denitrifying conditions (Powell et al. 2006b). Recently we isolated a hydrocarbon-degrading bacterium from ornithogenic soil identified as *Alkanindiges* CH71 (Aislabie et al. 2009). Like the type strain *Alkanindiges illinoisensis*, *Alkanindiges* CH71 grew weakly or not at all on rich media (e.g., R2A) without hydrocarbons (Bogan et al. 2003).

9.4.4 Effect of Hydrocarbon Contamination on *In Situ* Bacterial Communities

Because only a small percentage of microbes in the environment can be cultivated in the laboratory and typically differ markedly from the genera detected by culture-independent methods, the significance *in situ* of the hydrocarbon-degrading bacteria isolated from polar soils must be questioned (see Section 9.4.3). Hence culture-independent methods have been employed to determine the impacts of hydrocarbon contamination on the diversity of bacterial communities in polar soils. These methods utilize biomarker genes, such as the phylogenetic gene, that encode 16S rRNA or functional genes to identify bacteria or potential bacterial functions in naturally occurring microbial populations without the need for culturing them (Whyte et al. 1999a, 2001, Juck et al. 2000, Saul et al. 2005, Powell et al. 2006c).

9.4.4.1 *In Situ* Bacterial Diversity

The presence of hydrocarbons in polar soils can result in an increased abundance of culturable bacteria. Using cultivation-independent molecular tools, however, it has been revealed that the effect on the *in situ* bacterial communities is variable and may depend on soil type. In some soils, an increase in bacterial diversity has been observed (Juck et al. 2000, Aislabie et al. 2009) whereas other soils show a decrease (Saul et al. 2005). The cold-adapted bacterial communities in petroleum-contaminated and control soils from Alert, Nunavut were analyzed using the DGGE separation of 16S rDNA PCR fragments (Juck et al. 2000). A comparison of the overall microbial molecular diversity suggests diversity was maintained or increased in the oiled soil compared with the control soil. The sequencing of selected 16S rDNA bands from the gels revealed the prevalence of Actinobacteria and Proteobacteria. Many of the Actinobacteria were related to *Nocardioides*, some of which have subsequently been isolated from cold soils and shown to metabolize hydrocarbons (Vasileva-Tonkova and Gesheva 2005).

16S rRNA gene clone libraries were prepared from total community DNA extracted from two soil depths in hydrocarbon-contaminated and control mineral soils near Scott Base on Ross Island (Saul et al. 2005), and an ornithogenic soil sample from the site of a former research station at Cape Hallett (Aislabie et al. 2009). Clones numerically dominating the respective samples were sequenced. In the mineral soil, members of the Phlya Acidobacteria, Bacteroidetes, *Deinococcus/Thermus*, Firmicutes, and Candidate TM7 occurred exclusively in the control soils, whereas the contaminated soils were dominated by Alpha-, Beta-, and Gammaproteobacteria, specifically members of the genera *Sphingomonas*, *Sphingobium* (including former *Sphingomonas* isolates), *Pseudomonas*, or *Variovorax*. Members of the Actinobacteria were found in both oiled and control soils. However, whereas *Rubrobacter* were most prevalent in the control soil, *Rhodococcus* spp. were prevalent in the oiled soil. Similar results were obtained for both soil layers analyzed. Hydrocarbon-degrading bacteria isolated from the oiled mineral soil were identified as *Sphingomonas*, *Sphingobium*, *Pseudomonas*, and *Rhodococcus*.

In the ornithogenic soil, Gammaproteobacteria dominated the control and oiled soils (Aislabie et al. 2009). However, Gammaproteobacteria dominating the control soil were most closely related to *Rhodanobacter* or *Dokdonella*, while those in the oiled soils were related to *Psychrobacter* and *Alkanindiges*. Bacteroidetes and Gemmatimonadetes occurred exclusively in the control soil, whereas Alphaproteobacteria occurred exclusively in the oiled soil. The contamination of ornithogenic soil with hydrocarbons appears to have resulted in an increase in bacterial diversity in the high organic surface layers. In the subsurface mineral layer, however, a single ribotype related to *Alkanindiges* had a relative abundance of 0.39. *Alkanindiges* subsequently isolated from the subsurface layer of the ornithogenic soil was shown to metabolize alkanes.

In mineral soil, the spillage of oil can lead to a decrease in overall soil bacterial diversity (Saul et al. 2005) whereas in low diversity ornithogenic soils an increase in bacterial diversity was detected in the surface organic soil layer but a decrease in subsurface mineral layer (Aislabie et al. 2009). Many of the dominant bacteria detected in hydrocarbon-contaminated polar soils have been assigned to the Proteobacteria (e.g., *Pseudomonas*, *Sphingobium*, *Sphingomonas*, *Alkanindiges*) or Actinobacteria (*Rhodococcus* or *Nocardioides*), indicating that members of these genera (most of which are listed in Table 9.1) are likely to play a role in hydrocarbon degradation *in situ*.

9.4.4.2 Detection of Functional Genes

Functional genes encoding enzymes for hydrocarbon degradation from bacterial isolates, both mesophiles and psychrophiles, have been used to probe hydrocarbon-contaminated and control cold soils for the presence of microbes with a genetic potential to degrade hydrocarbon contaminants (Whyte et al. 1999a, 2001, 2002, Laurie and Lloyd-Jones 2000, Flocco et al., 2009) and to determine the impact of soil treatment methods on gene copy number (Powell et al. 2006c).

The genes encoding the alkane hydroxylase system (typically comprising a particulate, integral-membrane alkane monooxygenase, and a soluble rubredoxin and rubredoxin reductase) have homologues in *Pseudomonas putida* as well as in *Acinetobacter* and *Rhodococcus* spp. The hybridization of DNA extracted from polar soils with four alkane monooxygenase genotypes from *P. putida* (Pp *alkB*), *Rhodococcus* spp. (Rh *alkB1* and Rh *alkB2*), and *Acinetobacter calcoaceticus* (Ac *alkM*) revealed that Rh *alkB1* and Rh *alkB2* homologues are common in both contaminated and control polar soils, and Rh *alkB1* was more prevalent in culturable psychrotolerant bacteria. Pp *alkB* homologues were commonly detected in contaminated soil but Ac *alkM* homologues were rare (Whyte et al. 2002). Based on these results, Whyte et al. (2002) proposed that *Rhodococcus* is the predominant alkane degrader in both pristine and contaminated cold soils, whereas *Pseudomonas* may become enriched by the presence of contaminant hydrocarbons, and *Acinetobacter* is rare. Powell et al. (2006c) have recently developed a real-time PCR method to quantify the portion of microorganisms containing alkane monooxygenase. When applied to Antarctic soils undergoing bioremediation, it was revealed that the number of *alkB* genes in the soil increased in response to fertilizer application and then decreased as the concentration of *n*-alkanes in the soil decreased.

In contrast to alkane degradation genes present in cold tolerant bacteria, the genes used for aromatic degradation by cold-adapted bacteria do not appear to differ significantly from those identified in mesophilic isolates. Both Whyte et al. (1996) and Ma et al. (2006) found that catabolic genes from several aromatic-degrading psychrotolerant strains had homology to those described in mesophilic bacteria. The aromatic degradation genes may be either located on plasmids or the chromosome. Various PAH-degrading *Pseudomonas* strains isolated from Antarctic soil carried the *ndo* gene encoding naphthalene dioxygenase on a large self-transmissible plasmid that could be transferred to mesophilic strains (Ma et al. 2006). This indicates that horizontal gene transfer might play a role in the transfer of hydrocarbon degradation genes from outside polar regions to indigenous species. The probing of polar soil DNA extracts has revealed the presence of the archetypal catabolic genotypes *ndoB* and/or *xylE* (encoding 2,3-catchol dioxygenase) in various contaminated soils (Whyte et al. 1999a, 2001, Flocco et al. 2009). The *phnAc* gene encoding an aromatic dioxygenase subunit has been detected in pristine Antarctic soil following enrichment with naphthalene or phenanthrene (Laurie and Lloyd-Jones 2000).

9.5 SUMMARY

Hydrocarbon spills on soil result in higher levels of soil carbon, which may either serve as substrate for microbial growth or prove toxic to microbial growth and activity. The observed increase in hydrocarbon degraders in polar soils following spillage indicates that in these soils hydrocarbons can serve as a carbon source. While most research has considered hydrocarbon degradation in the active layer, a substantial number of hydrocarbon degraders have also been detected in permafrost soils. Soils conditions that may limit hydrocarbon degradation include low and fluctuating temperatures, and low levels of moisture and nutrient availability. A few studies indicate that hydrocarbon spills have a variable effect on the diversity of *in situ* bacterial communities. Many of the dominant bacteria detected in hydrocarbon-contaminated polar soils have been assigned to the Proteobacteria (e.g., *Pseudomonas*, *Sphingomonas*, or *Sphingobium*) or the Actinobacteria (*Rhodococcus* and *Nocardioides*). Hydrocarbon-degrading bacteria belonging to these phyla are commonly cultured from hydrocarbon-contaminated polar soil. As the genes used for aromatic degradation in cold-adapted bacteria are often located on plasmids, and do not appear to differ from those identified in mesophilic strains, it has been proposed that horizontal gene transfer may play a role in the transfer of hydrocarbon degradation genes from mesophilic to psychrophilic regions.

ACKNOWLEDGMENT

This work was supported by funding from the Foundation for Research, Science and Technology, New Zealand (C09X0307).

REFERENCES

Aislabie, J., McLeod, M., and R. Fraser. 1998. Potential for biodegradation of hydrocarbons in soil from the Ross Dependency, Antarctica. *Applied Microbiology and Biotechnology* 49:210–214.

Aislabie, J., Fraser, R., Duncan, S., and R. L. Farrell. 2001. Effects of oil spills on microbial heterotrophs in Antarctic soils. *Polar Biology* 24:308–313.

Aislabie, J. M., Balks, M. R., Foght, J. M., and E. J. Waterhouse. 2004. Hydrocarbon spills on Antarctic soils: Effects and management. *Environmental Science & Technology* 38:265–1274.

Aislabie, J., Saul, D. J., and J. M. Foght. 2006. Bioremediation of hydrocarbon-contaminated polar soil. *Extremophiles* 10:171–179.

Aislabie, J., Ryburn, J., and A. Sarmah. 2008. Hexadecane mineralization activity in ornithogenic soil from Seabee Hook, Cape Hallett, Antarctica. *Polar Biology* 31:421–428.

Aislabie, J., Jordan, S., Ayton, J., Klassen, J. L., Barker, G. M., and S. Turner. 2009. Bacterial diversity associated with ornithogenic soil of the Ross Sea region, Antarctica. *Canadian Journal of Microbiology* 55:21–36.

Balks, M. R., Paetzold, R. F., Kimble, J. M., Aislabie, J., and I. B. Campbell. 2002. Effects of hydrocarbon spills on the temperature and moisture regimes in the Ross Sea region. *Antarctic Science* 14:119–326.

Baraniecki, C. A., Aislabie, J., and J. M. Foght. 2002. Characterisation of *Sphingomonas* sp. Ant 17, an aromatic hydrocarbon-degrading bacterium isolated from Antarctic soil. *Microbial Ecology* 43:44–54.

Bej, A. K., Saul, D. J., and J. Aislabie. 2000. Cold tolerance of alkane-degrading bacteria isolated from soil near Scott Base, Antarctica. *Polar Biology* 23:100–105.

Biggar, K. W., Haidar, S., Nahir, M., and P. M. Jarrett. 1998. Site investigations for fuel spill migration into permafrost. *Journal of Cold Regions Engineering* 12:84–104.

Bogan, B. W., Sullivan, W. R., Kayser, K. J., Deer, K. D., Aldrich, H. C., and J. R. Paterek. 2003. *Alkanindiges illinoisensis* gen. nov., sp. nov., an obligately hydrocarbonoclastic, aerobic squalene-degrading bacterium isolated from oilfield soils. *International Journal of Systematic and Evolutionary Microbiology* 53:1389–1395.

Børessen, M. H. and A. G. Rike. 2007. Effects of nutrient content, moisture content and salinity on mineralization of hexadecane in Arctic soil. *Cold Regions Science and Technology* 48:129–138.

Børessen, M. H., Breedveld, G. D., and A. G. Rike. 2003. Assessment of the biodegradation potential of hydrocarbons in contaminated soils from a permafrost site. *Cold Regions Science and Technology* 37:37–149.

Børessen, M. H., Barnes, D. L., and A. G. Rike. 2007. Repeated freeze-thaw cycles and their effects on mineralization of hexadecane and phenanthrene in cold climate soils. *Cold Regions Science and Technology* 49:215–225.

Braddock, J. F., Ruth, M. L., Catterall, P. H., Walworth, J. L., and K. A. McCarthy. 1997. Enhancement and inhibition of microbial activity in hydrocarbon-contaminated arctic soils: Implications for nutrient-amended bioremediation. *Environmental Science & Technology* 31:2078–2084.

Chuvilin, E. M., Naletova, N. S., Miklyaeva, E. C., Kozlova, E. V., and A. Instanes. 2001. Factors affecting spreadability and transportation of oil in regions of frozen ground. *Polar Record* 37:229–238.

Delille, D. 2000. Response of Antarctic soil bacterial assemblages to contamination by diesel fuel and crude oil. *Microbial Ecology* 40:159–168.

Eckford, R., Cook, F. D., Saul, D., Aislabie, J., and J. Foght. 2002. Free-living nitrogen-fixing bacteria from Antarctic soils. *Applied and Environmental Microbiology* 68:5181–5185.

Eriksson, M., Ka, J.-O., and W. W. Mohn. 2001. Effects of low temperature and freeze-thaw cycles on hydrocarbon biodegradation in Arctic tundra soil. *Applied and Environmental Microbiology* 67:5107–5112.

Eriksson, M., Dalhammar, G., and W. W. Mohn. 2002. Bacterial growth and biofilm production on pyrene. *FEMS Microbiology Ecology* 40:21–27.

Eriksson, M., Sodersten, E., Yu, Z., Dalhammer, G., and W. W. Mohn. 2003. Degradation of polycyclic aromatic hydrocarbons at low temperature under aerobic and nitrate-reducing conditions in enrichment cultures from Northern soils. *Applied and Environmental Microbiology* 69:275–284.

Ferguson, S. H., Franzmann, P. D., Revill, A. T., Snape, I., and J. L. Rayner. 2003. The effects of nitrogen and water on mineralisation of hydrocarbons in diesel-contaminated terrestrial soils. *Cold Regions Science and Technology* 37:197–212.

Flocco, C. G., Gomes N. C. M., Mac Cormack, W., and Smalla, K. 2009. Occurrence and diversity of naphthalene dioxygenase genes in soil microbial communities from the maritime Antarctic. *Environmental Microbiology* 11:700–714.

Frysinger, G. S., Gaines, R. B., Xu, L., and C. M. Reddy. 2003. Resolving the unresolved complex mixture in petroleum-contaminated sediments. *Environmental Science & Technology* 37:1653–1662.

Gore, D. B., Revill, A. T., and D. Guille. 1999. Petroleum hydrocarbons ten years after spillage at a helipad in Bunger Hills, East Antarctica. *Antarctic Science* 11:427–429.

Green, G., Skerratt, J. H., Leeming, R., and P. D. Nichols. 1992. Hydrocarbon and coprostanol levels in seawater, sea-ice algae and sediments near Davis Station in Eastern Antarctica: A regional survey and preliminary results for a field fuel spill experiment. *Marine Pollution Bulletin* 25:293–302.

Hughes, K. A., Bridge, P., and M. S. Clark. 2007. Tolerance of Antarctic soil fungi to hydrocarbons. *Science of the Total Environment* 372:539–548.

Juck, D., Charles, T., Whyte, L. G., and C. W. Greer. 2000. Polyphasic microbial community analysis of peteroleum hydrocarbon-contaminated soils from two northern Canadian communities. *FEMS Microbiology Ecology* 33:241–249.

Kennicutt, M. C., McDonald, T. J., Denoux, G. J., and S. McDonald. 1992. Hydrocarbon contamination on the Antarctic Peninsula. I. Arthur Harbor- subtidal sediments. *Journal of Marine Pollution Bulletin* 24:499–506.

Kerry, E. 1990. Microorganisms colonizing plants and soil subjected to different degrees of human activity, including petroleum contamination in the Vestfold Hills and MacRobertson Land Antarctica. *Polar Biology* 10:423–430.

Kim, M., Kennicutt II, M. C., and Y. Qian. 2006. Molecular and stable carbon isotopic characterization of PAH contaminants at McMurdo Station, Antarctica. *Marine Pollution Bulletin* 52:1585–1590.

Laurie, A. D. and G. Lloyd-Jones. 2000. Quantification of *phnAc* and *nahAc* in contaminated New Zealand soils by competitive PCR. *Applied and Environmental Microbiology* 66:1814–1817.

Ma, Y., Wang, L., and Z. Shao. 2006. *Pseudomonas*, the dominant polycyclic aromatic hydrocarbon-degrading bacteria isolated from Antarctic soils and the role of large plasmids in horizontal gene transfer. *Environmental Microbiology* 8:455–465.

Macdonald, R. W., Barrie, L. A., Bidleman, T. F. et al. 2000. Contaminants in the Canadian Arctic: 5 years of progress in understanding sources, occurrence and pathways. *Science of the Total Environment* 254:93–234.

Master, E. R. and W. W. Mohn. 1998. Psychrotolerant bacteria isolated from Arctic soil that degrade polychlorinated biphenyls at low temperatures. *Applied and Environmental Microbiology* 64:4823–4829.

McCarthy, K., Walker, L., and L. Vigoren. 2004. Subsurface fate of spilled petroleum hydrocarbons in continuous permafrost. *Cold Regions Science and Technology* 38:43–54.

Mohn, W. W. and G. R. Stewart. 2000. Limiting factors for hydrocarbon degradation at low temperature in Arctic soils. *Soil Biology and Biochemistry* 32:1161–1172.

Poland, J. S., Riddle, M. J., and B. A. Zeeb. 2003. Contaminants in the Arctic and the Antarctic: A comparison of sources, impacts, and remediation options. *Polar Record* 39:369–384.

Powell, S. M., Ferguson, S. M., Snape, I., and S. D. Sicilano. 2006a. Fertilization stimulates anaerobic fuel degradation of Antarctic soils by denitrifying microorganisms. *Environmental Science & Technology* 40:2011–2017.

Powell, S. M., Ma, W. K., and S. D. Siciliano. 2006b. Isolation of denitrifying bacteria from hydrocarbon-contaminated Antarctic soil. *Polar Biology* 30:69–74.

Powell, S. M., Ferguson, S. H., Bowman, J. P., and I. Snape. 2006c. Using real-time PCR to assess changes in the hydrocarbon-degrading microbial community in Antarctic soils during bioremediation. *Microbial Ecology* 52:523–532.

Rike, A. G., Børessen, M., and A. Instanes. 2001. Response of cold-adapted microbial populations in permafrost profile to hydrocarbon contaminants. *Polar Record* 37:239–248.

Ritchie, G. D. 2003. Biological and health effects of exposure to kerosene-based jet fuels and performance additives. *Journal of Toxicology and Environmental Health, Part B* 6:357–451.

Saul, D. J., Aislabie, J., Brown, C. E., Harris, L., and J. M. Foght. 2005. Hydrocarbon contamination changes the bacterial diversity of soil from around Scott Base, Antarctica. *FEMS Microbiology Ecology* 53:141–155.

Sexstone, A. J. and R. M. Atlas. 1977. Response of populations in Arctic tundra soils to crude oil. *Canadian Journal of Microbiology* 23:1327–1333.

Tarnocai, C. and I. B. Campbell. 2002. Soils of the polar regions. In *Encyclopedia of Soil Science*, ed. R. Lal. Marcel Dekker Inc., New York, pp. 1018–1012.

Thomassin-Lacroix, E. J. M., Yu, Z., Eriksson, M., Reimer, K. J., and W. W. Mohn. 2001. DNA-based and culture-based characterization of hydrocarbon-degrading consortium enriched from Arctic soil. *Canadian Journal of Microbiology* 47:1107–1115.

Vasileva-Tonkova, E. and V. Gesheva. 2005. Glycolipids produced by Antarctic *Nocardioides* sp. during growth on n-paraffin. *Process Biochemistry* 40:2387–2391.

Walworth, J. L., Woolard, C. R., and K. C. Harris. 2003. Nutrient amendments for contaminated peri-glacial soils: Use of cod bone meal as a controlled release nutrient source. *Cold Regions Science and Technology* 37:81–88.

Wang, Z., Fingas, M., and Blenkinsopp, S., et al. 1998. Comparison of oil composition changes due to biodegradation and physical weathering in different oils. *Journal of Chromatography A* 809:89–107.

Whyte, L. G., Greer, C. W., and W. E. Inniss. 1996. Assessment of the biodegradation potential of psychrotrophic microorganisms. *Canadian Journal of Microbiology* 42:99–106.

Whyte, L. G., Bourbonnière, L., and C. W. Greer. 1997. Biodegradation of petroleum hydrocarbons by psychrotrophic *Pseudomonas* strains possessing both alkane (alk) and naphthalene (nah) catabolic pathways. *Applied and Environmental Microbiology* 63:3719–3723.

Whyte, L. G., Bourbonnière, L., Bellerose C., and C. W. Greer. 1999a. Bioremediation assessment of hydrocarbon-contaminated soils from the high Arctic. *Bioremediation Journal* 3:69–79.

Whyte, L. G., Slagman, S. J., Pietrantonio, F., et al. 1999b. Physiological adaptations involved in alkane assimilation at low temperatures by *Rhodococcus* sp. strain Q15. *Applied and Environmental Microbiology* 65:2961–2968.

Whyte, L. G., Goalen, B., Hawari, J., Labbé, D., Greer, C. W., and M. Nahir. 2001. Bioremediation treatability assessment of hydrocarbon-contaminated soils from Eureka, Nunavut. *Cold Regions Science and Technology* 32:121–132.

Whyte, L. G., Schultz, A., van Beilen, J. B., et al. 2002. Prevalence of alkane monooxygenase genes in Arctic and Antarctic hydrocarbon-contaminated and pristine soils. *FEMS Microbiology Ecology* 41:41–150.

Yu, Z., Stewart, G. R., and W. W. Mohn. 2000. Apparent contradiction: Psychrotolerant bacteria from hydrocarbon-contaminated Arctic tundra soils that degrade diterpenoids synthesized by trees. *Applied and Environmental Microbiology* 66:5148–5154.

10 Microbial Degradation of Petroleum Compounds in Cold Marine Water and Ice

Odd G. Brakstad, Andrew M. Booth, and Liv-Guri Faksness

CONTENTS

10.1 INTRODUCTION

The activities of the oil production and transport industries in the polar regions of the world represent a continuous risk for releases of petroleum to these marine environments. While oil production has progressed for several decades in the American Arctic, more recently focus has been placed on large oil and gas fields in the European Arctic (e.g., Barents Sea region). Despite strict governmental regulations being placed on the oil industry in these areas, small accidental releases from exploration and production may have impacts on the local marine environment. In addition, the possibilities of blowouts and oil tanker accidents represent a continuous threat to these vulnerable polar environments.

In polar regions, spilled oil may reach the marginal ice zone and become ice-infested. Owing to increasing temperatures in the Arctic, it is expected that more of the polar ice will become one-year ice. Subsequently, a second discharge of the oil to seawater occurs as this thin ice melts. During the ice-infested period, the oil may be transported for long distances away from the source area (Pfirman et al., 1995). If the oil reaches shallow water some of it may be sequestered in seabed sediments after association to suspended particles, and/or the oil may become stranded on the shoreline. However, considerable quantities of the oil will remain present in the seawater column as dissolved compounds or dispersions until removed through biotic or abiotic processes. Eventually, some of the emulsified oil may be incorporated in small tarballs that can be transported over large distances.

Biodegradation represents an important natural removal process for petroleum compounds released into the marine environment. When oil is spilled in open seawater, the process of biodegradation is initiated immediately. However, the rate and extent of the degradation depends strongly on a variety of factors, which include both oil characteristics and environmental conditions. Several reviews have described oil biodegradation in marine environments (e.g., Atlas, 1981, 1984; Leahy and Colwell, 1990; Prince, 1993; Swannell et al., 1996; Harayama et al., 1999, 2004; NRC, 2003; Van Hamme et al., 2003; Venosa and Zhu, 2003). In contrast, oil biodegradation in cold environments has only been reviewed in a few cases (e.g., Margesin and Schinner 1999, 2001; Brakstad 2008). These reviews have focused mainly on biodegradation and bioremediation in marine sediments rather than the seawater column. However, in this chapter we will mainly concentrate on the interactions between discharged oil and marine microorganisms in cold seawater and marine ice. We will also discuss some of the concerns related to persistent oil compounds in polar regions.

10.2 OIL CHARACTERISTICS IN COLD SEAWATER AND ICE

In order to describe microbial interactions with crude oil in cold marine environments it is fundamental to understand the physical–chemical characteristics of crude oils and oil compounds in these environments.

10.2.1 Oil Discharges

Following release to the environment, petroleum hydrocarbons will gradually weather due to natural physical and chemical processes, including spreading, evaporation, dispersion, emulsification, dissolution, biodegradation, and chemical modification by oxidative processes. The first processes significantly affecting a surface oil spill are evaporation, spreading, dissolution, and dispersion. These processes can significantly effect the composition and concentration of the seawater soluble fraction of spilled crude oil and thus determine its early impact on marine biota.

For example, evaporation is most important in the first few days following the spill and will determine how rapidly toxic volatile crude oil compounds (e.g., BTEX) are removed into the atmosphere. Conditions supporting rapid evaporation can lead to lower concentrations of volatile compounds in the seawater soluble fraction. In contrast, biodegradation is not considered a significant removal process (regarding oil spill contingency) until 1–2 weeks after the spill.

In polar regions, low temperatures and the absence of waves in waters with ice floes act to reduce oil spreading, evaporation, emulsification, and dispersion (e.g., Brandvik and Faksness, 2009). The fate of discharged oil compounds in these areas, therefore, differs from those in temperate or tropical regions. Wax and asphaltene contents, for example, in combination with temperature-dependent oil viscosity and weathering processes, are important for the fate of the oil in cold-water conditions (Brandvik et al., 2004; Brandvik and Faksness, 2009). Ultimately, the oil may become entrapped in the ice, although the time required for this to occur is dependent on conditions such as air–ice–water temperature, ice thickness, and quantity of oil to be encapsulated. Once trapped within the ice, ocean currents can transport spilled oil over large distances. A secondary discharge situation occurs during the spring melt season and, if the ice has been transported from the original spill site, this can result in the pollution of new and unexpected locations. In the spring and summer seasons, the chemical alteration of the crude oil through photooxidation may also become an important degradation process. Although the immediate impact of oil spills on ice has been studied (e.g., Fingas and Hollebone, 2003) and is fairly well understood, little is known about the long-term fate and effects of such pollutants on ecosystems in polar environments.

10.2.2 Spreading of Oil

The behavior of oil spilled in cold seawater or on ice is largely dependent on the oil type and therefore its chemical composition and properties. Owing to differences in the aqueous solubility and toxicity of individual oil components, the transport, migration, and environmental impact of different oils remain important areas of study. Field experiments conducted on Svalbard have shown that increased ice coverage reduces the rate of oil weathering processes due to higher film thickness and reduced wave energy (Brandvik and Faksness, 2009). The increased film thickness decreases the evaporative losses, and the lower energy input reduces the emulsification of water droplets into the oil. In addition, oil spilled in the Arctic marine environment can be rapidly frozen into the ice, effectively locking it away before it can spread. Once entrapped, the oil will become partly preserved as evaporation, dissolution, and biodegradation are expected to be significantly reduced. In such cases, the oil will retain much of its potential toxicity upon release from the ice, which is typically via transport in the brine channels and/or eventual break up and melting of the ice.

10.2.3 Whole Oil and Individual Compound Behavior in Ice

Sea ice in polar regions constitutes an important habitat for a variety of marine organisms ranging from bacteria to crustaceans. Specifically, the interstitial system of brine channels and pockets provides the habitats of sea ice organisms (Weissenberger et al., 1992). Brine channels in first-year ice occur throughout the growing season; however, their size and spatial frequency increases as the ice warms. The major brine channel growth period occurs in the spring when the air temperature increases and the ice warms toward the melting point. The most important oil transport mechanism is the opening of the top-to-bottom brine channels, which allows the oil to both rise through the ice to the surface, and to permeate the channel system within the ice (Martin, 1979; Dickins et al., 2008; Faksness and Brandvik, 2008b). Migration rates of oil in brine channels are thought to vary as a function of oil properties and chemical composition. The encapsulated oil will subsequently be released during the spring melt as the ice sheet deteriorates. Oil escapes from the ice sheet by a combination of two general processes: vertical rise of the oil through the brine channels and ablation of the ice surface down to the entrapped oil (Fingas and Hollebone, 2003). Exposed oil on the ice surface may further adsorb heat from the Sun and increase the surface water temperature (Fingas and Hollebone, 2003), propagating the ice-melting process.

To date, few studies have attempted to determine the transport and fate of individual water-soluble oil components in sea ice. However, data from some recent studies have shown that the more water-soluble compounds (mainly naphthalenes, phenanthrenes, and dibenzothiophenes) migrate through the brine channels in the ice. As a result, such compounds come into contact with sea ice microbes in the brine and the underlying water (Faksness and Brandvik, 2008a,b).

10.3 MARINE MICROBES INVOLVED IN COLD ENVIRONMENT CRUDE OIL BIODEGRADATION

Several studies over the last 30 years have shown that the growth of psychrophilic bacteria in seawater is stimulated by the presence of oil hydrocarbons (e.g., Cundell and Traxler, 1976; Delille and Vaillant, 1990; Delille and Siron, 1993, Siron et al., 1995; Brakstad and Bonaunet, 2006). Typical growth curves in marine culture media for heterotrophic (Marine Broth) and oil-degrading (Bushnell-Haas broth with fresh oil as carbon source) microbes at a temperature of 0°C are shown in Figure 10.1. In both cases, a significant increase in the number of bacterial cells was observed after only a few days. A few studies have also shown that bacterial growth in marine ice is stimulated by oil pollution. The long-term effects of diesel fuel and crude oil on microbial communities were investigated over 9 months in Antarctic land-fast ice (Terre Adélie). In these studies a three orders of magnitude increase in bacterial counts occurred in sea ice contaminated with diesel and crude oil, and the proportion of oil-degrading bacteria increased from <0.001% (uncontaminated ice) to 10% of the community after 30 weeks of contamination (Delille et al., 1997). In a recent 4 month winter field study at Svalbard (Van Mijen Fjord), Arctic fjord ice was contaminated with fresh crude oil in February and polluted ice cores sampled monthly from March to June. The total concentrations of bacteria-like particles were determined using epifluorescence microscopy after nucleic acid staining (DAPI). The concentrations were nearly five times higher in oil-contaminated sections of ice cores than in corresponding clean core sections at the end of the field study.

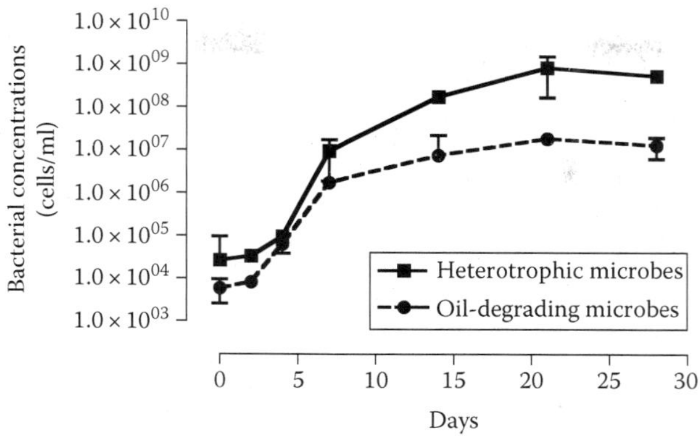

FIGURE 10.1 Typical growth curves for enrichment cultures grown at 0°C in seawater-based media for heterotrophic and oil-degrading microbes. Marine Broth 2216 was used for heterotrophic microbes and marine Bushnell-Haas with 1% paraffinic crude oil for oil-degrading microbes.

In addition, concentrations of heterotrophic microbes growing on nutritious media (Marine Agar 2216) were approximately 10 times higher in the oil-contaminated sections than in the clean core sections. However, the concentrations of oil-degrading bacteria (most probable number measurements in marine Bushnell-Haas broth with crude paraffinic oil as carbon source) were not clearly related to the presence of the oil in this study (Brakstad et al., 2008).

The ability to metabolize hydrocarbons has been detected in a wide variety of microbial genera, including bacteria, fungi, alga, and diatoms (Prince, 2005). Although archaea have typically been associated with the oxidation of small hydrocarbons such as methane, they have also been reported to be involved in the degradation of long-chain alkanes (Zengler et al., 1999; Anderson and Lovley, 2000). However, during oil spill remediation studies in temperate regions (marine intertidal flats, Somerset, United Kingdom), archaea were found to play an insignificant role in the degradation of petroleum hydrocarbons (Röling et al., 2004).

Hydrocarbon-degrading bacteria associated with temperate seawater usually include members of the classes Alphaproteobacteria and Gammaprotobacteria, typically represented by the genera *Roseobacter* and *Alcanivorax* (Harayama et al., 1999; Chang et al., 2000; Kasai et al., 2002; Brakstad and Lødeng, 2004). Table 10.1 shows the bacterial genera reported to be associated with hydrocarbon degradation in Arctic or Antarctic seawater and marine ice. The pollution of cold seawater or marine ice with crude oil or oil compounds typically results in shifts of bacterial communities with a relative increase in the abundance of comparatively few genera. When seawater collected from the Antarctic Ross Sea (Terra Nova Bay) was contaminated with evaporated crude oil, enrichment cultures grown at 4°C were predominated by members of *Colwellia*, *Oleispira*, and *Acinetobacter*, which accounted for 76% of sequenced 16S rDNA clones (Yakimov et al., 2004). A recent study comparing the bacterial diversity in clean and oil-amended sub-Antarctic seawater (collected outside Ushuaia, Argentina) revealed increases in the Gammaproteobacteria genus

TABLE 10.1
Polar Bacteria Associated with Biodegradation of Oil Compounds in Cold Seawater and Marine Ice

Phylogenetic Group	Genus	Bacterial Origin[a]		Sources
		Seawater	Marine Ice	
Alphaproteobacteria	*Sphingomonas*	Ant		Yakimov et al. (2004)
	Erythrobacter	Ant		Yakimov et al. (2004)
	Pelagibacter		Arct	Brakstad et al. (2008)
	Unidentified	Ant		Pradagaran et al. (2006), Yakimov et al. (2004)
Betaproteobacteria	*Variovorax*		Arct	Gerdes et al. (2005)
	Delftia		Arct	Gerdes et al. (2005)
	Unidentified		Arct	Gerdes et al. (2005), Brakstad et al. (2008)
Gammaproteobacteria	*Glaciecola*		Arct	Brakstad et al. (2008)
	Marinobacter	Ant/Arct	Arct	Gerdes et al. (2005), Deppe et al. (2005), Yakimov et al. (2004)
	Marinobacterium		Arct	Brakstad et al. (2008)
	Halomonas	Ant	Arct	Gerdes et al. (2005), Yakimov et al. (2004)
	Pseudomonas	Ant/Arct	Arct	Gerdes et al. (2005), Yakimov et al. (2004)
	Pseudoalteromonas	Ant/Arct	Arct	Deppe et al. (2005), Brakstad et al. (2008), Yakimov et al. (2004)
	Shewanella	Ant/Arct	Arct	Gerdes et al. (2005), Deppe et al. (2005), Yakimov et al. (2004)
	Colwellia	Ant	Arct	Brakstad et al. (2008), Yakimov et al. (2004)
	Marinomonas	Ant	Arct	Brakstad et al. (2008), Yakimov et al. (2004)
	Psychrobacter	Ant		Pradagaran et al. (2006)
	Oleispira	Ant	Arct	Pradagaran et al. (2006), Brakstad et al. (2008), Yakimov et al. (2004)
	Unidentified	Ant	Arct	Pradagaran et al. (2006), Brakstad et al. (2008), Yakimov et al. (2004)
Deltaproteobacteria	*Halangium*		Arct	Brakstad et al. (2008)

TABLE 10.1 (continued)
Polar Bacteria Associated with Biodegradation of Oil Compounds in Cold Seawater and Marine Ice

		Bacterial Origin[a]		
Phylogenetic Group	**Genus**	**Seawater**	**Marine Ice**	**Sources**
Epsilonproteobacteria	*Arcobacter*	Ant		Pradagaran et al. (2006), Yakimov et al. (2004)
Bacteroidetes	*Cytophaga*	Ant		Pradagaran et al. (2006), Yakimov et al. (2004)
	Polaribacter	Ant	Arct	Pradagaran et al. (2006), Gerdes et al. (2005)
	Ulvibacter	Ant		Pradagaran et al. (2006)
	Tenacibaculum	Ant		Pradagaran et al. (2006)
	Flavobacteria	Arct	Arct	Gerdes et al. (2005)
Actinobacteria	*Actinomycetales*	Arct	Arct	Gerdes et al. (2005)
	Agreia	Arct	Arct	Gerdes et al. (2005), Deppe et al. (2005)
Firmicutes	*Clostridium*	Arct	Arct	Gerdes et al. (2005)

Note: The genera were characterized by sequence analyses of the bacterial 16S rRNA gene.
[a] Ant = Antarctic; Arct = Arctic.

Psychrobacter and the *Bacteroidetes* genus *Polaribacter*. In contrast, the contribution of Alphaproteobacteria (*Roseobacter*, *Sufitobacter*) was reduced in oil-polluted seawater compared to clean seawater (Prabagaran et al., 2006). In a study of enrichment cultures from Arctic seawater (Svalbard), the genera *Pseudoalteromonas*, *Shewanella*, and *Marinobacte*r were found to be associated with crude oil biodegradation (Deppe et al., 2005).

Bacterial communities in Arctic marine ice stimulated by the presence of crude oil have been investigated in two recent studies (Gerdes et al., 2005; Brakstad et al., 2008). In one experiment, samples of marine ice were melted and incubated with crude oil at 1°C for 1 year. Strong shifts in community compositions were revealed in the oil-contaminated samples, with a predominance of the genera *Marinobacter*, *Shewanella*, and *Pseudomonas* (Gerdes et al., 2005). In the Svalbard field study described earlier, oil-contaminated ice cores were dominated by members of the genera *Colwellia* and *Marinomonas* toward the end of the study (April and June samples). While these genera constituted 77% of the 16S rDNA clone library in oil-contaminated samples, they were not detected in the library from corresponding clean ice cores (Brakstad et al., 2008). Several of the bacterial phylotypes described in these studies are typically psychrophilic and associated with polar seawater and ice in both the Arctic and Antarctic (Bowman et al., 1997; Junge et al., 2002; Brinkmeyer et al., 2003; Zeng et al., 2007).

10.4 BIODEGRADATION OF OIL IN COLD MARINE ENVIRONMENTS

In general, the biodegradation of oil compounds is expected to follow the order *n*-alkanes > branched alkanes > low molecular weight aromatics > cyclic alkanes (Perry, 1984). In cold seawater, the same order is expected, although degradation will be highly influenced by the physical–chemical characteristics of the oil. The low temperature affects both dissolution from the nonaqueous (crude oil) to the aqueous phase (Schluep et al., 2001), and the evaporation of volatile compounds. The reduced volatilization of toxic BTEX may result in delayed biodegradation at high concentrations due to their acute toxicity to microorganisms (Atlas and Bartha, 1972; Hokstad et al., 1999). Under ice conditions, BTEX may be transported by the ice brine to the seawater below the ice (Payne et al., 1991). For wax-rich oils with high pour points, dilution and dispersion may be reduced in cold seawater since precipitated wax may build a matrix that limits the internal mixing of the oil and acts as a diffusion barrier between the oil and the water. Important factors influencing the biodegradation of crude oil in cold marine environments include temperature, ice, light conditions, and the abilities of naturally occurring microbial communities to biodegrade oil compounds under these conditions.

10.4.1 SEAWATER

At temperatures above the freezing point of seawater (approx −1.8°C), the biodegradation of crude oil hydrocarbons is well documented. Figure 10.2 shows the

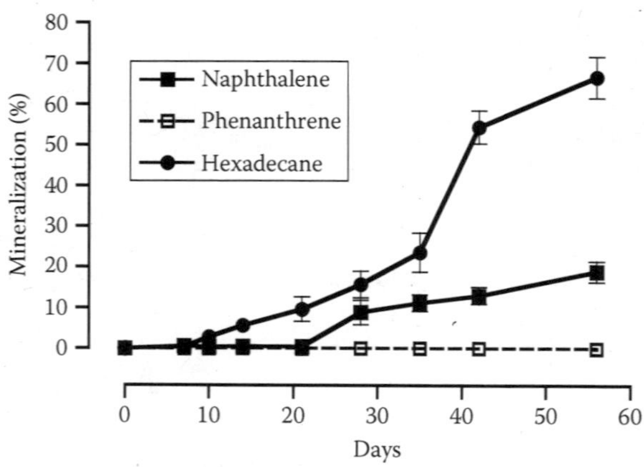

FIGURE 10.2 Mineralization in seawater at 0°C of ^{14}C-labeled hydrocarbons spiked in crude oil. No mineralization was measured in sterile controls (not shown). (From Brakstad, O.G. and Bonaunet, K., *Biodegradation*, 17, 71, 2006. With permission.)

mineralization curves of ^{14}C-labeled naphthalene, phenanthrene, and hexadecane in seawater at 0°C when the compounds were spiked into crude paraffinic oil. The degradation of the *n*-alkane (hexadecane) was faster than for the aromatic compounds, and the smaller aromatic (naphthalene; 2-ring) degraded faster than the larger homologue (phenanthrene; 3-ring). This pattern followed the generally accepted order of crude oil compound biodegradation described earlier. However, most biodegradation experiments with crude oil have been conducted at seawater temperatures more relevant for temperate regions than for the cold water of polar environments. The temperature-related plots have, therefore, been used to transpose results of bacterial metabolism and growth between different temperatures. Temperature-related bacterial growth rates may be determined by using modified Arrhenius plots (Arrhenius, 1889). Ideally, Arrhenius plots should show temperature-related linearity. However, in a study with psychrotrophic toluene-degrading strains of *P. putida* grown on toluene or benzoate, growth rates had to be fitted using two linear segments at a temperature range of 4°C–30°C, one segment above and one below 17°C–20°C (Chablain et al., 1997). In Arrhenius plots, the temperature range should therefore not be too broad. For water-soluble compounds, temperature-dependent biodegradation has been suggested to follow a Q_{10}-value, which is a factor describing the degradation rate increases when temperatures are raised by 10°C. The Q_{10}-values were determined in a seawater-based hydrocarbon biodegradation study with a heavy fuel oil (Bunker C), and with winter or summer water samples from the North Sea as inocula. When incubation temperatures of 4°C–18°C were used, Q_{10}-values of 2.4 and 2.1 were determined for waters in winter and summer, respectively, when biodegradation was measured as biological oxygen demand (Minas and Gunkel, 1995).

Temperature affects both physical and biochemical oil weathering characteristics. This is exemplified in Figure 10.3, which shows both the dissolution and biodegradation of compounds from thin oil films submerged in cold seawater at two temperatures, 5°C and 0°C (Brakstad and Bonaunet, 2006). The *n*-alkane fraction (n-C_{12}

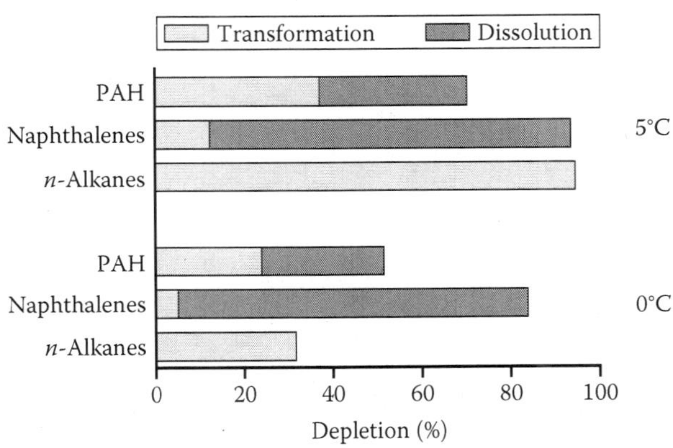

FIGURE 10.3 Depletion of the crude oil compounds *n*-alkanes, naphthalenes and PAH from immobilized thin oil films (~10 μm thickness) submerged in natural seawater at 0°C and 5°C for 56 days. (From Brakstad, O.G. and Bonaunet, K., *Biodegradation*, 17, 71, 2006. With permission.)

to n-C_{36}) was not water-soluble and biodegradation represented the only depletion mechanism from the oil film. This process showed a clear temperature-dependence with higher degradation at 5°C than at 0°C. In contrast, the C_0 to C_4-alkylated naphthalenes were water-soluble, and the depletion mechanism from the oil film was almost entirely through dissolution. As dissolution is not as temperature-dependent as biodegradation, the depletion of the water-soluble naphthalenes from the oil-phase was similar at both 5°C and 0°C. For a mixture of PAHs (3–5 aromatic rings), which exhibited lower water solubilities than the naphthalenes, depletion was again more temperature-dependent, reflecting an increased contribution from biodegradation than dissolution (Figure 10.3).

Photooxidation is an important process in degrading and transforming crude oil compounds after release to the environment. The polar regions exhibit vast seasonal differences in light conditions, and as a result, photooxidation varies significantly between the polar summer and winter. UV-illumination of crude oils has shown that aliphatic compounds are resistant to photodegradation, while aromatic compounds appear particularly sensitive to this process. In contrast to biodegradation, increased size and alkyl substitution result in an increased sensitivity of aromatic hydrocarbons to photochemical oxidation. As photooxidation leads to the inclusion of oxygen atoms in the structures of these compounds, photooxidized products appear mainly in the polar resin fractions of the oil (Garrett et al., 1998; Prince et al., 2003b). Furthermore, the average molecular weight of oil compounds is reduced and the oxygen content increased. Studies have also shown that these photooxidized compounds subsequently exhibit increased susceptibility to biodegradation (Ni'matuzahroh et al., 1999; Dutta and Harayama, 2000; Maki et al., 2001). Consistent with the physicochemical properties of the photooxidized compounds, both the dissolved organic carbon concentration and acute toxicity of the water-soluble fraction of oil increased during the irradiation period (Maki et al., 2001). Thus, photooxidation results in a greater proportion of oxidized compounds that exhibit increased water-solubility and subsequently more significant impacts on toxicity and biodegradation.

10.4.2 Marine Ice

The conditions in marine ice vary considerably from those in seawater. The temperature is decreased in the ice and may drop as low as the air temperature during the polar winter. Light levels in the ice are severely reduced compared to seawater, and may be absent when the ice is covered by a snowcap. In the first-year ice, brine inclusion networks are generated by salting-out processes. The brine channels of the ice provide saline liquid niches at subzero temperatures that enable microbial motility and respiration (Junge et al., 2003, 2004, 2006). The brine channels also act as a matrix for the transport of hydrocarbons (Fingas and Hollebone, 2003; Faksness and Brandvik, 2005) and soluble nutrients. Under such harsh conditions, where ice temperatures can be as low as −20°C and salinity conditions can be as high as 200 PSU (salinity in normal seawater is 35 PSU), laboratory experiments have shown that bacteria (*Colwellia psychroerythraea*) are able to incorporate amino acids into proteins (Junge et al., 2006). This respiration has been associated with particles or with the surfaces of the brine channels (Junge et al., 2003, 2004). In addition, the ability to grow at subzero temperatures has been demonstrated by the bacterium "*Psychromonas ingrahamii*," which has been cultured at temperatures as low as −12°C (Breezee et al., 2004).

Although some studies from oil-polluted Arctic soils have indicated slow biodegradation at subzero temperatures (Rike et al., 2003; Børresen et al., 2007), the biodegradation of petroleum hydrocarbons in marine ice under subzero temperatures still remains to be proven. However, a few studies have shown that oil in ice clearly stimulates bacterial growth (Delille et al., 1997; Brakstad et al., 2008). During the 2004 winter field study performed on Svalbard, changes in n-C_{17}/Pristane and naphthalene/phenanthrene ratios during the experimental period indicated slow biodegradation, but changes in these ratios were not significant (Brakstad et al., 2008).

10.4.3 Bacterial Adaptations to Metabolism at Low Temperatures

Bacteria have been shown to degrade hydrocarbons at low temperatures. The psychrotrophic *Rhodococcus* sp. strain Q15 was able to degrade a broad range of aliphatic hydrocarbons at 5°C, including n-C_{10} to n-C_{21} alkanes, branched alkanes, and a substituted cyclohexane. As expected, mineralization of the short-chain alkanes dodecane (C_{12}) and hexadecane (C_{16}) at 0°C and 5°C occurred to a greater extent than for the long-chain octacosane (C_{28}) and dotriacontane (C_{32}) (Whyte et al., 1998). However, hydrocarbon metabolism at very low temperatures requires adaptation to cold environments. Psychrophilic bacteria may produce extracellular polymeric substances (EPSs) that are associated with halotolerance, and which may have a cryoprotective role in sea ice brine channels, as well as binding essential cationic trace metals (Nichols et al., 2005). When *Pseudoalteromonas* sp. CAM025 was grown at 2°C to −10°C the EPS yield was increased 30 times when compared to growth at 20°C (Nichols, 2006). Further indication of the cryoprotective role of EPS has come from field studies, which showed that particulate EPS was present in Arctic ice, and that this EPS increased over time and with decreased ice temperatures (Collins et al., 2007). *Rhodococcus* sp. Q15 grown on hexadecane or diesel fuel at 5°C also produced EPS that connected flocs of cells. Interestingly, the EPS morphology differed between cultures grown on diesel fuel and acetate. In addition, the cells produced cell-surface associated biosurfactants and indicated more hydrophobic cell surfaces when grown on diesel fuel (Whyte et al., 1999).

Enzymes in bacteria isolated from sea ice may be cold active, with catalytic activities well below the freezing point of seawater (Groudieva et al. 2004). The temperature optima and stabilities of psychrophilic proteins typically occur at lower temperatures than for mesophilic proteins. Such cold-adaptive proteins are often characterized by the reduced strength of stabilizing interactions (Collins et al., 2008). Microbes exposed to large drops in temperature may produce cold shock and cold acclimation proteins (Berger et al. 1996). For example, cold-tolerant alkane-degrading strains of *Rhodococcus* from Antarctica contained the cold-shock gene *cspA*, encoding the cold-shock protein CS7.4 (Bej et al., 2000). Psychrophilic and halophilic bacteria are also capable of adjusting the fatty acid composition and the protein content composition of their membranes in such a way that the proton permeability at the respective growth temperature remains constant (Whyte et al., 1999; Chintalapati et al 2004).

This process may be of importance to microorganisms inhabiting the brine channels in marine ice. During ice generation, ice prokaryotes may exert a number of actions to avoid freezing. Indeed, several bacteria producing such cryoprotecting proteins have been reported. These proteins may protect more freeze-labile proteins from freezing and thawing denaturation, or they may have protein refolding activities. Other proteins are thought to control the generation of ice crystals (Kawahara, 2008).

The presence of hydrocarbon-catabolism genes have been reported from Arctic and Antarctic environments (Whyte et al., 2002; Luz et al., 2004). However, cold-adaptations of hydrocarbon-degrading enzymes have not been described so far. A study of aromatic ring hydroxylating dioxygenase peptide sequences from a cold marine environment (Patagonia) showed less than 70% identity to previously described dioxygenases (Lozada et al., 2008). However, it is not known if these differences were related to cold-adaptations.

10.5 ACCELERATED BIODEGRADATION OF OIL IN COLD SEAWATER AND ICE

Naturally occurring hydrocarbon-degrading bacteria are found almost everywhere in biotic environments. Although these bacteria are capable of initiating the biodegradation of spilled crude oil, attempts have been made to increase removal through bioremediation strategies. Most bioremediation attempts have focused on developing good biostimulation strategies, typically by applying dispersants, degradation rate-limiting nutrients, or the combination of these and other treatments, to accelerate the natural biodegradation processes. Bioremediation processes, if successful, are cost-effective and reduce the environmental impacts of marine oil spills (Prince 1993, 2005; Swannell et al., 1996; Prince and Clark, 2004; Prince and Atlas, 2005). An alternative to biostimulation is bioaugmentation that involves the inoculation of exogenous microbial cultures with high biodegradation potentials for contaminants. Bioaugmentation approaches have been reported to improve the biodegradation of hydrocarbons from oil spills in cold soil or marine sediments, and can be used in combination with fertilizers (Margesin and Schinner, 1997; Ruberto et al., 2003). Bacterial mats from marine oil-contaminated sites have also been suggested for use in the degradation of coastal oil spills, although these are of greater relevance for spills in temperate areas (Cohen, 2002).

To date, most experimental oil bioremediation studies in Arctic or Antarctic environments have been conducted on stranded oils, employing the application of fertilizers to stimulate the indigenous flora, and often in combination with mechanical treatments that improve oxygen and nutrient availability (e.g., Sveum and Ladousse, 1989; Prince et al., 2003a; Obbard et al., 2004; Pelletier et al., 2004). Bioremediation strategies have also been applied to real oil spill situations such as the Exxon Valdez accident in 1989, where it formed part of a beach cleaning strategy (Bragg et al., 1994).

10.5.1 SEAWATER

Efforts to stimulate crude oil biodegradation in seawater and ice have not been investigated to the same extent as for stranded oil. Most remediation strategies in oil-polluted seawater have focused on mechanical removal methods such as the use of oil booms and skimmers. In addition to mechanical methods, chemical dispersants have also been widely used as an alternative treatment. Dispersants are mixtures of surface-active chemicals that reduce the surface tension of the oil, resulting in the formation of small oil droplets of sizes less than those generated by mechanical wave action (Brandvik, 1997). These chemicals are primarily used to disperse spilled oil on the seawater surface into the water column. This approach aims to reduce the impacts of oil spills on seabird and mammal populations near the spill and help to prevent the oil reaching the coastline. In addition, this process increases the oil–water interface and generates more bioavailable surface area for microorganisms.

The efficiencies of chemical dispersants on crude oil degradation at low seawater temperatures have shown conflicting results. In a seawater mesocosm experiment ($3.5\,m^3$ flow-through tanks with natural seawater from the St. Lawrence Estuary, Quebec) with temperatures from −1.8°C to 5.5°C, Forties and Western Sweet Blend crude oils were treated with the dispersant Corexit 9527 or different surfactant mixtures. Over a 63 day period at water temperatures >0°C chemical dispersal was found to result in higher biodegradation rates than in untreated oil samples (Siron et al., 1995). In microcosm studies with seawater temperatures of 8°C, MacNaughton et al. (2003) added the dispersant Corexit 9500 to Alaskan North Slope (ANS) crude oil. The dispersant resulted in the rapid colonization of oil droplets by bacteria, and heterotrophic and oil-degrading microbes proliferated in the microcosms. However, when the total hydrocarbons, C_{11} to C_{35} *n*-alkanes, or the sum of selected aromatics were assessed over a degradation period of 35 days, the amendments of dispersants resulted in only a slow or negligible biodegradation of the oil when compared to naturally dispersed oil. In another biodegradation experiment, Lindstrom and Braddock (2002) exposed cultures of oil-degrading microbes to ANS (fresh or evaporated and spiked with radiolabeled hydrocarbons) dispersed with Corexit 9500 over a 2 month period at 8°C. Respirometric analyses were conducted to determine $^{14}CO_2$ mineralization of ^{14}C-labeled dodecane, hexadecane, 2-methyl-naphthalene, and phenanthrene. The dispersant was found to inhibit the degradation of some of the hydrocarbons (hexadecane and phenanthrene), while others (dodecane and 2-methyl-naphthalene) were unaffected when compared to the mineralization of the oil without dispersant. It was suggested that carbon mineralization, at least initially, was the result of dispersant mineralization rather than the degradation of the oil compounds. In a biodegradation comparison study of two dispersants, Corexit 9500 and JD2000, fresh Prudhoe Bay crude oil was mixed with the dispersants and a microbial culture originating from the shorelines contaminated during the Exxon Valdez accident. The biodegradation of *n*-alkanes and PAHs was measured over a period of 46 days at 5°C. The first-order degradation rates of most *n*-alkanes and PAHs were found to be higher with dispersants than without dispersants, although these data were not statistically significant. When the same experiment was completed at 20°C, the degradation rates of *n*-alkanes and PAHs were more rapid than at 5°C, showing temperature-related biodegradation rates both in non-dispersed and dispersed oil. Furthermore, the influence of the two dispersants on biodegradation differed between the two degradation temperatures (Venosa and Holder, 2007). Correlation studies between the droplet surface area of a dispersed oil and the resulting degradation indicated that both dispersed area and dispersant chemistry controlled the degradation and that the surfactant blend hydrophile–lipophile balance and treatment levels were also significant controlling factors (Varadaraj et al., 1995).

10.5.2 Marine Ice

Bioremediation of oil in ice is an intriguing prospect. If the biodegradation of crude oil could be stimulated in ice, especially for the most toxic compounds migrating out of the ice through the brine channels, this would be of benefit for organisms inhabiting the polluted ice or nearby areas. Studies have shown that fertilizers can stimulate the biodegradation of crude oils in cold seawater under controlled experimental conditions (Delille et al., 1998). The slow-release oleophilic fertilizer Inipol EAP 22 was added to Antarctic seawater contaminated with "Arabian light" crude oil in a mesocosm study. The experiment was completed over 5 weeks during the Austral summers of 1992/1993 and 1993/1994. In both ice-covered and ice-free seawater, the addition of the fertilizer enhanced both the concentrations of heterotrophic- and hydrocarbon-degrading bacteria and increased the rate of biodegradation during the experiments, measured as n-C_{17}/Pristane and n-C_{18}/Phytane ratios.

A winter field experiment was conducted at Svalbard in 2004 as part of the ARCOP program. Crude Statfjord oil with and without fertilizers (mixture of Inipol EAP 22 and fish meal) was placed in fjord ice (Van Mijen Fjord, Svea) for a period of 6 months (December 2004–June 2005). At subzero temperatures no significant degradation of oil hydrocarbons occurred with the addition of nutrients, but at 0°C melt pool samples fertilized with inorganic nutrients showed a significant change in bacterial diversity (Gerdes and Dieckmann, 2006). Importantly, many of the available slow-release fertilizers are not suitable for use in polar regions as they will solidify if used in ice at very low temperatures. For example, the pour point of the Inipol EAP 22 is 11°C, which makes it difficult to use effectively in Arctic conditions. As a result, slow-release fertilizers will require reformulation or new products will need to be developed specifically for use at very low temperatures.

A few studies have also been conducted to determine the impacts on ice protist communities after oil contamination and subsequent fertilizer treatment. One of these studies formed part of the ARCOP field trial on Svalbard. In oil-contaminated ice (no fertilizers), the protist communities were destroyed through the complete ice coverage. Upon the addition of fertilizers, a less pronounced decline of organisms in the ice interior was observed. Thus, the use of fertilizers (Inipol and fish meal) helped to maintain higher diversity and biomass of protists in the ice. In a separate study, heterotrophic flagellates appeared to escape or avoid the oil contamination by downward migration (Ikävelko et al., 2005). In an Antarctic field experiment conducted during the Austral winter of 1993, land-fast ice on the continental shelf of Terre Adélie was contaminated with crude oil (Arabian light) or diesel fuel, and negative effects on the ice microalgae determined by chlorophyll A measurements. In crude oil-contaminated ice, negative effects were induced that lasted throughout the ice coverage period. The diesel contamination studies were found to cause an even more rapid effect on the algae than the crude oil. However, the addition of the fertilizer Inipol EAP 22 resulted in clearly favorable effects on the sea ice microalgae (Fiala and Delille, 1999).

10.6 THE CHALLENGE OF PERSISTENT OIL COMPOUNDS

Most environmental studies of petroleum-derived chemicals have focused on effects related to specific and easily identified hydrocarbons such as *n*-alkanes, BTEX, and PAHs. These compounds are typically easy to identify and quantify using standard analytical chemical techniques such as gas chromatography mass spectrometry (GC-MS). However, in the case of environmentally weathered samples, most oil compounds appear as an unresolved complex mixture (UCM) in gas chromatograms, and are often referred to as the "hump" (see Figure 10.4). Having undergone a variety of weathering processes (e.g., evaporation, biodegradation, and photooxidation), this residual UCM is comprised of thousands of environmentally persistent compounds (Gough and Rowland, 1990; Killops and Al-Juboori, 1990). In fact, it has been established that the natural biodegradation of spilled crude oil leads to a significant increase in the UCM concentration relative to other crude oil components (e.g., Meredith et al., 2000), highlighting the persistence of these compounds. Fractionation and subsequent characterization studies have shown that both nonpolar (e.g., aliphatic and aromatic) and polar (resin and asphaltene) compounds contribute to crude oil UCMs. Both the aromatic hydrocarbon and polar UCMs have proven to bioaccumulate in marine organisms and elicit ecotoxicological responses and health impairments (e.g., Farrington et al., 1982; Widdows et al., 1995; Barron et al., 1999; Smith et al., 2001; Rowland et al., 2001; Donkin et al., 2003). Owing to significant analytical challenges in studying polar UCM fractions, most research has focused on studying the composition, fate, and effects of the nonpolar fractions of UCMs. A detailed analysis of the aromatic hydrocarbon UCM fraction indicates that it contains a variety of highly branched alkylated homologues of well-known aromatic

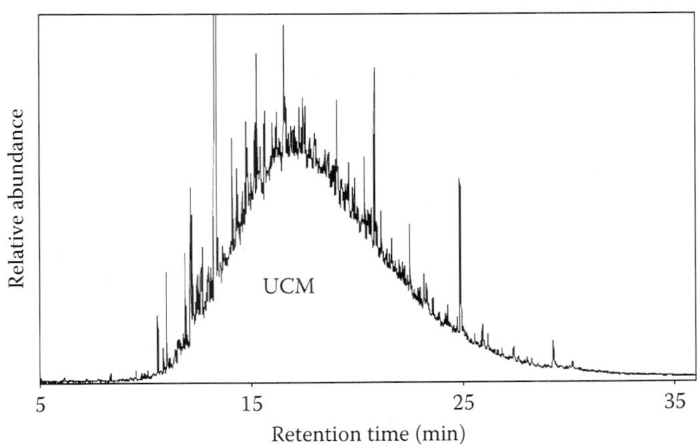

FIGURE 10.4 Gas chromatogram of the aromatic hydrocarbon UCM fraction extracted from mussels (*Mytilus edulis*) from a petroleum-impacted site (Southend) on the U.K. coast. (From Booth, A.M. et al., *Environ. Sci. Technol.*, 41, 457, 2007a. With permission.)

hydrocarbons (e.g., benzenes, tetralins, naphthalenes, and PAHs) that are not readily susceptible to biodegradation (Reddy et al., 2002; Booth et al., 2007a). The polar UCM fractions comprise compounds containing highly polar nitrogen, sulfur, and oxygen atoms in their structures (so-called NSO compounds). Despite their polarity, many of these compounds (e.g., phenols and naphthenic acids) are thought to be homologous in structure to compounds present in the nonpolar fraction of the UCM, hence their resistance to biodegradation. Due to their persistence, these compounds may reach polar regions both through local oil releases and as long-range transported pollutants.

The biodegradation potentials of these seemingly persistent UCM compounds are further complicated by the environmental conditions prevalent in polar regions. Lower ambient temperatures will result in significantly reduced biodegradation rates, and currently nothing is known about the abilities of psychrophilic or psychrotrophic bacteria to degrade these compounds. In temperate regions, microbial communities from previous oil-impacted sites have been shown to partially degrade model UCM compounds, such as alkylcyclohexyltetralins, alkylcyclohexylnaphthalene, and naphthenic acids (Scott et al., 2005; Booth et al., 2007b; Frenzel, 2008). Recent studies of naphthenic acids showed that it was the molecular structures rather than the number of carbon atoms that were important for determining biodegradation. Specifically, the most recalcitrant compounds included those with relatively high degrees of alkyl branching (Han et al., 2008). During future degradation studies of oil compounds in cold environments it is therefore of major importance to consider these environmentally persistent and toxic UCM-related compounds.

10.7 SUMMARY

At present, relatively little is known about the biodegradation of petroleum-derived compounds in cold marine waters and ice. A number of recent studies have indicated that the natural microbial biodegradation of some crude oil hydrocarbons can proceed in Arctic environments, but that low temperatures significantly reduce the biodegradation rate. In addition, the presence of sea ice can severely influence crude oil biodegradation by locking it away within the ice. The subsequent movement of the sea ice and the seasonal spring melt can then result in the occurrence of new pollution events when trapped crude oil is released in a new location. However, the presence of brine channels within the ice offers both a transport mechanism for trapped oil and a specialized environment for potential oil-degrading microbes to inhabit. Further studies have indicated that anthropogenic remediation processes may be useful in aiding the removal of compounds from oil spilled in the Arctic environment. However, under the climatic conditions in these regions recovery will be on much longer timescales than those observed for more temperate areas.

When considering crude oil reserves and exploration activities in Arctic regions, the impact of climate change can be viewed in different ways by people with different perceptions and interests. Due to a combination of technological advances and rising temperatures in the Arctic, oil reserves located in these areas are now becoming increasingly viable resources for exploitation. While this is a positive move for oil companies, the increased transport and the extraction of crude oil could also pose pollution problems for such relatively pristine and delicate environments. Interestingly, as the climate changes in polar regions the available data indicate that crude oil compounds contaminating these environments will become more readily degradable, especially due to longer ice-free seasons, resulting in reduced long-term impacts. However, this will clearly come at the cost of the polar environments we know today. Furthermore, an increase in polar warming will lead to the increased melting of existing ice and therefore to the release of pollutant chemicals that are currently locked away in the ice.

REFERENCES

Anderson, R. and Lovley, D. 2000. Hexadecane decay by methanogenesis. *Nature* 404: 722–723.

Arrhenius, S. 1889. Über die Reaktionsgeschwindigkeit bei der Inversion von Rohrzucker durch Säuren. *Zeitschrift für Physikalische Chemie* 139: 79–103.

Atlas, R.M. 1981. Microbial degradation of petroleum hydrocarbons: An environmental perspective. *Microbiological Reviews* 45: 180–209.

Atlas. R.M. 1984. *Petroleum Microbiology*. Macmillan, New York.

Atlas, R.M. and Bartha, R. 1972. Biodegradation of petroleum in seawater at low temperatures. *Canadian Journal of Microbiology* 18: 1851–1855.

Bej, A.K., Saul, D., and Aislabie, J. 2000. Cold-tolerant alkane-degrading *Rhodococcus* species from Antarctica. *Polar Biology* 23: 100–105.

Barron, M.G., Podrabsky, T., Ogle, S., and Ricker, R.W. 1999. Are aromatic hydrocarbons the primary determinant of petroleum toxicity to aquatic organisms? *Aquatic Toxicology* 46: 253–268.

Berger, F., Morellet, N., and Corrieu, G. 1996. Cold shock and cold acclimation proteins in the psychrophilic bacterium *Arthrobacter globiformis* SI55. *Journal of Bacteriology* 178: 2999–3007.

Booth, A.M., Sutton, P.A., Lewis, C.A., Lewis, A.C. Scarlett, A. Chau, W., Widdows, J., and Rowland, S.J. 2007a. Unresolved complex mixtures of aromatic hydrocarbons: Thousands of overlooked persistent, bioaccumulative, and toxic contaminants in mussels. *Environmental Science & Technology* 41: 457–464.

Booth, A.M., Aitken, C., Jones, D.M., Lewis, C.A., and Rowland, S.J. 2007b. Resistance of toxic alkylcyclohexyltetralins to biodegradation by aerobic bacteria. *Organic Geochemistry* 38: 540–550.

Børresen, M.H., Barnes, D.L., and Rike, A.G. 2007. Repeated freeze-thaw cycles and their effects on mineralization of hexadecane and phenanthrene in cold climate soils. *Cold Regions Science and Technology* 49: 215–225.

Bowman, J.P., McCammon, S.A., Brown, M,V., Nichols, D.S., and McMeekin, T.A. 1997. Diversity and association of psychrophilic baceria in Antarctic sea ice. *Applied Environmental Microbiology* 63: 3068–3078.

Bragg, J.R., Prince, R.C., Harner, E.J., and Atlas, R.M. 1994. Effectiveness of bioremediation for the Exxon Valdez oil spill. *Nature* 368: 413–418.

Brakstad, O.G. 2008. Natural and stimulated biodegradation of petroleum in cold marine environments. In *Psychrophiles. From Biodiversity to Biotechnology*, eds. R. Margesin, F. Schinner, J.-C. Marx, and C. Gerday. Springer-Verlag, Heidelberg, Germany, pp. 389–407.

Brakstad, O.G. and Bonaunet, K. 2006. Biodegradation of petroleum hydrocarbons in seawater at low temperatures (0–5°C) and bacterial communities associated with degradation, *Biodegradation* 17: 71–82.

Brakstad, O.G. and Lødeng, A.G.G. 2004. Microbial diversity during biodegadation oif crude oil in seawater from the North Sea. *Microbial Ecology* 49: 94–103.

Brakstad, O.G., Nonstad, I., Faksness, L.-G., and Brandvik, P.J. 2008. Responses of microbial communities in Arctic sea ice after contamination by crude petroleum oil. *Microbial Ecology*. 55: 540–552.

Brandvik, P.J. 1997. Optimisation of oil spill dispersants on weathered oils. PhD dissertation, Norwegian University of Science and Technology, Trondheim, Norway.

Brandvik, P.J. and Faksness, L.G. 2009. Weathering processes in Arctic oil spills. Meso-scale experiments with different ice conditions. *Cold Regions Science and Technology* 55: 160–166.

Brandvik, P.J., Singsaas, I., and Daling, P.S. 2004. Oil spill R & D in Norwegian Arctic waters with special focus on large-scale oil weathering experiments. Paper presented at The Inter Spill Conference, Trondheim, Norway, June 14–17, 2004.

Breezee, J., Cady, N., and Staley, J.T. 2004. Subfreezing growth of the sea ice bacterium "*Psychromonas ingrahamii*." *Microbial Ecology* 47: 300–304.

Brinkmeyer, R., Knittel, K., Jürgens, J., Weyland, H., Amann, R., and Helmke, E. 2003. Diversity and structure of bacterial communities in Arctic versus Antarctic pack ice. *Applied and Environmental Microbiology* 69: 6610–6619.

Chablain, P.A., Philippe, G., Groboillot, A., Truffaut, N., and Guespin-Michel, J.F. 1997. Isolation of soil psychrotrophic toluene-degrading *Pseudomonas* strain: Influence of temperature on the growth characteristics on different substrates. *Research in Microbiology* 148: 153–161.

Chang, Y-J., Stephen, J.R., Richter, A.P., Venosa, A.D., Brüggemann, J., Macnaughton, S.J., Kowalchuk, G.A., Haines, J.R., Kline, E., and White, D.C. 2000. Phylogenetic analysis of aerobic freshwater and marine enrichment cultures efficient in hydrocarbon degradation: effects of profiling method. *Journal of Microbiological Methods* 40: 19–31.

Chintalapati, S., Kiran, M.D., and Shivaji, S. 2004. Role of membrane lipid fatty acids in cold adaptions. *Cellular and Molecular Biology* 50: 631–642.

Cohen, Y. 2002. Bioremediation of oil by marine microbial mats. *International Microbiology* 5: 189–193.

Collins, R.E., Carpenter, S.D., and Deming, J.W. 2007. Spatial heterogeneity and temporal dynamics of particles, bacteria, and pEPS in Arctic winter sea ice. *Journal of Marine Systems* 74: 902–917.

Collins, T., Roulling, F., Piette, F., Marx, J.-C., Feller, G., Gerday, C., and D'Amico, S. 2008. Fundamental of cold-adapted enzymes. In *Psychrophiles. From Biodiversity to Biotechnology*, eds. R. Margesin, F. Schinner, J.-C. Marx, and C. Gerday. Springer-Verlag, Heidelberg, Germany, pp. 211–227.

Cundell, A.M. and Traxler, R.W. 1976. Psychrophilic hydrocarbon-degrading bacteria from Narragansett Bay, Rhode Island, U.S.A. *Material und Organismen* 11: 1–17.

Delille, D. and Siron, R. 1993. Effects of dispersed oil on heterotrophic bacterial communities in cold marine waters. *Microbial Ecology* 25: 263–272.

Delille, D. and Vaillant, N. 1990. The influence of crude oil on the growth of suantarctic marine bacteria. *Antarctic Science* 2: 123–127.

Delille, D., Basseres, A., and Dessommes, A. 1997. Seasonal variation of bacteria in sea ice contaminated by diesel fuel and dispersed crude oil. *Microbial Ecology* 33: 97–105.

Delille, D., Basseres, A., and Dessommes, A. 1998. Effectiveness of bioremediation for oil-polluted Antarctic seawater. *Polar Biology* 19: 237–241.

Deppe, U., Richnow, H.-H., Michaelis, W., and Antranikian, G. 2005. Degradation of crude oil by an arctic microbial consortium. *Extremophiles* 9: 461–470.

Dickins D., Brandvik, P.J., Bradford, J., Faksness, L.-G., Liberty, L., and Daniloff, R. 2008. Svalbard 2006 experimental oil spill under ice: remote sensing, oil weathering under arctic conditions and assessment of oil removal by in-situ burning. Paper accepted for presentation at the 2008 International Oil Spill Conference, May 4–8, 2008 Savannah, Georgia.

Donkin, P., Smith, E.L., and Rowland, S.J. 2003. Toxic effects of unresolved complex mixtures of aromatic hydrocarbons accumulated by mussels, *Mytilus edulis*, from contaminated field sites. *Environmental Science & Technology* 37: 4825–4830.

Dutta, T.K. and Harayama, S. 2000. Fate of crude oil by the combination of photooxidation and biodegradation. *Environmental Science & Technology* 34: 1500–1505.

Faksness, L.-G. and Brandvik, P.J. 2005. Dissolution of water soluble components from oil spills encapsulated in ice. *Proceedings to the 2005 Arctic and Marine Oilspill Program Technical Seminar*, June 7–9, 2005 Calgary, Canada, pp. 59–73.

Faksness, L.-G. and Brandvik, P.J. 2008a. Distribution of water soluble components from Arctic marine oil spills—A combined laboratory and field study. *Cold Regions Science and Technology* 54: 97–105.

Faksness, L.-G. and Brandvik, P.J. 2008b. Distribution of water soluble components from oil encapsulated in Arctic sea ice: Summary of three field seasons. *Cold Regions Science and Technology* 54: 106–114.

Farrington, J.W., Davis, A.C., Frew, N.M., and Rabin, K.S. 1982. No. 2 fuel oil compounds in *Mytilus edulis*: Retention and release after an oil spill. *Marine Biology* 66: 15–26.

Fiala, M. and Delille, D. 1999. Annual changes of microalgae biomass in Antarctic sea ice contaminated by crude oil and diesel fuel. *Polar Biology* 21: 391–396.

Fingas, M.F. and Hollebone, H.P. 2003. Review of the behaviour of oil in freezing environments. *Marine Pollution Bulletin* 47: 333–340.

Frenzel, M. 2008. Persistence, bioremediation and fate of unresolved complex mixtures (UCMs) of hydrocarbons in the environment. PhD dissertation, University of Exeter, U.K.

Garrett, R.M., Pickering, I.J., Haith, C.E., and Prince, R.C. 1998. Photooxidation of crude oils. *Environmental Science & Technology* 32: 3719–3723.

Gerdes, B. and Dieckmann, G. 2006. Biological degradation of crude oil in sea ice. Abstract to the report "Growth Project GRD-2000-30112 ARCOP Technology and Environment WP 6 Workshop Activities," 16.02.2006 (http://www.arcop.fi/reports/workshop_report8.pdf).

Gerdes, B., Brinkmeyer, R., Dieckmann, G., and Helmke, E. 2005. Influence of crude oil on changes of bacterial communities in Arctic sea-ice. *FEMS Microbiology Ecolology* 53: 129–139.

Gough, M.A. and Rowland, S.J. 1990. Characterisation of unresolved complex mixtures of hydrocarbons in petroleum. *Nature* 344: 648–650.

Groudieva, T., Kambourova, M., Yosuf, H., Royter, M., Grote, R., Trinks, H., and Antranikian, G. 2004. Diversity and cold-active hydrolytic enzymes of culturable bacteria associated with Arctic sea ice, Spitzbergen. *Extremophiles* 8: 475–488.

Han, X.M., Scott, A.C., Fedorak, P.M., Bataineh, M., and Martin, J.W. 2008. Influence of molecular structure on the biodegradability of naphthenic acids. *Environmental Science & Technology* 42: 1290–1295.

Harayama, S., Kishira, H., Kasai, Y., and Shutsubo, K., 1999. Petroleum biodegradation in marine environments. *Journal of Molecular Microbiology and Biotechnology* 1: 63–70.

Harayama S., Kasai, Y., and Hara, A. 2004. Microbial communities in oil-contaminated seawater. *Current Opinion in Biotechnology* 15: 205–214.

Hokstad, J.N., Daling, P.S., Buffagni, M., and Johnsen, S. 1999. Chemical and ecotoxicological characterization of oil-water systems. *Spill Science and Technology Bulletin* 5: 75–80.

Ikävelko, J., Gerdes, B., and Dieckmann, G. 2005. An experimental study on the effects of Statfjord crude oil, and application of Inipol and fish meal on the sea ice biota in Svalbard in February—April 2004. *Proceedings to the 2005 Arctic and Marine Oilspill Program Technical Seminar*, June 7–9, 2005 Calgary, Canada, pp. 993–1003.

Junge, K., Imhoff, F., Staley, T., and Deming, J.W. 2002. Phylogenetic diversity of numerically important Arctic sea-ice bacteria cultured at subzero temperature. *Microbial Ecology* 43: 315–328.

Junge, K., Eicken, H., and Deming, J.W. 2003. Motility of *Colwellia psychrerythraea* strain 34H at subzero temperatures. *Applied and Environmental Microbiology* 69: 4282–4284.

Junge, K., Eicken, H., and Deming, J.W. 2004. Bacterial activity at −2 and −20°C in Arctic wintertime sea ice. *Applied and Environmental Microbiology* 70: 550–557.

Junge, K., Eicken, H., Swanson, B.D., and Deming, J.W. 2006. Bacterial incorporation of leucine into protein down to −20°C with evidence for potential activity in sub-eutectic saline formations. *Cryobiology* 52: 417–429.

Kasai, Y., Kishira, H., Sasaki, T., Syutsubo, K., Watanabe, K., and Harayama, S. 2002. Predominant growth of *Alcanivorax* strains in oil-contaminated and nutrient-supplemented sea water. *Environmental Microbiology* 4: 141–147.

Kawahara, H. 2008. Cryoprotectants and ice-binding proteins. In: *Psychrophiles. From Biodiversity to Biotechnology*, eds. R. Margesin, F. Schinner, J.-C. Marx, and C. Gerday. Springer-Verlag, Heidelberg, Germany, pp. 2229–2246.

Killops, S.D. and Al-Juboori, M.A.H.A. 1990. Characterisations of unresolved complex mixtures in the gas chromatograms of biodegraded petroleums. *Organic Geochemistry* 15: 147–160.

Leahy, J. and Colwell, R. 1990. Microbial degradation of hydrocarbons in the environment. *Microbiological Reviews* 54: 305–315.

Lindstrom, J.E. and Braddock, J.F. 2002. Biodegradation of petroleum hydrocarbons at low temperature in the presence of the dispersant Corexit 9500. *Marine Pollution Bulletin* 44: 739–747.

Lozada, M., Mercadal, J.P.R., Guerrero, L.D., Di Marzio, W.D., Ferrero, M.A., and Dioniso, H.M. 2008. Novel aromatic ring-hydroxylating dioxygenase genes from coastal marine sediments of Patagonia. *BMC Microbiology* 8: 50. Open Access article (http://www.biomedcentral.com/1471-2180/8/50).

Luz, A.P., Pellizari, V.H., Whyte, L.G., and Greer, C.W. 2004. A survey of indigenous microbial hydrocarbon degradation genes in soils from Antarctica and Brazil. *Canadian Journal of Microbiology* 50: 323–333.

MacNaughton, S.J., Swannell, R., Daniel, F., and Bristow, L. 2003. Biodegradation of dispersed Forties crude and Alaskan North Slope oils in microcosms under stimulated marine conditions. *Spill Science & Technology* 8: 179–186.

Maki, H., Sasaki, T., and Harayama, S. 2001. Photo-oxidation of biodegraded crude oil and toxicity of the photo-oxidized products. *Chemosphere* 44: 1145–1151.

Margesin, R. and Schinner, F. 1997. Efficiency of indigenous and inoculated cold-adapted soil microorganisms for biodegradation of diesel oil in Alpine soils. *Applied and Environmental Microbiology* 63: 2660–2664.

Margesin, R. and Schinner, F. 1999. Biological decontamination of oil spills in cold environments (review). *Journal of Chemical Technology and Biotechnology* 74: 1–9.

Margesin R. and Schinner, F. 2001. Biodegradation and bioremediation of hydrocarbons in extreme environments (review). *Applied Microbiology and Biotechnology* 56: 650–663.

Martin, S. 1979. A field study of brine drainage and oil entrainment in first-year sea ice. *Journal of Glaciology* 22: 473–502.
Meredith, W., Kelland, S.J., and Jones, D.M. 2000. Influence of biodegradation on crude oil acidity and carboxylic acid composition. *Organic Geochemistry* 31: 1059–1073.
Minas, W. and Gunkel, W. 1995. Oil pollution in the North Sea—a microbiological point of view. *Helgoländer Meeresuntersuchungen* 49: 143–158.
Nichols, C.M. 2006. Exopolysacchararides from Antarctic sea-ice bacteria. PhD dissertation, University of Tasmania, Hobart.
Nichols, C.M., Bowman, J.P., and Guezennec, J. 2005. Effects of incubation temperature on growth and production of exopolysaccharides by an Antarctic sea ice bacterium grown in batch culture. *Applied and Environmental Microbiology* 71: 3519–3523.
Ni'matuzahroh, M., Gilewicz, M., Guiliano, M., and Bertrand, J.C. 1999. In-vitro study of interaction between photooxidation and biodegradation of 2-methylphenanthrene by *Sphingomonas* sp. 2MPII. *Chemosphere* 38: 2501–2507.
NRC. 2003. Oil in the Sea III: Inputs, Fates and Effects, National Research Council.
Obbard, J.P., Ng, K.L., and Xu, R. 2004. Bioremediation of petroleum contaminated beach sediments: Use of crude palm oil and fatty acids to enhance indigenous biodegradation. *Water Air and Soil Pollution* 157: 149–161.
Payne, J.R., Hachmeister, L.E., McNebb, G.D., Sharpe, H.E., Smith, G.S., and Manen, C.A. 1991. Brine-induced advection of dissolved aromatic hydrocarbons to arctic bottom waters. *Environmental Science and Technology* 25: 940–951.
Pelletier, E., Delille, D., and Delille, B. 2004. Crude oil bioremediation in sub-Antarctic intertidal sediments: Chemistry and toxicity of oiled residue. *Marine Environmental Research* 57: 311–327.
Perry, J.J. 1984. Microbial metabolism of cyclic alkanes. In *Petroleum Microbiology*, ed. R.M. Atlas. Macmillan Publ. Co., New York, pp. 61–98.
Pfirman, S.L., Eicken, H., Bauch, D., and Weeks, W.F. 1995. The potential transport of pollutants by Arctic sea-ice. *Sciences of the Total Environment* 159: 129–146.
Prabagaran, S.R., Manorama, R., Delille, D., and Shivaji, S. 2006. Predominance of *Roseobacter, Sulfitobacter*, *Glaciecola* and *Psychrobacter* in seawater collected off Ushuaia, Argentina, Sub-Antarctica. *FEMS Microbiology Ecology* 59: 342–355.
Prince, R.C. 1993. Petroleum spill bioremediation in marine environments. *Critical Reviews in Microbiology* 19: 217–242.
Prince, R.C. 2005. The microbiology of marine oil spill bioremediation. In *Petroleum Microbiology*, eds. B. Ollivier, and M. Magot. ASM Press, Washington D.C., pp. 317–335.
Prince, R.C. and Atlas, R.M. 2005. Bioremediation of marine oil spills. In *Bioremediation: Applied Microbial Solutions for Real-World Environmental Cleanup*, eds. R.M. Atlas and J. Philp. ASM Press, Washington D.C., pp. 269–292.
Prince, R.C. and Clark, J.R. 2004. Bioremediation of marine oil spills. In *Petroleum Biotechnology: Development and Perspectives. Studies in Surface Science and Catalysis, vol 151*, eds. R. Vasquez-Duhalt, and R. Quintero-Ramirez. Elsevier B.V, Amsterdam, the Netherlands, pp. 495–512.
Prince, R.C., Bare, R.E., Garrett, R.M., Grossman, M.J., Haith, C.E., Keim, L.G., Lee, K., Holtom, G.J., Lambert, P., Sergy, G.A., Owens, E.H., and Guenette, C.C. 2003a. Bioremediation of stranded oil on an Arctic shoreline. *Spill Science & Technology Bulletin* 8: 303–312.
Prince, R.C., Garrett, R.M., Bare, R.E., Grossman, M.J., Townsend, T., Suflita, J.M., Lee, K., Owens, E.H., Sergy, G.A., Braddock, J.F., Lindstrom, J.E., and Lessard, R.R. 2003b. The roles of photooxidation and biodegradation in long-term weathering of crude and heavy fuel oils. *Spill Science & Technology Bulletin* 8: 145–156.

Reddy, C.M., Eglinton, T.I., Hounshell, A., White, H.K., Xu, L., Gaines, R.B., and Frysinger, G.S. 2002. The West Falmouth oil spill after thirty years: The persistence of petroleum hydrocarbons in marsh Sediments. *Environmental Science & Technology* 36: 4754–4760.

Rike, A.G., Haugen, K.B., Børresen, M., Engene, B., and Kolstad, P. 2003. *In situ* biodegradation of petroleum hydrocarbons in frozen arctic soils. *Cold Regions Science and Technology* 37: 97–120.

Rowland, S., Donkin, P., Smith, E., and Wraige, E. 2001. Aromatic hydrocarbon "humps" in the marine environment: unrecognized toxins? *Environmental Science & Technology* 35: 2640–2644.

Ruberto, L., Vazquez, S.C., Mac Cormack, W.P. 2003. Effectiveness of the natural bacterial flora, biostimulation and bioaugmentation on the bioremediation of a hydrocarbon contaminated Antarctic soil. *International Biodeterioration and Biodegradation* 52: 115–125.

Röling, W.F.M., de Brito, I.R.C., Swannell, R.P.J., and Head, I.M. 2004. Response of archaeal communities in beach sediments to spilled oil and bioremediation. *Applied and Environmental Microbiology* 70: 2614–2620.

Schluep, M., Imboden, D.M., GallI, R., and Zeyer, J. 2001. Mechanisms affecting the dissolution of nonaqueous phase liquids into the aqueous phase in slow stirring batch systems. *Environmental Toxicology and Chemistry* 20: 459–466.

Scott, A.C., MacKinnon, M.D., and Fedorak, P.M. 2005. Naphthenic acids in Athabasca oil sands tailings waters are less biodegradable than commercial naphthenic acids. *Environmental Science & Technology* 39: 8388–8394.

Siron, R., Pelletier, É., and Brochu, C. 1995. Environmental factors influencing the biodegradation of petroleum hydrocarbons in cold seawater. *Archives of Environmental Contamination and Toxicology* 28: 406–416.

Smith, E., Wraige, E., Donkin, P., and Rowland S. 2001. Hydrocarbon humps in the marine environment: Synthesis, toxicity, and aqueous solubility of monoaromatic compounds. *Environmental Toxicology and Chemistry* 20: 2428–2432.

Sveum, P. and Ladousse, A. 1989. Biodegradation of oil in the Arctic: Enhancement by oil-soluble fertilizer application. In: *Proceedings of the 1989 International Oil Spill Conference.* San Antonia, February 13–16, 1989 American Petroleum Institute, Washington D.C., pp. 439–445.

Swannell, R.P.J., Lee, K., and McDonagh, M.1996. Field evaluation of marine spills bioremediation. *Microbiological Reviews* 60: 242–365.

Van Hamme, J.D., Singh, A., and Ward, O.P. 2003. Recent advances in petroleum microbiology. *Microbiology and Molecular Biology Reviews* 67: 503–549.

Varadaraj, R., Robbins, M.L., Bock, J., Pace, S., and MacDonald, D. 1995. Dispersion and biodegradation of oils spills on water. In: *Proceedings to the 1995 International Oil Spill Conference*, American Petroleum Institute, Washington D.C., pp. 101–106.

Venosa, A.D. and Holder, E.L. 2007. Biodegradability of dispersed crude oil at two different temperatures. *Marine Pollution Bulletin* 54: 545–553.

Venosa, A.D. and Zhu, X. 2003. Biodegradation of crude oil contaminating marine shorelines and freshwater wetlands. *Spill Science & Technology Bulletin* 8: 163–178.

Weissenberger, J., Dieckmann, G., Gradinger, R., and Spindler, M. 1992. Sea ice: A cast technique to examine and analyze brine pockets and channel structure. *Limnology and Oceanography* 37: 179–183.

Widdows, J., Donlkin, P., Brinsley, M.D., Evans, S.V., Salkeld, P.N., Franklin, A., Law, R.J., and Waldock, M.J. 1995. Scope for growth and contaminant levels in North Sea mussels *Mytilus edulis. Marine Ecology Progress Series* 127: 131–148.

Whyte, L.G., Hawari, J., Zhou, E., Bourbonniere, L., Inniss, W.E., and Greer, C.W. 1998. Biodegradation of variable chain-length alkanes at low temperature by a psychrotrophic *Rhodococcus* sp. *Applied and Environmental Microbiology* 64: 2578–2584.

Whyte, L.G., Slagman, S.J., Pietrantonio, F., Bourbonniére, L., Koval, S.F., Lawrence, J.R., Inniss, W.E., and Greer, C.W. 1999. Physiological adaptations involved in alkane assimilation at a low temperatures by *Rhodococcus* sp. strain Q15. *Applied and Environmental Microbiology* 65: 2961–2968.

Whyte, L.G., Schultz, A., van Beilen, J.B., Luz, A.P., Pellizari, V., Labbe, D., and Greer, C.W. 2002. Prevalence of alkane monooxygenase genes in Arctic and Antarctic hydrocarbon-contaminated and pristine soils. *FEMS Microbiology Ecology* 41: 141–150.

Yakimov, M.M., Gentile, G., Bruni, V., Capello, S., D'Auria, G., Golyshin, P.N., and Giuliano, L. 2004. Crude oil-induced structural shift of coastal bacterial communities of rod bay (Terra Nova Bay, Ross Sea, Antarctica) and characterization of cultured cold-adapted hydrocarbonoclastic bacteria. *FEMS Microbiology Ecology* 49: 419–432.

Zeng, Y.X., Li, H.R., Yong, Y., Boz, Z., and Zheng, T.L. 2007. Phylogenetic diversity and phenotypic characterization of cultivable bacterioplankton isolated from polar oceans. *Acta Oceanologica Sinica* 26: 93–103.

Zengler, K., Richnow, H., Rossello-Mora, R., Michaelis, W., and Widdel, F. 1999. Methane formation from long-chain alkanes by anaerobic microorganisms. *Nature* 401: 266–269.

11 Potential for Microbial Biodegradation of Polychlorinated Biphenyls in Polar Environments

Angelina Lo Giudice, Vivia Bruni, and Luigi Michaud

CONTENTS

11.1 INTRODUCTION

Polychlorinated biphenyls (PCBs) belong to the so-called dirty dozen, a list of 12 persistent organic pollutants (POPs), which include dioxins, furans, and some pesticides (e.g., aldrin, chlordane, and DDT). The list was compiled in 2001 in the framework of the Stockholm Convention, whose aim was to protect human health and the environment by banning the production and reducing the release of some of the most hazardous chemicals known to humankind.

Contamination with PCBs is not only of great concern to industrialized countries. It has also been demonstrated that a continuous deposition/evaporation process favors combined atmospheric and ocean-borne long-range transport to areas far from known PCB sources.

Conventional cleanup technologies (e.g., land filling, deep-well injection, and high temperature incineration), currently adopted and/or recommended for the remediation of PCB-contaminated natural environments and waste, are generally expensive and often inefficient. Moreover, they may be a source of secondary pollution, e.g., the formation of dioxins. As a result, the development of bioremediation processes has been proposed as a promising and cost-saving alternative, since indigenous microorganisms could be used and *in situ* treatments are allowed. This trend has extended to cold regions such as the Arctic and the Antarctic in spite of the substantial limitations related to the extreme climate conditions (Mohn et al., 1997; Tumeo and Guinn, 1997). The assessment of the PCB biodegradative potential of psychrotolerant Arctic and Antarctic microorganisms has been previously reported, although rarely. Psychrotolerant microbes, which have adapted to a wider temperature range, low water activity, and low nutrient availability, may have important advantages in biotechnological applications (Gounot, 1991). It should be noted, however, that the introduction of nonnative species to Antarctica is forbidden by the Antarctic Treaty, and therefore the eventual utilization of autochthonous microorganisms is required. Conversely, in addition to Arctic microorganisms PCB-degrading Antarctic microorganisms could also be exploited for the remediation in other cold systems.

After a brief review of PCBs and their biodegradation, this chapter will discuss their occurrence in polar environments and the potential for their removal exhibited by Arctic and Antarctic bacteria, as well as by psychrotolerant microbes from temperate environments.

11.2 POLYCHLORINATED BIPHENYLS: AN OVERVIEW

11.2.1 Physicochemical Properties, Industrial Uses, and Environmental Concerns

PCBs are part of a larger group of persistent organic pollutants (POPs), which are released into the environment through a variety of human activities and have a range of harmful effects on the health of ecosystems, wildlife, and people. Theoretically, the family of PCBs consists of 209 related compounds, called congeners, consisting of a biphenyl carbon ring substituted with 1–10 chlorine atoms (chemical formula: $C_{12}H_{10-n}Cl_n$). PCB mixtures are resistant to combustion, nonexplosive, hydrophobic and lipophilic, and soluble in organic solvents. They exhibit an excellent chemical stability under acid and alkali conditions, and a thermal stability up to 350°C, insulation features, and have desirable dielectric properties. Many PCB features are strongly affected by the degree of chlorination, e.g., their viscosity, melting point, and lipophilicity increase with an increasing degree of chlorination, whereas vapor pressure and water solubility decrease. The half-life of PCBs in the environment also depends on the number of chlorines in the biphenyl structure. It can range from a few days to approximately 10 years for mono- to pentachlorobiphenyls, reaching more than 20 years for higher substituted congeners (Borja et al., 2005; Ceccarini and Giannarelli, 2006).

Due to their excellent physicochemical features, more than 1.5 million tons of PCBs were synthesized during the mid-twentieth century and sold for industrial and commercial purposes under different trade names, e.g., Acelor (France), Aroclor (United States), Clophen (Germany), Delor (Czechoslovakia), Fenclor (Italy), and Kanechlor (Japan). They were used in a variety of ways, as dielectric fluid in electrical transformers and capacitors, pesticide extenders, flame retardants, and as components in heat transfer and hydraulic systems. Additional uses also included inks, hydraulic lubricants, sealants, paints, waxes, carbonless copy paper, adhesives, and dust control agents. Each commercial mixture is characterized by a typical congener composition derived from the different reaction and separation conditions used during production. They are generally made by the isomeric forms of 4–5 congeners (predominantly trihexachlorinated) and have a chlorine content ranging from 21% to 68% (Pieper, 2005; Ceccarini and Giannarelli, 2006).

Large amounts of PCBs have legally or illegally entered the environment for decades through accidental spills, inappropriate disposal techniques, or a variety of thermal and chemical processes with no consideration for their impact on biota. Since the 1970s, when their strong toxicity and recalcitrant nature were recognized, PCBs have been regulated or banned in Western nations because of the serious health and ecological problems potentially due to their widespread utilization. This has led to worldwide contamination, even in remote areas.

The high environmental persistence, and the low biodegradability and lipophilicity of PCBs contribute to their bioaccumulation potential and biomagnification along the trophic chain. This can cause adverse physiological effects on the reproductive, nervous, and endocrine systems, mainly on top predators, including humans. Although their acute toxicity is rather low, the chronic effects of PCBs are severe as certain congeners are teratogenic, carginogenic, and/or immunogenic (for detailed discussion, see Ross, 2004 and Safe, 2004).

11.2.2 Microbial Transformation of PCBs

Despite their recalcitrant nature, PCB congeners can be biotransformed by different microbial species inhabiting both contaminated (Master and Mohn, 1998; Lambo and Patel, 2006a) and uncontaminated sites (Macedo et al., 2007). Microbial biodegradation depends on specific features of the PCBs. This includes the position and number of the chlorine substituents in the congeners, the concentration, bioavailability, and water solubility of the pollutant, and environmental factors such as pH, temperature, and the presence of toxic and inhibitory substances or competitive substrates (Borja et al., 2005).

The biodegradation of PCBs consists of two main processes: anaerobic reductive dechlorination and aerobic oxidative degradation. These processes have been mainly reported for mesophilic microorganisms inhabiting soils and sediments, where such xenobiotic compounds generally accumulate. Both anaerobic and aerobic degradation pathways have previously been described in detail (Furukawa, 2000; Wiegel and Wu, 2000; Abraham et al., 2002; Borja et al., 2005; Vasilyeva and Strijakova, 2007).

The degree of chlorination of the single congeners is a major factor affecting the biodegradation of PCBs (Bedard and Haberl, 1990). Highly chlorinated biphenyls are substrates for anaerobes. Therefore, as their toxicity decreases their degradability increases. Congeners with five or more chlorine atoms remain generally recalcitrant to aerobic degradation, whereas they are substrates for anaerobic microorganisms. The lowly chlorinated PCB congeners resulting from the anaerobic transformation can subsequently become suitable substrates for the oxidative attack of aerobic bacteria. Consequently, a sequential anaerobic/aerobic biotransformation is required to achieve the complete biodegradation of higher chlorinated congeners, along with the action of different microorganisms possessing specific congener preferences.

Typically, the reductive dechlorination of PCBs occurs in both freshwater and marine anaerobic sediments as has been previously demonstrated for several anaerobes, e.g., *Dehalococcoides*, *Desulfomonile*, and *Desulfitobacterium* and the white rot fungus *Phanerochaete chrysosporium* (Holliger et al., 1999; Bedard et al., 2006). The anaerobic biodegradation of PCBs consists of the removal of the chlorines as halogen ions and their concurrent replacement by hydrogen as electrons and protons (hydrogenolysis). Thus, the biphenyl structure is kept intact while the number of substituted chlorine atoms is reduced, mainly from the *para* and *meta* positions. This leaves mono- to trichlorobiphenyls with mainly *ortho* chlorines. At least six distinct microbial dechlorination processes have been identified in contaminated sediments through the comparison of the pattern of congener loss and reaction products. These differences may be dependent on several variables, such as the occurrence of different microbial populations (having dehalogenases exhibiting distinct congener specificities) or environmental factors (e.g., temperature, pH, and partial pressure of H_2), as well as on the presence/absence of suitable electron donors and acceptors, carbon sources, and micronutrients (Mohn and Tiedje, 1992; Wu et al., 1997; Wiegel and Wu, 2000).

The aerobic biodegradation of PCBs has mainly been reported for bacteria and only rarely for fungi (e.g., *Aspergillus niger*), yeasts, and cyanobacteria, which are able to metabolize biphenyl and low-chlorinated biphenyls (Furukawa, 2000). To date, among PCB-degrading Gram-negative bacteria, members of the genera *Pseudomonas*, *Acinetobacter*, *Alcaligenes*, *Burkholderia*, and *Sphingomonas* have been isolated, along with some Gram-positives, such as *Micrococcus*, *Janibacter*, *Dietzia*, *Arthrobacter*, and *Rhodococcus*. The aerobic degradation pathways of PCBs appear to be very similar among these microorganisms. The first step, known as biphenyl "upper" pathway, is initiated by the multicomponent enzyme biphenyl-2,3-dioxygenase (BPDO) and through a series of intermediates leads to the formation of benzoate and 2-hydroxypenta-2,4-dienoate as reaction products. BPDO comprises a reductase, a ferredoxin, and an oxygenase component of $\alpha_3\beta_3$ constitution (iron-sulfur protein, ISP). Other enzymes involved in the process are the *cis*-2,3-dihydro-2,3-dihydroxybiphenyl dehydrogenase (BDDH), the 2,3-dihydroxybiphenyl-1,2-dioxygenase (DBDO), and the 2-hydroxy-6-oxo-6-phenylhexa-2,4-dienoate hydrolase (HOPDA hydrolase). PCB degraders exhibit significantly different activity spectra as their congener patterns are generally strongly dependent on the substrate specificity and regioselectivity of the biphenyl-2,3-dioxygenase. Some bacteria are also assumed to possess biphenyl-3,4-dioxygenase enzyme, which attacks the molecule at the positions 3,4 (Bedard et al., 1987).

The genes (termed *bph*) encoding for the catabolic enzymes involved in the biodegradation of PCBs are organized in an operon (*bphABCD*) that can be localized on chromosomes, plasmids, or transposons and whose arrangement depends on the bacterial species. In this regard, the occurrence of similar catabolic genes in different strains suggests that even chromosomal *bph* genes have, or previously had, a mechanism for mobilization to other strains (Furukawa, 2000).

In particular, five open reading frames (ORFs) have been detected, i.e., *bphA1*, *bphA2*, *bphA3*, and *bphA4* encoding for a large subunit and a small subunit of the terminal dioxygenase, a ferredoxin and a ferredoxin reductase, respectively. The function of another *orf3* found between *bphA2* and *bphA3* remains unclear (Furukawa et al., 2004). For recent reviews on the genetic and biochemical aspects of aerobic degradation of PCBs see Furukawa et al. (2004) and Pieper (2005).

Biphenyl-degrading microorganisms are generally able to degrade several PCB congeners using the same enzymatic system. Microbial PCB degradation generally requires biphenyl as the carbon source and inducer of catabolic enzymes. However, other growth substrates (e.g., linoleic acid, pyruvate, and acetate) have been demonstrated to support the removal of PCB (Master and Mohn, 1998). These alternative substrates could be used for field applications to induce the biphenyl catabolic pathway as replacements for biphenyl, which is toxic.

11.3 CONTAMINATION OF POLAR REGIONS WITH PCBs

Geographically the Antarctic and the Arctic differ greatly. The Antarctic is a mostly ice-covered continent surrounded by the Southern Ocean and very distant from populated areas. The Arctic is an ice-covered ocean, surrounded by tree-less frozen land often covered with snow and ice. In particular, the boundaries of the Arctic region, generally considered to be north of the Arctic Circle (66° 33′ N), can be traced out in different ways based on social, political, and ecological criteria. In this chapter, we will consider as "Arctic" the area within the 10°C July isotherm including the Arctic Ocean (overlying the North Pole) and parts of the Northern territories of Canada, Russia, the United States (Alaska), Denmark (Greenland), Norway, and Iceland.

Since the 1960s, despite their remoteness, the presence of anthropogenic pollutants such as PCBs and other organochlorine compounds has been reported in both polar regions. The two regions were previously considered to be pristine (Risebrough et al., 1968, 1976). Several pathways have been suggested as the source of PCB contamination in polar regions. These include long-range transport by air (mainly for volatile contaminants) and water, local contamination due to improper disposal practices (e.g., old electrical equipment) and/or the incineration of waste produced at research bases, riverine inputs (for the Arctic) and local accumulation due to biotic activities (Negoita et al., 2003; Montone et al., 2005).

Sources proximity and the chemical properties of PCBs (along with other POPs) are among factors which determine their occurrence and deposition in these remote regions (Corsolini et al., 2002a). Atmospheric and oceanic currents are the main vectors of persistent pollutants from continental areas to oceans, which may be considered as a major sink for these compounds. Cold polar regions may become sinks themselves for several POPs due to the global distillation mechanism. This consists of the evaporation of pollutants at low warm latitude and the deposition at cold higher latitudes (Burkow and Kallenborn, 2000; Negoita et al., 2003; Gambaro et al., 2005; Montone et al., 2005). Migratory organisms, in particular seabirds and whales, can also contribute to pollutant transportation to polar regions through their excrement and carcasses. Corsolini et al. (2002a) found the highest concentrations of PCBs in south polar skua tissues (*Catharacta maccormicki*). This suggests that Antarctic birds migrating to sub-Antarctic or North Pacific Islands carry heavy amounts of PCBs from their wintering grounds.

PCB concentrations measured in Antarctic and sub-Antarctic Regions are generally lower than those determined for the rest of the world, including the Arctic (Fuoco and Ceccarini, 2001). This finding is reasonable as the latter is highly impacted by neighboring industrialized and urbanized regions, which act as the major sources of a variety of organic pollutants (Bustnes et al., 2006). Moreover, the atmospheric transportation of semivolatile compounds from the Northern to the Southern hemisphere is severely limited by the slow interhemispheric air exchange that generally occurs over a period of 1–2 years. The occurrence of PCBs has been reported for several Arctic and Antarctic abiotic matrices such as air, water, soil, snow, and sediment (Tanabe et al., 1983; Risebrough et al., 1990; Larsson et al., 1992; Bright et al., 1995a,b; Kennicutt et al., 1995; Gregor et al., 1996; Fuoco et al., 1996, 1999, 2005; Kallenborn et al., 1998; Ockenden et al., 2001; Montone et al., 2001b, 2005; Crockett and White, 2003; Negoita et al., 2003; Gambaro et al., 2005; Hung et al., 2005; Negri et al., 2006). Both winter darkness and low temperature slow down the degradation of PCBs that may be entrapped in pack ice and snow for long periods. Subsequent ice melting then releases them into terrestrial and marine systems where they enter the food webs, bioaccumulating in organism tissues and biomagnifying with increasing trophic level from planktons to top predators (Fuoco et al., 1996, 1999).

In the Arctic, indigenous people are highly vulnerable to contaminant exposure due to their dietary reliance on marine mammals. The concentrations of PCBs measured in people from the Canadian and the Norwegian Arctic are among the highest detected worldwide (AMAP, 1997). Therefore PCB contamination in these areas is of great environmental concern.

Arctic and Antarctic marine and maritime ecosystems are characterized by a low species diversity, with a predominance of marine mammals and birds, thus resulting in simple predator–prey relationships. Investigations into PCB concentrations in Arctic wildlife have mainly focused on the polar bear (*Ursus maritimus*) that represents the top predator of the plankton–fish–seal food web, and the ringed seal (*Phoca hispida*), which is the most abundant Arctic pinniped with a circumpolar distribution and is a top predator in the nearshore pelagic food webs (Muir et al., 2000; Haave et al., 2003; Braathen et al., 2004). Very high PCB concentrations have also been detected in Arctic glaucous gulls (*Larus hyperboreus*) and sub-Arctic great black-backed gulls (*Larus marinus*) in comparison to Antarctic skuas (Bustnes et al., 2006). In addition, several authors have measured PCBs in microbial loop-involved Arctic organisms (e.g., bacteria and zooplankton) and investigated their accumulation with increasing trophic levels (Borgå et al., 2005; Sobek et al., 2006a,b). The global transport mechanisms of anthropogenic pollutants including PCBs and the susceptibility of Arctic food webs have been reviewed by Bard (1999) and Braune et al. (2005; for the Canadian Arctic).

Mosses, lichens, algae, benthos, krill, fish, seabirds, and mammals are among biotic PCB-contaminated Antarctic matrices (Focardi et al., 1991, 1992, 1995; Weber and Goerke, 1996, 2003; Court et al., 1997; van den Brink et al., 1997; Montone et al., 2001a; Corsolini et al., 2002b, 2003, 2006; Chiuchiolo et al., 2004; Borghini et al., 2005; Negri et al., 2006; Bustnes et al., 2006, 2007). PCBs have also been detected in seabird eggs (penguins and skuas). This proves their transfer from the mother to eggs, thereby exposing the future generations (Corsolini et al., 2002a).

PCB levels in Antarctic seawater are generally several orders of magnitude lower than in organisms. The biomagnification of PCBs along the Antarctic marine food webs was demonstrated by Focardi et al. (1995) and Corsolini et al. (2002a). Both observed a clear relationship between PCB concentrations (ng/g wet weight) in marine organisms and their trophic level (fish < Adélie penguin < Weddel seal, and krill < crocodile icefish < sharp-spined notothen < silverfish < Weddel seal, respectively).

11.4 BIODEGRADATION OF PCBs BY PSYCHROTOLERANT MICROBES

11.4.1 PCB-Degrading Microorganisms from Arctic Environments

The first attempt to establish the biodegradative potential of psychrotolerant bacteria from the Arctic was reported by Whyte et al. (1995), who screened isolates from Arctic and sub-Arctic environments for the utilization of several commonly occurring pollutants (i.e., toluene, naphthalene, dodecane, hexadecane, 2-chlorobiphenyl, and pentachlorophenol). However, none of the 66 isolates obtained from soil, water, and sediment samples collected from the relatively pristine Resolute Bay (Canadian Arctic) were able to mineralize 2-chlorobiphenyls used as a substrate.

Since 1997, Mohn and coworkers have made great advances in the comprehension of PCB biodegradation by psychrotolerant microorganisms from Arctic soils. Mohn et al. (1997) first demonstrated the capability of Canadian Arctic soil microflora to aerobically degrade biphenyls and the commercial PCB mixture Aroclor 1221 examining both the response of the soil microflora to PCB pollution in Arctic soil slurries and the biodegradation potential of the microflora. In soil slurries, the rate and extent of biphenyl mineralization in the Arctic and temperate soils were comparable. Generally, biphenyl mineralization in Arctic soils was stimulated slightly more at 30°C (maximum mineralization rates were typically from 1.2 to 1.4 mg of biphenyl g of dry soil^{-1} day^{-1}) than at 7°C (maximum mineralization rates were typically from 0.52 to 1.0 mg of biphenyl g of dry soil^{-1} day^{-1}). These small differences were explained by the potential coexistence in the soils of biphenyl degraders with different temperature optima. Biphenyl mineralization in PCB-contaminated Arctic soils was generally faster than in those uncontaminated Arctic soils (at both 7°C and 30°C). This suggests that an *in situ* selection for biphenyl-mineralizing microorganisms can occur in Arctic polluted soils, due to (1) a selective pressure that may have been for organisms using PCBs as growth substrates or against organisms inhibited by those compounds or their metabolites, or (2) a genetic and physiological state that is more prone to biphenyl mineralization by biphenyl degraders.

Results obtained when adding Aroclor 1221 to soil slurries showed that the removal of the substrate by indigenous microorganisms was more effective at 30°C (in the range 71%–76%) than at 7°C (in the range 14%–40%) (Mohn et al., 1997). The authors suggested that the limited removal of PCBs observed at 7°C in soils might be due to (1) the cometabolism of PCBs by biphenyl degradation enzymes, which may be more sensitive to low temperature than biphenyl degradation. Thus PCB-degrading psychrotolerant microbes could be metabolically less active at temperatures lower than the optimum growth temperature, (2) the presence of distinct microorganisms responsible for biphenyl degradation at low temperature and lacking the capability to cometabolize PCBs, (3) the effect of temperature on the bioavailability of PCBs, strongly dependent on the adsorption of PCBs into the soil, and (4) the toxicity of the high concentrations of the substrates (biphenyl or PCBs) that can inhibit biphenyl-degrading microorganisms active at 7°C.

Mohn et al. (1997) isolated both psychrophilic and psychrotolerant biphenyl-degrading bacteria from enrichment cultures incubated at 7°C. Among them, two psychrotolerant isolates (denominated Sag-50A and Sag-50G, both Gram-negatives) removed Aroclor 1221 in the range of 54%–60% and showed similar patterns of congener removal at 7°C. Their capability to aerobically degrade Aroclor 1221 congeners with up to three chlorine substituents at low temperature may make them useful in a sequential process in which PCBs are first dechlorinated by anaerobic organisms. Conversely, none of the Gram-positive psychrophilic isolates substantially degraded PCBs at 7°C. Based on the fact that PCB degradation was proven to be more efficient at low temperature in pure cultures rather than in soil slurries, Mohn et al. (1997) first suggested the utilization of indigenous microorganisms for the bioremediation at Arctic sites.

Further results on psychrotolerant PCB degraders isolated from PCB-contaminated Arctic soil samples (Northwest Territories, Canada) are reported by Master and Mohn (1998). Among approximately 50 biphenyl-degrading microorganisms obtained from enrichment cultures, only three isolates, called Cam-1, Sag-1, and Iqa-1, were selected as they could remove Aroclor 1221 congeners at 7°C by batch cultures after 5 weeks, although to different extents (total removal: 56%, 82.1%, and 87.3%, respectively). Their congener ranges were similar to that exhibited by the isolate Sag-50G, which had been previously reported by Mohn et al. (1997). None of the isolates could remove 2,2′-chlorobiphenyl, although Cam-1 and Sag-1 removed 4,4′-chlorobiphenyl, suggesting the expression of 2,3-biphenyl dioxygenase rather than 3,4-biphenyl dioxygenase (Bedard and Haberl, 1990).

The phylogenetic analysis of the 16S rDNA sequences of these isolates, including Sag-50G (Mohn et al., 1997), revealed that they were most closely related to the genus *Pseudomonas*. The degradation of Aroclor 1242 at both 7°C and 15°C by batch cultures of each isolate was also tested by Master and Mohn (1998). The total removal of the substrate by the isolates Cam-1 and Sag-1 was higher at 15°C (14.4% and 10.2%, respectively) than at 7°C (11.3% and 8.1%, respectively), whereas isolate Iqa-1 degraded Aroclor 1242 more efficiently at 7°C (7.1%) rather than at 15°C (5.3%). However, the range of Aroclor 1242 congeners removed by each isolate at 7°C and 15°C were similar, with the exception of Cam-1 that appeared to degrade some tetrachlorobiphenyls.

The extent and rate of removal of a PCB mixture by each Arctic soil isolate at different temperatures (7°C, 37°C, and 50°C) were then compared with those of the mesophile *Burkholderia xenovorans* LB400, which is known to preferentially degrade *ortho*-substituted PCB congeners with up to six chlorines (Bopp, 1986). At 7°C similar total removal values of Aroclor 1242 were observed for LB400 and most Arctic soil isolates (range 3.2%–10.1%), although the removal rates of some individual PCB congeners by psychrotolerant isolates were up to 10 times higher than the corresponding rates of removal by the mesophile. At 37°C the total removal of Aroclor 1242 by the Arctic isolates was higher (range 5.5%–17.6%) than, or the same as, that obtained at 7°C but as much as 90% lower at 50°C (range 1%–3.6%) than at 37°C. In addition, the removal fell to zero for some PCB congeners at 50°C. In contrast, the rates of PCB removal by LB400 were greater at 50°C than at 37°C. Higher rates of PCB removal by the psychrotolerant isolates than by the mesophile LB400 at 7°C and their decreased PCB removal at high temperatures suggested that either the PCB-degrading enzyme systems of the bacteria isolated from Arctic soils were cold adapted or the higher biodegradation rate was due to the better growth and development of the strains at the optimal growth temperature.

Subsequently, experiments were carried out with the aim of elucidating the genetic and biochemical aspects of PCB biodegradation by the Arctic bacterium *Pseudomonas* sp. Cam-1, previously isolated by Master and Mohn (1998). The regulation of *bph* genes in a chromosomal context was investigated by Master and Mohn (2001) by constructing a *bphA-lacZ* reporter in Cam-1 and *Burkholderia* sp. strain LB400 (thus generating Cam-10 and LB400-1). The induction of beta-galactosidase activity by several potential inducers (e.g., aromatic compounds, terpenoids, soil extracts and flavenoids, and chlorinated biphenyls), that have been previously shown to stimulate PCB degradation or that are structurally similar to biphenyl, was correlated to the induction of *bphA*. In particular, results suggested that the regulation of *bphA* in Cam-1 was highly specific: biphenyl induced beta-galactosidase activity in Cam-10 to a level approximately six times greater than the basal level of expression in cells grown with pyruvate. Other growth substrates, such as naphthalene, salicylicate, 2-chlorobiphenyl, and 4-chlorobiphenyl, appeared to be the inducers of *bphA* in Cam-1, but at levels that were no more than 30% of the levels induced by biphenyl. The same authors, in order to investigate the role of *bphA* induction in the efficiency of PCB removal at low temperatures, compared the beta-galactosidase activities in cell suspensions of Cam-10 incubated at 7°C with pyruvate or biphenyl plus pyruvate. After 24 h, the beta-galactosidase activity of cells incubated with biphenyl was four times higher than that of cells incubated with pyruvate. This suggests that *bphA* is induced by biphenyl in Cam-1 at low temperature, being consistent with the psychrotolerant nature of this bacterium isolated from Arctic soil. Conversely, the constitutive level of beta-galactosidase activity in LB400-1 grown at 15°C was about five times less than the level in LB400-1 grown at 30°C. The author concluded that physical and chemical environmental variables could diversely affect the *bphA* induction in bacteria and that therefore there is a requirement to determine the optimal conditions for PCB bioremediation.

The comprehension of factors involved in the transmembrane transport of biphenyl and PCB for catalysis is also important to improve the bioremediation process. Master et al. (2005) first described the kinetics and investigated the energy dependence of bacterial biphenyl uptake through the use of *Pseudomonas* sp. Cam-1 as a model. The results on the effects of growth substrates and temperature were compared with those obtained for *Burkholderia* LB400. Biphenyl uptake by both Cam-1 and LB400 was induced by growth on biphenyl, whereas it was inhibited by dinitrophenol and carbonyl cyanide *m*-chlorophenylhydrazone that are both metabolic uncouplers.

At 22°C, biphenyl uptake by Cam-1 was induced by growth on biphenyl and showed saturable kinetics with respect to biphenyl concentration; the rate for biphenyl uptake by Cam-1 was approximately 1.5 times higher than that shown by LB400. At 15°C, the rate of biphenyl uptake by Cam-1 and LB400 decreased when high concentrations of biphenyl were used. This suggests that changes in the membrane fluidity at low temperature may increase the toxicity of biphenyl; the maximum rate of biphenyl uptake by Cam-1 resulted about three times higher than that of LB400 under the same conditions. Results suggested that the transport of biphenyl at low temperature was facilitated by the cold adaptation of Cam-1.

The congener preference of *Pseudomonas* sp. strain Cam-1 is similar to that of *Pseudomonas pseudoalcaligenes* KF707, preferentially degrading *para*-substituted PCBs with up to four chlorine substituents. In addition, the oxygenase component of Cam-1 biphenyl dioxygenase ($BPDO_{Cam-1}$) shares 99% amino acid sequence identities with that of $BPDO_{KF707}$, as the α-subunits of ISP_{Cam-1} and ISP_{KF707} differ by a single residue at position 178 (Master and Mohn, 2001). $BPDO_{Cam-1}$ has been recently purified and its temperature dependence was analyzed by performing standard activity assays at different temperatures (Master et al., 2008). Between 4°C and 45°C, $BPDO_{Cam-1}$ activity was higher than that of $BPDO_{LB400}$ (of *B. xenovorans* LB400). Furthermore, the highest activity of $BPDO_{Cam-1}$ occurred at 47°C, at a temperature lower than that of $BPDO_{LB400}$ (55°C). At 57°C, the half-life on the $BPDO_{Cam-1}$ oxygenase was less than half that of the $BPDO_{LB400}$ oxygenase.

Although data suggest that $BPDO_{Cam-1}$ is not cold adapted, differences between $BPDO_{Cam-1}$ and $BPDO_{LB400}$ account for the differences between their parental strains (Cam-1 and LB400, respectively). At 7°C, certain PCB congeners are removed by *Pseudomonas* sp. Cam-1 more efficiently than *Burkholderia* sp. LB400 (Master and Mohn, 1998). Moreover, high temperatures were shown to inhibit PCB removal by Cam-1, whereas that by LB400 increased.

The first attempt to evaluate the exploitation of Arctic soil bacteria for bioremediation purpose is reported by Ahn et al. (2001). When a microbial amendment is used to enhance the bioremediation process, it is useful to know if it is viable and survives during the treatment. Ahn et al. (2001) used the psychrotolerant *Pseudomonas* spp. Cam-1 (Master and Mohn, 1998) and Sag-50G (Mohn et al., 1997) in soil microcosms spiked with 2,3-dichlorobiphenyl to study their survival, persistence, and ability to degrade PCBs at 4°C and 22°C. To allow the detection of Cam-1 and Sag-50G and therefore to distinguish them from the indigenous microflora in microcosm bioremediation experiments, the two strains were marked by inserting the green fluorescent protein gene (*gfp*) into the chromosome, thus generating *gfp*-transformants, which were stable and had no adverse effects on metabolism and growth. The psychrotolerant *gfp*-transformants survived in high numbers over 16 weeks at 22°C and 18 weeks at 4°C, respectively. Their abundance was similar to those of indigenous microorganisms at 4°C, but resulted lower at 22°C. This suggests that the psychrotolerant nature of parental PCB-degrading bacteria may offer them advantages in the competition against the indigenous population for nutrients or in their environmental survival over a longer period of time. Therefore, Ahn et al. (2001) proposed the utilization of *gfp*-transformants for PCB bioremediation applications at cold temperatures.

Finally, although aerobic PCB-degrading bacteria have been isolated from Arctic sites (Mohn et al., 1997; Master and Mohn, 1998), highly chlorinated congeners in Aroclor 1260, a contaminant present in considerable quantities in Arctic soils, make it very recalcitrant to aerobic biodegradation. Kuipers et al. (2003) investigated the microbial reductive dechlorination of both weathered and nonweathered Aroclor 1260 in soils from Resolution Island (Nunavut, Canada). Experiments were carried out in soil slurry cultures incubated at 30°C for 8 weeks, after which dechlorination ceased. In all cultures, an extensive dechlorination was observed, as indicated by the reduction in the average number of chlorines *per* biphenyl molecule (from 6.6 to 5.1 and from 6.2 to 4.5 for weathered and nonweathered Aroclor 1260, respectively); mainly for nonweathered Aroclor 1260, a significant decrease in highly chlorinated PCB congeners (with five or more chlorine substituents) was accompanied by the increase or appearance of less highly chlorinated PCBs (mainly tetrachlorobiphenyls).

Kuipers et al. (2003) compared the dechlorination at 21°C in soils from Resolution Island and Saglek (Labrador, Canada), which differed in the initial composition of weathered Aroclor 1260. The overall trends and the extent of changes in major dechlorination substrates and products were consistent with those of the treatment of Resolution Island soil at 30°C. Weathered Aroclor 1260 was dechlorinated to a greater extent in Saglek soil than in Resolution Island soil (the average number of chlorines per biphenyl molecule was reduced from 6.5 to 4.6, and from 6.7 to 5.1, respectively). The authors concluded that anaerobic PCB degradation, producing less chlorinated PCB congeners, is a feasible bioremediation option for contaminated aerobic Arctic soils and that the weathering of the PCBs (possibly reducing their bioavailability) may inhibit dechlorination to a varying extent with different soils.

11.4.2 PCB-Degrading Microorganisms from Antarctic Environments

Research focusing on the microbial PCB degradation in Antarctica remains quite scarce. However, contrary to the Arctic, PCB-degrading microorganisms have been detected in Antarctic marine sediments and seawater, as well as in soils.

Tumeo and Guinn (1997) first investigated the existence of PCB-degrading indigenous microorganisms in soil at McMurdo Station (the main U.S. research station in Antarctica) and in marine sediments from beneath Winter Quarters Bay in order to evaluate easy-to-implement, low-cost remediation alternatives for application in this cold remote area. Active PCB-degraders occurred only in sediments (having an average temperature of −2°C), where they accounted for approximately 10^4 microbes g^{-1} of wet sediment after 3 weeks of incubation at 4°C on nutrient agar plates and PCB as a carbon source. However, no information on the isolation and identification of these PCB-degrading microorganisms has been documented.

Yakimov et al. (1999) and De Domenico et al. (2004) reported on two psychrotolerant hydrocarbon-degrading *Rhodococcus* strains (namely, B11 and B15). They were previously isolated from the marine water column at Terra Nova Bay in the Ross Sea (Maugeri et al., 1996) and were able to grow in the presence of both biphenyl and Aroclor 1242 as the sole carbon sources. The strains grew on biphenyl at temperatures ranging from 4°C to 35°C, with an optimal between 15°C and 20°C; the maximum yield of biomass obtained was 11.3 mg of dry weight *per* liter of culture. Interestingly, during the cultivation on *n*-alkanes, isolates produced a mixture of surface active trehalose lipids (both cell-bound and extracellular, and structurally genus-typical for *Rhodococcus*). This is potentially exploitable for bioremediation purposes as it strongly enhanced the growth on PCBs of the well-known PCB-degrader *Burkholderia cepacia* LB400 (Yakimov et al., 1999). For the estimation of Aroclor 1242 removal by *Rhodococcus* spp. B11 and B15, biodegradation assays were performed by incubating bacterial cultures at 4°C and 20°C for a month and a week, respectively (De Domenico et al., 2004). For the strain B11, total PCB removal was 31.9% and 62.8% at 4°C and 20°C, respectively; *Rhodococcus* sp. B15 appeared to be less effective than B11 as it removed 22.8% and 30.8% of Aroclor 1242 at 4°C and 20°C, respectively. Lower incubation temperature was noted to severely limit PCB degradation. This was observed in the total substrate removal, which was approximately twofold higher after only 1 week of incubation at 20°C than after 1 month at 4°C.

Michaud et al. (2007) investigated the biodegradation of Aroclor 1242 by marine psychrotolerant Antarctic bacteria belonging to the genera *Pseudoalteromonas*, *Psychrobacter*, and *Arthrobacter* (strains no. 19, 15, and 74, respectively). After 3 weeks of incubation at 4°C, Aroclor 1242 could sustain bacterial populations between 1.5 and 6.8×10^6 CFU mL^{-1}, and between 2.8 and 4.6×10^6 CFU mL^{-1} at 15°C. The isolates were able to reduce most chromatographic peaks by more than 50%; however, none totally disappeared (range 0.4%–82.8%). Contrary to the observations previously made for *Rhodococcus* spp. B11 and B15 (De Domenico et al., 2004), PCB removal did not prove to be strongly dependent on incubation temperature. Additionally, it was also not related to the phylogenetic affiliation of isolates. At the two tested temperatures, PCB congeners were generally removed to a similar extent, even if the isolates were slightly more efficient at 15°C. *Arthrobacter* sp. 74 exhibited similar biodegradative potential when growing at 4°C and 15°C, whereas *Pseudoalteromonas* sp. 19 more effectively removed PCBs at 15°C; *Psychrobacter* sp. 15 weakly removed PCB congeners exclusively at 15°C (range 0.4%–25.2%). At 4°C, the removal of di-, tri- and tetrachlorobiphenyls was similar for *Pseudoalteromonas* sp. 19 and *Arthrobacter* sp. 74, whereas pentachlorobiphenyls were more significantly removed by *Arthrobacter* sp. 74 (range 50.7%–70.6%). In comparison, at 15°C *Pseudoalteromonas* sp. 19 was more efficient than *Arthrobacter* sp. 74 in the degradation of almost all PCB congeners.

Finally, Luz et al. (2004) assessed the biodegradative potential of indigenous microbial populations in 16 sub-Antarctic (from South Shetland Islands) and 13 tropical soils (Brazil), which were both impacted and not impacted by anthropogenic activities. A survey of catabolic genes encoding for alkane monooxygenase and aromatic dioxygenases was performed (including *alk*, *ndo*, *tod*, *xyl*, *cat*, and *bph*). All catabolic genes were present in higher frequency in Antarctic soil samples, collected from the Admiralty Bay (King George Island) and the Maxwell Bay (Nelson Island), than in Brazilian soils, with *todC1* and *bphA1* (encoding for the large subunit of the terminal dioxygenase) as the most common. In particular, the *bphA1* genotype was detected by PCR in 10 contaminated (at both Admiralty Bay and Maxwell Bay) and 3 uncontaminated (all at Admiralty Bay) soils, respectively. A DNA fragment with a size comparable (830-bp) to that of *P. pseudoalcaligenes* KF707 was produced and results were confirmed by hybridization to the *bphA1* gene probe. Results revealed that catabolic gene systems involved in biodegradative pathways mainly described in mesophiles also occur in microbial populations inhabiting very cold climates. The culture-independent molecular approach adopted by Luz et al. (2004) appears to be a useful tool in predicting the biodegradation potential of soil microbial communities in both cold and tropical ecosystems through the provision of basic knowledge on the distribution of degradative genotypes.

11.4.3 PCB Biodegradation by Psychrotolerant Bacteria from Temperate Environments

PCB-degrading psychrotolerant microorganisms deriving from environments different to the Poles, or more generally microorganisms able to degrade PCBs at low temperature, may be used for bioremediation purposes in the Arctic where, in contrast to Antarctica, nonindigenous microorganisms could be introduced. This is despite a lack of evidence to date on the level of usefulness deriving from their utilization in these areas.

Initial experiments dealt with the response of the whole microbial community to PCB contamination (without giving further information on the microorganisms involved in the process) in relation to the effects of temperature on the substrate removal in both aerobic and anaerobic conditions. Williams and May (1997) demonstrated the occurrence of microbial aerobic degradation of PCBs in Hudson River sediment samples, which were spiked with Aroclor 1242 and incubated at 4°C for several months. Certain di- and trichlorobiphenyls began to disappear from the surface sediment layer after 6 weeks, and >50% loss occurred within 5 months. In the same year, Wu et al. (1997) reported about the effect of incubation temperature (range 4°C–66°C) on the anaerobic reductive dechlorination of 2,3,4,6-tetrachlorobiphenyl in freshwater pond sediments from Athens (Greece) and Massachusetts (United States). Temperature influenced both the timing and the relative predominance of parallel pathways of dechlorination; e.g., *meta* dechlorination of 2,3,4,6-tertachlorobiphenyl to 2,4,6-tetrachlorobiphenyl resulted predominant at all but two of the temperatures tested (18°C and 34°C); *para* dechlorination dominated at 20°C, and *ortho* dechlorination dominated at 15°C. Wu et al. (1997) concluded that field temperatures can be expected to play a significant role in controlling the extent and the nature of PCB dechlorination occurring at a given site.

Subsequently, Lambo and Patel (2006a,b; 2007a,b) reported on a psychrotolerant PCB-degrading bacterium, namely, *Hydrogenophaga taeniospiralis* IA3-A that was isolated from a long-term PCB-contaminated soil in the Western Newfoundland (Canada), characterized by its temperate climate. The strain IA3-A was able to remove mono-, di- and trichlorobiphenyls in Aroclor 1221 at 5°C and 30°C after 48 h of incubation (Lambo and Patel, 2006a). Unlike cultures incubated at 30°C, none of the congeners were totally removed at 5°C. Interestingly, certain congeners (i.e., 3,3′-dichlorobiphenyls, 2,3,6- and 2,3′,6-trichlorobiphenyls) that were not removed at 30°C partially disappeared at 5°C. In particular, monochlorobiphenyls were removed in the range of 63%–89% at 5°C, whereas they totally disappeared at 30°C, suggesting that low temperature probably slows down the cometabolism of these congeners. Dichlorobiphenyl removal was between 30% and 78% at 5°C, and between 30% and 100% at 30°C and finally, the removal of trichlorobiphenyls was in the range of 30%–75% and 27%–59% at 5°C and 30°C, respectively. Several congeners with two or more *ortho* chlorines were generally resistant to the degradation by *H. taeniospiralis* IA3-A. The removal of di-, *ortho*-, and *para*-substituted congeners at 30°C rather than at 5°C indicated that the cometabolism of these congeners is probably enhanced by the presence of *para*-chlorines at the higher temperature tested. Results suggested that the incubation temperature could be a major determining factor in the biodegradation process and that the specificity of cells for some congeners was probably different at different temperatures. Furthermore, Lambo and Patel (2006a) investigated the removal of individual congeners (i.e., 2,3- and 2,4′-dichlorobiphenyl) in an attempt to assess the extent of removal of PCBs and the metabolites produced at different temperatures. After 72 h, the strain IA3-A removed 68% and 83% of 2,4′-dichlorobiphenyl, and 35% and 44% of 2,3-dichlorobiphenyl at 5°C and 30°C, respectively. In particular, 2,3-dichlorobiphenyl was totally transformed into 2,3-chlorobenzoic acid (2,3-CBA) at both temperatures, whereas most of the 2,4′-dichlorobiphenyl was not recovered as CBAs.

The influence of temperature on the removal of PCB congeners was further investigated by Lambo and Patel for the strain *H. taeniospiralis* IA3-A through the use of Aroclor 1248 and Aroclor 1232 as substrates. After 48 h, di- and trichlorobiphenyls in Aroclor 1248 were degraded at both temperatures, whereas a unique congener with four chlorine substituents (more likely 2,3,4′,6-tetrachlorobiphenyl) was removed at 30°C (Lambo and Patel, 2006b). When studying the removal of congeners in Aroclor 1232, Lambo and Patel (2007a) observed that tetrachlorobiphenyls were also removed at 5°C (although to a lesser extent than at 30°C), whereas none of the pentachlorobiphenyls and the hexachlorobiphenyls disappeared. However, with regard to the chlorination of congeners in Aroclor 1248 and Aroclor 1232, the removal patterns were generally similar at both temperatures, with few exceptions (e.g., congeners in Aroclor 1248 with 2-*ortho*-, plus 5-*meta*- or 2-*ortho*, 3-*meta*, plus 4-*para* substitution that were more slightly removed at 30°C).

The cold adaptation of chlorobiphenyl-degrading upper pathway enzymes of *H. taeniospiralis* IA3-A was confirmed by analyzing the production of metabolites from 2,4′-dichlorobiphenyl at 5°C after 72 h. Both the rings of 2,4′-dichlorobiphenyl were oxidized as the production of 2- and 4-CBAs (both dead-end metabolites as confirmed by oxygen uptake experiments) was recorded (Lambo and Patel, 2006b). It has been suggested that *ortho*- and *para*-substituted congeners, including 2,4′-dichlorobiphenyl, could be used for both identifying bacteria that are able to more efficiently transform PCBs and discriminating the specificity of attack of PCBs by monitoring the accumulation of chlorobenzoates and *meta*-cleavage products (i.e., 2-hydroxy-6-oxo-6-phenyl-hexa-2,4-dienoic acids, HOPDAs) (Maltseva et al., 1999). With this aim, a further investigation on the temperature effect (in the range 10°C–45°C) on the pattern of transformation of 2,4′-dichlorobiphenyl by *H. taeniospiralis* IA3-A was performed by Lambo and Patel (2007b). After 48 h, the isolate transformed 2,4′-dichlorobiphenyl to 2- and 4-CBA, 2-hydroxy-6-oxo-6-phenyl-hexa-2,4-dienoic acids (HOPDAs, some of which are precursors of toxic compounds) and inorganic chloride. The extent of degradation of 2,4′-dichlorobiphenyl resulted quite similar at low and moderate temperatures (10°C, 25°C, and 37°C; range 190–196 μmol/L), whereas it was much lower at 45°C (46 μmol/L). This suggests that 2,4′-dichlorobiphenyl-degrading enzymes are sensitive to or inhibited by high temperature. Between 10°C and 37°C, the recovery of degraded 2,4′-dichlorobiphenyl as CBAs was in the range of 23%–27%; in contrast, at the higher temperature 87% of the degraded compound was recovered as CBAs. All data suggested that the pattern of transformation was different at 45°C and, contrary to low and moderate temperatures, it did not result in high level of HOPDAs. Finally, the detection of several isomeric intermediates suggested that multiple pathways may be involved in the transformation of 2,4′-dichlorobiphenyl by the strain IA3-A and that these were regulated differently at various temperatures.

Overall, the results obtained by Lambo and Patel (2007b) support the idea that *H. taeniospiralis* IA3-A could be exploited in the aerobic phase of the anaerobic–aerobic bioremediation process. During this process, highly chlorinated congeners are anaerobically dechlorinated to produce lightly chlorinated congeners that can be degraded aerobically.

11.5 SUMMARY

Previous research focusing on the biodegradation of PCBs in polar environments has mainly addressed the detection and isolation of PCB-degrading microorganisms and the influence of incubation temperature on both the rate and extent of PCB removal by bacterial isolates. However, other important abiotic factors (e.g., salinity, pH, occurrence of other contaminants, and the real contamination levels in the natural environment) should be taken into consideration in future research. It is well known that the differences in physical and chemical environmental variables, in addition to microbial physiological conditions, can strongly affect many biological processes, including those involving degradation. Coculture experiments should also be implemented in order to illustrate, even if on a small scale, the effect of microbial interactions on the biodegradation of PCBs, as different microorganisms share the same natural environment and could possess different congener specificities.

Therefore, the treatment of a PCB-polluted cold area with the enrichment cultures of PCB-degrading microorganisms could not result in a real environmental decontamination. Further analyses should be performed in an attempt to optimize the biodegradation of PCBs by Arctic and Antarctic microbes at laboratory scale. An attempt to closely reflect the field conditions could result in more effective subsequent bioremediation practices.

Further analyses with the aim of elucidating several biochemical and genetic aspects of PCB transformation in polar environments are required. This is especially true with regard to Antarctic microorganisms as studies on the subject have focused predominantly on the identification and estimation of their biodegradative potentials.

Finally, PCB degradation in polar regions has been reported for only a small number of bacterial genera obtained from soil and seawater (i.e., *Pseudomonas* for the Arctic, and *Pseudoalteromonas*, *Psychrobacter*, *Rhodococcus*, and *Arthrobacter* for the Antarctic). Thus, research efforts should be directed at the isolation from different matrices of novel microorganisms (both biotic and abiotic) involved in the degradation of PCBs in these remote areas, particularly in Antarctica where the introduction of allochthonous organisms is not permitted.

In conclusion, the research on the degradation of PCBs in polar environments is in its infancy, and further small and large scale investigations are required to achieve a more in-depth understanding of the biotechnological potential of Arctic and Antarctic PCB-degrading microorganisms as bioremediation players in cold and temperate systems.

REFERENCES

Abraham, W.R., B. Nogales, P.N. Golyshin, D.H. Pieper, and K. Timmis. 2002. Polychlorinated-biphenyl-degrading microbial communities in soils and sediments. *Current Opinion in Microbiology* 5:246–253.

Ahn, Y.-B., L.A. Beaudette, H. Lee, and J.T. Trevors. 2001. Survival of a GFP-labeled polychlorobiphenyl degrading psychrotolerant *Pseudomonas* spp. in 4 and 22°C soil microcosms. *Microbial Ecology* 42:614–623.

AMAP. 1997. Arctic pollutant issues: A state of the Arctic environment report. *Arctic Monitoring and Assessment Program*. Oslo, Norway. ISBN 82-7655-060-6.

Bard, S.M. 1999. Global transport of anthropogenic contaminants and the consequences for the Arctic marine ecosystem. *Marine Pollution Bulletin* 38:356–379.

Bedard, D.L. and M.L. Haberl. 1990. Influence of chlorine substitution pattern on the degradation of polychlorinated biphenyls by eight bacterial strains. *Microbial Ecology* 20:87–102.

Bedard, D.L., M.L. Haberl, R.J. May, and M.J. Brennan. 1987. Evidence for novel mechanisms of polychlorinated biphenyl metabolism in *Alcaligenes eutrophus* H850. *Applied and Environmental Microbiology* 53:1103–1112.

Bedard, D.L., J.J. Bailey, L.R. Brandon, and G. van Slyke Jerzak. 2006. Development and characterization of stable sediment-free anaerobic bacterial enrichment cultures that dechlorinate Aroclor 1260. *Applied and Environmental Microbiology* 72:2460–2470.

Bopp, L.H. 1986. Degradation of highly chlorinated PCBs by *Pseudomonas* strain LB400. *Journal of Industrial Microbiology and Biotechnology* 1:23–29.

Borghini, F., J.O. Grimalt, J.C. Sanchez-Hernandez, and R. Bargagli. 2005. Organochlorine pollutants in soils and mosses from Victoria Land (Antarctica). *Chemosphere* 58:271–278.

Borgå, K., A. Fisk, B. Hargrave, P.F. Hoekstra, D. Swackhamer, and D.G. Muir. 2005. Bioaccumulation factors for PCB revisited. *Environmental Science and Technology* 39:4523–4532.

Borja, J., D.M. Taleon, J. Auresenia, and S. Gallardo. 2005. Polychlorinated biphenyls and their biodegradation. *Process Biochemistry* 40:1999–2013.

Braathen, M., A.E. Derocher, Ø. Wiig, et al. 2004. Relationships between PCBs and tyroid hormones and retinol in female and male Polar bears. *Environmental Health Perspectives* 112:826–833.

Braune, B.M., P.M. Outridge, A.T. Fisk, et al. 2005. Persistent organic pollutants and mercury in marine biota of the Canadian Arctic: An overview of spatial and temporal trends. *The Science of the Total Environment* 351–352:4–56.

Bright, D.A., W.T. Dushenko, S.L. Grundy, and K.J. Reimer. 1995a. Evidence for short-range transport of polychlorinated biphenyls in the Canadian Arctic using congener signatures of PCBs in soils. *The Science of the Total Environment* 160–161:251–263.

Bright, D.A., W.T. Dushenko, S.L. Grundy, and K.J. Reimer. 1995b. Effects of local and distant contaminant sources: Polychlorinated biphenyls and other organochlorines in bottom-dwelling animals from an Arctic estuary. *The Science of the Total Environment* 160–161:265–283.

Burkow, I.C. and R. Kallenborn. 2000. Sources and transport of persistent organic pollutants to the Arctic. *Toxicology Letters* 112–113:87–92.

Bustnes, J.O., T. Tveraa, J.A. Henden, Ø. Varpen, K. Janssen, and J.U. Skaare. 2006. Organochlorines in Antarctic and Arctic avian top predators: A comparison between the South Polar skuas and two species of Northern hemisphere gulls. *Environmental Science and Technology* 40:2826–2831.

Bustnes, J.O., T. Tveraa, Ø. Varpe, J.A. Henden, and J.U. Skaared. 2007. Reproductive performance and organochlorine pollutants in an Antarctic marine top predator: The south Polar skua. *Environmental International* 33:911–918.

Ceccarini, A. and S. Giannarelli. 2006. Polychlorobiphenyls. In *Chromatographic Analysis of the Environment*, ed. L.M.L. Nollet, pp. 667–710. CRC Press, Boca Raton, FL.

Chiuchiolo, A.L., R.M. Dickhut, M.A. Cochran, and H.W. Ducklow. 2004. Persistent organic pollutants at the base of the Antarctic marine food web. *Environmental Science and Technology* 38:3551–3557.

Corsolini, S., K. Kannan, T. Imagawa, S. Focardi, and J.P. Giesy. 2002a. Polychloronaphthalenes and other dioxin-like compounds in Arctic and Antarctic marine food webs. *Environmental Science and Technology* 36:3490–3496.

Corsolini, S., T. Romeo, N. Ademollo, S. Greco, and S. Focardi. 2002b. POPs in key species of marine Antarctic ecosystem. *Microchemical Journal* 73:187–193.

Corsolini, S., N. Ademollo, T. Romeo, S. Olmastroni, and S. Focardi. 2003. Persistent organic pollutants in some species of a Ross Sea pelagic trophic web. *Antarctic Science* 15:95–104.

Corsolini, S., A. Covaci, N. Ademollo, S. Focardi, and P. Schepens. 2006. Occurrence of organochlorine pesticides (OCPs) and their enantiomeric signatures, and concentrations of polybrominated diphenyl ethers (PBDEs) in the Adélie penguin food web, Antarctica. *Environmental Pollution* 140:371–382.

Court, G.S., L.S. Davis, S. Focardi, et al. 1997. Chlorinated hydrocarbons in the tissues of South Polar skuas (*Catharacta maccormicki*) and Adélie penguins (*Pygoscelis adeliea*) from Ross Sea, Antarctica. *Environmental Pollution* 97:295–301.

Crockett, A.B. and G.J. White. 2003. Mapping sediment contamination and toxicity in Winter Quarter Bay, McMurdo Station, Antarctica. *Environmental Monitoring and Assessment* 85:257–275.

De Domenico, M., A. Lo Giudice, L. Michaud, M. Saitta, and V. Bruni. 2004. Diesel oil and PCB-degrading bacteria isolated from Antarctic seawater (Terra Nova Bay, Ross Sea). *Polar Research* 23:141–146.

Focardi, S., C. Gaggi, G. Chemello, and E. Bacci. 1991. Organochlorine residues in moss and lichen samples from two Antarctic areas. *Polar Records* 162:241–244.

Focardi, S., L. Lari, and L. Marsili. 1992. PCB congeners, DDTs and hexachlorobenzene in Antarctic fish from Terra Nova Bay (Ross Sea). *Antarctic Science* 4:151–154.

Focardi, S., R. Bargagli, and S. Corsolini. 1995. Isomer-specific analysis and toxic potential evaluation of polychlorinated biphenyls in Antarctic fish, seabirds and Weddell seals from Terra Nova Bay (Ross Sea). *Antarctic Science* 7:31–35.

Fuoco, R. and A. Ceccarini. 2001. Polychlorobiphenyls in Antarctic matrices. In *Environmental Contamination in Antarctica: A Challenge for Analytical Chemistry*, ed. S. Caroli, P. Cescon, and D.W.H. Walton, pp. 237–274. Elsevier Science, Oxford.

Fuoco, R., M.P. Colombini, A. Ceccarini, and C. Abete. 1996. Polychlorobiphenyls in Antarctica. *Microchemical Journal* 54:384–390.

Fuoco, R., S. Giannarelli, C. Abete, M. Onor, and M. Termine. 1999. The effect of seasonal pack ice melting on the sea water polychlorobiphenyl contents at Gerlache Inlet and Wood Bay (Ross Sea–Antarctica). *International Journal of Environmental Analytical Chemistry* 75:367–375.

Fuoco, R., S. Giannarelli, Y. Wei, C. Abete, S. Francesconi, and M. Termine. 2005. Polychlorobiphenyls and polycyclic aromatic hydrocarbons in the sea-surface microlayer and the water column at Gerlache Inlet, Antarctica. *Journal of Environmental Monitoring* 7:1313–1319.

Furukawa, K. 2000. Biochemical and genetic bases of microbial degradation of polychlorinated biphenyls (PCBs). *Journal of General and Applied Microbiology* 46:283–296.

Furukawa, K., H. Suenaga, and M. Goto. 2004. Biphenyl dioxygenases: Functional versatilities and directed evolution. *Journal of Bacteriology* 186:5189–5196.

Gambaro, A., L. Manodori, R. Zangrando, A. Cincinelli, G. Capodaglio, and P. Cescon. 2005. Atmospheric PCB concentrations at Terra Nova Bay, Antarctica. *Environmental Science and Technology* 39:9406–9411.

Gounot, A.M. 1991. Bacterial life at low temperature: Physiological aspects and biotechnological implications. *Journal of Applied Bacteriology* 71:386–397.

Gregor, D., C. Teixeira, and R. Roswell. 1996. Deposition of atmospherically transported polychlorinated biphenyls in the Canadian Arctic. *Chemosphere* 33:227–244.

Haave, M., E. Ropstad, A.E. Derocher, et al. 2003. Polychlorinated biphenyls and reproductive hormones in female Polar bears at Svalbard. *Environmental Health Perspectives* 111:431–436.

Holliger, C., G. Wohlfarth, and G. Diekert. 1999. Reductive dechlorination in the energy metabolism of anaerobic bacteria. *FEMS Microbiology Review* 22:383–398.

Hung, H., P. Blanchard, C.J. Halsall, et al. 2005. Temporal and spatial variabilities of atmospheric polychlorinated biphenyls (PCBs), organochlorine (OC) pesticides and polycyclic aromatic hydrocarbons (PAHs) in the Canadian Arctic: Results from a decade of monitoring. *The Science of the Total Environment* 342:119–144.

Kallenborn, R., M. Oehme, D.D. Wynn-Williams, M. Schlabach, and J. Harris. 1998. Ambient air levels and atmospheric long-range transport of persistent organochlorines to Signy Island, Antarctica. *The Science of the Total Environment* 220:167–180.

Kennicutt, M.C., S.J. McDonald, J.L. Sericano, et al. 1995. Human contamination of the marine environment: Arthur Harbor and McMurdo Sound, Antarctica. *Environmental Science and Technology* 29:1279–1287.

Kuipers, B., W.R. Cullen, and W.W. Mohn. 2003. Reductive dechlorination of weathered Aroclor 1260 during anaerobic biotreatment of Arctic soils. *Canadian Journal of Microbiology* 49:9–14.

Lambo, A.J. and T.R. Patel. 2006a. Isolation and characterization of a biphenyl-utilizing psychrotrophic bacterium, *Hydrogenophaga taeniospiralis* IA3-A, that cometabolize dichlorobiphenyls and polychlorinated biphenyl congeners in Aroclor 1221. *Journal of Basic Microbiology* 46:94–107.

Lambo, A.J. and T.R. Patel. 2006b. Cometabolic degradation of polychlorinated biphenyls at low temperature by psychrotolerant bacterium *Hydrogenophaga* sp. IA3-A. *Current Microbiology* 53:48–52.

Lambo, A.J. and T.R. Patel. 2007a. Biodegradation of polychlorinated biphenyls in Aroclor 1232 and production of metabolites from 2,4,4′-trichlorobiphenyl at low temperature by psychrotolerant *Hydrogenophaga* sp. strain IA3-A. *Journal of Applied Microbiology* 102:1318–1329.

Lambo, A.J. and T.R. Patel. 2007b. Temperature-dependent biotransformation of 2,4′-dichlorobiphenyl by psychrotolerant *Hydrogenophaga* strain IA3-A: Higher temperatures prevent excess accumulation of problematic meta-cleavage products. *Letters in Applied Microbiology* 44:447–453.

Larsson, P., C. Järnmark, and A. Södergren. 1992. PCBs and chlorinated pesticides in the atmosphere and aquatic organisms of Ross Island, Antarctica. *Marine Pollution Bulletin* 25:281–287.

Luz, A.P., V.H. Pellizari, L.G. Whyte, and C.W. Greer. 2004. A survey of indigenous microbial hydrocarbon degradation genes in soils from Antarctica and Brazil. *Canadian Journal of Microbiology* 50:323–333.

Macedo, A.J., K.N. Timmis, and W.A. Abraham. 2007. Widespread capacity to metabolize polychlorinated biphenyls by diverse microbial communities in soils with no significant exposure to PCB contamination. *Environmental Microbiology* 9:1890–1897.

Maltseva, V.V., T.V. Tsoi, J.F. Quensen III, M. Fukuda, and J.M. Tiedje. 1999. Degradation of anaerobic reductive dechlorination products of Aroclor 1242 by four aerobic bacteria. *Biodegradation* 10:363–371.

Master, E.R. and W.W. Mohn. 1998. Psychrotolerant bacteria isolated from Arctic soil that degrade polychlorinated biphenyls at low temperatures. *Applied and Environmental Microbiology* 64:4823–4829.

Master, E.R. and W.W. Mohn. 2001. Induction of *bphA*, encoding biphenyl dioxygenase, in two polychlorinated biphenyl-degrading bacteria, psychrotolerant *Pseudomonas* strain Cam-1 and mesophilic *Burkholderia* strain LB400. *Applied and Environmental Microbiology* 67:2669–2676.

Master, E.R., J.J. McKinlay, G.R. Stewart, and W.W. Mohn. 2005. Biphenyl uptake by psychrotolerant *Pseudomonas* sp. strain Cam-1 and mesophilic *Burkholderia* sp. strain LB400. *Canadian Journal of Microbiology* 51:399–404.

Master, E.R., N.Y.R. Agar, L. Gòmez-Gil, J.B. Powlowski, W.W. Mohn, and L.D. Eltis. 2008. Biphenyl dioxygenase from an Arctic isolate is not cold-adapted. *Applied and Environmental Microbiology* 74:3908–3911.

Maugeri, T.L., C. Gugliandolo, and V. Bruni. 1996. Heterotrophic bacteria in the Ross Sea (Terra Nova Bay, Antarctica). *New Microbiologica* 19:67–76.

Michaud, L., G. Di Marco, V. Bruni, and A. Lo Giudice. 2007. Biodegradative potential and characterization of psychrotolerant polychlorinated biphenyl-degrading marine bacteria isolated from a coastal station in the Terra Nova Bay (Ross Sea, Antarctica). *Marine Pollution Bulletin* 54:1754–1761.

Mohn, W.W. and J.M. Tiedje. 1992. Microbial reductive dehalogenation. *Microbiological Reviews* 56:482–507.

Mohn, W.W., K. Westerberg, W.R. Cullen, and K.J. Reimer. 1997. Aerobic biodegradation of biphenyl and polychlorinated biphenyls by Arctic soil micro-organisms. *Applied and Environmental Microbiology* 63:3378–3384.

Montone, R.C., S. Taniguchi, J. Sericano, R.R. Weber, and W.H. Lara. 2001a. Determination of polychlorinated biphenyls in Antarctic macroalgae *Desmaretia* sp. *The Science of the Total Environment* 277:181–186.

Montone, R.C., S. Taniguchi, and R.R. Weber. 2001b. Polychlorinated biphenyls in marine sediments of Admiralty Bay, King George Island, Antarctica. *Marine Pollution Bulletin* 42:611–614.

Montone, R.C., S. Taniguchi, C. Boian, and R.R. Weber. 2005. PCBs and chlorinated pesticides (DDTs, HCHs and HCB) in the atmosphere of the southwest Atlantic and Antarctic oceans. *Marine Pollution Bulletin* 50:778–786.

Muir, D., F. Riget, M. Cleemann, et al. 2000. Circumpolar trends of PCBs and organochlorine pesticides in the Arctic marine environment inferred from levels in ringed seals. *Environmental Science and Technology* 34:2431–2438.

Negoita, T.G., A. Covaci, A. Gheorghe, and P. Schepens. 2003. Distribution of polychlorinated biphenyls (PCBs) and organochlorine pesticides in soils from the East Antarctic coast. *Journal of Environmental Pollution* 5:281–286.

Negri, A., K. Burns, S. Boyle, D. Brinkman, and N. Webster. 2006. Contamination in sediments, bivalves and sponges of McMurdo Sound, Antarctica. *Environmental Pollution* 143:456–467.

Ockenden, W.A., R. Lohmann, J.R. Shears, and K.C. Jones. 2001. The significance of PCBs in the atmosphere of the Southern hemisphere. *Environmental Science and Pollution Research International* 8:189–194.

Pieper, D.H. 2005. Aerobic degradation of polychlorinated biphenyls. *Applied Microbiology and Biotechnology* 67:170–191.

Risebrough, R.W., P. Reiche, D.B. Peakall, S.G. Herman, and M.N. Kirven. 1968. Polychlorinated biphenyls in the global ecosystem. *Nature* 220:1098–1102.

Risebrough, R.W., W. Walker, T.T. Scmidt, B.W. de Lappe, and C.W. Connors. 1976. Transfer of chlorinated biphenyls to Antarctica. *Nature* 264:738–739.

Risebrough, R.W., B.W. de Lappe, and C. Younghans-Haug. 1990. PCB and PCT contamination in Winter Quarters Bay, Antarctica. *Marine Pollution Bulletin* 21:523–529.

Ross, G. 2004. The public health implications of polychlorinated biphenyls (PCBs) in the environment. *Ecotoxicology and Environmental Safety* 59:275–291.

Safe, S. 2004. Toxicology, structure-function relationship, and human and environmental health impacts of polychlorinated biphenyls: Progress and problems. *Environmental Health Perspectives* 100:259–268.

Sobek, A., K. Olli, and Ö. Gustafsson. 2006a. On the relative significance of bacteria for the distribution of polychlorinated biphenyls in Arctic ocean surface waters. *Environmental Science and Technology* 40:2586–2593.

Sobek, A., M. Reigstad, and O. Gustafsson. 2006b. Partitioning of polychlorinated biphenyls between Arctic seawater and size-fractionated zooplankton. *Environmental Toxicology and Chemistry* 25:1720–1728.

Tanabe, S., H. Hidaka, and R. Tatsukawa. 1983. PCBs and chlorinated hydrocarbon pesticides in Antarctic atmosphere and hydrosphere. *Chemosphere* 12:277–288.

Tumeo, M.A. and D.A. Guinn. 1997. Evaluation of bioremediation in cold regions. *Journal of Cold Regions Engineering* 11:221–231.

van den Brink, N.W., J.A. van Franeker, and E.M. de Ruiter-Dijkman. 1997. Fluctuating concentrations of organochlorine pollutants during a breeding season in two Antarctic seabirds: Adélie penguin and Southern fulmar. *Environmental Toxicology and Chemistry* 17:702–709.

Vasilyeva, G.K. and E.R. Strijakova. 2007. Bioremediation of soils and sediments contaminated by polychlorinated biphenyls. *Microbiology* 76:639–653.

Weber, K. and H. Goerke. 1996. Organochlorine compounds in fish off the Antarctic Peninsula. *Chemosphere* 33:377–392.

Weber, K. and H. Goerke. 2003. Persistent organic pollutants (POPs) in Antarctic fish: Levels, patterns, changes. *Chemosphere* 53:667–678.

Whyte, L.G., C.W. Greer, and W.E. Inniss. 1995. Assessment of the biodegradation potential of psychrotrophic microorganisms. *Canadian Journal of Microbiology* 42:99–106.

Wiegel, J. and Q. Wu. 2000. Microbial reductive dehalogenation of polychlorobiphenyls. *FEMS Microbiology Ecology* 32:1–15.

Williams, W.A. and R.J. May. 1997. Low temperature microbial aerobic degradation of polychlorinated biphenyls in sediment. *Environmental Science and Technology* 31:3491–3496.

Wu, Q., D.L. Bedard, and J. Wiegel. 1997. Effect of incubation temperature on the route of microbial reductive dechlorination of 2,3,4,6-tetrachlorobiphenyl in polychlorinated biphenyl (PCB)-contaminated and PCB-free freshwater sediments. *Applied and Environmental Microbiology* 63:2836–2843.

Yakimov, M.M., L. Giuliano, V. Bruni, S. Scarfì, and P.N. Golyshin. 1999. Characterization of Antarctic hydrocarbon-degrading bacteria capable of producing bioemulsifiers. *New Microbiologica* 22:249–259.

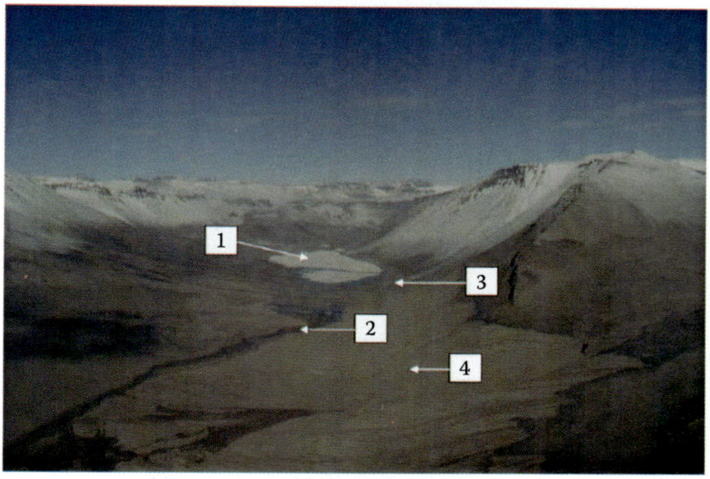

FIGURE 1.1 The Miers Valley, Eastern Antarctica. This valley incorporates all the typical terrestrial microbial habitats of the Dry Valleys including lakes (1), lake-derived streams (2), glacial meltwater streams (3), and desert soils (4).

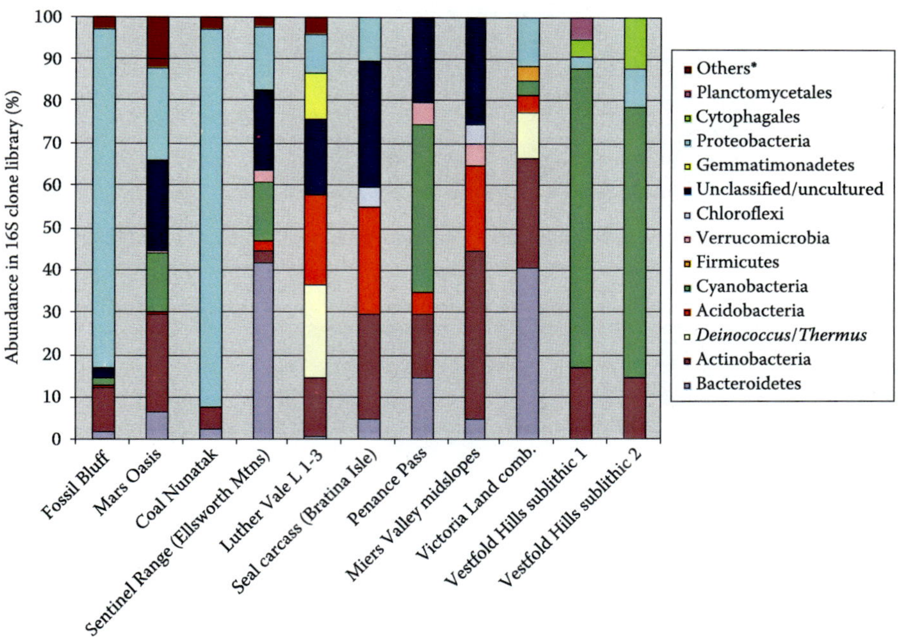

FIGURE 1.3 A snapshot of prokaryotic diversity across a range of Antarctic arid soil habitats. *NB: For Mars Oasis, Fossil Bluff, Coal Nunatak, and Sentinel Range, "Others" category includes Chloroflexi, Firmicutes, Fusobacteria, Gemmatimonadetes, Planctomycetes, and the candidate phyla OP10 and TM7. For Luther Vale, "Others" category includes Chloroflexi, Firmicutes, Bacteroidetes, Verrucomicrobia, Nitrospira, and Cyanobacteria.

FIGURE 1.4 Filamentous, fungal-dominated hypolithic community loosely attached to the underside of translucent quartz (Miers Valley, Eastern Antarctica).

FIGURE 1.5 Unattached, moss-dominated community beneath a quartz rock (Miers Valley, Eastern Antarctica).

FIGURE 1.6 Zonation visible in a typical cyanobacteria-dominated hypolithon with orange-pigmented organisms around the periphery and green cyanobacteria directly underneath the quartz rock (Miers Valley, Eastern Antarctica).

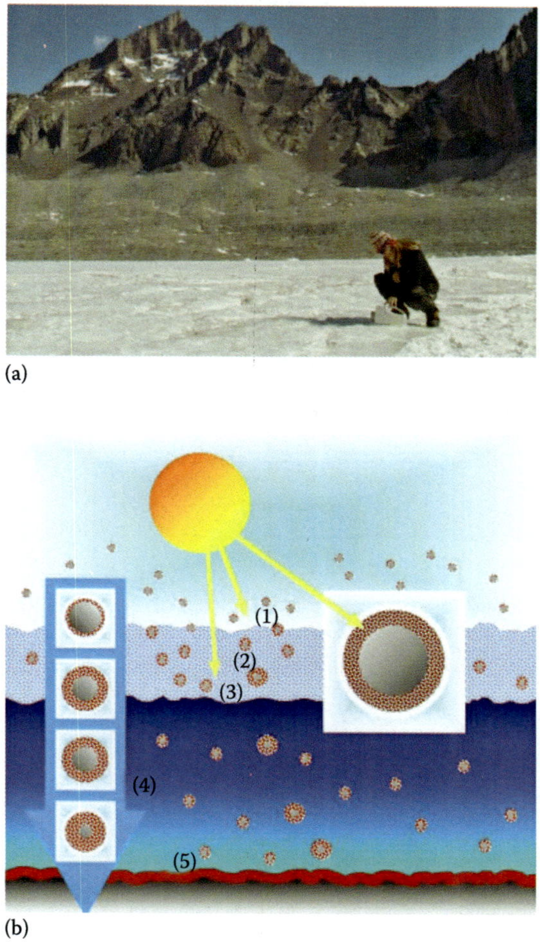

FIGURE 4.1 One of the authors (BS) collecting air samples (a) for 16s RNA analysis of airborne assemblages of microbes on the ice-covered Lake Untersee, Dronning Maud Land, Antarctica. The microbial ecosystem of the lake begins as airborne particles of microbes, humic material, and minerals adhere (b1) to the icy surface. Warmed by the sun the dark mixture of organic and inorganic materials sinks into the ice (b2) and is covered by melt water that soon freezes. Solar energy heats the assemblage sufficiently to produce a thin biofilm on the particle surface comprised of liquid water and multiple microbial species (insert). Photosynthetic cyanobacteria, protected from predators and provided with solar energy for photosynthesis, multiply (b3) until the particle surface is maximally covered (b4). The assemblage passes into the lake water and finally joins the cyanobacteria-dominated mat community (b5) at the lake floor.

(a)

(b)

FIGURE 4.2 Cryoconites come in all sizes from accumulations at the base of massive boulders (a) floating on the ice cover of Lake Untersee to millimeter scale particles wrapped in refractive crystals of ice spread across the lake surface (b).

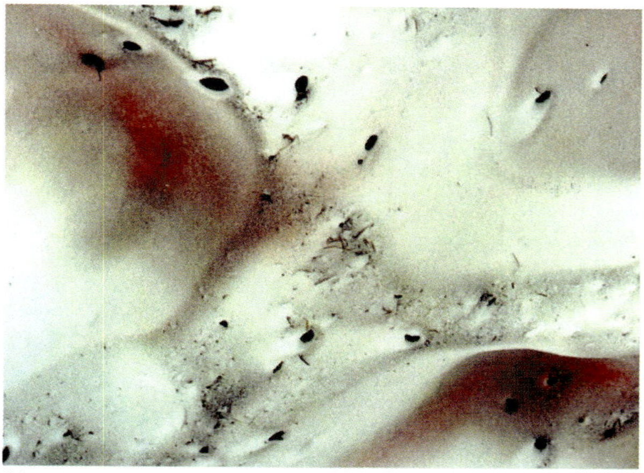

(a)

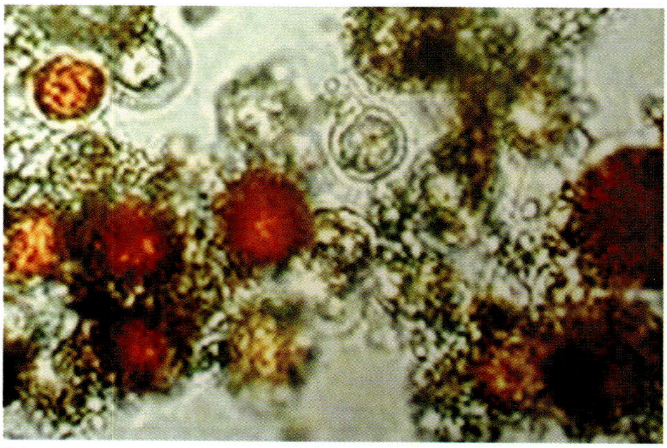

(b)

FIGURE 5.1 (a) *C. nivalis* on a snowbank in the Kolyma Lowlands, North Siberia. (b) Photomicrograph of red spores of *C. nivalis*.

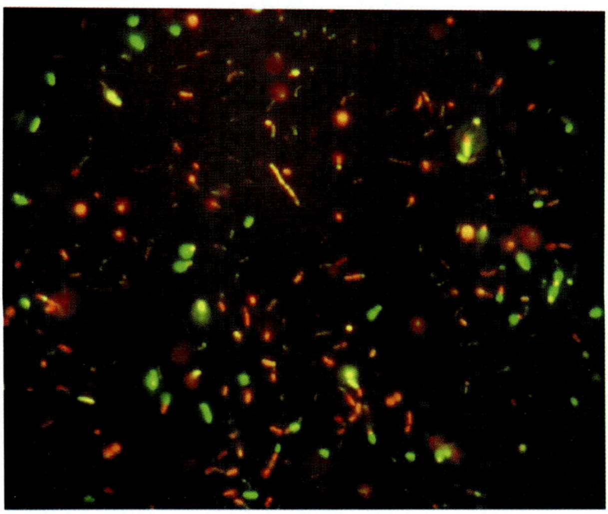

FIGURE 5.4a Images of strain FTR1^{T} with BacLite live/dead stain (Photo by Prof. A.K. Bej, University of Alabama at Birmingham, Birmingham, AL.)

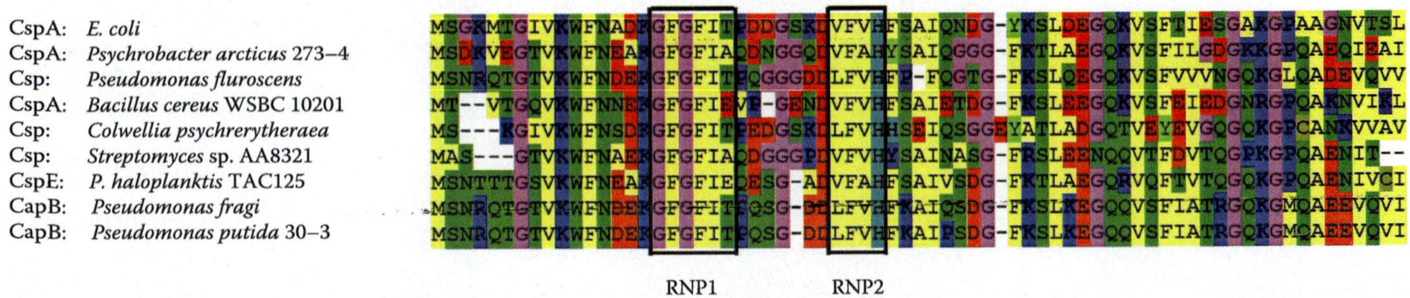

FIGURE 6.1 Multiple alignment of cold-shock proteins (Csps) and cold acclimation protein (Caps) from Antarctica with CspA of *E. coli* with the help of CLUSTAL X (http://www.clustal.org/). The RNP1/RNP2 motifs shown in square boxes are conserved domains in Csps and Caps isolated from various bacterial species.

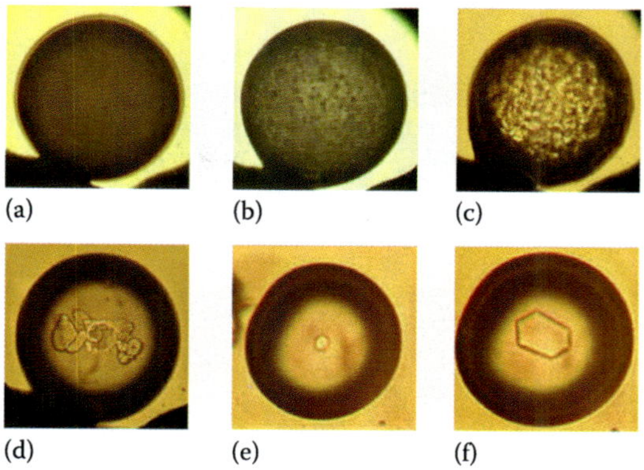

FIGURE 6.3 A sequence of cryomicroscopic images (a–f) in a Clifton nanoliter osmometer monitoring the growth of a single ice crystal when temperature from −40°C is elevated to just below the freezing point. The effect of the antifreeze function of the crude extract of the total cellular protein from a biodegradative bacterium *P.* sp. 30/3 isolated from oil-contaminated soil from Antarctica is shown by the growth along the *c*-axis of a stable hexagonal ice crystal.

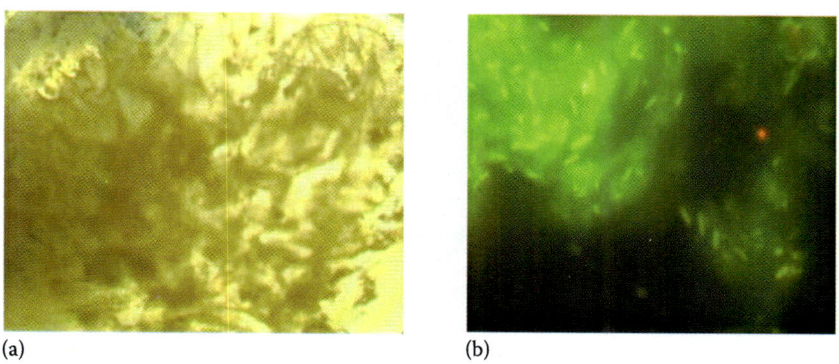

FIGURE 6.4 Extracellular polymeric substance (EPS) produced by *Janthinobacterium* sp. strain Ant 5-2 isolated from a Proglacial lake P9 located at the Schirmacher Oasis of East Antarctic Dronning Maud Land: (a) "slimy" EPS secreted by *Janthinobacterium* sp. Ant 5-2 in a stationary culture after 7 days of incubation at 15°C; (b) microscopic examination of the bacillus *Janthinobacterium* sp. Ant 5-2 aggregated flocs of cells stained with Live/Dead *Bac*Light fluorescent staining kit (Molecular Probes) embedded in thick EPS moiety.

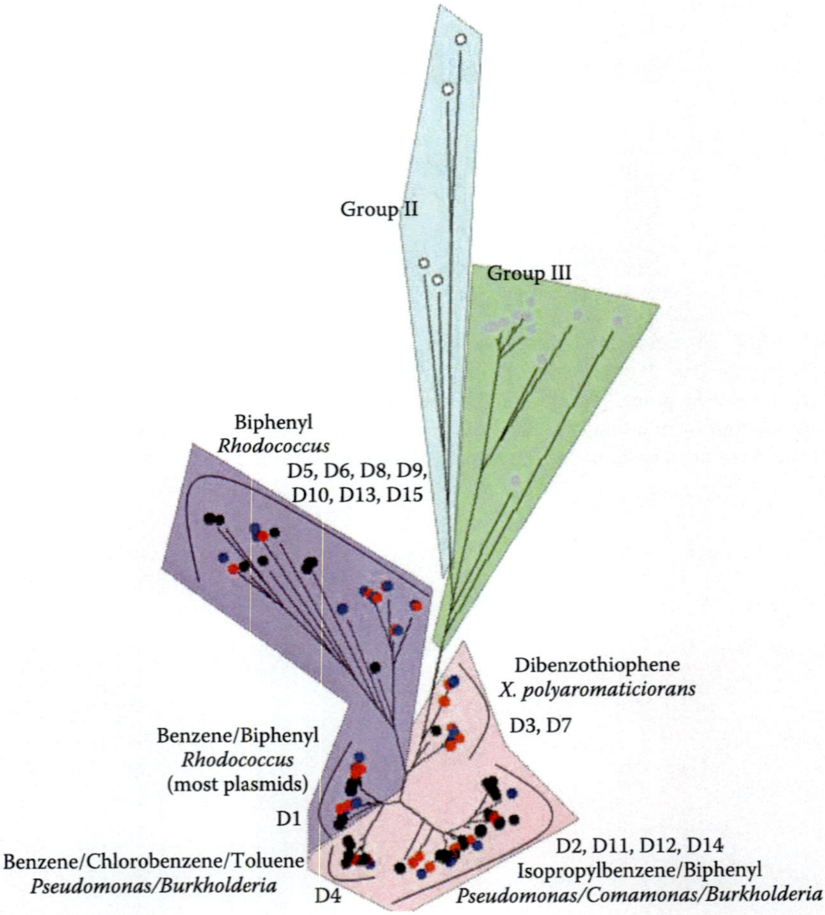

FIGURE 15.5 ARHDs radial tree showing the clustering according to the substrates oxidized by the members of Group IV dioxygenases. Antarctic clone sequences were related to five distinct substrate groups previously described in Gram Positive (purple color) bacteria and Gram Negative (pink color) bacteria such as *Rhodococcus*, *Pseudomonas*, *Comamonas*, *Burkholderia*, and *Xanthobacter*. Group I was not included in this tree because of the low homology of the sequences related to the other three Groups of dioxygenases.

12 Tolerance of Antarctic Soil Fungi to Hydrocarbons and Their Potential Role in Soil Bioremediation

Kevin A. Hughes and Paul Bridge

CONTENTS

12.1 ANTARCTIC ENVIRONMENT

Antarctica, which is the area south of 60°S, is the most pristine and least populated continent on Earth. It is separated from the other continents by the Southern Ocean, which, at its narrowest between the Antarctic Peninsula and South America, is 1000 km wide (Hughes et al., 2006). When Antarctica broke away from South America ~30 million years ago, the formation of the circumpolar Southern Ocean and atmospheric current initiated a cooling process that has resulted in the frozen continent found today. Permanent ice sheets cover 99.6% of the land surface and the remaining ice-free land is composed of rocky nunataks that rise through the surrounding ice sheet and coastal areas, including the expansive continental Dry Valley region. In general, as latitude increases, mean annual temperature and therefore liquid water availability decreases, ultraviolet (UV; 280–400 nm) radiation flux increases and seasonal variations in daylight duration become more pronounced. Stratospheric ozone depletion during the austral spring may also increase the levels of incident UV radiation, particularly at high latitudes.

Antarctica can be divided into two terrestrial biogeographical zones: (1) the maritime Antarctic (which largely encompasses the western Antarctic Peninsula down to ~72°S and the South Shetland, South Sandwich, and South Orkney Islands) and (2) the much larger continental Antarctic region (which includes all of the vast East Antarctic, most of the West Antarctic and the eastern side of the Antarctic Peninsula). The fauna of the terrestrial maritime Antarctic is limited to two higher invertebrates (the Diptera *Belgica antarctica* and *Parochlus steinenii*) and around 130 species of hexapods, mites, nematodes, and tardigrades. The flora comprises of many species of moss and lichens but only two vascular plants (*Colobanthus quitensis* [Antarctic pearlwort] and *Deschampsia antarctica* [Antarctic hair grass]). Continental Antarctica has a reduced diversity with probably fewer than 100 species of lower invertebrates and nematodes, and is dominated by simple organisms that can withstand the extreme environmental conditions, such as lichens (fungal-eukaryotic algal/cyanobacterial symbionts) and microorganisms including bacteria, fungi, protozoa, eukaryotic algae, and cyanobacteria. In the maritime Antarctic thick peat banks can be found (e.g., Signy Island, South Orkney Islands), but with increasing latitude, and certainly within continental Antarctica, mineral soils, which are characterized by low organic content, poor water-holding capacity, and low nutrient content, dominate. The sub-Antarctic islands that encircle Antarctica make up an additional biogeographic zone that experiences warmer and wetter conditions than Antarctica and as a result has much richer biodiversity, including many higher plants and invertebrates, and soils with more organic material.

12.2 ANTARCTIC FUNGI

The native Antarctic microbiota is poorly understood in comparison to other geographic regions. Since the first records in the mid-nineteenth century, more than 750 currently accepted species of fungi have been reported from the broader Antarctic region, with around 500 of these being reported from either the maritime or continental regions (Bridge et al., 2008a,b). Surprisingly, relatively few of these are species that are considered true psychrophiles, and the majority are normally mesophilic species that have some degree of cold tolerance. The larger fungi have been reported to be "remarkably similar to that of North Temperate-Arctic regions" (Pegler et al., 1980), and the same is largely true of the microfungi. Onofri et al. (2007) suggested that some 22 species described from the region may be endemic, but some of these have since been isolated from elsewhere in the world including food cold-storage facilities, and arctic and alpine environments (e.g., Margesin et al., 2005; Frisvad et al., 2006). It is difficult to compare the limited information on Antarctic soil fungi with results from elsewhere due to a number of constraints. The figure of 750+ accepted species is derived from the 1000+ fungal species names that have been reported over 150 years. However, there has been relatively little specific targeted fungal collection and isolation during that period, and this list is probably derived from around 100 individual field trips. In contrast, the best-recorded site for fungi in the United Kingdom is probably Esher Common, where some 3300 fungal species have been recovered from 380 ha due to a continuous extensive campaign by mycologists over some 30 years involving at least several hundred collection activities (see Hawksworth, 2001; Spooner and Roberts, 2005). Isolation methods and approaches also seem to affect the assessment of Antarctic microfungal diversity. In culture-based studies the most isolated fungi are filamentous ascomycetes and these often account for >80% of all fungi recovered (e.g., Baublis et al., 1991). Molecular diversity studies, however, have suggested that there may be an equal number of basidiomycete and zoosporic fungi, although these are much rarer in culture-based studies (Lawley et al., 2004). In one of the few studies to use both culture and molecular methods, Arenz et al. (2006) identified 71 fungal taxa from a series of sites, but of these only 16 were detected by both methodologies. Another constraint in determining fungal diversity is the lack of homogeneity between often-isolated sampling sites. Studies on the Antarctic Dry Valleys have reported soil moisture contents varying from 0.3% to 4.9% from nearby sites (Fell et al., 2006), while differences of 2.5%–8% moisture content can occur over a few hundred yards at some maritime Antarctic sites (Newsham, personal communication). Given these constraints, it would seem unlikely that there is an extensive specialist Antarctic fungal population, and the species present in the region appear to be largely a restricted subset of those found in similar environments elsewhere.

12.3 HUMAN ACTIVITY

Since before the onset of International Geophysical Year of 1957–1958, humans have had a continuous presence in Antarctica and, in that time, there has been a heavy reliance on hydrocarbons to sustain human existence in the extreme polar environment. Hydrocarbons, in the form of fuel oils, are used by national scientific operators (such as, the British Antarctic Survey) for transportation of personnel and cargo (via ships, aircraft, and overland vehicles) and operation of stations and infrastructure for scientific research (Platt et al., 1981). In recent years, tourism has increased in Antarctica to some 33,000 tourists landing in the region during 2007–2008 from around 60 different vessels (International Association of Antarctica Tour Operators,

2008), with tourist operators relying on fuels for both ship and land-based tourism. Within Antarctica, inappropriate fuel oil transport and storage and/or human error have resulted in spills on land (Kerry, 1993), ice shelves (Wilkness, 1990), and at sea (Janiot et al., 2003). When oil spills occur, whether in the form of diesel (marine diesel oil), paraffin, avtur, petrol, engine oils, or lubricants, they can be highly visible against the backdrop of the pristine and often barren Antarctic terrestrial environment. Some spills that occurred several decades ago are still visible because nonbiological weathering can be slow and the indigenous microbiota have failed to sufficiently degrade the spilled hydrocarbons due to factors such as a lack of nutrients, the short periods each year when soil temperatures are high enough to permit biological activity, and low availability of liquid water (Kerry, 1993; Aislabie et al., 2004).

Under the Protocol on Environmental Protection to the Antarctic Treaty (1991) operators are obliged to clean up contaminated sites (Annex III to the Protocol on Environmental Protection to the Antarctic Treaty; Antarctic Treaty Consultative Parties, 1991). Removing contaminated soil is prohibitively expensive, so there has been considerable interest in finding *in situ* methods of cleaning up impacted sites using bioremediation. To complicate the situation to some extent, the Antarctic Treaty also states that the introduction of microbial species, including microorganisms that are not native to Antarctica is prohibited (Annex II to the Protocol on Environmental Protection to the Antarctic Treaty, 1991). Therefore, well-characterized hydrocarbon-degrading bacteria and fungi suitable for bioaugmentation purposes that were isolated outside Antarctica, and in particular other cold environments, probably cannot be used, although the difficulties in determining what constitutes a nonnative fungus further complicates this issue.

12.4 NATURAL SOURCES OF HYDROCARBONS IN THE ANTARCTIC

So far, natural seeps of hydrocarbons have not been reported from terrestrial Antarctic environments, although some investigation has been made into some Antarctic sedimentary basins (e.g., the Larsen basin) for hydrocarbon potential (Macdonald and Butterworth, 1990). Nevertheless, Antarctic fungi will have come across hydrocarbons associated with plants, cyanobacteria, algae, invertebrates, and meteorites (Crespo et al., 2000; Aislabie et al., 2004; Bridge et al., 2005). Fungi, in general, have a wide range of extracellular enzyme activities (see Paterson and Bridge, 1994) and those found associated with bird feathers may also be able to degrade or utilize the associated oils. This may account for the report of the normally lignicolous *Pycnoporus coccineus* from Antarctic penguin cadavers (Baublis et al., 1991).

12.5 FUNGAL DEGRADATION OF HYDROCARBONS

Fungi have different means of degrading hydrocarbons depending upon the environmental conditions, strain type, and hydrocarbon chemistry.

12.5.1 Aliphatic Hydrocarbons

Aliphatic hydrocarbons often make up the largest proportion of hydrocarbons released during a spill, e.g., petrol and diesel consist of up to 90% C_{14}–C_{20} aliphatics, the majority of which are alkanes (Stroud et al., 2007). The physiochemical properties of aliphatics, such as insolubility and low volatility, particularly at chain lengths >C_{14}, mean that, once spilled, they may not be readily removed from soils by physical processes such as evaporation or leaching. Nevertheless, both Antarctic bacteria and fungi may produce biosurfactants, which can aid in the degradation of hydrocarbons by increasing the concentration of aliphatic hydrocarbons in the aqueous phase available for biodegradation (Yakimov et al., 1999; Kitamoto et al., 2001; Vasileva-Tonkova and Gesheva, 2004). Fungi have long been known to degrade the saturated aliphatic fraction of oils, with different species able to degrade different chain lengths (Markovetz et al., 1968). April et al. (2000) isolated filamentous fungi, including some from northern Canada, that could oxidize aliphatic fractions of crude oil, but there was no evidence of aromatic hydrocarbon degradation. Most studies on the fungal degradation of aliphatic compounds have been undertaken for alkanes and there is little information available for alkenes. However, the yeast *Candida lipolytica* has been shown to degrade hexadec-1-ene and heptadec-1-ene to produce a variety of products (Klug and Markovetz, 1968).

12.5.2 BTEX and Monoaromatic Compounds

Monoaromatic compounds, including the BTEX hydrocarbons (benzene, toluene, ethylbenzene, and *m*-, *o*-, and *p*-xylene), may enter the Antarctic environment through petroleum or fuel oil spills and leakages. Many metabolic routes exist to convert benzenoid compounds into readily assimilable molecules. Monoaromatic hydrocarbons can be degraded by enzymes commonly involved in both lignin degradation and in methylation. However, aerobic degradation by both intracellular and extracellular enzymes is involved using a wide variety of metabolic routes. Before ring cleavage can occur, the ring must have two hydroxyl groups in place, which may involve the removal or modification of existing groups. The hydroxylation is performed by monooxygenases or enzymes associated with lignin degradation. Dioxygenases cleave the modified aromatic ring by the insertion of a molecule of oxygen, after which the fission product is further degraded by passing through standard well-characterized catabolic pathways (Buswell, 2001). Some fungi can grow directly on one or more of the BTEX compounds, examples include some *Exophiala* isolates that utilize ethylbenzene and *Cladosporium sphaerospermum* that can grow on toluene. Other compounds in this group that are known to be metabolized by fungal oxidation or hydroxylation include some propyl- and dialkyl-benzenes, xylene isomers, and styrene (see Sutherland, 2004, for a review).

12.5.3 Polyaromatic Hydrocarbons (PAHs)

The major natural sources of PAHs are crude oil and coal tar, but they can be anthropogenically released into the environment by petroleum hydrocarbon spills, incomplete combustion of fossil fuels from engine exhausts, and inadequate incineration of waste. PAHs can be accumulated by some soil invertebrates, but are not generally taken up by plants (Wilcke, 2000). PAHs show little solubility in water and have a strong binding affinity with organic material in soils, which may decrease their availability to potential degrading microorganisms. PAHs are toxic to mammalian cells by a number of processes that generally result in damage to cellular DNA. PAH metabolites and degradation products may have a direct effect on the cell function by binding directly to DNA, RNA, and proteins (reviewed by Harvey, 1992; Harvey et al., 1999; Luch, 2005). As eukaryotes, fungi have many similar metabolic pathways to mammalian cells and as a result are likely to be susceptible to DNA damage by PAHs.

12.5.3.1 Oxidation of PAHs by Lignin-Degrading Fungi

The white-rot fungi are normally wood degrading fungi that utilize lignin and degrade it to CO_2, largely through the action of extracellular peroxidase enzymes (Holzbaur et al. 1991). Some fungi may be able to degrade PAHs when degrading other substrates (Marquez-Rocha et al., 2000), while white-rot fungi can degrade a wide range of PAHs and other pollutants due to their use of nonspecific radical oxidation (Aust, 1995). White-rot fungi produce a range of oxidative enzymes including catalases, laccases (polyphenol oxidases), and lignin peroxidases that act together in the degradation of lignin (Thurston, 1994; Carlisle et al., 2001). These enzymes act to form short-lived free radicals that can oxidatively degrade a wide range of environmental pollutants including those found in petroleum hydrocarbons (Yateem et al., 1998). Lignin peroxidase (LiP) and manganese-dependent peroxidase (MnP) are both heme-containing extracellular enzymes that require hydrogen peroxide, produced by the fungus, to complete the catalytic cycle. Lignin peroxidases have a pH optimum of around 3, as the environment within wood is normally around 4 and decreases further due to oxalic acid production by the fungus. The enzymes react with hydrogen peroxide to release highly reactive oxygen-containing free radicals that break the covalent bonds in hydrocarbon chains (Carlisle et al., 2001). The combination of LiP, MnP, and laccases secreted by fungi depends upon the culture conditions and substrate. Unusually for carbon cycle enzymes, the production of these peroxidases are largely unaffected by the presence of utilizable carbon, but their production may be stimulated under nitrogen-limited conditions (Fenn and Kirk, 1981). This may not be a major problem in the Antarctic where the largely mineral soils have a very low organic content, and generally low nitrogen levels (see Bridge and Newsham, 2009).

12.5.3.2 Oxidation of PAHs by Non-Lignin-Degrading Fungi

The first step in the degradation of PAHs by non-lignin-degrading fungi is an oxidation of aromatic rings by a cytochrome P450 monooxygenase. Subsequent breakdown can involve hydroxylases and conjugation to produce a wide range of breakdown products (see Sutherland, 2004). Some species of *Penicillium* are able to degrade the four-ringed PAH pyrene (Saraswathy and Hallberg, 2002). Many Antarctic soils have properties that permit good levels of aeration and oxygenation, i.e., little organic carbon, low water retention, and coarse texture. Conversely, the aeration of some polluted subsurface soils may be reduced when spilled hydrocarbons become immobilized above the permafrost layer (Walworth et al., 2008). Such soils may become oxygen-limited or depleted, and molecular oxygen may not be available for oxygenases to operate, so limiting catabolism of different hydrocarbons of whatever chemistry. Oxygen is not always essential for hydrocarbon degradation and, mono-aromatic and PAHs may be metabolized by microbial consortia by sulfation, methylation, or conjugation with some monosaccharides or glucuronic acid (Zeyer et al., 1986; Mihelcic and Luthy, 1988; Cerniglia and Sutherland, 2001). In general terms, although PAHs are readily transformed and degraded by fungi, they are only rarely used as growth substrates, one notable exception being the growth on phenanthrene by *Rhodotorula glutinis* (Romero et al., 1998; Sutherland, 2004).

12.6 OCCURRENCE OF HYDROCARBON-DEGRADING FUNGI IN ANTARCTICA

There have been some discussions on the number of fungi that are considered to be truly endemic to the Antarctic (de Hoog et al., 2005; Onofri et al., 2007; Bridge et al., 2008b). Onofri et al. (2007) listed some 22 species of microfungi that they considered endemic to continental Antarctica, although some of these, such as *Penicillium antarcticum* have subsequently been found to occur in other low temperature environments. Given that around 1000 fungal species have been described from the broad Antarctic area, the 20 or so considered endemic would be around 2% of the diversity. How the other 98% have arrived in the region has also been the subject of some speculation. Chalmers et al. (1996) have shown that air movements could transport fungi from South America to the Antarctic, and that the meteorological conditions required would occur some three times in every 2 years (Chalmers et al., 1996; Marshall, 1996). Airborne distribution would be unlikely for many other species, such as *Pythium* spp. that are normally waterborne taxa, and de Hoog et al. (2005) have demonstrated that some species of *Thelebolus* found in Antarctic lakes and deposits are also associated with birds, and such vectors may be involved in the routine introduction of fungi. Some studies have suggested that fungi may be introduced by human activity, as diversity and numbers are higher in areas associated with Antarctic bases (e.g., Bial Lasiewicz and Czarnecki 1999), but there is probably insufficient data to determine if this is due to the introduction of fungi or the environmental changes (heat, substrate availability, etc.) associated with the bases.

Table 12.1 lists genera and species of hydrocarbon-degrading fungi that have been isolated from the Antarctic region.

12.6.1 Basidiomycete Fungi and Yeasts

Lignin-degrading fungi (generally basidiomycetes) are known to produce enzymes that degrade aromatic hydrocarbons. The almost ubiquitous distribution of lignin-degrading fungi in other parts of the world, including cold regions, may not extend, to any significant degree, to Antarctica as the climatic conditions preclude the growth of woody plants, and therefore the capacity to degrade lignin would confer little

TABLE 12.1
Nonlignolytic Fungi Capable of Polycyclic Aromatic Hydrocarbon Metabolism and Genera of Hydrocarbon-Degrading Fungi That Have Been Reported from the Antarctica and the Sub-Antarctic Islands

Class	Name	Details	Antarctic Biogeographic Zone		
			Sub-Antarctic Islands	Maritime Antarctica	Continental Antarctica
Zygomycetes	*Mortierella ramanniana* (Syn. *Mucor ramannianus*)[a,b]. Current name *Umbelopsis ramanniana*	Soil			+
	Mucor racemosus[a]	Soil, ice core		+	+
	Rhizopus arrhizus[a] Current name *Rhizopus oryzae*	Environmental isolations		+	+
	Syncephalastrum racemosum[a]	Air		+	
Filamentous Basidiomycetes	*Rhizoctonia solani*[a]	Single report from soil		+	
Basidiomycetous yeasts	*Candida krusei*[a]	Current name *Issatchenkia orientalis*. Soil			+
	Cryptococcus albidus[a]	Numerous reports. Soil and moss	+	+	+
	R. glutinis[a,b]	Soil		+	+
	Rhodotorula minuta[a,b]	Numerous reports. Soil, moss, air, and freshwater	+	+	+
	Trichosporon penicillatum[a]	Eight species of *Trichosporon* reported. Soil		+	+
	Sporobolomyces[b]	Four species including one synonym of *P. antarctica*. Soil and water		+	+
Filamentous Ascomycetes	*Neurospora crassa*[a]	As *Neurospora* sp. soil			+
	Sporormiella australis[a]	As *Sporormiella* sp. air		+	

Acremonium[b]	Twelve species reported from soil, lichens, *D. antarctica*, water foam, sea water, mud, soil, and midge gut	+	+	+
Aureobasidium[b]	Most collections identified as *A. pullulans*. Soil, air, and *Campylopus pyriformis*	+	+	+
Cochliobolus[b]	As *C. heliconiae*. Soil			+
Cylindrocarpon[b]	Lichens and moss	+		
Geotrichum[b]	Three species reported. Wide range of moss and lichen habitats, water foam, sea water, soil, air, and wood	+	+	+
Gliocladium[b]	As *G. roseum*. Soil	+		
Paecilomyces[b]	Five species reported. Soil, mosses, and Collembola	+	+	+
Tolypocladium[b]	As *T. cylindrosporum* and *T. nubicola*. Soil and moss		+	
Aspergillus niger[a,b]	Soil, freshwater, and air		+	+
Aspergillus versicolor[a]	Soil, freshwater, and air. Also isolated from within ice-cores	+	+	+
Beauveria alba[a,b]	*B. bassiana* only species reported in region. Soil and lake sediment	+	+	
Botrytis cinerea[a,b]	Air and soil	+	+	
Chrysosporium pannorum[a,b]	Numerous reports from wide range of substrates including soil, air, freshwater, lake mats and sediments, lichen, mosses, introduced wood, and hydrocarbon-contaminated ground	+	+	+
Cladosporium herbarum[a,b]	Air, soil, lichens, mosses, and from buildings	+	+	+
Curvularia lunata[a]	*Curvularia* sp. reported from microbial mats in freshwater lakes		+	

(*continued*)

TABLE 12.1 (continued)
Nonlignolytic Fungi Capable of Polycyclic Aromatic Hydrocarbon Metabolism and Genera of Hydrocarbon-Degrading Fungi That Have Been Reported from the Antarctica and the Sub-Antarctic Islands

			Antarctic Biogeographic Zone		
Class	Name	Details	Sub-Antarctic Islands	Maritime Antarctica	Continental Antarctica
	Curvularia tuberculata[a]	As above		+	
	Embellisia annulata[a]	*Embellisia* sp. and *E. telluster*. Soil and moss		+	+
	Fusarium subglutinans[a,b]	*Fusarium* sp. and six species reported. Soil, freshwater, *Deschampsia*, and imported wood	+	+	+
	Monosporium olivaceum[a]	*Monosporium* sp. air	+		
	Penicillium chrysogenum[a,b] (and as the later synonym *P. notatum*[a,b])	Numerous reports. Soil air, microbial mats, freshwater, and imported foods. Also found in ice cores	+	+	+
	Penicillium simplicissimum[a,b] (and as the later synonym *P. janthinellum*[a,b])	Moss, air, and ice cores	+	+	
	Pestalotia palmarum[a]	Only report is of *Pestalotia* sp. moss associated soil			+
	Lecythophora hoffmannii[a]	*Lecythophora lignicola* from soil			+
	Scopulariopsis brumptii[a]	*Scopulariopsis* sp. and *S. brevicaulis* (syn. *Microascus brevicaulis*). Soil and air		+	
	Scytalidium lignolitica[a]	*Scytalidium* sp. and *S. thermophilum* soil and lichen		+	+
	Sporothrix cyanescens[a]	*Sporothrix* sp. and *S. ramosissima*. Soil and contaminated soil			+

	Trichoderma harazianum[a,b]	Soil		+	+
	Trichoderma viride[a,b]	Soil		+	+
	Verticillium lecanii[a,b] (current name *Lecanicillium lecanii*)	Rocks and mosses		+	+
	Humicola[b]	*H. alopallonella* and *H. marvinii*. Soil and wood baits in sea water	+	+	
	Scolecobasidium[b]	*S. salinum*. Soil, mosses, and petroleum contaminated soil		+	+
	P. herbarum[a,b]	Soil and mosses		+	+
Ascomycetous yeasts	*Saccharomyces cerevisiae*[a,b]	Soil, *Schistidium antarctici* and *Hennediella heimii*		+	+
	Debaryomyces[b]	As *Debaryomyces hansenii* or synonyms. Soil		+	+

Sources: Data from Species Lists, Cerniglia, C.E. and Sutherland, J.B., Bioremediation of polycyclic aromatic hydrocarbons by ligninolytic and non-ligninolytic fungi, in *Fungi in Bioremediation*, Gadd, G.M. (ed), Cambridge University Press, Cambridge, U.K., 2001, 136–187; Bossert, I. and Bartha, R., The fate of petroleum in soil ecosystems, in *Petroleum Microbiology*, Atlas, R.M. (ed), MacMillan Publishing Co., New York, 1984, 435–473; Bridge, P.D. et al., *Mycotaxon*, 106, 485, 2008.

[a] Cerniglia and Sutherland (2001).

[b] Bossert and Bartha (1984).

advantage. Trees did persist in Antarctica until the Eocene, and fossil remains can still be found with evidence of ancient wood decaying fungal activity (Cantrill and Poole, 2005). Today, lignin-availability is limited to wood debris washed ashore by ocean currents (Barnes et al., 2004), or timber imported to Antarctica for construction purposes during the last 100 years. Typical white-rot fungi, such as *Phanerochaete chrysosporium*, have not been recorded from the true Antarctic, although *Trametes* species have been reported from the sub-Antarctic island of South Georgia. Within the maritime and continental Antarctic, brown rot or white rot wood-decaying basidiomycetes may be rare due to their sensitivity to exceedingly dry environmental conditions common at high latitudes (Blanchette et al., 2004). There are few records of specialist lignin-degrading fungi from the Antarctic including the single record of a filamentous specialist wood-degrader *Hohenbuehelia* sp. from dry continental soil, a *Trametes* sp. from imported wood on the maritime Antarctic Signy Island and the wood rotting *P. coccineus* from a mummified penguin cadaver (Pegler et al., 1980; Baublis et al., 1991; Fell et al., 2006).

Where found, basidiomycetous yeasts are generally more common than filamentous basidiomycetes. The yeasts reported from the Antarctic are largely species that are cosmopolitan, or have also been reported from other cold environments, although most species are psychrotolerant rather than psychrophilic (Ludley and Robinson, 2008). Atlas et al. (1978) isolated *Cryptococcus*, *Rhodotorula*, *Candida*, *Sporobolomyces*, *Torulopsis*, and *Aureobasidium* spp. from continental Antarctica, and in some soils, yeasts were found in numbers equal to the total enumerable heterotrophic microbial population. Pigmentation may reduce the susceptibility of yeasts and filamentous fungi to solar UV damage, and enhance their ability to survive in the Antarctic terrestrial environment (Butler and Day, 1998; Hughes et al., 2003). *Cryptococcus* spp. that have been isolated frequently from Victoria Land mosses (Tosi et al., 2002) and *Trichosporon* spp. that have been recovered in the McMurdo Dry Valleys (Fell et al., 2006). In contrast, basidiomycetes are relatively common in Arctic and sub-Antarctic soils, probably because many higher plants and cryptogamic species are found within these biogeographic zones and climatic conditions may not be as extreme as at equivalent latitude in the Antarctic (Pegler et al., 1980; Smith, 1994; Ludley and Robinson, 2008).

From Table 12.1 it is clear that there are many yeasts and soil fungi in Antarctica belonging to genera with isolates known to degrade hydrocarbons, including PAHs. These fungi may be an unexploited source of cold-tolerant hydrocarbon-degrading enzymes. Yeasts are common components of the Antarctic soil microbial community and may play an important role in facilitating hydrocarbon degradation for other microbial groups and undertaking degradation themselves. Much research has focused on the basidiomycetous yeast-like fungus *Pseudozyma antarctica* (previously known as *Candida antarctica*) that could convert C_{12}–C_{18} alkanes into glycolipid biosurfactants (e.g., mannosylerythritol lipids), which, in turn, could enhance the emulsification and degradation of kerosene *in vitro* (Kitamoto et al., 2001; Hua et al., 2004). Cold-tolerant alpine yeasts were shown to degrade monoaromatic hydrocarbons (Bergauer et al., 2005); strains included *Cryptococcus*, *Rhodosporidium*, *Rhodotorula*, and *Sporobolomyces* spp.—all of which are present in Antarctic soils. The cold-tolerant yeast *Yarrowia lipolytica* (which has been isolated from Antarctic microbial mats) proved effective at degrading diesel oil in soils at low temperatures (Margesin and Schinner, 1997). Indeed, the capacity of yeasts to degrade hydrocarbons may be greater than that of bacteria, for example, of 61 cold-adapted bacteria and 28 yeast isolates, 79% of yeast could degrade hexadecane (cf. 6% bacteria), up to a third could use phenol, phenanthrene, or anthracene for growth (cf. 13% of bacteria), and optimal temperatures for phenol degradation were generally lower than for bacteria (Margesin et al., 2003).

12.6.2 Ascomycete Fungi

The majority of the fungi listed in Table 12.1 are filamentous ascomycetes. Many of these, such as *Aspergillus* sp., *Penicillium* sp., and *Trichoderma* sp., are cosmopolitan fungi that occur on a wide range of substrates worldwide. In addition to the species and genera included in Table 12.1, a number of other fungi that have been reported from the Antarctic are known to be able to utilize aliphatic compounds. These include the ascomycete yeast *Y. lipolytica* that can utilize both alkanes and alkenes and normally invertebrate associated *Beauveria bassiana* that has been grown on a range of alkane substrates (Sutherland, 2004), both of which have been isolated from microbial mats in Antarctic lakes (Göttlich et al., 2003).

In the last 100 years, man has colonized Antarctica and imported wood for the construction of huts and larger buildings. Studies of the decomposition of wood in huts erected in the "Heroic Age" by Scott and Shackleton have shown that Ascomycete soft-rot *Cladospora* spp. that are probably endemic to Antarctica, are largely responsible for the recorded decay (Blanchette et al., 2004). Ascomycetes do not have the PAH-degrading enzymes common in basidiomycetes. Arenz et al. (2006) suggested that fungi found associated with wood and artefacts in continental Antarctica are likely to be indigenous and, as opportunists, have taken advantage of the anthropogenic introduction of organic materials into their environment. In contrast, nonindigenous fungi imported on non-treated wood to research station in the climatically less extreme northern Antarctic Peninsula, may be able to colonize soil ecosystems, where they may play a part in the degradation of any hydrocarbons spills associated with the research stations (Hughes and Stallwood, 2006). A potentially nonnative (alien) hydrocarbon-degrading fungus may already have been introduced to Antarctica; Line (1988) and Kerry (1990) found *Hormoconis resinae* in Antarctic soils, but only those that had experienced significant human impact including contamination with oil. *H. resinae* is well known in the aviation industry for contaminating fuel and blocking fuel lines through the rapid conversion of alkanes into fungal filamentous biomass. It is likely that this fungus was introduced to Antarctica in contaminated fuel and then colonized impacted soils.

12.7 TOLERANCE AND UTILIZATION OF HYDROCARBONS BY ANTARCTIC SOIL FUNGI

12.7.1 *In Vitro* Experiments

Few studies have investigated the tolerance of resident Antarctic fungi to hydrocarbons, or their ability to utilize them as a carbon source. In one recent study (based around Rothera Research Station, Adelaide Island, Antarctic Peninsula) soils were collected from (a) relatively pristine sites and (b) areas that had experienced oil contamination, and attempts were made to culture microorganisms in the presence of seven separate aromatic hydrocarbons (toluene, xylene, naphthalene, phenol, biphenyl, benzoic acid, and hydroxybenzoic acid), two aliphatic hydrocarbons (dodecane and hexadecane), and marine gas oil (MGO; a complex mixture of aliphatic, aromatic hydrocarbons, and fuel additive) (Hughes et al., 2007). Generally, soil bacteria were better able to grow in the presence of a wider range of hydrocarbons than fungi, with only toluene preventing bacterial growth in all soils tested. Fungi were not readily isolated from soils in the presence of most aromatic hydrocarbons compared to the aliphatic hydrocarbons and MGO. Six filamentous fungi, namely, *Mollisia* sp., *Penicillium commune*, *Mortierella* sp., *Trichoderma koningii*, *Trichoderma* sp., and *Phoma herbarum* isolated from Rothera Point soil were further examined for their ability to grow on a minimal media agar-containing glucose in the presence of the hydrocarbons. All the aromatic hydrocarbons reduced the growth rates of all six fungi, while aliphatics generally either increased or had no significant effect on the growth rate compared to the no hydrocarbons control (Figure 12.1). When glucose was removed from the minimal media, leaving no readily utilizable carbon source, the addition of aromatic hydrocarbons usually reduced hyphal extension rate, while the alkanes had no significant effect. MGO generally inhibited growth, possibly because the complex mixture of aromatic components makes it toxic to some fungal strains even though the alkane component, which makes up the majority of the oil, may not be toxic to these organisms. When fungal isolates were tested for their ability to assimilate hydrocarbons as a sole carbon source, *Mortierella* sp. was able to utilize dodecane. *Mortierella* sp. biomass was greater when additional nutrients were supplied in the form of yeast extract, indicating that adequate nutrient availability may be necessary for rapid hydrocarbon degradation in the field (as shown for bacteria by Stallwood et al. (2005) and Kerry (1993)). These experiments were performed *in vitro*, with high or saturated concentrations of hydrocarbons, so the

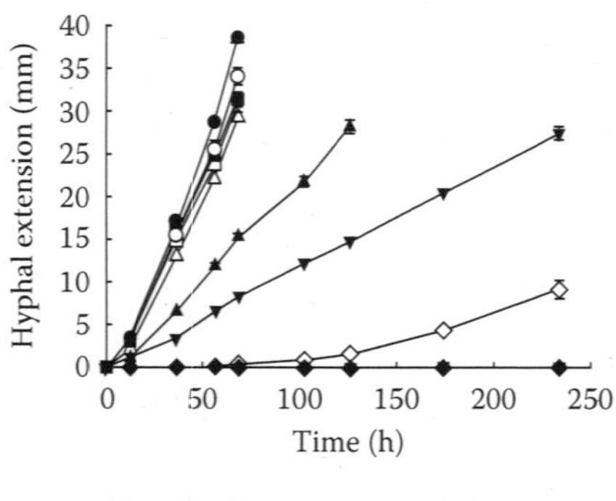

FIGURE 12.1 Growth rates of *Trichoderma* sp. on minimal media containing 1% glucose in addition to marine gas oil (■), hydroxybenzoic acid (❑), dodecane (●), hexadecane (❍), *m*- and *p*-xylenes (◆), naphthalene (◊), benzoic acid (▼), toluene (∇), biphenyl (▲), or a no hydrocarbon control (Δ). (Taken from Hughes, K.A. et al., *Sci. Total Environ.*, 372, 539, 2007. With permission.)

degradation and utilization of a wider range of hydrocarbons may have been possible at lower concentrations.

12.7.2 Potential *In Vivo* Impacts

Spilled hydrocarbons may have two opposing effects on indigenous microbial populations. (1) Hydrocarbons may increase soil microbial biomass by providing an additional carbon source for microbes, with the appropriate physiological and biochemical capabilities, to undertake aliphatic and/or aromatic compound degradation and assimilation. (2) Hydrocarbons (especially BTEX and PAHs) may have toxic effects on microbiota, thereby reducing the opportunity for potential hydrocarbon-degraders to utilize any hydrocarbonoclastic capabilities, and potentially reducing soil biodiversity by killing hydrocarbon-sensitive microbial strains. Which effect dominates will depend on many factors and interactions; however, the data available to make informed predictions are scanty. For example, we have relatively little information on true levels of fungal biodiversity across Antarctic soils. Nor do we know how the functional capabilities of that biodiversity fluctuates under different conditions of (1) hydration, (2) natural nutrient input, (3) oxygen availability, (4) low or rapidly fluctuating temperatures, (5) pH, (6) salinity, (7) soil organic or humus content, or (8) environmental variability throughout the polar year. The addition of hydrocarbons to this already complex system would cause changes that we are not currently able to predict with accuracy. Shifts in soil community composition may occur due to physiological and biochemical characteristics of individual component species. Kerry (1990) found that *Geomyces pannorum* and *Thelebolus microsporus* were less common in contaminated soils compared to less impacted sites, while Leahy and Colwell (1990) reported that *Trichoderma* and *Mortierella* spp. were the most common isolates in hydrocarbon-polluted soils, with *Penicillium* spp. also frequently isolated. Four of the six species isolated by Hughes et al. (2007) from impacted soil belonged to these genera, which may point to a degree of ubiquity in fungal species found associated with hydrocarbon-contaminated soils in different regions. Aislabie et al. (2001) reported that culturable fungi increased in hydrocarbon-contaminated Antarctic soils compared to the pristine control soils, and the dominant species changed from *Chysosporium* sp. in control soils to *Phialophora* sp. in hydrocarbon-contaminated soils. The declining levels of fungal biodiversity found as climatic condition become more severe (and typically with increasing latitude [Lawley et al., 2004]) mean that for some inland continental sites there may not be adequate enzymatic capacity to degrade all the complex mixture of hydrocarbons found in commonly used Antarctic fuels and oils. However, some general outcomes of fungal populations might be tentatively suggested. Following an oil spill, volatile compounds, including BTEX, are likely to inhibit fungal activity until the concentrations are reduced to subinhibitory levels by evaporation, microbial degradation, or dispersal/seepage through the soil or ice. The time that this process takes will depend on the characteristics of the soil, the environmental conditions, and the volume/type of hydrocarbon spilt. Humic substances can reduce the toxicity of spilled hydrocarbons by adsorbing/absorbing some of the toxic components; however, many Antarctic soils have few organic components making this outcome less likely.

Low organic content in soils may allow a rapid vertical seepage of hydrocarbons to considerable depths, which may also impair the removal of volatile fractions. Fungal strains (a) with sufficient inoculum in the soil, (b) that are less sensitive to BTEX or PAHs, and (c) that can degrade accessible aliphatic hydrocarbons may become dominant within the fungal community, assuming there is sufficient water and oxygen levels, nutrient availability, and temperatures are high enough to allow them to capitalize on their physiological advantage.

12.8 FUNGAL BIOREMEDIATION OF CONTAMINATED ANTARCTIC SOILS

The use of filamentous fungi as bioremediative agents in Antarctic soils may have several benefits:

- Fungal extracellular low temperature enzyme systems can permit access to hydrocarbon substrates of low solubility.
- Use of extracellular enzymes may allow hydrocarbon-degrading fungi to tolerate higher levels of toxic substrates than would be possible if the substrate had to be internalized before catabolism.
- Depending upon the strain, fungi may be more tolerant to high concentrations of some pollutants than bacteria.
- Although bacteria are generally faster PAH degraders, fungi may show the ability to degrade a wider selection of aromatic hydrocarbons, and therefore play an important part in soil bioremediation (Saraswathy and Hallberg, 2002).
- Filamentous fungi may rapidly penetrate the soil matrix and access hydrocarbons, including PAHs that are initially inaccessible to bacteria (Novotny et al., 1999).
- Fungi can tolerate low nutrient concentrations and pHs.
- Some fungi show considerable desiccation tolerance (Boddy et al., 1985).

Despite these benefits, reports suggest that hydrocarbon-degrading basidiomycete filamentous fungi are rare in Antarctica. However, as mentioned earlier, the option of introducing a well-characterized hydrocarbon-degrading fungal species to Antarctica from another part of the globe to help to remediate contaminated soils is not available, as importation of nonnative species (including microorganisms) is not permitted under The Protocol on Environmental Protection to the Antarctic Treaty (Annex II). If fungal bioremediation is considered to have future potential, research emphasis should focus on finding native Antarctic low-temperature hydrocarbon-degrading fungi, as has already been done for Antarctica bacteria (Baraniecki et al., 2002; Panicker et al., 2002; Stallwood et al., 2005).

12.8.1 Biostimulation of Indigenous Fungi with Nutrients

Biostimulation is the activation of indigenous bacteria and fungi through the addition of nutrients to degrade a pollutant. Native microorganisms are generally well adapted to the local environmental conditions and therefore the addition of nutrients to stimulate growth should, in theory, permit the degradation of the pollutants. Problems arise when the indigenous microorganisms do not have the necessary enzymatic capabilities, or they are not represented to an effective level in the soil to bioremediate the soil at an acceptable rate. For example, indigenous microorganisms may not be able to degrade high molecular weight molecules or emulsify insoluble compounds (Mancera-Lopez et al., 2008). It is also notoriously difficult to optimize the levels of nutrients and hydration in contaminated soils so that bioremediation is enhanced, and until now, most research has focused on optimizing conditions for bacterial bioremediators (Walworth and Ferguson, 2008).

12.8.2 Bioaugmentation with Artificially Added Fungi (Mycoremediation)

Bioaugmentation is the addition of microorganisms, previously cultured in the laboratory, to assist in the degradation of pollutants in the soil. Much of the research that has been performed on mycoremediation has used basidiomyces (e.g., Radtke et al., 1994; Pozdnyakova et al., 2008), which have little known presence in Antarctica (Bridge et al., 2008a). If a suitable fungus was isolated from the Antarctic environment, which showed acceptable hydrocarbon-degrading properties in laboratory and field experiments, it may be suitable for *in situ* bioremediation in Antarctic soils by augmenting the bioremediative action of the indigenous microbial population. Two possible candidates for this may be *Y. lipolytica* and *P. antarctica* as both are present in the Antarctic and are known to be capable of growth on hydrocarbons and high lipase production *in vitro*, respectively. There is little evidence as yet as to how effective these activities are *in vivo* in the Antarctic soil environment. Bioaugmentation is unlikely to be effective if the fungus does not grow to a sufficient biomass in the soil to have a significant bioremediative effect (Reddy, 1995). Problems may include: (1) competition and predation by well-adapted indigenous Antarctic soil species (Federici et al., 2007), particularly if biostimulation is used simultaneously, (2) a slowing of fungal growth due to an inability to predict or supply adequate nutrients, and (3) unsuitable growing conditions at the contaminated site (temperature, pH, hydration, and aeration). The interactions of the artificially added fungus with the indigenous population can determine the extent of bioremediation. In nonpolar soils, microorganisms can inhibit the introduced fungus (Radtke et al., 1994; Andersson et al., 2003) or enhance bioremediation by the further degradation of the pollutants partially metabolized by the fungi (Mougin et al., 1997; Kotterman et al., 1998; Federici et al., 2007; Mancera-Lopez et al., 2008). Some fungi can undertake the initial oxidation of pollutants, which may increase water solubility and increase hydrocarbon availability to bacterial species for further degradation (Sasek, 2003). The initial application and care of the allochthonous fungus during the bioremediative process have yet to receive sufficient research. Although several small-scale bioremediation projects using fungi have been successful outside Antarctica (Lamar and White, 2001), so far there is no reliable large-scale soil bioremediation technology available or appropriate for use in Antarctica. Individual bacterial strains may only have the ability to degrade a limited range of hydrocarbons (Farinazleen et al., 2004); however, the addition of a consortium of microbial species to soils may greatly increase the enzymatic capability of the population and increase the range, extent, and rate of degradation, particularly if the consortium includes fungi that produce nonspecific hydrocarbon-degrading enzymes, such as the white-rot fungi (Colombo et al., 1996; Mancera-Lopez et al., 2008). For example, Boonchan et al. (2000) showed that cocultures of *Penicillium janthinellum* and a consortium of bacteria isolated from a contaminated site showed enhanced levels of PAH bioremediation compared with the indigenous microorganisms and soil amended with only axenic inocula.

12.8.3 Mycorrhizal Bioremediation

The potential for using fungi associated with plant roots to bioremediate hydrocarbon spills in Antarctica has received little attention. Plant roots and associated microorganisms, including fungi, may accelerate the rate of pollutant degradation due to (1) the stimulation of microbial activity at the plant roots, and (2) the selection of microorganisms that can degrade plant secondary metabolites and, therefore, many hydrocarbons that may be co-metabolized alongside plant material (Kuiper et al., 2004). Fungi commonly found as mycorrhizal in the Northern Hemisphere have been found in the Antarctic associated with lower plants and the two vascular plant species (Upson et al., 2007, 2008). Recent increases in temperature on the Antarctic Peninsula (Vaughan et al., 2003) and concomitant increases in water availability may increase plant distribution and biodiversity (Convey and Smith, 2006) with associated increases in mycorrhizal biomass (e.g., *Hymenoscythus ericae*). Nevertheless, the current scanty distribution of higher plants and other mycorrhiza-forming cryptogams in Antarctica, and the fact that they are largely absent at existing contaminated sites, mean that mycorrhizal fungi are unlikely to be of direct practical use in the bioremediation of Antarctic hydrocarbon spills in the near future, although this may not be true for the Arctic.

12.9 SUMMARY

Antarctic environmental conditions make the bioremediation of hydrocarbon-contaminated soils technically challenging. Potentially, there are many cold-adapted fungal enzymes for hydrocarbon degradation that could be exploited for bioremediation purposes. The biostimulation of local microbial population, and bioaugmentation with a consortium of native hydrocarbon-degrading bacteria and fungi may produce the best opportunity for successful *in situ* bioremediation in Antarctica. However, there are no guarantees of success, as much is still unknown about the physical and biological processes involved in Antarctic soil bioremediation. To complicate matters further, lignin-degrading fungal species, traditionally chosen for hydrocarbon and PAH bioremediation elsewhere, are largely absent from Antarctic soils (although not in the sub-Antarctic), and use of nonnative fungi for bioremediation is not permitted under the Antarctic Treaty. As a result, if mycoremediation is to become a cost-effective reality, further investigations into Antarctic fungal biodiversity, bacterial–fungal interactions, and hydrocarbon-degradation capacity are required.

REFERENCES

Aislabie, J., R. Fraser, S. Duncan, and R. L. Farrell. 2001. Effects of oil spills on microbial heterotrophs in Antarctic soils. *Polar Biology* 24: 308–313.

Aislabie, J. M., M. R. Balks, J. M. Foght, and E. J. Waterhouse. 2004. Hydrocarbon spills on Antarctic soils: Effects and management. *Environmental Science and Technology* 38: 1265–1274.

Andersson, B. E., S. Lundstedt, K. Tornberg, Y. Schnuerer, L. G. Oeberg, and B. Mattiasson. 2003. Incomplete degradation of polycyclic aromatic hydrocarbons in soil inoculated with wood-rotting fungi and their effects on the indigenous soil bacteria. *Environmental Toxicology and Chemistry* 22: 1238–1243.

Antarctic Treaty Consultative Parties. 1991. Protocol on Environmental Protection to the Antarctic Treaty. CM 1960. London: Her Majesty's Stationery Office.

April, T. M., J. M. Foght, and R. S. Currah. 2000. Hydrocarbon-degrading filamentous fungi isolated from flare pit soils in northern and western Canada. *Canadian Journal of Microbiology* 46: 38–49.

Arenz, B. E., B. W. Held, J. A. Jurgens, R. L. Farrell, and R. A. Blanchette. 2006. Fungal diversity in soils and historic wood from the Ross Sea Region of Antarctica. *Soil Biology and Biochemistry* 38: 3057–3064.

Atlas, R. M., M. E. di Menna, and R. E. Cameron. 1978. Ecological investigations of yeasts in Antarctic soils. *Antarctic Research Series* 30: 27–34.

Aust, S. D. 1995. Mechanisms of degradation by white-rot fungi. *Environmental Health Perspective* 103: 59–61.

Baraniecki, C. A., J. Aislabie, and J. M. Foght. 2002. Characterization of *Sphingomonas* sp. Ant 17, an aromatic hydrocarbon-degrading bacterium isolated from Antarctic soil. *Microbial Ecology* 43: 44–54.

Barnes, D. K. A., N. L. Warren, K. Webb, B. Phalan, and K. Reid. 2004. Polar pedunculate barnacles piggy-back on pycnogona, penguins, pinniped seals and plastics. *Marine Ecology-Progress Series* 284: 305–310.

Baublis, J. A., R. A. Wharton, and P. A. Volz. 1991. Diversity of micro-fungi in an Antarctic dry valley. *Journal of Basic Microbiology* 31: 3–12.

Bergauer, P., P. A. Fonteyne, N. Nolard, F. Schinner, and R. Margesin. 2005. Biodegradation of phenol and phenol-related compounds by psychrophilic and cold-tolerant alpine yeasts. *Chemosphere* 59: 909–918.

Bial Lasiewicz, D. and B. Czarnecki. 1999. Microfungi in the aerosphere of the Arctowski polar station. *Polish Polar Research* 20: 319–324.

Blanchette, R. A., B. W. Held, J. A. Jurgens, D. L. McNew, T. C. Harrington, S. M. Duncan, and R. L. Farrell. 2004. Wood-destroying soft rot fungi in the historic expedition huts of Antarctica. *Applied and Environmental Microbiology* 70: 1328–1335.

Boddy, L., D. M. Gibbon, and M. A. Grundy. 1985. Ecology of *Daldinia concentrica*: Effects of abiotic variables on mycelial extension and interspecific interactions. *Transactions of the British Mycological Society* 85: 201–211.

Boonchan, S., M. L. Britz, and G. A. Stanley. 2000. Degradation and mineralization of high-molecular-weight polycyclic aromatic hydrocarbons by defined fungal-bacterial cocultures. *Applied and Environmental Microbiology* 66: 1007–1019.

Bossert, I. and R. Bartha. 1984. The fate of petroleum in soil ecosystems. In *Petroleum Microbiology*, ed. R. M. Atlas, pp. 435–473. New York: MacMillan Publishing Co.

Bridge, P. D. and K. K. Newsham. 2009. Soil fungal community composition at Mars Oasis, a southern maritime Antarctic site, assessed by PCR amplification and cloning. *Fungal Ecology* doi:10.1016/j.funeco.2008.10.008.

Bridge, P. D., M. S. Clark, and D. A. Pearce. 2005. A new species of Paecilomyces isolated from the Antarctic springtail *Cryptopygus antarcticus*. *Mycotaxon* 92: 213–222.

Bridge, P. D., B. M. Spooner, and P. J. Roberts. 2008a. List of non-lichenized fungi from the Antarctic region. http://www.antarctica.ac.uk//bas_research/data/access/fungi/.

Bridge, P. D., B. M. Spooner, and P. J. Roberts. 2008b. Non-lichenized fungi from the Antarctic region. *Mycotaxon* 106: 485–490.

Buswell, J. A. 2001. Fungal biodegradation of chlorinated monoaromatics and BTEX compounds. In *Fungi in Bioremediation*, ed. G. M. Gadd, pp. 113–135. Cambridge, U.K.: Cambridge University Press.

Butler, M. J. and A. W. Day. 1998. Fungal melanins: A review. *Canadian Journal of Microbiology* 44: 1115–1136.

Cantrill, D. J. and I. Poole. 2005. A new Eocene Araucaria from Seymour Island, Antarctica: Evidence from growth form and bark morphology. *Alcheringa* 29: 341–350.

Carlisle, M. J., S. C. Watkinson, and G. W. Gooday. 2001. *The Fungi*, 2nd edition. London, U.K.: Academic Press.

Cerniglia, C. E. and J. B. Sutherland. 2001. Bioremediation of polycyclic aromatic hydrocarbons by ligninolytic and non-ligninolytic fungi. In *Fungi in Bioremediation*, ed. G. M. Gadd, pp. 136–187. Cambridge, U.K.: Cambridge University Press.

Chalmers, M. O., M. A. Harper, and W. A. Marshall. 1996. *An Illustrated Catalogue of Airborne Microbiota from the Maritime Antarctic*. Cambridge, U.K.: British Antarctic Survey.

Colombo, J., M. Cabello, and A. M. Arambarri. 1996. Biodegradation of aliphatic and aromatic hydrocarbons by natural soil microflora and pure cultures of imperfect and lignolitic fungi. *Environmental Pollution* 94: 355–362.

Convey, P. and R. I. L. Smith. 2006. Responses of terrestrial Antarctic ecosystems to climate change. *Plant Ecology* 182:1–10.

Crespo, R., M. P. Juarez, and L. F. R. Cafferata. 2000. Biochemical interaction between entomopathogenous fungi and their insect-host-like hydrocarbons. *Mycologia* 92: 528–536.

de Hoog G.S., E. Göttlich, G. Platas, O. Genilloud, G. Leotta, and J. van Brummelen. 2005. Evolution, taxonomy and ecology of the genus *Thelebolus* in Antarctica. In *Fungi of the Antarctic: Evolution under Extreme Conditions*, ed. G. S. de Hoog; *Studies in Mycology* 51: 33–76.

Farinazleen, M. G., Z. A. Raja Noor, B. S. Abu, and B. Mahiran. 2004. Biodegradation of hydrocarbons in soils by microbial consortium. *International Biodeterioration and Biodegradation* 54: 61–67.

Federici, E., V. Leonardi, M. A. Giubilei, D. Quaratino, R. Spaccapelo, A. D'Annibale, and M. Petruccioli. 2007. Addition of allochthonous fungi to a historically contaminated soil affects both remediation efficiency and bacterial diversity. *Applied Microbiology and Biotechnology* 77: 203–211.

Fell, J. W., G. Scorzetti, L. Connell, and S. Craig. 2006. Biodiversity of micro-eukaryotes in Antarctic dry valley soils with <5% soil moisture. *Soil Biology and Biochemistry* 38: 3107–3119.

Fenn, P. and T. K. Kirk. 1981. Relationship of nitrogen to the onset and suppression of ligninolytic activity and secondary metabolism in *Phaerochaete chrysosporium. Archive of Microbiology* 130: 59–65.

Frisvad, J. C., T. O. Larsen, P. W. Dalsgaard, K. A. Seifert, G. Louis-Seize, E. K. Lyhne, B. B. Jarvis, J. C. Fettinger, and D. P. Overy. 2006. Four psychrotolerant species with high chemical diversity consistently producing cycloaspeptide A, *Penicillium jamesonlandense* sp. nov., *Penicillium ribium* sp. nov., *Penicillium soppii* and *Penicillium lanosum. International Journal of Systematic and Evolutionary Microbiology* 56: 1427–1437.

Göttlich, E., G. S. de Hoog, O. Genilloud, B. E. Jones, and F. Marinelli. 2003. MICROMAT: Culturable fungal diversity in microbial mats of Antarctic lakes. In *Anatarctic Biology in a Global Context*, eds. A. H. L. Huiskes, W. W. C. Gieskes, J. Rozema, R. M. L. Schorno, S. M. van der Vies, and W. J. Wolff, pp. 251–254. Leiden: Backhuys Publishers.

Harvey, R. G. 1992. *Polycyclic Aromatic Hydrocarbons: Chemistry and Carcinogenicity*. Cambridge, U.K.: Cambridge University Press.

Harvey, R. G., T. M. Penning, J. Jarabak, and F. J. Zhang. 1999. Role of quinine metabolites in PAH carcinogenesis. *Polycyclic Aromatic Compounds* 16: 13–20.

Hawksworth, D. L. 2001. The magnitude of fungal diversity: The 1–5 million species estimate revisited. *Mycological Research* 105: 1422–1432.

Holzbaur, E. L. F., A. Andrawis, and M. Tien. 1991. Molecular biology of lignin peroxidases from *Phanerochaete chrysosporium*. In *Molecular Industrial Mycology*, eds. S. A. Leong and R. M. Berka, pp. 197–223. New York: Marcel Dekker Publishing.

Hua, Z. Z., Y. Chen, G. C. Du, and J. Chen. 2004. Effects of biosurfactants produced by *Candida antarctica* on the biodegradation of petroleum compounds. *World Journal of Microbiology and Biotechnology* 20: 25–29.

Hughes, K. A. and B. Stallwood. 2006. Oil pollution in the Antarctic terrestrial environment. *Polarforschung* 75: 141–144.

Hughes, K. A., B. Lawley, and K. K. Newsham. 2003. Solar UV-B radiation inhibits the growth of Antarctic terrestrial fungi. *Applied and Environmental Microbiology* 69: 1488–1491.

Hughes, K. A., S. Ott, M. Bolter, and P. Convey. 2006. Colonisation processes. In *Trends in Antarctic Terrestrial and Limnetic Ecosystems*, eds. D. Bergstrom, P. Convey, and A. H. L. Huiskes. Dordrecht, the Netherlands: Kluwer.

Hughes, K. A., P. Bridge, and M. S. Clark. 2007. Tolerance of Antarctic soil fungi to hydrocarbons. *Science of the Total Environment* 372: 539–548.

International Association of Antarctica Tour Operators (IAATO). 2008. Tourism statistics. http://www.iaato.org/tourism_stats.html.

Janiot, L. J., J. L. Sericano, and O. Marcucci. 2003. Evidence of oil leakage from the *Bahai Paraiso* wreck in Arthur Harbour, Antarctica. *Marine Pollution Bulletin* 46: 1619–1622.

Kerry, E. 1990. Microorganisms colonising plants and soils subjected to different degrees of human activity, including petroleum contamination in the Vestforld Hills and MacRobertson Land, Antarctica. *Polar Biology* 10: 423–430.

Kerry, E. 1993. Bioremediation of experimental petroleum spills on mineral soils in the Vestfold Hills and MacRobertson Land, Antarctica. *Polar Biology* 13: 163–170.

Kitamoto, D., T. Ikegami, G. T. Suzuki, A. Sasaki, Y. Takeyama, Y. Idemoto, N. Koura, and H. Yanagishita. 2001. Microbial conversion of n-alkanes into glycolipid biosurfactants, mannosylerythritol lipids, by Pseudozyma (*Candida antarctica*). *Biotechnology Letters* 23: 1709–1714.

Klug, M. J. and A. J. Markovetz. 1968. Degradation of hydrocarbons by members of the genus Candida. III. Oxidative intermediates from 1-hexadecene and 1-heptadecene by *Candida lipolytica*. *Journal of Bacteriology* 96: 1115–1123.

Kotterman, M. J. J., E. H. Vis, and J. A. Field. 1998. Successive mineralization and detoxification of benzo[*a*]pyrene by the white-rot fungus *Bjerkandera* BOS55 and indigenous microflora. *Applied and Environmental Microbiology* 64: 2853–2858.

Kuiper, I., E. L. Lagendijk, G. V. Bloemberg, and B. J. J. Lugtenberg. 2004. Rhizoremediation: A beneficial plant-microbe interaction. *Molecular Plant-Microbe Interactions* 17: 6–15.

Lamar, R. T. and R. B. White. 2001. Mycoremediation—Commercial status and recent developments. In *Bioremediation Symposium: Wood-Treating and Phenolic Wastes, vol. 6(6)*, eds. V. S. Magar, F. M. von Fahnestock, and A. Leeson, pp. 263–278. Columbus, OH: Battelle Press.

Lawley, B., S. Ripley, P. Bridge, and P. Convey. 2004. Molecular analysis of geographic patterns of eukaryotic diversity in Antarctic soils. *Applied and Environmental Microbiology* 70: 5963–5972.

Leahy, J. G. and R. R. Colwell. 1990. Microbial degradation of hydrocarbons in the environment. *Microbiological Reviews* 54: 305–315.

Line, M. A. 1988. Microbial-flora of some soils of Mawson Base and the Vestfold Hills, Antarctica. *Polar Biology* 8: 421–427.

Luch, A. 2005. *The Carcinogenic Effects of Polycyclic Aromatic Hydrocarbons.* London, U.K.: Imperial College Press.

Ludley, K. E. and C. H. Robinson. 2008. 'Decomposer' Basidiomycota in Arctic and Antarctic ecosystems. *Soil Biology and Biochemistry* 40: 11–29.

Macdonald, D. I. M. and P. J. Butterworth. 1990. The stratigraphy, setting and hydrocarbon potential of the Mesozoic sedimentary basins of the Antarctic Peninsula. In *Antarctica as an Exploration Frontier, Hydrocarbon Potential, Geology, and Hazards. Studies in Geology*, vol. 31, ed. B. St. John, pp. 102–125. Tulsa, OK: American Association of Petroleum Geologists.

Mancera-Lopez, M., F. Esparza-Garcia, B. Chavez-Gomez, R. Rodriguez-Vazquez, G. Saucedo-Castaneda, and J. Barrera-Cortes. 2008. Bioremediation of an aged hydrocarbon-contaminated soil by a combined system of biostimulation-bioaugmentation with filamentous fungi. *International Biodeterioration and Biodegradation* 61: 151–160.

Margesin, R. and F. Schinner. 1997. Effect of temperature on oil degradation by a psychrotrophic yeast in liquid culture and in soil. *FEMS Microbiology* 24: 243–249.

Margesin, R., S. Gander, G. Zacke, A. M. Gounot, and F. Schinner. 2003. Hydrocarbon degradation and enzyme activities of cold-adapted bacteria and yeasts. *Extremophiles* 7: 451–458.

Margesin, R., V. Fauster, and P. A. Fonteyne. 2005. Characterization of cold-active pectate lyases from psychrophilic *Mrakia frigida. Letters in Applied Microbiology* 40: 453–459.

Markovetz, A. J., J. R. Cazin, and J. E. Allen. 1968. Assimilation of alkanes and alkenes by fungi. *Applied Microbiology* 16: 487–489.

Marquez-Rocha, F. J., V. Z. Hernandez-Rodriguez, and R. Vazquez-Duhalt. 2000. Biodegradation of soil-adsorbing polycyclic aromatic hydrocarbons by the white rot fungus *Pleurotus ostreatus. Biotechnology Letters* 22: 469–472.

Marshall, W. A. 1996. Biological particles over Antarctica. *Nature* 383: 680.

Mihelcic, J. R. and R. G. Luthy. 1988. Degradation of polycyclic aromatic hydrocarbons under various redox conditions in soil-water systems. *Applied and Environmental Microbiology* 54: 1182–1187.

Mougin, C., C. Pericaud, J. Dubroca, and M. Asther. 1997. Enhanced mineralization of lindane in soils supplemented with white rot basidiomycete *Phanerochaete chrysosporium. Soil Biology and Biotechnology* 29: 1321–1324.

Novotny, C., P. Erbanova, V. Sasek, A. Kubatova, T. Cajthaml, E. Lang, J. Krahl, and F. Zadrazil. 1999. Extracellular oxidative enzyme production and PAH removal in soil by exploratory mycelium of white rot fungi. *Biodegradation* 10: 159–168.

Onofri, S., L. Zucconi, and S. Tosi. 2007. *Continental Antarctic Fungi.* Eching, Germany: IHW-Verlag.

Panicker, G., J. Aislabie, D. Saul, and A. K. Bej. 2002. Cold tolerance of *Pseudomonas* sp 30-3 isolated from oil-contaminated soil, Antarctica. *Polar Biology* 25: 5–11.

Paterson, R. R. M. and P. D. Bridge. 1994. *Biochemical Techniques for Filamentous Fungi. IMI Technical Handbooks Vol. 1.* Wallingford, U.K.: CAB International.

Pegler, D. N., B. M. Spooner, and R. I. L. Smith. 1980. Higher fungi of Antarctica, the sub-Antarctic zone and Falkland Islands. *Kew Bulletin* 35: 499–561.

Platt, H. M., P. R. Mackie, and A. Clarke. 1981. Sources of Antarctic hydrocarbons. *Marine Pollution Bulletin* 12: 407–410.

Pozdnyakova, N. N., V. E. Nikitina, and O. V. Turovskaya. 2008. Bioremediation of oil-polluted soil with an association including the fungus *Pleurotus ostreatus* and soil microflora. *Applied Biochemistry and Microbiology* 44: 60–65.

Radtke, C, W. S. Cook, and A. Anderson. 1994. Factors affecting antagonism of the growth of *Phanerochaete chrysosporium* by bacteria isolated from soils. *Applied Microbiology and Biotechnology* 41: 274–280.

Reddy, C. A. 1995. The potential for white rot fungi in the treatment of pollutants. *Current Opinions in Biotechnology* 6: 320–328.

Romero, M. C., M. C. Cazau, S. Giorgieri, and A. M. Arambarri. 1998. Phenanthrene degradation by microorganisms isolated from a contaminated stream. *Environmental Pollution* 101: 355–359.

Saraswathy, A. and R. Hallberg. 2002. Degradation of pyrene by indigenous fungi from a former gasworks site. *FEMS Microbiology Letters* 210: 227–232.

Sasek, V. 2003. Why mycoremediations have not yet come into practice. In *Utilization of Bioremediation to Reduce Soil Contamination: Problems and Solutions*, eds. V. Sasek, J. A. Glaser, and P. Baveye, pp. 247–266. NATO Science Series, sub-series IV: Earth and Environmental Sciences: 19. Berlin, Germany: Springer.

Smith, R. I. L. 1994. Species-diversity and resource relationship of South Georgia fungi. *Antarctic Science* 6: 45–52.

Spooner, B. and P. Roberts. 2005. *Fungi.* London, U.K.: Collins New Naturalist Library.

Stallwood, B., J. Shears, P. A. William, and K. A. Hughes. 2005. Low temperature bioremediation of oil-contaminated soils using biostimulation and bioaugmentation of oil-contaminated with a *Pseudomonas* sp. from maritime Antarctica. *Journal of Applied Microbiology* 99: 794–802.

Stroud, J. L., G. I. Paton, and K. T. Semple. 2007. Microbe-aliphatic hydrocarbon interactions in soil: Implications for biodegradation and bioremediation. *Journal of Applied Microbiology* 102: 1239–1253.

Sutherland, J. B. 2004. Degradation of hydrocarbons by yeasts and filamentous fungi. In *Fungal Biotechnology in Agricultural, Food and Environmental Applications*, ed. D. K. Arora, pp. 443–455. New York: Marcel Dekker.

Thurston, C. F. 1994. The structure and function of fungal laccases. *Microbiology* 40: 19–26.

Tosi, S., B. Casado, R. Gerdol, and G. Caretta. 2002. Fungi isolated from Antarctic mosses. *Polar Biology* 25: 262–268.

Upson, R., D. J. Read, and K. K. Newsham. 2007. Widespread association between the ericoid mycorrhizal fungus *Rhizoscyphus ericae* and a leafy liverwort in the sub- and maritime Antarctic. *New Phytologist* 176: 460–471.

Upson, R., K. K. Newsham, and D. J. Read. 2008. Root-fungal associations of *Colobanthus quitensis* and *Deschampsia antarctica* in the maritime and sub-Antarctic. *Arctic, Antarctic, and Alpine Research* 40: 592–599.

Vasileva-Tonkova, E. and V. Gesheva. 2004. Potential for biodegradation of hydrocarbons by microorganisms isolated from Antarctic soils. *Zeitschrift fur Naturforschung C-A Journal of Biosciences* 59: 140–145.

Vaughan, D. G., G. J. Marshall, W. M. Connolley, C. L. Parkinson, R. Mulvaney, D. A. Hodgson, J. C. King, C. J. Pudsey, and J. Turner. 2003. Recent rapid regional climate warming on the Antarctic Peninsula. *Climatic Change* 60: 243–274.

Walworth, J. L. and S. Ferguson. 2008. Nutrient requirements for bioremediation. In *Bioremediation of Petroleum Hydrocarbons in Cold Regions*, ed. D. M. Filler, I. Snape, and D. L. Barnes, pp. 154–169. Cambridge, U.K.: Cambridge University Press.

Walworth, J. L., C. M. Reynolds, A. Rutter, and I. Snape. 2008. Landfarming. In *Bioremediation of Petroleum Hydrocarbons in Cold Regions*, ed. D. M. Filler, I. Snape, and D. L. Barnes, pp. 170–189. Cambridge, U.K.: Cambridge University Press.

Wilcke, W. 2000. Polycyclic aromatic hydrocarbons (PAHs) in soil—Review. *Journal of Plant Nutrition and Soil Science* 163: 229–248.

Wilkness, P. 1990. Fuel spill clean up in the Antarctic. *Antarctic Journal of the US* 25: 3–8.

Yakimov, M. M., L. Giuliano, V. Bruni, S. Scarfi, and P. N. Golyshin. 1999. Characterization of Antarctic hydrocarbon-degrading bacteria capable of producing bioemulsifiers. *Microbiologica* 22: 249–256.

Yateem, A., M. T. Balba, N. Al-Awadhi, and A. S. El-Nawawy. 1998. White rot fungi and their role in remediating oil-contaminated soil. *Environment International* 24: 181–187.

Zeyer, J., E. P. Kuhn, and R. P. Schwarzenbach. 1986. Rapid microbial mineralization of toluene and 1,3-dimethylbenzene in the absence of molecular-oxygen. *Applied and Environmental Microbiology* 52: 944–947.

13 Do Fertilizers Help in Effective Bioremediation of Polycyclic Aromatic Hydrocarbons in Polar Soils?

Daniel Delille, Emilien Pelletier, and Frédéric Coulon

CONTENTS

13.1 INTRODUCTION

Naturally occurring petroleum hydrocarbons are complex mixtures of linear, polycyclic, and heterocyclic organic compounds with molecular composition known to change overtime and location when released into the environment due to biotic and abiotic processes. Petroleum hydrocarbons are substantial components of a wide range of commercial fuels, lubricants, sealants, and tars, including mineral oils, greases, aviation and vehicles fuels, heavy fuel oils, asphalt, and are incorporated in a wide variety of other materials such as paints, solvents, and creosotes. The two main classes of the major constituents of crude oil are the saturated hydrocarbons (linear and cyclic alkanes and alkenes) and the polycyclic aromatic hydrocarbons (referred hereafter as PAHs). The remaining two other classes contain polar nonhydrocarbons, heterocycles, and heavy nonvolatile compounds, and are forming classes of asphaltenes (including polyphenols, ketones, esters, fatty acids, and porphyrins) and resins (pyridines, quinolines, carbazoles, sulfoxides, and amides). The mutagenic and carcinogenic effects of some high molecular weight PAHs have been clearly established (Cavalieri and Rogan 1998). In addition, even PAHs showing low mutagenic effects can produce, as part of their catabolism by different organisms, metabolic intermediates with significantly higher toxic effects compared to original compounds (Hall and Grover 1990).

Human activities in high-latitude regions rely heavily on fossil fuels for transportation and generation of heat and electricity. In the northern hemisphere, petroleum industry activities are rapidly growing in the Arctic Barents Sea. In the South, the Antarctic can no longer be considered as a pristine environment because hydrocarbon contamination has affected many coastal marine and terrestrial areas (Karl 1992, Mazzera et al. 1999, Bej et al. 2000, Delille and Pelletier 2002, Martins et al. 2004). A number of accidental fuel spills on land occurred mainly near scientific stations where the storage and refueling of vehicles can result in small and large spills (Aislabie et al. 1999). While the fuels used in polar regions are predominantly composed of aliphatic hydrocarbons ranging from C_9 to C_{30}, the relatively smaller fraction of aromatic hydrocarbons are pollutants of great environmental concern due to their toxic, mutagenic, and carcinogenic properties (Snape et al. 2001, Ferguson et al. 2003a,b, Coulon et al. 2004, Coulon and Delille 2006). For example, toxic PAHs have been shown to accumulate in soils from Palmer Station (Kennicut et al. 1992), Scott Base (Aislabie et al. 1998, 1999), and Grytviken Whaling Station (Cripps and Priddle 1991). Of all the different types of contamination reported up to now in the Antarctic Ocean and on the continent, petroleum has been identified as the most significant problem to be solved to preserve the environmental integrity of Antarctica (Snape et al. 2001).

The microbial decontamination (bioremediation) of oil-polluted soils has been claimed as an efficient, economic, and versatile alternative to physicochemical treatments (Atlas 1981, Joergensen et al. 1995, Møller et al. 1996). Once a hydrocarbon spill has occurred in Antarctica, the recovery of oiled residues and the remediation of the site represent a major challenge, as many techniques deployed in temperate regions are either unsuitable or difficult to implement in this extreme environment. Bioremediation has been proposed in early 1990s for cleanup of oil spills in Antarctic soils (Kerry 1993) and is increasingly viewed as an appropriate remediation technology for hydrocarbon-contaminated polar soils (Aislabie et al. 2006). The biodegradation of petroleum components has been reported in a variety of cold terrestrial systems, including sub-Arctic and Arctic soils (Griffiths et al. 1981, Horowitz et al. 1983), alpine soils (Margesin and Schinner 1997, Margesin 2000), and sub-Antarctic and Antarctic soils (Delille 2000, Coulon et al. 2004, 2005, Delille et al. 2004a,b, 2007a,b). Bioremediation has been proposed as the only viable management option that can be implemented on a large scale in polar areas (Snape et al. 2001).

The rate of the microbial degradation of hydrocarbons in soils is affected by several physical, chemical, and biological parameters including the abundance and diversity of microorganisms present in the pristine environment before the spill, the conditions for microbial activity (e.g., the concentration of nutrients, oxygen, and temperature) and the nature, quantity, and bioavailability of the contaminants. It is well established that nutrients are among the major limiting factors of hydrocarbon biodegradation. In spite of harsh weather conditions, a number of studies indicated that hydrocarbon degradation rates in soils from Arctic regions can be enhanced by the addition of nutrients (Braddock et al. 1997, Whyte et al. 1999, Walworth et al. 2001). The aim of this chapter is therefore to summarize scientific information about factors favoring the biodegradation of hydrocarbons under severe conditions and discuss how the addition of fertilizers might play a determining role on the degradation of PAH in polar soils.

13.2 PROPERTIES AND PROCESSES INFLUENCING THE BEHAVIOR AND FATE OF AROMATIC COMPOUNDS IN POLAR SOILS

Aromatic compounds are broadly defined as a collection of related compounds with two or more benzene fused rings that exhibit similar chemical behavior (Farrell-Jones 2003). Due to their high hydrophobicity, PAHs tend to interact with non-aqueous phase and soil organic and mineral matter (geosorbents), consequently, becoming potentially recalcitrant for future biodegradation. The field and laboratory biodegradation of PAHs, particularly those of low molecular weight (<3-rings fused benzene), has been widely investigated. It has been shown that PAHs can be totally degraded or partially transformed either by a community of microorganisms or by a single microorganism (Whyte et al. 1997, Grishchenkov et al. 2000, Andersson et al. 2003, Ruberto et al. 2005).

Microbial degradation is thought to be the major process involved in the effective bioremediation of contaminated soils and sediments. However, the degradation of PAHs in soil is not only dependent on the presence of microorganisms with the enzymatic capacity to attack hydrocarbons but also on the bioavailability of the oil compounds, especially PAHs within soil aggregates (Amellal et al. 2001, Hamdi et al. 2007). In many cases, high residual PAH concentrations are caused by a limited availability to microorganisms of medium (3 and 4 rings) and high molecular weight (>5 rings) PAHs (Cuypers 2001, Coulon et al. 2004, Coulon and Delille 2006). When deposited on soil, PAHs have a number of possible fates including volatilization, photooxidation, bioaccumulation, and microbial degradation. While volatilization and photodegradation are two physical processes mainly controlled by air and soil temperature and light intensity, their influence is only important on the fate and degradation of the semivolatile PAHs including 2- and 3-rings PAHs.

Biodegradation is quite a complex process depending on biotic conditions. The rate and the degree of degradation depend on several chemical, physical, and biological factors that may differ significantly among ecosystems and their location on the surface of the earth. Furthermore, like many other organic compounds, PAHs show a declining availability to organisms with their increasing residence time in soil (Hatzinger and Alexander 1995). This aging process, often named sequestration, is responsible for changes in availability and can be attributed to a slow migration of

PAH molecules into a condensed organic matter and inaccessible nanopores (Doick et al. 2005, Allan et al. 2006). When sequestrated, molecules are inaccessible to organisms and extracellular microbial enzymes, as diffusion out of these sites is mainly governed by the slow rate of molecular desorption (Brion and Pelletier 2005). Soil texture, that is the particle size distribution of the soil, has a great effect on the structure and the activity of microbial communities (Coulon et al. 2004, Hyun et al. 2008). The proportion of small particles and their dynamic behavior in the soil matrix represent crucial factors in determining the fate and distribution of PAHs in Antarctic soils (Curtosi et al. 2007). The proportion of clay in a soil is a key factor in bioremediation because very fine clay particles modify soil porosity and more complex chemical factors such as surface activation and adsorption coefficients (Kan et al. 1994). Clays can catalyze humic acid formation and protect organic material from decomposition within aggregates (Stott and Martin 1990). The presence of an active soil layer and a permafrost layer is a unique feature of high-latitude cold areas that could influence the fate of PAHs (Biggar et al. 1998, McCarthy et al. 2004a, Curtosi et al. 2007). When repetitive cycles of freezing and thawing occur, soil components are submitted to a slow sieving process where the smallest particles migrate from surface to depth while pebbles and stones tend to migrate from depth to surface (Anderson et al. 1978). PAHs are mainly adsorbed to the smallest soil particles but it has been found that they are usually metabolized to a much greater extent in soils dominated by larger particles and a low content of clay and silt (Krauss and Wilcke 2002).

A number of conditions must be satisfied for a microbial transformation to occur:

1. Microorganisms with the necessary enzymatic capacity to catalyze the specific transformation must be present. Some nonspecific enzymes can attack several organic substrates while others only catalyze the breakdown of one specific bond in specific compounds such as alkanes or cyclic hydrocarbons. Different bacterial strains may also degrade the same compound by different degradation routes, depending on the type of enzymes used (Cutright and Lee 1994, Whyte et al. 1997, Jenisch-Anton et al. 1999). Little is known about the relative importance of these different pathways in the conversion of an organic compound in the environment. However, the proliferation success of some degrading strains will depend on their ability to compete for organic substrates and oxygen or other limiting environmental factors. Many degradation pathways are made efficient only by the joint activity of several species (Green et al. 2000, Röling et al. 2002, Del Panno et al. 2005). Many different mechanisms for such synergistic relationships have been suggested and seem to vary between communities (Alexander 1994, Chavez-Gomez et al. 2003). Some bacterial strains have the capability to degrade substances with multiple pathways (Whyte et al. 2002). In these cases as well, environmental factors like oxygen level usually control the degradation pathway.
2. The microorganisms must live in the same environment as the chemicals to be degraded. Although microorganisms are present essentially in all

environments even the most severe ones, not all types of microorganisms exist everywhere. Because bacterial strains differ in their occurrence and their ability to use particular substrates, the same group of organic compounds can be readily degraded in a given environment and is highly persistent in another one.

3. The targeted chemicals must be available to the microbial community. Even if microorganisms with the necessary enzymes coexist in the same microenvironment, a heterogeneous distribution of microorganisms or a lack of chemical availability or both may inhibit biodegradation. Inaccessibility may also result from the chemical existing in a phase different from the bacteria (e.g., in a liquid phase immiscible with water, or sorbed to a solid phase).
4. The chemicals to be removed must be sensitive to biodegradation. If the enzyme catalyzing the transformation is intracellular, the target molecule must penetrate the surface of the cell. Molecular weight, shape, polarity, and other properties of the chemical may prevent permeability. If the initial-degrading enzyme is extracellular, the specific bond to be broken in the molecule must be exposed. Likewise, the active sites of the performing enzymes must not be sterically hindered by some substituents or by sorption to a solid surface. Some long-chain hydrocarbons are attacked at the terminal bonds of the molecule. If these structures are folded or coiled, resistance to biodegradation may occur.
5. Environmental growth conditions must be fulfilled to allow bacterial proliferation. Environmental factors such as inorganic and organic nutrient levels (Wang et al. 1994, Liebeg and Cutright 1999), pH (Hambrick et al. 1980), redox-potential (Hambrick et al. 1980), oxygen level (Huesemann and Truex 1996), pressure, temperature (Margesin and Schinner 1997), salinity (Chuvilin and Miklyaeva 2003), and water availability (McGroddy et al. 1996) dictate the conditions under which bacteria can synthesize enzymes and degrade organic substances. Even though anaerobic mechanisms of PAH biodegradation have been described (Eriksson et al. 2003, Steinbach et al. 2004) only aerobic processes, which occur in the upper layer of the soil where oxygen concentrations is not a limiting factor, can lead to a significant removal rate of hydrocarbons (Morgan and Watkinson 1992).

The structure of a chemical substance highly affects its biodegradability. For polyaromatic hydrocarbons, their environmental persistence increases as the molecular size increases over three fused benzene rings (Bossert and Bartha 1986, Heitkamp and Cerniglia 1989, Cerniglia 1992). The slower degradation rate of larger molecules has been attributed to their limited transport through cell membranes and the higher hydrophobicity of larger PAHs, increasing their sorption to the soil matrix and inaccessibility for degradation (Hatzinger and Alexander 1995). Substituents at the benzene ring such as methyl groups, halogens, NO_2–, NH_2–, SO_3H–, and CN groups have a slowing effect on the degradation compared to unsubstituted benzene rings (Sutherland et al. 1995, Meyer and Steinhart 2000), and monosubstituted benzenes are more rapidly degraded than disubstituted or trisubstituted ones (Bossert and Bartha 1986, Coulon et al. 2004). Moreover, the position of substituents at the ring could have an impact on the degradation (Bossert and Bartha 1986). Because bacteria degrade aromatic compounds through either ortho- or meta-cleavage pathways, the effect of the position of the substituent may not be the same in all molecular locations. Generally, highly branched hydrocarbons are degraded much more slowly than unbranched hydrocarbons (Morgan and Watkinson 1992, Coulon et al. 2004) and unsaturated hydrocarbons slower than saturated analogues. Exceptions to these generalizations exist. For instance, many degradation patterns are different in anaerobic compared with aerobic environments, and a chemical substance can be resistant in one environment and highly degradable in another one depending on the availability of different electron acceptors (Heider et al. 1999, Grishchenkov et al. 2000, Chang and Shiung 2002). The biodegradation efficiency of a given compound can also differ between soils with a previous chemical exposure of that compound compared to a pristine soil. This observation could be explained by a difference in microbial community adaptation between soils (Spain and van Veld 1983, Macleod and Semple 2002).

Only very few features are known about the fate of PAHs in polar soils. Naphthalene and methylnaphthalenes predominate in soil samples collected around Antarctic stations (Kennicut et al. 1992, Aislabie et al. 1999, Curtosi et al. 2007). An increase in PAH concentration with depth has also been reported (Aislabie et al. 1999, Curtosi et al. 2007). It has been hypothesized that PAHs first deposited on surface soils are rapidly associated to fine soil particles and hence transported downward by runoff from snow and ice melt as well as by rain waters (Mazzera et al. 1999). Temperature is an important parameter that influences sorption and desorption. For most compounds sorption decreases and desorption increases with increasing temperature (Hulscher and Cornelissen 1996).

Low air temperature may be considered as one of the determining rate factors in soil bioremediation in polar regions (Sutherland et al. 1995, Shi et al. 2001). Oil bioremediation in cold climates is frequently regarded with scepticism (Rike et al. 2008) because scientists generally recognize that biodegradation rate is mainly related to temperature. However, ambient temperatures close to 0°C do not completely stop oil biodegradation in seawater and sea-ice (Delille et al. 1997, 1998), and petroleum hydrocarbon biodegradation has been observed at a temperature as low as −1.1°C (Huddleston and Cresswell 1976). Actually, microbial metabolism increases as temperature increases (Leahy and Colwell 1990), usually doubling for each 10°C increase from 10°C to 40°C (Bossert and Bartha 1984). However, there is no evidence that the microbial potential for degrading hydrocarbons is lower in cold regions than in warm regions (Aislabie and Foght 2008). Biodegradation efficacy can be similar at both cold and temperate sites during comparable treatment periods (Filler et al. 2001). In the course of previous studies conducted on the soil and intertidal sediments of Kerguelen Archipelago it was observed that temperature was not a determining factor in the development and activity of a hydrocarbon-degrading community in oiled sands and soils (Delille et al. 2002, 2004b, 2007a, 2008, Coulon et al. 2005). One important remaining challenge is to optimize biodegradation in cold climate and for this we need to determine the temperature range at which remediation can be performed with efficiency within a reasonable time frame.

13.3 PAH-DEGRADING BACTERIA

For the application of bioremediation techniques in Antarctica, only indigenous microbes are considered because the Antarctic Treaty prohibits the introduction of foreign organisms. Hydrocarbon-degrading microorganisms have been isolated from both Arctic (Atlas 1986, Braddock et al. 1999, Whyte et al. 1999, Yu et al. 2000, Walworth et al. 2001, Thomassin-Lacroix et al. 2001, 2002, Eriksson et al. 2003) and Antarctic soils (Kerry 1990, MacCormack and Fraile 1997, Delille 2000, Aislabie et al. 2001, Baraniecki et al. 2002, Eckford et al. 2002, Saul et al. 2005, Hughes et al. 2007).

Aromatic-degrading bacteria isolated from polar soils typically belong to the genera *Pseudomonas* or *Sphingomonas* [see Aislabie et al. (2006) for a review]. A prime example of a readily isolated group of soil bacteria is given by *Pseudomonas* and *Pseudomonas*-like strains isolated for their ability to degrade naphthalene and its close analogues (Eckford et al. 2002). Numerous examples of catabolic genes from such strains have been described that show a surprising (>90%) degree of conservation, and a broad distribution (Whyte et al. 1997, Wilson et al. 1999). Gram-positive bacterial strains of the genus *Rhodococcus* are widely distributed in both pristine and polluted soils from cold environments (Whyte et al. 2002, Ruberto et al. 2005) and show an enormous catabolic versatility (Warhurst and Fewson 2004). The phylogenetic analysis of 16S rRNA genes from alkane-degrading bacterial isolates indicates that *Rhodococcus* spp. from cold soils group with *R. erythropolis* or *R. fascians*. In laboratory assays, these strains were able to degrade a broad range of environmental pollutants including both aliphatic and aromatic hydrocarbons (Whyte et al. 1997, Ruberto et al. 2005). Psychrotolerant *Rhodococcus* strains are therefore an interesting target to develop bioremediation techniques in permanently cold environments such as Antarctic soils. Of the alpha proteobacteria, cold-tolerant PAH-degrading isolates groups with *Sphingomonas chlorophenolica* and are related to aromatic-degrading *Sphingomonas* spp. from globally distributed sources (Aislabie et al. 2000). Further in their review, Aislabie et al. (2006) highlighted that *Pseudomonas* isolates tend to degrade a narrower range of aromatic substrates than *Sphingomonas* spp.

13.4 TOXICITY OF PAHs

PAHs have been recognized as chemicals of concern owing to their intrinsic chemical stability and their potentially deleterious effects on living organisms and human health (Alexander 1994). The carcinogenicity of polycyclic aromatic hydrocarbons has been assessed by several studies. PAHs and related compounds have been recently reviewed (Xue and Warshawsky 2005). As the aqueous solubility of PAHs decreases almost logarithmically with increasing molecular mass, high molecular weight PAH ranging in size from five to seven rings are of special environmental concern (Johnsen et al. 2005). Aromatic hydrocarbons inhibited microbial growth more than aliphatic hydrocarbons in Antarctic soils (Hughes et al. 2007).

The exposure of organisms to mixtures of xenobiotics represents an emerging environmental issue because interactions between chemicals and their multiple biological effects are largely unknown (Carpenter et al. 2002). Investigations on the effects of chemical mixtures are particularly lacking for the Antarctic ecosystem (Regoli et al. 2005). The study of Benedetti et al. (2007) revealed a complex pathway of interactions between trace metals (Cd, Cu, Hg, Ni, Pb) and a risk driver PAH, benzo[*a*]pyrene, in the Antarctic fish *Trematomus bernacchii*.

Two recent papers addressed the environmental risks associated to PAH and oxidized by-products present in contaminated soils (Lemieux et al. 2008, Park et al. 2008). The authors examined the mutagenicity of fractions extracted from the soil collected from former manufactured gas plants and confirmed mutagenic responses in polar fractions (Hughes et al. 1998). During a weathering process, contaminated soils tend to lose low molecular weight PAH (volatilization and biodegradation) and polar fractions tend to become enriched. Several studies found a direct link between mutagenicity and polar fractions but the chemical identification of mutagens is still unsolved. Park et al. (2008) observed PAH quinones and ketones in their mutagenic polar fractions, but none of these compounds by itself were mutagenic in the bioassay. The authors suggested that mutagenicity might result from polymeric quinone conjugates or interactions between fraction constituents as often observed in the toxicity of complex mixtures.

In our studies of oiled residues after crude oil bioremediation in sub-Antarctic intertidal sediments (Pelletier et al. 2004) and soils (Delille et al. 2007b) we currently observed an increase in soil toxicity as the bioremedition process was in progress. The toxicity of soils was monitored by Microtox® assay, which is a sensitive but nonspecific toxicity test using the bioluminescence of a bacterial strain. We also conducted an exhaustive analytical study of polar fractions extracted from the most toxic soils using a high temperature pressurized water extraction method as proposed by Kubátová et al. (2004) and Lundstedt et al. (2006). We obtained a number of oxidized PAHs including quinones, ketones, and naphthenic acids but it was impossible to relate the concentrations of these molecules to the observed toxicity in whole soils (unpublished results). Our understanding of the residual toxicity of weathered oiled soils is still incomplete and formulating fertilizers that could reduce lasting toxicity is a real challenge for those working on oil bioremediation in high latitudes (Pelletier et al. 2004).

13.5 BIOREMEDIATION EXPERIMENTS

While petroleum can supply a good deal of metabolic carbon, most hydrocarbon products are deficient in other nutrients, such as nitrogen and phosphorus. However, the effects of N and P addition are at times conflicting (Bossert and Bartha 1984). The suitable ratios of C:N:P are not readily apparent for petroleum degradation. Reported optimal C to N ratios range from 200:1 to 9:1 (Morgan and Watkinson 1989). Best ratios are likely to be contaminant specific and related to carbon availability. Relatively low fertilizer additions may lead to toxic effects on microbial populations due to osmotic stress (Walworth et al. 1997). Under long decomposition periods, bacteria and especially fungi, recycle nitrogen very efficiently (Piehler et al. 1999). A slowly released carbon source can be degraded with wider C:N:P ratios because N and P will be recycled during decomposition. For example, McMillen et al. (1995)

found that C:N:P ratio of 100:5:1.7 worked well for degrading petroleum wastes. Investigations with Antarctic soil confirmed this point and showed that the extent of ^{14}C-octadecane mineralization in soil from Casey Station incubated at 10°C peaked between 1000 and 1600 mg N/kg soil water (Ferguson et al. 2003a,b).

13.5.1 *In Situ* Treatment

The addition of significant amounts of plant materials and manures to petroleum-contaminated soils can be used to initiate a "soil composting" process. Due to the scarcity of the vegetation, such techniques eventually used in Arctic and sub-Arctic regions are not valuable for Antarctica. Biostimulation treatments using the addition of nutrients have been shown to stimulate the biodegradation of oil on a number of contaminated subpolar shorelines (Bragg et al. 1994, Delille et al. 1997, 2002, Delille and Delille 2000). In 1995, Braddock et al. (1997) conducted a treatability study for determining the effectiveness of a bioremediation treatment strategy for a coarse sand pad near Barrow (Alaska) that supported storage tanks containing various refined fuel products. Microbial activity and populations were stimulated relative to controls, but the stimulation was inversely related to the added concentration of fertilizer. The pad material has little water-holding capacity, and increasing fertilizer addition led to decreasing soil–water potential. The following year, the same authors conducted a small field plot study to determine the effect of fertilization on losses of aliphatic versus aromatic hydrocarbons in the same site (Braddock et al. 1999). Even at this cold and dry Arctic site, biological processes were effective in removing both aromatic and aliphatic hydrocarbons from these soils. Both surface and subsurface aliphatic losses were enhanced by fertilizer addition. In contrast, PAH declined in all surface soils, including non-fertilized plots, but decreases in the subsurface were significantly greater in the fertilized soil than in the unfertilized one.

Some bioremediation experiments have been conducted in Antarctica on hydrocarbon-contaminated soils (Kerry 1993, Delille 2000, Delille et al. 2003, Snape et al. 2003). All of them demonstrated the potential efficacy of bioremediation under these extreme environments but none of them provided specific information concerning the long-term behavior of PAHs.

An accidental oil spill occurred in 1997 in the sub-Antarctic Crozet Island leaving thousands of liters of diesel fuel trapped in soil (Delille and Pelletier 2002). On October 1998, the oiled area visible from air was estimated to approximately 1200 m^2. To examine the effects of bioremediation on natural assemblages of soil bacteria and hydrocarbon biodegradation, biostimulation treatments were conducted in wood enclosures settled in the contaminated area. Two years of field survey demonstrated that the nature of substrate and humidity level influenced greatly the degradation rate of both aliphatic and aromatic compounds (Coulon and Delille 2006). After the addition of fertilizer a nearly complete degradation of the alkyl homologues of naphthalene and fluorene was already observed within the arid areas. In contrast, the degradation rate decreased drastically with increasing number of aromatic rings.

In order to determine the best bioremediation approach to be developed for hydrocarbon-contaminated sub-Antarctic soils, a controlled field study was initiated in December 2000 in soils of a sub-Antarctic island (Grande Terre, Kerguelen

Archipelago) with better research facilities than Crozet Island (Delille et al. 2004a,b, Delille et al. 2007a). The total data set provided evidence of the presence of indigenous hydrocarbon-degrading microorganisms in sub-Antarctic soils and their high potential for a bioremediation action on both alkanes and PAHs. Alkanes were degraded more quickly than PAHs in most experiments. The degradation of *n*-alkanes was enhanced significantly in the presence of a fertilizer, while the degradation of PAHs was only barely enhanced (Coulon et al. 2004). Despite the very small amount of remaining nondegraded molecules, 4 years of biostimulation treatment were not sufficient to obtain a complete return to pristine conditions. As discussed earlier, the residual toxicity can represent a limiting factor in the bioremediation of sub-Antarctic soils. The identification of potentially toxic by-products has to be done and represents a major analytical challenge (Delille et al. 2007b).

13.5.2 *Ex Situ* Treatment

Mesocosm studies using sub-Antarctic soils artificially contaminated with diesel or crude oil were conducted in Kerguelen Archipelago to evaluate the potential of a bioremediation approach in high-latitude environments (Coulon et al. 2005, Delille et al. 2007b, Delille and Coulon 2008). From an applied perspective, heating soil and nutrient amendment appeared to be efficient means to accelerate the biodegradation of total petroleum hydrocarbons (TPH) in contaminated sub-Antarctic soils. The benefit of adding nutrient differs from one soil to another one, but fertilizer application always induced a significant increase in the rate of degradation of both aliphatic and aromatic fractions. However, the mass fraction of aromatics relative to aliphatic hydrocarbons increased by more than 20% after 42 treatment days (Delille and Coulon 2008). These observations are in good agreement with those reported by Whyte et al. (2001) demonstrating that fertilizer supplementation markedly increased naphthalene mineralization in soil microcosm studies conducted in Arctic soils.

Biopiles have been successfully used for the bioremediation of diesel-contaminated Arctic soils using various combinations of biostimulation (heating, nutrients, and aeration) and bioaugmentation (Filler et al. 2001, Mohn et al. 2001). The vapor pressure and solubility of most diesel constituents preclude their remediation by bioventing alone (Hinchee 1994). As the vapor pressure of a substance is a function of air temperature, an increase in biopile temperature has a potential dual effect of enhancing microbial activity and contaminant volatility. In summer 1999, Filler and collaborators combined mechanical heating and fertilization with bioventing of a large (61 m-long) diesel-contaminated biopile in the Arctic at Prudhoe Bay, Alaska (Filler et al. 2001). Their results evidenced a significant reduction in the overall level of petroleum compounds in the top half of the biopile through two enhanced bioremediation seasons. It is difficult to appreciate the specific effect of fertilization alone in such a global study. At a much smaller scale, we studied the effect of fertilization in pilot biopiles in sub-Antarctic area (Delille et al. 2008). Biopile technique developed in this study was highly efficient and all nutrient enriched biopiles reached the TPH regulatory USEPA standard of 100 μg/g (less than 0.2% of initially added diesel hydrocarbons) in less than 12 months. The huge decrease of contaminants was observed into both aliphatic and aromatic compounds. However, after one year, a significantly larger amount of PAHs was remaining in fertilized biopiles without a physical treatment compared to ventilated or agitated ones.

Landfarming could be a valuable alternative in some polar regions (Reynolds et al. 1994). As recently reported, 3600 m^3 of diesel-contaminated sandy soil was successfully treated at a field scale by on-site landfarming at Barrow, Alaska (McCarthy et al. 2004b). The soil was amended with a commercial fertilizer manually applied. The site was managed with intensive tilling and selective fertilization to exploit the short-treatment season, and the soil moisture content was maintained by light precipitation. This work demonstrates that even under harsh climatic conditions, soils moderately contaminated with petroleum hydrocarbons can be effectively and economically remediated within a reasonable time frame via landfarming. Concerning the more specific problem of high molecular weight PAH biodegradation, we demonstrated that manual tilling decreased the persistence of pyrene in fertilized diesel-contaminated sub-Antarctic soils (Delille et al. 2008).

13.6 CONCLUSION

As shown by the presence of significant populations of cold-adapted microorganisms, in general, and cold-adapted hydrocarbon-degrading organisms, in particular, the bioremediation of polar soils contaminated with hydrocarbons containing cyclic compounds is feasible in polar regions. The biodegradation of petroleum hydrocarbons, and PAHs, in particular, is markedly enhanced by the addition of simple and inexpensive fertilizers such as fish meal and agriculture fertilizers. However, the degradation of heavy polycyclic compounds is usually slower than aliphatic hydrocarbons and a residual toxicity remains generally present.

ACKNOWLEDGMENTS

Our work in Antarctica region has been supported by the French Polar Institute (IPEV), the National Sciences and Engineering Research Council of Canada, and the Canada Research Chair program.

REFERENCES

Aislabie, J. and Foght, J. 2008. Hydrocarbon-degrading bacteria in contaminated soils. In: *Bioremediation of Petroleum Hydrocarbons in Cold Regions*. Filler, D.F., Snape, I., and Barns, D.L. eds. Cambridge University Press, Cambridge, U.K., pp. 69–83.

Aislabie, J., Balks, M., Astori, N., Stevenson, G., and Symons, R. 1999. Polycyclic aromatic hydrocarbons in fuel-oil contaminated soils, Antarctica. *Chemosphere* 39:2201–2207.

Aislabie, J., Foght, J., and Saul, D. 2000. Aromatic hydrocarbon-degrading bacteria from soil near Scott Base. *Polar Biology* 23:183–188.

Aislabie, J., Fraser, R., Duncan, S., and Farrell, R.L. 2001. Effects of oil spills on microbial heterotrophs in Antarctic soils. Potential for biodegradation of hydrocarbons in soil from the Ross dependency, Antarctica. *Polar Biology* 24:308–313.

Aislabie, J., McLeod, M., and Fraser, R. 1998. Potential of biodegradation of hydrocarbons in soil from Ross dependensy, Antarctica. *Applied Microbiology and Biotechnology* 49:210–214.

Aislabie, J., Saul, D.J., and Foght, J.M. 2006. Bioremediation of hydrocarbon-contaminated polar soils. *Extremophiles* 10:171–179.

Alexander, M. 1994. *Biodegradation and Bioremediation*. Academic Press, San Diego, CA.

Allan, I.J., Semple, K.T., Hare, R., and Reid, B.J. 2006. Prediction of mono- and polycyclic aromatic hydrocarbon degradation in spiked soils using cyclodextrin extraction. *Environment Pollution* 144:562–571.

Amellal, N., Portal, J.M., Vogel, T., and Berthelin, J. 2001. Distribution and location of polycyclic aromatic hydrocarbons (PAH) and PAH degrading bacteria within polluted soil aggregates. *Biodegradation* 12:49–57.

Anderson, D.M., Pusch, R., and Penner, E. 1978. Physical and thermal properties of frozen ground. In: *Geotechnical Engineering for Cold Regions*, vol. 2. Andersland, O.B. and Anderson, D.M. eds. McGraw-Hill, New York, pp. 37–102.

Andersson, B.E., Lundstedt, S., Tornberg, K., Schnurer, Y., Oberg, L.G., and Mattiasson, B. 2003. Incomplete degradation of polycyclic aromatic hydrocarbons in soil inoculated with wood-rotting fungi and their effect on the indigenous soil bacteria. *Environmental Toxicology and Chemistry* 22(6):1238–1243.

Atlas, R.M. 1981. Microbial degradation of petroleum hydrocarbons: An environmental perspective. *Microbiological Review* 45:180–209.

Atlas, R.M. 1986. Fate of petroleum pollutants in Arctic ecosystems. *Water Science and Technology* 18:59–67.

Baraniecki, C.A., Aislabie, J., and Foght, J.M. 2002. Characterisation of *Sphingomonas* sp. Ant 17, an aromatic hydrocarbon-degrading bacterium isolated from Antarctic soil. *Microbial Ecology* 43:44–54.

Bej, A.K., Saul, D., and Aislabie, J. 2000. Cold-tolerant alkane-degrading *Rhodococcus* species from Antarctica. *Polar Biology* 23:100–105.

Benedetti, M., Martuccio, G., Fattorini, D., Capana, A., Barucca, M., Nigro, M., and Regoli, F. 2007. Oxidative and modulatory effects of trace metals on metabolism of polycyclic aromatic hydrocarbons in the Antarctic fish *Trematomus bernacchii*. *Aquatic Toxicology* 85:167–175.

Biggar, K.W., Haidar, S., Nahir, M., and Jarrett, P.M. 1998. Site investigations of fuel spill migration into permafrost. *Journal of Cold Regions Engineering* 12:84–104.

Bossert, I. and Bartha, R. 1984. The fate of petroleum in soil ecosystems. In: *Petroleum Microbiology*. Atlas, R. ed. McMillan, New York, pp. 435–474.

Bossert, I.D. and Bartha, R. 1986. Structure biodegradability relationships of polycyclic aromatic hydrocarbons in soil. *Bulletin of Environmental Contamination and Toxicology* 37:490–495.

Braddock, J.F., Ruth, M.L., Catterall, P.H., Walworth, J.L., and McCarthy, K.A. 1997. Enhancement and inhibition of microbial activity in hydrocarbons-contaminated Arctic soils: Implications for nutrient-amended bioremediation. *Environmental Science and Technology* 31:2078–2084.

Braddock, J.F., Walworth, J.L., and McCarthy, K.A. 1999. Biodegradation of aliphatic vs. aromatic hydrocarbons in fertilized arctic soils. *Bioremediation Journal* 3:105–116.

Bragg, J.R., Prince, R.C., Harner, E.J., and Atlas, R.M. 1994. Effectiveness of bioremediation for the Exxon Valdez oil spill. *Nature* 368:413–418.

Brion, D. and Pelletier, E. 2005. Modelling PAH adsorption and sequestration in freshwater and marine sediments. *Chemosphere* 61:867–876.

Carpenter, D.O., Arcaro, K., and Spink, D.C. 2002. Understanding the human health effects of chemical mixtures. *Environmental Health Perspective* 110:25–42.

Cavalieri, E. and Rogan, E. 1998. Mechanisms of tumour initiation by polycyclic aromatic hydrocarbons in mammals. In: *The Handbook of Environmental Chemistry*, vol. 3, *PAHs and Related Compounds*. Nelson, A.H. ed. Springer-Verlag, Heidelberg, Germany, pp. 81–117.

Cerniglia, C.E. 1992. Biodegradation of polycyclic aromatic hydrocarbons. *Biodegradation* 3:351–368.

Chang, B.V. and Shiung, L.C. 2002. Anaerobic biodegradation of polycyclic aromatic hydrocarbon in soil. *Chemosphere* 48:717–724.

Chavez-Gomez, B., Quintero, R., Esparza-Garcia, F., Mesta-Howard, A.M., de la Serna, F.J.Z.D., and Hernandez-Rodriguez, C.H. 2003. Removal of phenanthrene from soil by co-culture of bacteria and fungi pregrown on sugarcane bagasse pith. *Bioresource Technology* 89:177–183.

Chuvilin, E.M. and Miklyaeva, E.S. 2003. An experimental investigation of the influence of salinity and cryogenic structure on the dispersion of oil and oil products in frozen soils. *Cold Regions Science and Technology* 37:89–95.

Coulon, F. and Delille, D. 2006. Influence of substratum on the degradation processes in diesel polluted subAntarctic soils (Crozet Archipelago). *Polar Biology* 29:806–812.

Coulon, F., Pelletier, E., Gourhant, L., and Delille, D. 2005. Effects of nutrient and temperature on degradation of petroleum hydrocarbons in contaminated sub-Antarctic soil. *Chemosphere* 58:1439–1448.

Coulon, F., Pelletier, E., St. Louis, R., Gourhant, L., and Delille, D. 2004. Degradation of petroleum hydrocarbons in two sub-Antarctic soils: Influence of an oleophilic fertilizer. *Environmental Toxicology and Chemistry* 23:1893–1901.

Cripps, G.C. and Priddle, J. 1991. Hydrocarbons in the Antarctic environment. *Antarctic Science* 3:233–250.

Curtosi, A., Pelletier, E., Vodopivez, C.L., and McCormack, W.P. 2007. Polycyclic aromatic hydrocarbons in soil and surface marine sediment near Jubany Station (Antarctica). Role of permafrost as a low-permeability barrier. *Science Total Environment* 383:193–204.

Cutright, T.J. and Lee, S. 1994. Microorganisms and metabolic pathways for remediation of PAH in contaminated soil. *Fresenius Environmental Bulletin* 3:413–421.

Cuypers, C., Clemens, R., Grotenhuis, T., and Rulkens, W. 2001. Prediction of petroleum hydrocarbon bioavailability in contaminated soils and sediments. *Soil and Sediment Contamination* 10:459–482.

Delille, D. 2000. Response of Antarctic soil bacterial assemblages to contamination by diesel fuel and crude oil. *Microbial Ecology* 40:159–168.

Delille, D. and Coulon, F. 2008. Comparative mesocosm study of biostimulation efficiency in two different oil-amended subAntarctic soils. *Microbial Ecology* 56:243–252.

Delille, D. and Delille, B. 2000. Field observations on the variability of crude oil impact on indigenous hydrocarbon-degrading bacteria from subAntarctic intertidal sediments. *Marine Environmental Research* 49:403–417.

Delille, D. and Pelletier, E. 2002. Natural attenuation of diesel-oil contamination in a subantarctic soil (Crozet Island). *Polar Biology* 25:682–687.

Delille, D., Bassères, A., and Dessommes, A. 1997. Seasonal variation of bacteria in sea ice contaminated by diesel fuel and dispersed crude oil. *Microbial Ecology* 33:97–105.

Delille, D., Bassères, A., Dessommes, A., and Rosiers, C. 1998. Influence of daylight on potential biodegradation of diesel and crude oil in Antarctic seawater. *Marine Environmental Research* 45:259–268.

Delille, D., Coulon, F., and Pelletier, E. 2004a. Biostimulation of natural microbial assemblages in oil-amended vegetated and desert sub-Antarctic soils. *Microbial Ecology* 47:407–415.

Delille, D., Coulon, F., and Pelletier, E. 2004b. Effects of temperature warming during a bioremediation study of natural and nutrient-amended hydrocarbon-contaminated subAntarctic soils. *Cold Regions Science and Technology* 40:61–70.

Delille, D., Coulon, F., and Pelletier, E. 2007a. The influence of temperature on bacterial assemblages during bioremediation of a diesel fuel contaminated subAntarctic soil. *Cold Regions Science and Technology* 48:74–83.

Delille, D., Coulon, F., and Pelletier, E. 2007b. Long term changes of bacterial abundance, hydrocarbon concentration and toxicity during a biostimulation treatment of oil-amended organic and mineral sub-Antarctic soils. *Polar Biology* 30:925–933.

Delille, D., Delille, B., and Pelletier, E. 2002. Effectiveness of bioremediation of crude oil contaminated subAntarctic intertidal sediment: The microbial response. *Microbial Ecology* 44:118–126.

Delille, D., Duval, A., and Pelletier, E. 2008. Highly efficient pilot biopiles for on-site fertilization treatment of diesel oil-contaminated sub-Antarctic soil. *Cold Regions Science and Technology* 54:7–18.

Delille, D., Pelletier, E., Delille, B., and Coulon, F. 2003. Effect of nutrient enrichments on the bacterial assemblage of Antarctic soils contaminated by diesel or crude oil. *Polar Record* 39:1–10.

Del Panno, M.T., Morelli, I.S., Engelen, B., and Berthe-Corti, L. 2005. Effect of petrochemical sludge concentrations on microbial communities during soil bioremediation. *FEMS Microbial Ecology* 53:305–316.

Doick, K.J., Dew, N.M., and Semple, K.T. 2005. Linking catabolism to cyclodextrin extractability: Determination of the microbial availability of PAHs in soil. *Environmental Science and Technology* 39:8858–8864.

Eckford, R., Cook, F.D., Saul, D., Aislabie, J., and Foght, J. 2002. Free-living nitrogen-fixing bacteria from fuel-contaminated Antarctic soils. *Applied and Environmental Microbiology* 68:5181–5185.

Eriksson, M., Sodersten, E., Yu, Z., Dalhammer, G., and Mohn, W.W. 2003. Degradation of polycyclic aromatic hydrocarbons at low temperature under aerobic and nitrate-reducing conditions in enrichment cultures from Northern soils. *Applied and Environmental Microbiology* 69:275–284.

Farrell-Jones, J. 2003. Petroleum hydrocarbons and polyaromatic hydrocarbons. In: *Chemical Analysis of Contaminated Land.* Thompson, C.K. and Nathanail, P.C. eds. Blackwell, Oxford, U.K., pp. 132–176.

Ferguson, S.H., Franzmann, P.D., Revill, A.T., Snape, I., and Rayner, J.L. 2003a. The effects of nitrogen and water on mineralisation of hydrocarbons in diesel-contaminated terrestrial Antarctic soils. *Cold Regions Science and Technology* 37:197–212.

Ferguson, S.H., Franzmann, P.D., Snape, I., Revill, A.T., Trefry, M.G., and Zappia, L.R. 2003b. Effects of temperature on mineralization of petroleum in contaminated Antarctic terrestrial sediments. *Chemosphere* 52:975–987.

Filler, D.M., Lindstrom, J.E., Braddock, J.F., Johnson, R.A., and Nickalaski, R. 2001. Integral biopile components for successful bioremediation in the Arctic. *Cold Regions Science and Technology* 32:143–156.

Green, E.A., Kay, J.G., Jaber, K., Stehmeier, L.G., and Voordouw, G. 2000. Composition of soil microbial communities enriched on a mixture of aromatic hydrocarbons. *Applied and Environmental Microbiology* 66:5282–5299.

Griffiths, R.P., MacNamara, T.M., Cadwell, B.A., and Morita, R.Y. 1981. Field observations on the acute effect of crude oil on glucose and glutamate uptake in samples collected from Arctic and subArctic waters. *Applied and Environmental Microbiology* 41:1400–1406.

Grishchenkov, V.G., Townsend, R.T., McDonald, T.J., Autenrieth, R.L., Bonner, J.S., and Boronin, A.M. 2000. Degradation of petroleum hydrocarbons by facultative anaerobic bacteria under aerobic and anaerobic conditions. *Process Biochemistry* 35:889–896.

Hall, M. and Grover, P.L. 1990. Polycyclic aromatic hydrocarbons: Metabolism activation and tumour initiation. In: *Chemical Carcinogenesis and Mutagenesis*, vol. 1. Cooper, C.S. and Grover, P.L. eds. Spinger-Verlag, Berlin, Germany, pp. 327–372.

Hambrick, G.A.I., Delaune, R.D., and Patrick, W.H.J. 1980. Effect of estuarine sediment pH and oxidation reduction potential on microbial hydrocarbon degradation. *Applied and Environmental Microbiology* 40:365–369.

Hamdi, H., Benzarti, S., Manusadzianas, L., Aoyama, I., and Jedidi, N. 2007. Solid-phase bioassays and soil microbial activities to evaluate PAH-spiked soil ecotoxicity after a long-term bioremediation process simulating landfarming. *Chemosphere* 70:135–143.

Hatzinger, P.B. and Alexander, M. 1995. Effect of aging of chemicals in soil on their biodegradability and extractability. *Environmental Science and Technology* 29:537–545.

Heider, J., Spormann, A.M., Beller, H.R., and Widdel, F. 1999. Anaerobic bacterial metabolism of hydrocarbons. *FEMS Microbiology Review* 22:459–473.

Heitkamp, M.A. and Cerniglia, C.E. 1989. Polycyclic aromatic hydrocarbon degradation by a *Mycobacterium* sp. in microcosms containing sediment and water from a pristine ecosystem. *Applied and Environmental Microbiology* 55:1968–1973.

Hinchee, R.E. 1994. Bioventing of petroleum hydrocarbons. In: *Handbook of Bioremediation*. Norris, R.D. ed. Robert S. Kerr Environmental Research Laboratory, Lewis Publishers, Boca Raton, FL, pp. 39–59.

Horowitz, A., Krichevsky, M.I., and Atlas, R.M. 1983. Characteristics and diversity of subArctic marine oligotrophic, stenoheterotrophic, and euryheterotrophic bacterial populations. *Canadian Journal of Microbiology* 29:527–535.

Huesemann, M.H. and Truex, M.J. 1996. The role of oxygen diffusion in passive bioremediation of petroleum contaminated soils. *Journal Hazardous Materials* 51:93–113.

Huddleston, R.L. and Cresswell, L.W. 1976. Environmental and nutritional constraints of microbial hydrocarbon utilization in the soil. *Proceedings of 1975 Engineering Foundation Conference: The Role of Microorganisms in the Recovery of Oil*. NSF/RANN, Washington, DC, pp. 71–72.

Hughes, K.A., Bridges, P., and Clark, M.S. 2007. Tolerance of Antarctic soil fungi to hydrocarbons. *Science Total Environment* 372:539–548.

Hughes, T.J., Claxton, L.D., Brooks, L., Warren, S., Brenner, R., and Kremer, F. 1998. Genotoxicity of bioremediated soils from the Reilly tar site, St. Louis Park, Minnesota. *Environmental Health Perspective* 106:1427–1433.

Hulscher, T. and Cornelissen, G. 1996. Effect of temperature ion sorption equilibrium and sorption kinetics of organic micropollutants, a review. *Chemosphere* 32:609–626.

Hyun, S., Ahn, M.-Y., Zimmerman, A.R., Kim, M., and Kim, J.-G. 2008. Implication of hydraulic properties of bioremediated diesel-contaminated soil. *Chemosphere* 71:1646–1653.

Joergensen, R.G., Schmadeke, F., Windhorst, K., and Meyer, B. 1995. Biomass and activity of microorganisms in a fuel oil contaminated soil. *Soil Biology & Biochemistry* 27:1137–1143.

Jenisch-Anton, A., Mikolajczak, A., Rabenstein, A., Klindworth, J., Fischer, U., and Michaelis, W. 1999. Biodegradation of a high molecular weight aliphatic ether—Indications of an unusual biodegradation pathway. *Biodegradation* 10:383–389.

Johnsen, A.R., Wick, L.Y. and Harms, H. 2005. Principles of microbial PAH-degradation in soil. *Environmental Pollution* 133:71–84.

Kan, A.T., Fu, G., and Tomson, M.B. 1994. Adsorption/desorption hysteresis in organic pollutant and soil/sediment. *Environmental Science and Technology* 28:859–867.

Karl, D.M. 1992. The grounding of the *Bahia Paraiso*: Microbial ecology of the 1989 Antarctic oil spill. *Microbial Ecology* 24:77–89.

Kennicut, M.C. II, McDonald, T.J., Denoux, J., and McDonald, S.J. 1992. Hydrocarbon contamination on the Antarctic Peninsula. I. Arthur harbor-subtidal sediments. *Marine Pollution Bulletin* 24:499–506.

Kerry, E. 1990. Microorganisms colonizing plants and soil subjected to different degrees of human activity, including petroleum contamination in the Vestfold Hills and MacRobertson Land, Antarctica. *Polar Biology* 10:423–430.

Kerry, E. 1993. Bioremediation of experimental petroleum spills on mineral soils in the Vesfold Hills, Antarctica. *Polar Biology* 13:163–170.

Krauss, M. and Wilcke, W. 2002. Sorption strength of persistent organic pollutants in particle-size fractions of urban soils. *Soil Science Society America Journal* 66:430–437.

Kubátová, A., Steckler, T.S., Gallagher, J.R., Hawthorne, S.B., and Picklo, M.J. 2004. Toxicity of a wide-range polarity fractions from wood smoke and diesel exhaust particulate obtained using hot pressurized water. *Environmental Toxicology and Chemistry* 23:2243–2250.

Leahy, J.G. and Colwell, R.R. 1990. Microbial degradation of hydrocarbons in the environment. *Microbiological Review* 54:305–315.

Lemieux, C.L., Lambert, I.B., Lundstedt, S., Tysklind, M., and White, P.A. 2008. Mutagenic hazards of complex polycyclic aromatic hydrocarbon mixtures in contaminated soil. *Environmental Toxicology and Chemistry* 27:978–990.

Liebeg, E.W. and Cutright, T.J. 1999. The investigation of enhanced bioremediation through the addition of macro and micro nutrients in a PAH contaminated soil. *International Biodeterioration and Biodegradation* 44:55–64.

Lundstedt, A., Haglund, P., and Öberg, L. 2006. Simultaneous extraction and fractionation of polycyclic aromatic hydrocarbons and their oxygenated derivatives in soil using selective pressurized liquid extraction. *Analytical Chemistry* 78:2993–3000.

MacCormack, W.P. and Fraile, E.R. 1997. Characterization of a hydrocarbon degrading psychrotrophic Antarctic bacterium. *Antarctic Science* 9:150–155.

Macleod, C.J. and Semple, K.T. 2002. The adaptation of two similar soils to pyrene catabolism. *Environmental Pollution* 119:357–364.

Margesin, R. 2000. Potential of cold-adapted microorganisms for bioremediation of oil-polluted Alpine soils. *International Biodeterioration and Biodegradation* 46:3–10.

Margesin, R. and Schinner, F. 1997. Laboratory bioremediation experiments with soil from a diesel-oil contaminated site: Significant role of cold-adapted microorganisms and fertilizers. *Journal Chemistry Technology Biotechnology* 70:92–98.

Martins, C.C., Bicego, M.C., Taniguchi, S., and Montone, S.C. 2004. Aliphatic and polycyclic aromatic hydrocarbons in surface sediments in Admiralty Bay, King George Island, Antarctica. *Antarctic Science* 16:117–122.

Mazzera, D., Hayes, T., Lowenthal, D., and Zielinska, B. 1999. Quantification of polycyclic hydrocarbons in soil at McMurdo Station, Antarctica. *Science Total Environment* 229:65–71.

McCarthy, K., Walker, L., and Vigoren, L. 2004a. Subsurface fate of spilled petroleum hydrocarbons in continuous permafrost. *Cold Regions Science and Technology* 38:43–54.

McCarthy, K., Walker, L., Vigoren, L. and Bartel, J. 2004b. Remediation of spilled hydrocarbons by in situ landfarming at an Arctic site. *Cold Regions Science and Technology* 40:31–39.

McGroddy, S.E., Farrington, J.W., and Gschwend, P.M. 1996. Comparison of the *in situ* and desorption sediment-water partitioning of polycyclic aromatic hydrocarbons and polychlorinated biphenyls. *Environmental Science and Technology* 30:172–177.

McMillen, S.J., Gray, N.R., Kerr, J.M., Requejo, A.G., McDonald, T.J., and Douglas, G.S. 1995. Assessing bioremediation of crude oil in soils and sludges. In: *Monitoring and Verification of Bioremediation*, vol. 3. Hinchee, R.E., Douglas, G.S., and Ong, S.K. eds. Battelle Press, Columbus, OH, pp. 1–9.

Meyer, S. and Steinhart, H. 2000. Effects of heterocyclics PAH (N, S, O) on the biodegradation of typical tar oil PAH in a soil/compost mixture. *Chemosphere* 40:359–367.

Mohn, W.W., Radziminski, C.Z., Fortin, M.-C., and Reimer, K.J. 2001. On site bioremediation of hydrocarbon-contaminated Arctic tundra soils in inoculated biopiles. *Applied Microbiology and Biotechnology* 57:242–247.

Møller, J., Winther, P., Lund, B., Kirkebjerg, K., and Westermann, P. 1996. Bioventing of diesel oil-contaminated soil: Comparison of degradation rates in soil based on actual oil concentration and respirometric data. *Journal of Industrial Microbiology* 16:110–116.

Morgan, R. and Watkinson, R.J. 1989. Hydrocarbon degradation in soils and methods for soil treatment. *CRC Critical Review Biotechnology* 8:305–333.

Morgan, P. and Watkinson, R.J. 1992. Factors limiting the supply and efficiency of nutrient and oxygen supplements for the in situ biotreatment of contaminated soil and groundwater. *Water Research* 26:73–78.

Park, J., Ball, L.M., Richardson, S.D., Zhu, H.-B., and Aitken, M.D. 2008. Oxidative mutagenecity of polar fractions from polycyclic aromatic hydrocarbon-contaminated soils. *Environmental Toxicology and Chemistry* 27:2207–2215.

Pelletier, E., Delille, D., and Delille, B. 2004. Crude oil bioremediation in sub-Antarctic intertidal sediments: Chemistry and toxicity of oiled sediments. *Marine Environmental Research* 57:311–327.

Piehler, M.F., Swistak, J.G., Pinckney, J.L., and Paerl, H.W. 1999. Stimulation of diesel fuel biodegradation by indigenous nitrogen fixing bacterial consortia. *Microbial Ecology* 38:69–78.

Regoli, F., Nigro, M., Benedetti, M., Gorbi, S., Pretti, C., Gervasi, P.G., and Fattorini, D. 2005. Interactions between metabolism of trace metals and xenobiotics agonists of the Ah receptor in the Antarctic fish *Trematomus bernacchii*: Environmental perspectives. *Environmental Toxicology and Chemistry* 24:1475–1482.

Reynolds, C.M., Travis, M.D., Braley, W.A., and Scholze, R.J. 1994. Applying field-experiment bioreactors and landfarming in Alaskan climates. In: *Hydrocarbon Bioremediation*. Hinchee, R.E., Alleman, B.C., Hoeppel, R.E., and Miller, R.N. eds. Lewis Publishers, Boca Raton, FL, pp. 100–106.

Rike, A.G., Schiewer, S., and Filler, D.M. 2008. Temperature effects on biodegradation of petroleum contaminants in cold soils. In: *Bioremediation of Petroleum Hydrocarbons in Cold Regions*. Filler, D.F., Snape, I., and Barns, D.L. eds. Cambridge University Press, Cambridge, U.K., pp. 84–108.

Röling, W.F.M., Milner, M.G., Jones, D.M., Lee, K., Daniel, F., Swannell, R.J.P., and Head, I.M. 2002. Robust hydrocarbon degradation and dynamics of bacterial communities during nutrient enhanced oil spill bioremediation. *Applied and Environmental Microbiology* 68:5537–5548.

Ruberto, L.A.M., Vasquez, S., Lobalbo, A., and MacCormak, W.P. 2005. Psychrotolerant hydrocarbon-degrading *Rhodococcus* strains isolated from polluted Antarctic soils. *Antarctic Science* 17:47–56.

Saul, D.J., Aislabie, J., Brown, C.E., Harris, L., and Foght, J.M. 2005. Hydrocarbon contamination changes the bacterial diversity of soil from around Scott Base, Antarctica. *FEMS Microbial Ecology* 53:141–155.

Shi, T., Frederickson, J.K., and Balkwill, D.L. 2001. Biodegradation of polycyclic aromatic hydrocarbons by *Sphingomonas* strains isolated from the terrestrial subsurface. *Journal of Industrial Microbiology Biotechnology* 226:283–289.

Snape, I., Ferguson, S., and Revill, A. 2003. Constraints on rates of natural attenuation and in situ bioremediation of petroleum spills in Antarctica. In: *Assessment and Remediation of Contaminated Sites in Arctic and Cold Climates* (*Proceedings*). Nahir, M., Biggar, K., and Cotta, G. eds. St. Joseph's Print Group Inc., Edmonton AB, Canada, pp. 257–261.

Snape, I., Riddle, M.J., Stark, J.S., Cole, C.M., King, C.K., Duquesne, S., and Gore, D.B. 2001. Management and remediation of contaminated sites at Casey Station, Antarctica. *Polar Record* 37:199–214.

Spain, J.C. and van Veld, P.A. 1983. Adaptation of natural microbial communities to degradation of xenobiotic compounds effects of concentration exposure time inoculum and chemical structure. *Applied and Environmental Microbiology* 45:428–435.

Steinbach, A., Seifert, R., Annweiler, E., and Michaelis, W. 2004. Hydrogen and carbon isotope fractionation during anaerobic biodegradation of aromatic hydrocarbons. A field study. *Environmental Science and Technology* 38:609–616.

Stott, D.E. and Martin, J.P. 1990. Synthesis and degradation of natural and synthetic humic material in soil. In: *Humic Substances in Soil and Crop Sciences: Selected Readings*. International Humic Substances Society, Chicago, IL, pp. 37–63.

Sutherland, J.B., Rafii, F., Khan, A.A., and Cerniglia, C.E. 1995. Mechanisms of polycyclic aromatic hydrocarbon degradation. In: *Microbial Transformation and Degradation of Toxic Organic Chemicals*. Young, L.Y. and Cerniglia, C.E. eds. Willey-Liss, New York, pp. 269–306.

Thomassin-Lacroix, E.J.M., Eriksson, M., Reimer, K.J., and Mohn, W.W. 2002. Biostimulation and bioaugmentation for on-site treatment of weathered diesel fuel in Arctic soil. *Applied Microbiology and Biotechnology* 59:551–556.

Thomassin-Lacroix, E.J.M., Yu, Z., Eriksson, M., Reimer, K.J., and Mohn, W.W. 2001. DNA-based and culture-based characterization of hydrocarbon-degrading consortium enriched from Arctic soil. *Canadian Journal of Microbiology* 47:1107–1115.

Walworth, J., Braddock, J., and Woolard, C. 2001. Nutrient and temperature interactions in bioremediation of cryic soils. *Cold Regions Science and Technology* 32:85–91.

Walworth, J.L., Woolard, C.R., Braddock, J.F., and Reynolds, C.M. 1997. Enhancement and Inhibition of Soil Petroleum Biodegradation Through the Use of Fertilizer Nitrogen: An Approach to Determining Optimum Levels. *Journal of Soil Contamination* 6:465–480.

Wang, Z.D., Fingas, M., and Sergy, G. 1994. Study of 22-year old Arrow oil samples using biomarker compounds by GCMS. *Environmental Science and Technology* 28:1733–1746.

Warhurst, A.M. and Fewson, C.A. 2004. Biotransformations catalysed by the genus *Rhodococcus*. *Critical Reviews in Biotechnology* 14:29–73.

Whyte, L.G., Bourbonnière, C., Bellrose, C., and Greer, C.W. 1999. Bioremediation assessment of hydrocarbon contaminated soils from the high Arctic. *Bioremediation Journal* 3:69–79.

Whyte, L.G., Bourbonnière, L., and Greer, C.W. 1997. Biodegradation of petroleum hydrocarbons by psychrotrophic *Pseudomonas* strains possessing both alkane (*alk*) and the naphthalene (*nah*) catabolic pathways. *Applied Environmental Microbiology* 63:3719–3723.

Whyte, L.G., Goalen, B., Hawari, J., Labbé, D., Greer, C.W., and Nahir, M. 2001. Bioremediation treatability assessment of hydrocarbon-contaminated soils from Eureka, Nunavut. *Cold Regions Science and Technology* 32:121–132.

Whyte, L.G., Smits, T.H.M., Labbé, D., Greer, C.W., and van Beilen, J.B. 2002. Gene cloning and characterization of multiple alkane hydroxylase systems in *Rhodococcus* strains Q15 and NRRL B-16531. *Applied and Environmental Microbiology* 68:5933–5942.

Wilson, M.S., Bakermans, C., and Madsen, E.L. 1999. In situ, real-time catabolic gene expression: Extraction and characterization of naphthalene dioxygenase mRNA transcripts from groundwater. *Applied and Environmental Microbiology* 65:80–87.

Xue, W. and Warshawsky, D. 2005. Metabolic activation of polycyclic and heterocyclic aromatic hydrocarbons and DNA damage: A review. *Toxicological Applied Pharmacology* 206:73–93.

Yu, Z., Stewart, G.R., and Mohn, W.W. 2000. Apparent contradiction: Psychrotolerant bacteria from hydrocarbon-contaminated Arctic tundra soils that degrade diterpenoids synthesized by trees. *Applied and Environmental Microbiology* 66:5148–5154.

14 Bioremediation of Contaminated Sites in the Canadian Arctic: Monitoring Performance and the Effects of Biostimulation Using Molecular Methods

Charles W. Greer

CONTENTS

14.1 INTRODUCTION

The bioremediation of petroleum hydrocarbon–contaminated soil in polar, remote environments presents numerous challenges. In addition to extreme cold temperatures, short periods of above zero temperatures, low water and nutrient content of the soils, the distance and the cost of material, and equipment transport limit bioremediation options. Nevertheless, bioremediation remains a feasible alternative to treat hydrocarbon-contaminated soils in these regions because more invasive technologies (such as excavation and off-site treatment, incineration, etc.) are unlikely to be cost effective.

A number of studies have demonstrated that microorganisms, in particular bacteria, are capable of degrading hydrocarbons at the extreme temperatures typically encountered in polar environments (Margesin and Schinner, 1997, 2001; Whyte et al., 1997; Whyte et al., 1999; Aislabie et al., 2006). In fact, representatives from all the three major domains of life have demonstrated the ability to metabolize hydrocarbons to various degrees and there are currently in excess of 200 bacterial, algal, and fungal genera that are known to use hydrocarbons as carbon and energy sources (Prince et al., 2003; Head et al., 2006).

The extent to which microorganisms participate in the biodegradation of hydrocarbons appears to be a function of the ecosystem and the local environmental conditions (as reviewed by Leahy and Colwell, 1990). Prior exposure of a microbial community to anthropogenic and/or natural sources of hydrocarbons is an important factor in determining the rate of biodegradation.

In addition to the low temperature, limiting factors for the bioremediation of contaminated soils in polar environments are also related to the availability of other essential nutrients and the low levels of available water. Soils are often very low in organic carbon content, they can be very coarse textured, and also have low water holding capacities (Aislabie et al., 2006). Although there is no strict ratio of carbon to nitrogen to use when adding nitrogen fertilizers for biostimulation, the physicochemical characteristics of the site must be considered. The most important relationship is the mass of nitrogen per mass of soil water, because it is possible to overfertilize in relatively dry environments (Braddock et al., 1997; Walworth et al., 1997). It has been known for some time that biostimulation using fertilizers with readily available nitrogen and phosphorus have a positive impact on microbial hydrocarbon degradation activity in virtually all types of environments (Braddock et al., 1997; Macnaughton et al., 1999; Whyte et al., 1999; Margesin and Schinner, 2001; Whyte et al., 2001; Röling et al., 2002; Greer et al., 2003; Aislabie et al., 2004; Xu et al., 2004; Aislabie et al., 2006).

This chapter presents elements of two case studies from hydrocarbon-contaminated sites in northern Canada. Contaminated soils from each site were initially evaluated in a laboratory scale feasibility study to determine whether biotreatment was possible under the conditions normally encountered at the sites. In addition to determining the potential to treat contaminated soil using biological processes, this laboratory feasibility analysis also provides important data on possible treatment scenarios to optimize biodegradation rates. Results are presented from a number of culture-dependent and culture-independent microbiological techniques to monitor bioremediation performance during field trials and full-scale treatment to ensure that the treatment system is meeting desired objectives.

14.2 CHARACTERIZATION OF THE COLOMAC MINE SITE

Colomac, Northwest Territories, Canada, is the site of a former gold mine that was closed in 1996. Over the course of its operational life, a variety of fuel spills and leakage incidents occurred around the main fuel storage tank area. As part of the decommissioning, the empty fuel tanks were disassembled, and the fuel-contaminated soils were excavated and moved into a constructed biopile, that was equipped with aeration and irrigation systems, and the soils were fertilized with monoammonium phosphate (MAP) while being placed in the biopile. Prior to the removal of all the soils, samples were collected from the contaminated storage tank area and a feasibility study was conducted to determine if the soils contained hydrocarbon-degrading bacteria and to identify fertilizer conditions that would enhance degradation performance.

14.2.1 Indigenous Microbial Degradation Potential: Mineralization Analysis

The degradation potential of the indigenous soil bacterial population was examined using mineralization assays, in which, the degradation of a radiolabeled substrate, a representative of the hydrocarbon contaminants, is monitored by recovering radiolabeled carbon dioxide (Greer et al., 2003). These assays are performed on a small scale and can be used to evaluate a variety of factors, such as fertilizer type and concentration, on degradation. In this study, the degradation potential in unamended soil, and the effects of fertilizers at different concentrations were evaluated in the soils at 4°C using ^{14}C-hexadecane as a substrate (Figure 14.1). Without fertilizer, there was essentially no hexadecane mineralization, but with the addition of 20-20-20 (a commercial liquid fertilizer) or MAP, both at a concentration of 500 mg N/kg soil, hexadecane was mineralized effectively following a lag period of approximately 2 weeks. Adding half the initial concentration of MAP resulted in a shortening of the lag phase and a higher extent of mineralization, indicating that fertilizer concentration was an important factor in optimizing biodegradation performance.

14.2.2 Polymerase Chain Reaction Analysis

Total DNA was extracted (Fortin et al., 2004) from the initial contaminated composite soils (A and B) and the noncontaminated control soil (C) and subjected to polymerase chain reaction (PCR) analysis using oligonucleotide primers designed to detect bacterial genes involved in the biodegradation of alkanes (*alkB*) and polycyclic aromatic (*ndoB*) hydrocarbons (Figure 14.2). The *alkB* gene encodes alkane hydroxylase, the initial hydroxylating enzyme in the bacterial degradation pathway for aliphatic hydrocarbons (van Beilen et al., 1994). The *ndoB* gene encodes naphthalene dioxygenase, the initial hydroxylating gene in the bacterial pathway for

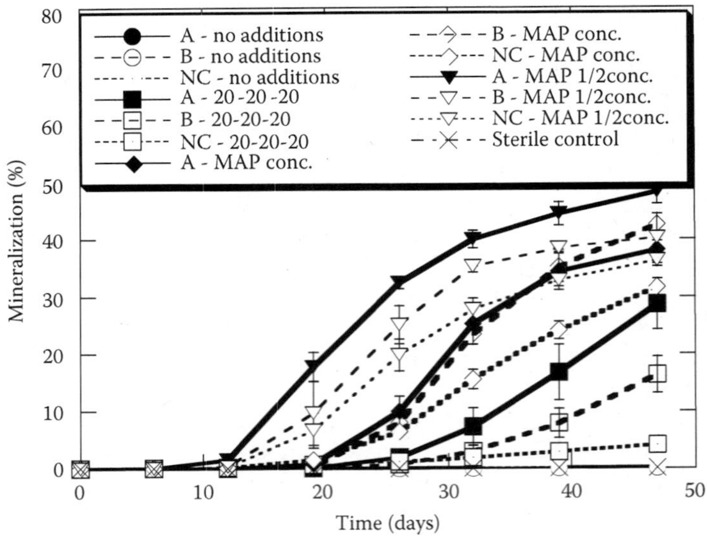

FIGURE 14.1 Mineralization of ^{14}C-hexadecane in Colomac soils at 4°C. Mineralization in initial composite soils (A and B) and the noncontaminated control soil (NC), without (no additions) and with amendment with the fertilizers 20-20-20 and MAP at two concentrations.

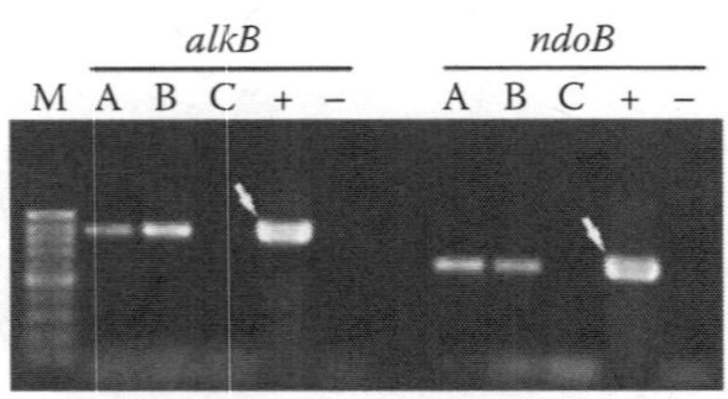

FIGURE 14.2 Agarose gel electrophoresis of the PCR amplified functional genes, *alkB* and *ndoB*, from total DNA extracts from the initial composite-contaminated soils (A and B) and from the noncontaminated control soil (C) from Colomac. The expected positive fragment size is indicated by an arrow in the + lane (positive PCR controls, *P. putida* ATCC 29347 (*alkB*) and *P. putida* ATCC17484 (*ndoB*)) and the negative PCR control is indicated by – (water). The molecular weight marker (100 base pair ladder) is in lane M.

polycyclic aromatic hydrocarbons. These two genes are commonly encountered in fuel-contaminated soils and are good indicators for the presence of bacteria with the genetic capacity to degrade the components of a variety of hydrocarbon fuels. Both contaminated soils showed the presence of the expected PCR fragment indicating that they contained indigenous bacteria possessing genes for alkane and polycyclic aromatic hydrocarbon degradation. The noncontaminated control soil did not show the presence of these genes suggesting that they were present at levels too low to be detected. As the fertilizer amended noncontaminated control soil did demonstrate mineralization after a long lag time (Figure 14.1), this confirms that hydrocarbon degraders were present in the indigenous microbial population in soil with no evidence of hydrocarbon contamination. Although PCR is a rapid, qualitative method for gene detection (sample extraction and analysis can be performed in 1 day), it does not provide direct evidence of gene expression in the studied systems. Another advantage of PCR detection is that the analysis can be performed at various levels of stringency to increase or decrease detection specificity.

14.2.3 Denaturing Gradient Gel Electrophoresis Analysis

Denaturing gradient gel electrophoresis (DGGE), a method to separate PCR-amplified DNA fragments (Muyzer et al., 1993), is used extensively as a rapid, high-throughput screening method for assessing the dominant members of a microbial community. When used to monitor the same system over time or with different treatments, DGGE analysis can provide extremely useful information on changes in the community structure. There are limitations to using DGGE as a method of community analysis, including PCR amplification biases, the choice of primers, detection sensitivity, and the lack of standardized databases containing the data for comparative analyses. Marzorati et al. (2008) presented a method to normalize DGGE data so it can be compared between laboratories and environments. This could be a major step forward in the use of screening data (as an alternative to large-scale sequencing) to evaluate different environments.

The mineralization study on the Colomac samples was followed more closely by examining the microbial population composition under different fertilizer regimens using DGGE analysis (Figure 14.3). The microbial population showed considerable variation in the presence of the different fertilizers (20-20-20, MAP, or urea). The banding patterns were the most similar (88%) for 20-20-20 and MAP, the two fertilizers that resulted in the best mineralization activity, and these were significantly different from the soil without fertilizer addition. The urea treated soil was the most different in terms of banding pattern, and no mineralization was observed in the soil using this fertilizer (data not shown). Bands were cut from the DGGE and the nucleotide sequences were determined, analyzed, and clustered to determine the phylogenetic distribution of bacteria under the different treatment regimens (Figure 14.4). The soils differed considerably with respect to the relative proportions of major bacterial phyla. In the nonamended contaminated soil, Alphaproteobacteria

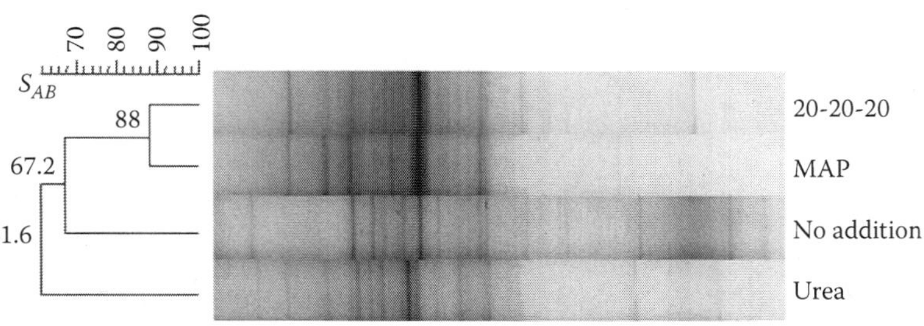

FIGURE 14.3 DGGE analysis of Colomac soils without (no addition) or following biostimulation with different fertilizers (20-20-20, MAP, urea).

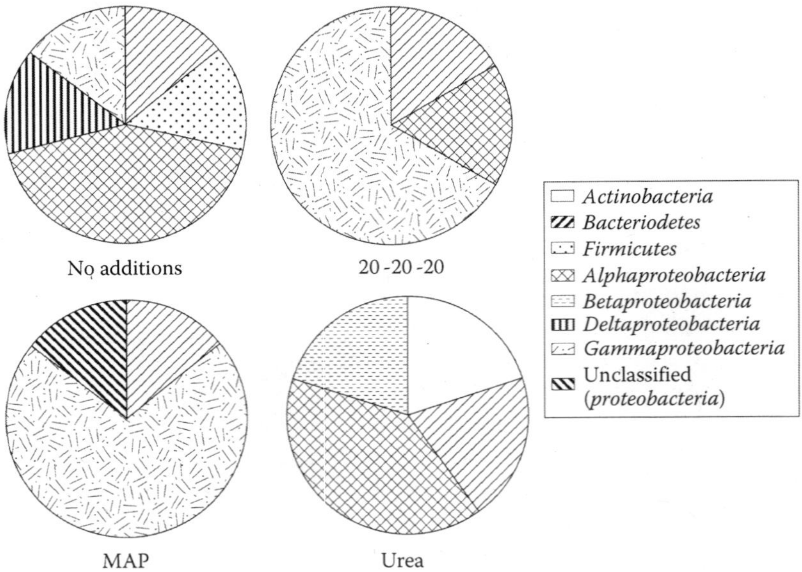

FIGURE 14.4 Relative distribution of microbial phyla in Colomac soils following incubation without (no additions) and with the fertilizers 20-20-20, MAP, or urea.

were dominant (ca. 50%), with roughly equal proportions of Bacteroidetes, Firmicutes, Deltaproteobacteria, and Gammaproteobacteria. Following MAP or 20-20-20 application, the Gammaproteobacteria became dominant, representing ca. 70% of the total microbial population. This coincided with the best hydrocarbon degradation conditions suggesting that the Gammaproteobacteria were key members of the hydrocarbon-degrading microbial community. In the urea-amended soil, the Alphaproteobacteria remained dominant, but the Betaproteobacteria and Actinobacteria also became significant members of the microbial community. The increase in numbers of Betaproteobacteria following nutrient amendment has been observed in other systems treating hydrocarbon-contaminated soils (Viñas et al., 2005). These results clearly show the effect of different fertilizer treatments on the microbial community structure, and suggest that fertilizer selection could be a significant factor in assuring optimum biodegradation performance.

14.2.4 Quantitative-PCR

Quantitative-PCR (Q-PCR) is a powerful technique to determine the relative number of target gene copies in a sample (Powell et al., 2006), which can be used to assess whether the target genes have increased or decreased in number in response to a bioremediation treatment. When Q-PCR is combined with reverse transcriptase (Q-RT-PCR), the level of expression of the target gene can be determined providing direct evidence of the targeted degradation pathway genes. The constructed biopiles at the Colomac mine site were examined after approximately 1 year of on-site treatment using Q-PCR.

Random composite samples were collected from several locations throughout the biopile, total microbial community DNA was extracted and analyzed using primers designed from several homologues of the *alkB* gene (Figure 14.5). The *alkB* (Pp) gene is derived from the well-characterized alkane degrader *Pseudomonas putida* ATCC 293487, the *alkB* (pp5) was derived from a *P. putida* strain isolated from a contaminated site at Eureka, Ellesmere Island in the high Arctic, and the *alkB1* (Q15) was derived from a *Rhodococcus* sp. Q15 containing at least four different *alk* gene homologues (Whyte et al., 2002). The results showed that these three genes were virtually undetectable in the original composite soils and in the noncontaminated control soil. Following 1 year of on-site treatment, there was a small increase in the quantity of the *alkB* (pp5) gene, but a very large increase in the quantity of the *alkB1* (Q15) gene in all samples collected from the biopile. These results indicate that the increased biodegradation potential in the biopile soil was largely attributable to *alkB* type genes found in *Rhodococcus* sp.

The feasibility study for the Colomac mine site demonstrated that the contaminated soil did possess indigenous hydrocarbon-degrading bacteria, and that fertilizer amendment was necessary to ensure an adequate supply of other essential nutrients (nitrogen, phosphorus) to the system. Of importance in the remediation of remote sites, the amount of fertilizer necessary for the optimal stimulation of

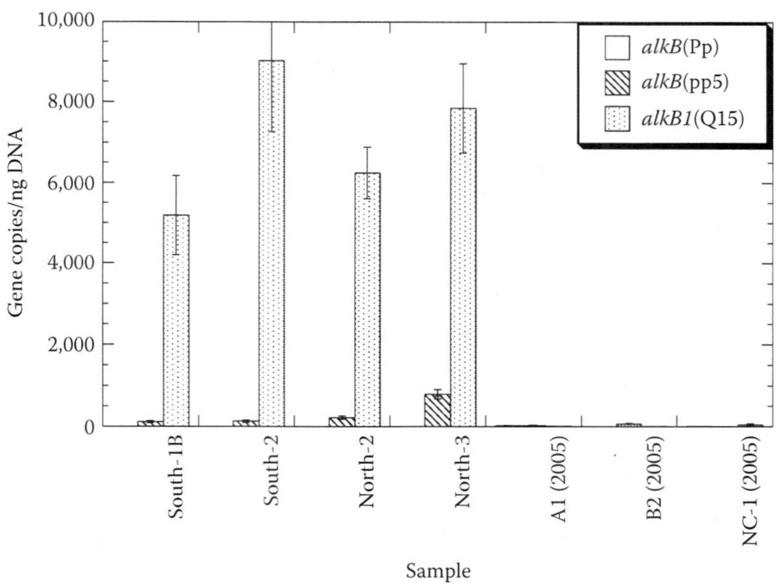

FIGURE 14.5 Q-PCR for genes involved in alkane (*alk* genes) degradation in soils from the Colomac mine site. Total DNA extracts from the initial-contaminated composite soils (A1, B1), from a noncontaminated control soil (NC-1), and from the biopile soils after 12 months of treatment (South-1B, South-2, North-2, North-3) were examined for the *alkB* gene from the well-characterized alkane-degrading *P. putida* ATCC 29347 (*alkB*(Pp)), from a high Arctic *P. putida* isolate (*alkB*(pp5)), and from the *alkB1* gene from *Rhodococcus* sp. Q15 (*alkB1*(Q15)).

the indigenous hydrocarbon-degrading bacteria was considerably lower than the amount that would have been added based on the contamination level in the soil. For the practitioner in this field, this means lower costs for fertilizer purchase and transportation to the site.

14.3 CHARACTERIZATION OF CANADIAN FORCES STATION-ALERT

Canadian Forces Station-Alert (CFS-Alert) is the most northerly, permanently inhabited community on the planet. It has been in operation for approximately 60 years, and during that time, due to the large amounts of fuel that must be stored and transported around the site, numerous spill, leakage or pipeline break incidents have occurred. As a result, several areas have been constructed on-site to contain and remediate fuel-contaminated soils. The on-site treatment included the application of fertilizer and yearly mixing of the soil. This section relates to the initial characterization of contaminated soil, the determination of its bioremediation potential, and the subsequent on-site monitoring after it had been relocated into one of the constructed areas. Two composite samples of the initial contaminated soil were examined to assess the biotreatability potential, using a number of microbiological and molecular analyses. These same analytical procedures were also used to monitor bioremediation performance using samples collected from the on-site biopiles at different times after biotreatment was initiated.

14.3.1 Microbiological Characteristics

The total heterotrophic bacterial population and the hydrocarbon (diesel)-degrading bacterial population were determined in the two composite contaminated soils and in the noncontaminated soil, as well as in the biopile soil samples collected after 2 and 15 months of on-site treatment (Table 14.1). The total viable heterotrophic population in the initial soil was relatively high for a nutrient-poor soil and the cold-adapted population (4°C) was comparable to, or in some cases, higher (i.e., composite B-1) than the population size observed at room temperature (22°C). The same trend was observed in the hydrocarbon-degrading bacterial population, with comparable population sizes determined at 4°C in the contaminated soils. The noncontaminated control soil typically had lower bacterial populations of both total heterotrophic bacteria and hydrocarbon-degrading bacteria. These results indicated that the bacterial populations were higher and well adapted to the low temperature conditions present in the contaminated soils, where the nutrient status (carbon source from the hydrocarbons) would be more favorable for growth. This inference also indicated that the soil was nutrient deficient, and the presence of the hydrocarbons partially and temporarily alleviated the poor nutrient status.

The total heterotrophic and hydrocarbon-degrading bacterial populations in the biopile increased by approximately one order of magnitude after 2 months of on-site treatment. After 15 months of on-site treatment, the total heterotrophic bacterial population density remained comparably high, but the hydrocarbon-degrading bacterial population had decreased by almost an order of magnitude.

TABLE 14.1
Viable Bacterial Population Density in Alert Soils

	CFU/g Wet Soil			
	Total Heterotrophs (YTS)		Hydrocarbon Degraders (MSM-Diesel)	
Sample	22°C	4°C	22°C	4°C
Samples from Initial Soil Samples				
A-1	1.52e+7	7.69e+6	1.16e+6	1.14e+6
B-1	2.14e+7	4.18e+7	5.69e+6	1.16e+6
C-1	9.68e+5	9.68e+5	9.69e+3	1.45e+4
Random Samples from Biopile after 2 Months (2005)				
Bio1	2.14e+8	1.28e+8	6.61e+7	4.92e+7
Bio2	1.70e+8	1.52e+7	1.18e+7	2.14e+7
Bio3	2.34e+8	1.99e+8	4.92e+7	1.96e+7
Bio4	4.18e+8	3.06e+7	1.18e+7	6.97e+7
Random Samples from Biopile after 15 Months (2006)				
Bio1	3.05e+8	1.16e+8	3.06e+6	2.53e+6
Bio2	4.92e+8	1.96e+8	2.14e+6	1.52e+6
Bio3	2.75e+8	1.52e+8	3.06e+6	3.72e+6
Bio4	4.18e+8	1.18e+8	2.32e+6	3.06e+6

Note: Total heterotrophic bacteria were determined on nutrient media (YTS) and hydrocarbon degraders were determined on mineral salts-diesel medium (MSM-diesel). Samples included the initial soil composites (A-1, B-1), a noncontaminated control soil (C-1), and random composite soils collected from the biopile during treatment after 2 months (Bio1–Bio4, 2005) and 15 months of treatment (Bio1–Bio4, 2006).

14.3.2 Indigenous Microbial Degradation Potential: Mineralization Analysis

The assays were performed on the untreated soils, on the nutrient amended soils and on soil samples collected from the biopiles after 2 and 15 months of treatment. The nutrient amendments were designed to improve microbial activity by supplying the nutrients that were most likely limiting, nitrogen and phosphorus. The assays were conducted at 4°C, the average summer temperature at the site. In the absence of nutrient amendments, the contaminated soil composite (AB) showed a negligible mineralization of hexadecane at 22°C (data not shown) and at 4°C (Figure 14.6). When the soils were amended with MAP, at 500 mg/kg (high) or 250 mg/kg (low) nitrogen, hexadecane mineralization was high (ca. 60%) after 7 weeks incubation. The amendment of the composite soil with 20-20-20 at 250 mg/kg resulted in slightly lower levels of mineralization compared to those observed with MAP (Figure 14.6). Urea was not

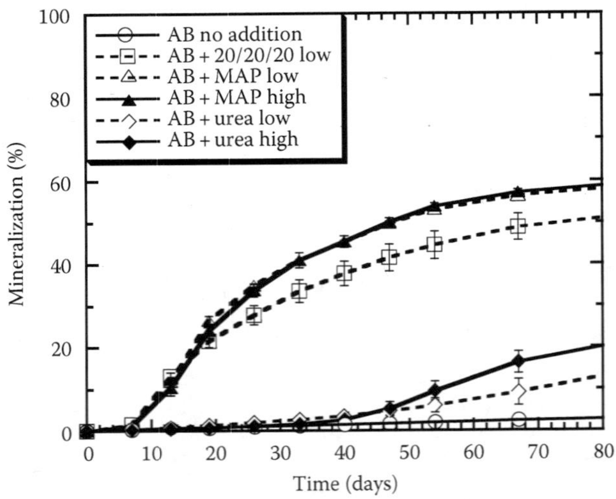

FIGURE 14.6 Mineralization of ^{14}C-hexadecane in Alert soils incubated at 4°C. Mineralization in the initial composite soil (AB) with no addition, following amendment with 20-20-20 fertilizer at a low concentration, or MAP or urea at low and high concentrations.

effective at stimulating the indigenous activity to any appreciable extent in comparison with the other two fertilizers, although weak mineralization was observed after an extensive lag phase. The results demonstrated that the initial composite-contaminated soil contained cold-adapted aliphatic hydrocarbon-degrading microbial populations, but to obtain activity, nutrient amendment was required. The specific nutrient amendment used also had a noticeable effect, with urea as a very poor candidate, while both 20-20-20 and MAP stimulated mineralization to comparable extents.

The soil samples from the biopiles were collected after approximately 2 and 15 months of on-site treatment and analyzed for indigenous hexadecane mineralization activity at 4°C (Figure 14.7). All the biopile samples collected after 2 months of treatment showed a rapid rate of hexadecane mineralization, and achieved the maximum extent only after 4 weeks of incubation (Figure 14.7A). The control soil showed almost no mineralization in the same time frame. After 15 months of on-site treatment, the rates of hexadecane mineralization were still very high, but somewhat slower than after 2 months (Figure 14.7B). This was not surprising considering that the samples were collected in October when the ambient air temperatures were already approximately −20°C on-site. Nevertheless, in both cases, the biopile samples showed essentially no lag times in hexadecane mineralization, somewhat better than the initial soil, which showed a lag time before mineralization began.

14.3.3 Laboratory Mesocosms

Laboratory mesocosms amended and incubated under identical conditions to the mineralization microcosms were analyzed for residual hydrocarbons (C10–C50) after

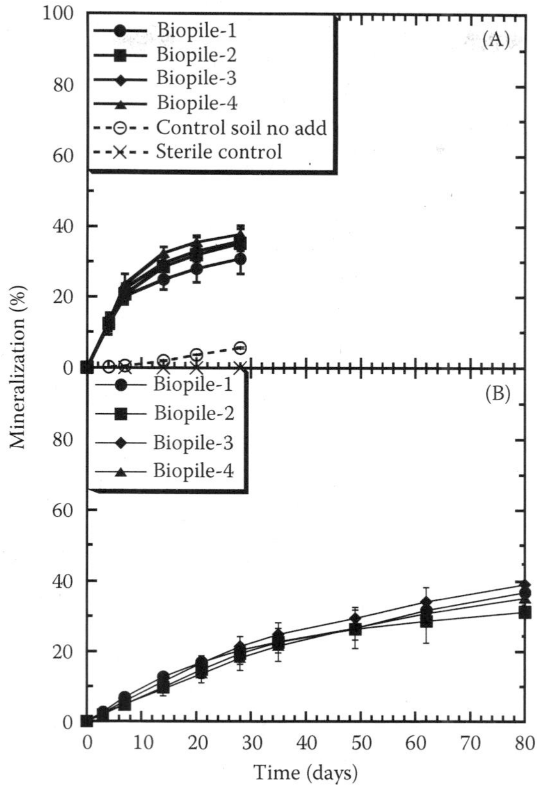

FIGURE 14.7 Mineralization of ^{14}C-hexadecane in Alert biopile soils incubated at 4°C. Four random soil samples were collected on site after 2 months (A) and 15 months (B) of on-site treatment. A noncontaminated soil from the site was included as a control, in addition to a sterile control.

3 months of incubation at either 22°C or 4°C (Table 14.2). The results showed that even without amendments, considerable degradation of the hydrocarbons had occurred at both incubation temperatures. The most significant decrease was in soil A amended with 20-20-20 and incubated at 22°C (>90% decrease). The results demonstrated that nutrient amendment significantly increased the degradation of hydrocarbons, especially at the higher incubation temperature (22°C). A typical GC-FID chromatogram showing the results of several composite soil B mesocosms in comparison to weathered diesel and several alkane standards is shown in Figure 14.8. The results show that the degradation proceeds more rapidly at the higher incubation temperature.

14.3.4 Monitoring Biopile Performance

A large biopile was constructed with the contaminated soils in a bermed area on-site in July, 2005. The soils were amended with MAP at a concentration equivalent to

TABLE 14.2
Total Petroleum Hydrocarbon Concentration in Alert Mesocosm Composite Soils (A and B) or Noncontaminated Control Soil (C) without and with Biostimulation (20-20-20 or MAP Fertilizers) and Incubation at 4°C or 22°C for 3 Months

	Total Petroleum Hydrocarbons (C10–C50) (mg/kg)		
		After 3 Months Incubation At	
Sample	Time = 0	22°C	4°C
A	3400		
A – no additions		1850	2650
A + 20-20-20		275	1400
A + MAP		650	1250
B	3250		
B – no additions		800	1150
B + 20-20-20		400	1150
B + MAP		300	950
C	<100		

500 mg/kg contaminated soil. After approximately 2 months of on-site treatment, random composite soil samples were collected for the analysis of residual hydrocarbons (Table 14.3) and for indigenous microbial population density (Table 14.1) and mineralization activity (Figure 14.7A). As indicated previously, the microbial population density (Table 14.1) and the hexadecane mineralization activity (Figure 14.7A) increased significantly in biopile soils relative to the starting contaminated soil. In addition, chemical analysis demonstrated that the residual hydrocarbon concentration had been reduced by almost 60% (Table 14.3). A more detailed analysis of the hydrocarbon fractions in the biopile soils collected after 12 and 15 months of on-site treatment demonstrated that the residual hydrocarbon concentration after 15 months of treatment was below the Canadian Council of Ministers of the Environment (CCME) criteria for industrial coarse soils, indicating that the soils can now be considered for release (Table 14.3). The results show that a minimal on-site biostimulation treatment system was sufficient to treat the on-site contamination within a 2 year time frame. This data indicates that should future contamination events occur on site, it would be possible to treat the contaminated soil within a very reasonable time without the need for significant treatment infrastructure.

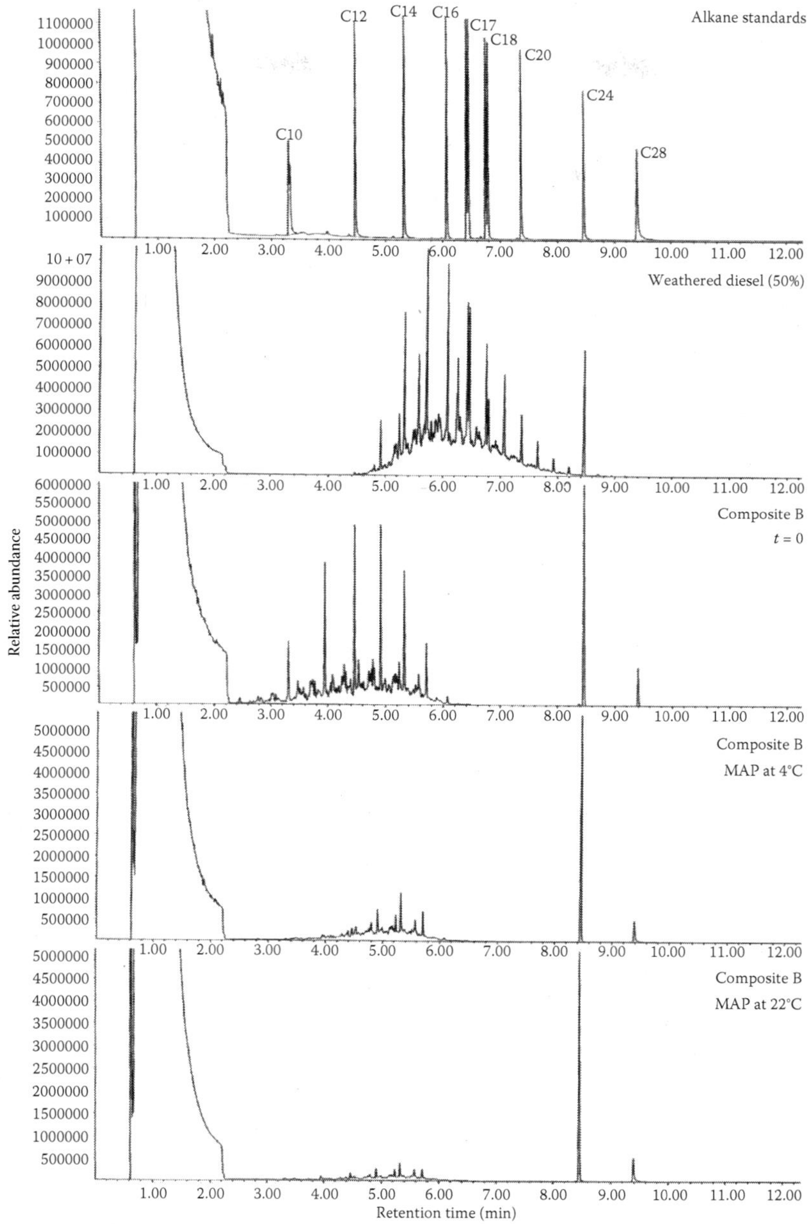

FIGURE 14.8 GC-FID chromatogram of C10–C50 hydrocarbons in Alert composite soil B laboratory mesocosms, initially, and following MAP amendment and incubation at either 4°C or 22°C for 3 months. Chromatograms showing standard reference alkanes and weathered diesel (50%) are presented for comparison.

TABLE 14.3
Total Petroleum Hydrocarbon Analysis in Alert Biopile Soils

		Residual Hydrocarbon Concentration (mg/kg)			
		Hydrocarbon Fraction			
Sample	**Total C10–C50**	**F1 (C6–C10)**	**F2 (C10–C16)**	**F3 (C16–C34)**	**F4 (C34–C50)**
Criteria—industrial soil		310	760	1700	3300
	Random Samples after 2 Months of On-Site Treatment (2005)				
Biopile-1	1433				
Biopile-2	1333				
Biopile-3	1267				
Biopile-4	1400				
Control (NC)	<100				
	Random Samples after 12 Months On-Site Treatment (2006)				
A1		18	680	37	ND
A2		22	1000	46	ND
B1		22	500	26	ND
B2		24	810	36	ND
B3		18	640	47	ND
C1		27	790	38	ND
D1		32	550	29	ND
D2		20	740	36	ND
D3		20	720	41	ND
	Random Samples after 15 Months On-Site Treatment (2006)				
Biopile-1		13	40	30	ND
Biopile-2		16	400	40	ND
Biopile-3		29	340	50	ND
Biopile-4		16	270	20	ND

Note: Samples initially analyzed using the C10–C50 method (CEAEQ, 1997), and subsequently using the F1–F4 method (Turle et al., 2007). Samples included a noncontaminated control soil (control) and random composite soils collected from the biopile during treatment after 2 months (biopile-1–biopile-4, 2005), 12 months (A1–D3), and 15 months of treatment (biopile-1–biopile-4, 2006). ND, not detected.

14.3.5 Catabolic Gene Microarray Analysis

A catabolic gene microarray containing amplicons from more than 100 bacterial genes involved in the degradation of hydrocarbon pollutants, chlorinated compounds, heavy metal resistance, and genes from several major nutrient cycles (nitrogen, carbon, sulfur) was used to monitor the total microbial communities in hydrocarbon-contaminated soil undergoing treatment at CFS-Alert. Microarray analysis has been developed and used by several researchers to detect specific target genes primarily using oligonucleotides derived from those genes (Wu et al., 2001; Zhou, 2003; Rhee et al., 2004). Several target genes involved in the degradation of aliphatic (*alkB*) and aromatic (*ndoB*) hydrocarbons were found to increase significantly following biostimulation with the fertilizer MAP (Figure 14.9). When the intensities of the hybridization signals relative to the internal controls (16S rRNA gene, lambda DNA, and the *luxA* gene) were analyzed at different times during the treatment, substantial increases in the frequencies of specific *alkB* genes, and the *ndoB* gene were observed, coinciding with increased hydrocarbon-degradation activity in the soil. In particular, two different *alkB* genes from *Rhodococcus* spp. were found to be the predominant alkane degradation genes in the first and second years of treatment. During the first year, the *ndoB* gene was detected at an elevated frequency, but its level during the second year of treatment was below the detection limit. Other well-known alkane-degradation genes from *Alcanivorax borkumensis* (*alkB1Ab*) and *Acinetobacter calcoaceticus* (*alkM*) were not detected in the initial-contaminated soil, or in the soil under treatment. The catabolic gene microarray proved to be sensitive enough to detect changes in the relative frequencies of several important target genes over the course of monitoring biodegradation performance in the biopiles. With further refinements to the technique to increase sensitivity and to render the method more quantitative, microarray analysis could be a very valuable, high-throughput approach for a long-term monitoring of bioremediation performance.

14.3.6 Biopile Monitoring Using Q-PCR

Q-PCR is a powerful technique to determine the relative number of target gene copies in a sample, which can be used to assess whether the target genes have increased or decreased in response to a bioremediation treatment (Powell et al., 2006). In the present study, Q-PCR was used to validate the results observed with the catabolic gene microarray on samples collected during the bioremediation of contaminated soil at CFS-Alert. The same samples that were examined using the microarray were assessed for the presence of several key target genes involved in the degradation of aliphatic and aromatic hydrocarbons. The Q-PCR results (Figure 14.10) supported the results seen with the catabolic microarray in that the alkane hydroxylase encoding gene from *Rhodococcus* Q15 (*alkB1*(Q15)) became a dominant alkane-degrading gene in the hydrocarbon-contaminated soil during treatment. This domination was observed through more than 1 year of on-site treatment. Another alkane hydroxylase-encoding gene from an indigenous strain of *P. putida* (*alkB*(Pp5)) was a major gene in the initial soil and during the first year of treatment, but its numbers decreased substantially during the second year of on-site treatment. The best known *alkB* gene, derived from *P. putida* (ATCC29347) was never detected in this Arctic soil. One of the key genes in the biodegradation of polycyclic aromatic hydrocarbons, the *ndoB* gene (*nahAc*) from *Pseudomonas* sp. was detected in the starting contaminated soil, and throughout the 2 year monitoring program, but it gradually decreased in concentration during the second year of treatment.

Q-PCR was shown to be a very effective technique to assess and monitor the number of target genes in the biopile samples, and to follow any

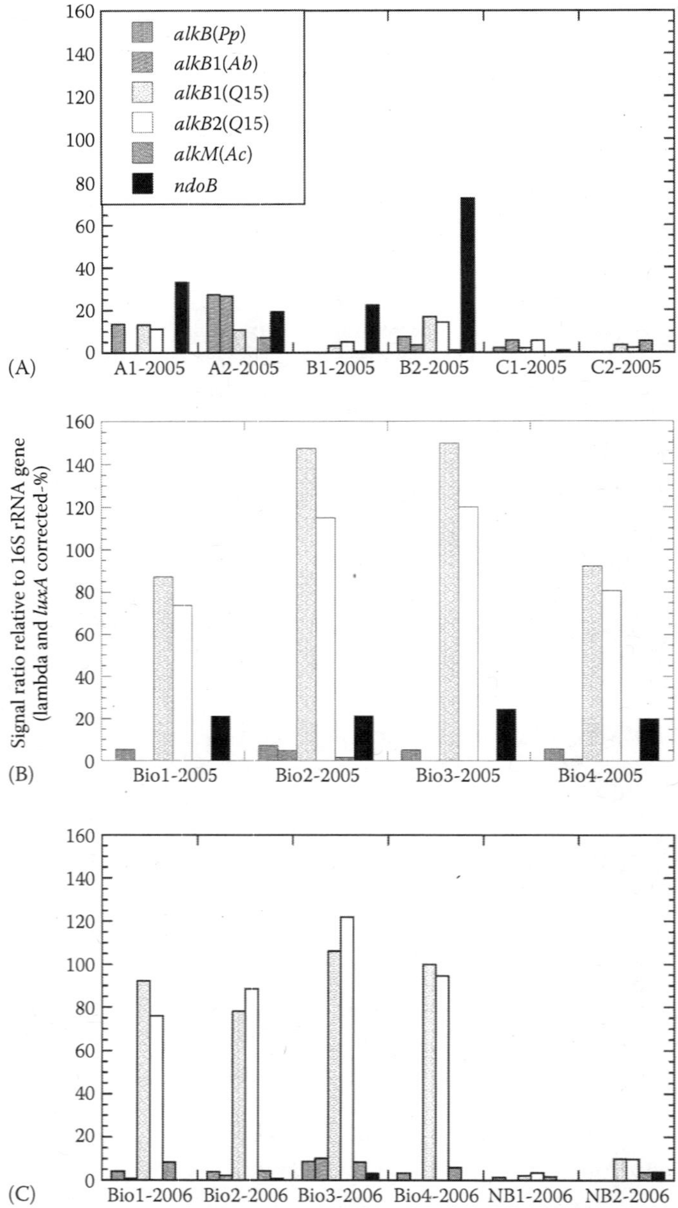

FIGURE 14.9 Catabolic gene microarray analysis of Alert soils. Gene frequencies were determined relative to total 16S rRNA genes, corrected for using internal lambda DNA for the *luxA* gene. Target genes involved in aliphatic (*alk* genes) or polycyclic aromatic (*ndoB* gene) hydrocarbon degradation are shown for total DNA extracted from the initial-contaminated composite soils (A1, A2, B1, B2), or noncontaminated control soil (C1, C2) in 2005 (panel A), from biopile samples collected in 2005 (Bio1–Bio4), 2 months after treatment started (panel B), and from biopile soils collected in 2006, 15 months after treatment started (Bio1–Bio4), in addition to contaminated soil from a fresh hydrocarbon spill (NB1, NB2) (panel C).

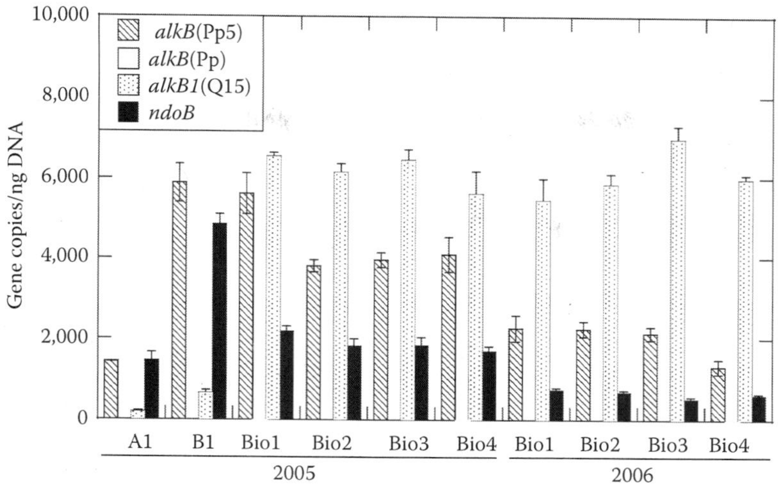

FIGURE 14.10 Q-PCR for genes involved in alkane (*alk* genes) and polycyclic aromatic hydrocarbon (*ndoB* gene) degradation in soils from CFS-Alert. Total DNA extracts from the initial-contaminated composite soils (A1, B1), and from the biopile soils after 2 months of treatment (Biol–Bio4, 2005) and from biopile soils after 15 months of treatment (Biol–Bio4, 2006) were examined for the *alkB* gene from a high Arctic *P. putida* isolate (*alkB*(Pp5)), from the well-characterized alkane-degrading *P. putida* ATCC 29347 (*alkB*(Pp)), the *alkB1* gene from *Rhodococcus* sp. Q15 (*alkB1*(Q15), and the *ndoB* gene from *P. putida* ATCC17484 (*ndoB*).

change in numbers during bioremediation. This technique was valuable in determining which treatment had the greatest stimulatory effect on the indigenous hydrocarbon-degrading microbial population. Importantly, the same trends of increased numbers of specific *alkB* genes that were observed using the catabolic gene microarray were observed during Q-PCR analysis. This suggests that microarray analysis can be semiquantitative when using appropriate quantitated standards, which could be a major advantage when it is desirable to monitor multiple target genes simultaneously.

14.4 CONCLUSIONS

The feasibility of biotreating hydrocarbon-contaminated soils in the Canadian high Arctic was evaluated by initial laboratory analyses of the hydrocarbon degradation potential and the activity of the indigenous microbial population and by biostimulation using several readily available fertilizer amendments.

An initial laboratory biotreatability study was performed on excavated fuel hydrocarbon-contaminated soils at a former gold mine site in Colomac, Northwest Territories, and at CFS-Alert, Nunavut, Canada. The study evaluated the presence of hydrocarbon-degrading viable bacterial populations, the genetic potential of the indigenous soil bacteria to degrade fuel hydrocarbons, the degradation and mineralization activity of soils with and without fertilizer amendments, and the use of

Q-PCR and a catabolic gene microarray to evaluate the initial indigenous microbial population and to monitor biopile soils during the on-site treatment phase. The laboratory results demonstrated good hydrocarbon-degradation activity by the indigenous microbial population following nutrient amendment, and this activity was sustained at 4°C. Chemical analyses of soils during laboratory or on-site treatment confirmed degradation activity and the positive influence of the fertilizer amendments. At CFS-Alert, an on-site biopile treatment system was constructed to treat contaminated soils in the field and provided comparable results to those that had been observed in the laboratory. These results demonstrated that a simple treatment design that includes controlled fertilizer amendment and soil tilling resulted in an effective treatment for fuel hydrocarbon-contaminated soil under ambient climatic conditions at the Colomac and CFS-Alert sites. The initial laboratory biotreatability study was a crucial step because each site presents specific conditions and challenges that will influence the ultimate success of any bioprocess. This approach of minimal intervention for the biotreatment of contaminated sites is essential in remote areas with extreme climates, and demonstrates the potential for the on-site remediation of other Arctic sites.

The contaminated soils from the two study sites contained indigenous hydrocarbon-degrading microbial populations that responded positively to fertilizer amendment and were active at 4°C, a temperature that might be achieved in the high Arctic during the summer months. In the laboratory studies, hexadecane mineralization activity was increased significantly using the fertilizers 20-20-20 and MAP, whereas urea had no significant effect on degradation activity. The microbial populations in contaminated soils from both sites responded to the fertilizer amendment by increasing the population density and relative hexadecane mineralization activity, and by increasing the relative numbers of key genes involved in the degradation of aliphatic (*alk* genes) and polycyclic aromatic (*ndoB gene*) hydrocarbons, determined by Q-PCR.

At the CFS-Alert site, an on-site biopile treatment system was set up in a bermed area in the spring of 2005, and monitored for microbial activity and residual hydrocarbon concentrations over a 15 month period. In the field, the microbial population density and hydrocarbon-degradation activity increased in a manner that was consistent with the results that had been observed in the laboratory feasibility study. Monitoring data over the 15 month treatment period demonstrated that the hydrocarbon concentration had been reduced to levels that allowed release of the soil based on the criteria for an industrial use, coarse grain soil. These results were supported by an observed increase in the numbers of key hydrocarbon-degradation genes by Q-PCR and by analyzing the genes semiquantitatively using a catabolic gene microarray.

The initial laboratory feasibility study was instrumental in establishing the treatment parameters for the on-site treatment system, which included the best fertilizer amendment and the amount necessary to obtain optimum-degradation activity. Performance monitoring in the field using culture-dependent and culture-independent analyses, including mineralization assays, Q-PCR, and a catabolic gene microarray, validated the efficacy of the designed treatment system and confirmed the endpoints necessary for the release of the treated soil.

ACKNOWLEDGMENTS

Thanks are extended to Sylvie Sanschagrin, Danielle Beaumier, Claude Masson, and Danielle Ouellette for their technical support, and to Serge Delisle, Lyle Whyte, and David Juck for their contributions to this work.

REFERENCES

Aislabie, J.M., M.R. Balks, J.M. Foght, and E.J. Waterhouse. 2004. Hydrocarbon spills on Antarctic soils: Effects and management. *Environ. Sci. Technol.* 38:1265–1274.

Aislabie, J.M., D.J. Saul, and J.M. Foght. 2006. Bioremediation of hydrocarbon-contaminated polar soils. *Extremophiles* 10:171–179.

Braddock, J.F., M.L. Ruth, P.H. Catterall, J.L. Walworth, and K.A. McCarthy. 1997. Enhancement and inhibition of the microbial activity in hydrocarbon-contaminated Arctic soil: Implications for nutrient-amended bioremediation. *Environ. Sci. Technol.* 31:2078–2084.

Centre d'expertise en analyse environnementale du Québec (CEAEQ). 1997. Méthode d'analyse-Sols-Dosage des hydrocarbures pétroliers C10 à C50, 1e édition. Ministère de l'environnement et de la faune du Québec, MA.410-HYD. 1.0, 10 pp.

Fortin, N., D. Beaumier, K. Lee, and C.W. Greer. 2004. Soil washing improves the recovery of total community DNA from polluted and high organic content sediments. *J. Microbiol. Methods* 56:181–191.

Greer, C.W., N. Fortin, R. Roy, L.G. Whyte, and K. Lee. 2003. Indigenous sediment microbial activity in response to nutrient enrichment and plant growth following a controlled oil spill on a freshwater wetland. *Bioremed. J.* 7:69–80.

Head, I.M., D. Martin Jones, and W.F.M. Röling. 2006. Marine microorganisms make a meal of oil. *Nat. Rev. Microbiol.* 4:173–182.

Leahy, J.G. and R.R. Colwell. 1990. Microbial degradation of hydrocarbons in the environment. *Microbiol. Rev.* 54:305–315.

Macnaughton, S.J., J.R. Stephen, A.D. Venosa, G.A. Davis, Y.-J. Chang, and D.C. White. 1999. Microbial population changes during bioremediation of an experimental oil spill. *Appl. Environ. Microbiol.* 65:3566–3574.

Margesin, R. and F. Schinner. 1997. Biological decontamination of oil spills in cold environments. *J. Chem. Technol. Biotechnol.* 74:381–389.

Margesin, R. and F. Schinner. 2001. Biodegradation and bioremediation of hydrocarbons in extreme environments. *Appl. Microbiol. Biotechnol.* 56:650–663.

Marzorati, M., L. Wittebolle, N. Boon, D. Daffonchio, and W. Verstraete. 2008. How to get more out of molecular fingerprints: Practical tools for microbial ecology. *Environ. Microbiol.* 10:1571–1581.

Muyzer, G., E.C. De Waal, and A.G. Uitterlinden. 1993. Profiling of complex microbial populations by denaturing gradient gel electrophoresis analysis of polymerase chain reaction-amplified genes coding for 16S rRNA. *Appl. Environ. Microbiol.* 59:695–700.

Powell, S.M., S.H. Ferguson, J.P. Bowman, and I. Snape. 2006. Using real-time PCR to assess changes in the hydrocarbon-degrading microbial community in Antarctic soil during bioremediation. *Microb. Ecol.* 52:523–532.

Prince, R.C., R.R. Lessard, and J.R. Clark. 2003. Bioremediation of marine oil spills. *Oil Gas Sci. Technol.* 58:463–468.

Rhee, S.-K., X. Liu, L. Wu, S.C. Chong, X. Wan, and J. Zhou. 2004. Detection of genes involved in biodegradation and biotransformation in microbial communities by using 50-mer oligonucleotide microarrays. *Appl. Environ. Microbiol.* 70:4303–4317.

Röling, W.F.M., M.G. Milner, D. Martin Jones, K. Lee, F. Daniel, R.J.P. Swannell, and I.A. Head. 2002. Robust hydrocarbon degradation and dynamics of bacterial communities during nutrient-enhanced oil spill bioremediation. *Appl. Environ. Microbiol.* 68:5537–5548.

Turle, R., T. Nason, H. Malle, and P. Fowlie. 2007. Development and implementation of the CCME reference method for the Canada-wide standard for petroleum hydrocarbons (PHC) in soil: A case study. *Anal. Bioanal. Chem.* 387:957–964.

van Beilen, J.B., M.G. Wubbolts, and B. Witholt. 1994. Genetics of alkane oxidation by *Pseudomonas oleovorans. Biodegradation* 5:161–174.

Viñas, M., J. Sabaté, M.J. Espuny, and A.M. Solanas. 2005. Bacterial community dynamics and polycyclic aromatic hydrocarbon degradation during bioremediation of heavily creosote-contaminated soil. *Appl. Environ. Microbiol.* 71:7008–7018.

Walworth, J.L., C.R. Woolard, J.F. Braddock, and C.M. Reynolds. 1997. Enhancement and inhibition of soil petroleum biodegradation through the use of fertilizer nitrogen: An approach to determining optimum levels. *J. Soil Contam.* 6:465–480.

Whyte, L.G., L. Bourbonniere, and C.W. Greer. 1997. Biodegradation of petroleum hydrocarbons by psychrotrophic *Pseudomonas* strains possessing both alkane (alk) and naphthalene (nah) catabolic pathways. *Appl. Environ. Microbiol.* 63:3719–3723.

Whyte, L.G., L. Bourbonniere, C. Bellerose, and C.W. Greer. 1999. Bioremediation assessment of hydrocarbon-contaminated soils from the high Arctic. *Biorem. J.* 3:69–79.

Whyte, L.G., B. Goalen, J. Hawari, D. Labbé, C.W. Greer, and M. Nahir. 2001. Bioremediation treatability assessment of hydrocarbon-contaminated soils from Eureka, Nunavut. *Cold Regions Sci. Technol.* 32:121–132.

Whyte, L.G., T. Smits, D. Labbé, B. Witholt, C.W. Greer, and J. van Beilen. 2002. Cloning and characterization of multiple alkane hydroxylases in *Rhodococcus* spp. strains Q15 and 16531. *Appl. Environ. Microbiol.* 68:5933–5942.

Wu, L., D.K. Thompson, G. Li, R.A. Hurt, J.M. Tiedje, and J. Zhou. 2001. Development and evaluation of functional gene arrays for detection of selected genes in the environment. *Appl. Environ. Microbiol.* 67:5780–5790.

Xu, R., N.L.A. Lau, K.L. Ng, and J.P. Obbard. 2004. Application of a slow-release fertilizer for oil bioremediation in beach sediment. *J. Environ. Qual.* 33:1210–1216.

Zhou, J. 2003. Microarrays for bacterial detection and microbial community analysis. *Curr. Opin. Microbiol.* 6:288–294.

15 Occurrence, Distribution, and Nature of Hydrocarbon-Degrading Genes in Biodegradative Microorganisms from the Antarctic Environment

Adriana Philippi Luz, Emanuele Kuhn, and Vivian Helena Pellizari

CONTENTS

15.1 INTRODUCTION

Hydrocarbons are the principal components of petroleum and fossil fuel, and their input into the environment can occur by geological process or by anthropogenic sources. The contamination of the environment by petroleum hydrocarbons is a widespread problem (Margesin and Schinner, 1999).

Polar environments usually and naturally contain low concentration of hydrocarbons that are overwhelmingly biogenic in origin, but human activities and oil spills have increased their concentration in previously pristine locations. The local sources of hydrocarbons in the Antarctic marine ecosystem are biogenic, petrogenic, and anthropogenic (Cripps and Priddle, 1991). The biogenic sources refer mainly to the aliphatic hydrocarbons present in the lipid fraction of all marine fauna and flora, which has been the largest potential source in the Antarctica. The petrogenic source comes from natural spills of submarine hydrocarbon reserves. The anthropogenic source comes from the human activities related to logistic and tourism in the area. The marine sediment is an important aquatic system, as it is the main destiny of the organic and inorganic particles that exist in the water column. Being hydrophobic, hydrocarbons tend to be incorporated into the particulate material and deposited in the subsurface marine sediment (Volkman et al., 1992).

In most areas in Antarctica, the soil may not be exposed to contamination, as it is permanently covered by snow. However, there are situations, mainly in the areas around research stations, where the soil is regularly exposed for a certain period of the year and in these cases, the soil is an important sink of contamination. The bioremediation of contaminated sites by microorganisms has become an accepted technology for the restoration of these environments outside the Antarctic, since it is considered to be quite safe and harmless, providing a more cost-effective alternative than conventional physical or chemical clean-up procedures (Milcic-Terzic et al., 2001). The biodegradation of a wide range of petroleum hydrocarbons has been reported *in vitro* and *in situ* experiments at low temperatures, such as the Arctic, Alaskan (Lindstrom et al., 1991; Pritchard et al., 1992; Master and Mohn, 1998; Whyte et al., 1999a,b), Alpine (Margesin et al., 2003), Patagonian (Peressutti et al., 2003), and Antarctic environments (Delille et al., 1997; Mac Cormack and Fraile, 1997; Aislabie et al., 1998; Margesin and Schinner, 1999; Bej et al., 2000; Whyte et al., 2002a; Stallwood et al., 2005; Luz et al., 2006; Ruberto et al., 2006), providing essential information on the biology, genetics, and ecology of involved microbial communities. From these studies, it is clear that cold-adapted autochthonous microorganisms may play a significant role in the biological decontamination of hydrocarbon-contaminated soils in cold climates.

The Protocol of Environmental Protection of the Antarctic Treaty (Madrid, 1991) requires that activities in the Antarctic continent be planned and conducted so as to limit adverse impacts on this pristine environment and that previous damage to the environment by human activities should be cleaned up. Thus, the importance of monitoring and minimizing anthropogenic impacts is of great value since Antarctica is an important resource to be preserved.

Microbial diversity and biogeography studies can help to identify whether microorganisms or specific genotypic characteristics are widely distributed or if they tend to be endemic only to specific areas (Staley, 1997; Staley and Gosink, 1999).

15.2 BACTERIAL HYDROCARBON-DEGRADING GENES

15.2.1 Classification

15.2.1.1 Alkane Monooxygenases

Alkane monooxygenases (*Alks*), enzymes that belong to the Alkane Hydroxylating System, are the key enzymes in the biodegradation of alkanes. They are membrane-bound enzymes that catalyze the initial oxidation of the alkane substrate to a 1-alkanol (Wyatt, 1984). The bacterial oxidation of alkanes is very common in nature and has been reported worldwide (Watkinson and Morgan, 1990; Sotsky et al., 1994; Whyte et al., 1996; Whyte et al., 2002a; van Beilen et al., 2003; Head et al., 2006; van Beilen and Funhoff, 2007). Genes encoding these enzymes (*alk* genes) have been characterized in *Pseudomonas putida* ATCC 29347 (*alkB*) (formerly *Pseudomonas oleovorans*, see van Beilen et al., 1994), *Rhodococcus* sp. strain Q15 (*alkB1*, *alkB2*) (Whyte et al., 2002b), *Acinetobacter* sp. ADP-1 (*alkM*) (Ratajczak et al., 1998), and various Gram Positive and Gram Negative bacteria (Smits et al., 1999, 2002; van Beilen et al., 2002). More than 60 Alk homologs are presently known, showing high sequence diversity (van Beilen et al., 2003; van Beilen and Funhoff, 2007). In Eubacteria, almost all alkane degraders belong to α-, β-, and γ-Proteobacteria Class and Actinomycetales Order (van Beilen et al., 2003; Coleman et al., 2006) (Figure 15.1).

Alks are identified as containing six transmembrane (TM) helices and four conserved histidine-containing sequence motifs (motif A, HEXXHK; motif B, EHXXGHH; motif C; NYXEHYG, and motif D, LQRHXDHHA), located near the ends of TM helices 4 and 6 (van Beilen et al., 2005). The eight conserved histidines in motifs A, B, and D are indeed essential for the activity of Alks (Shanklin et al., 1994; Shanklin and Whittle, 2003). The single conserved histidine in motif C (NYXEHYG) was identified as an additional potential ligand because it is conserved in all alkane monooxygenases sequences (Smits et al., 1999; van Beilen et al., 2003).

15.2.1.2 Aromatic Ring Hydroxylating Dioxygenases

Aromatic ring hydroxylating dioxygenases (ARHDs) are multicomponent enzyme systems that add dioxygen to the aromatic nucleus of aromatic substrates, breaking the resonance of the aromatic ring (Mishra et al., 2001). All members of the dioxygenase family have two or more proteins involved in the electron transport. Usually, they are a family of multicomponent mononuclear (nonheme) iron oxygenases composed of a reductase, a ferrodoxin, a terminal dioxygenase large (α) subunit, and a terminal dioxygenase small (β) subunit. Two prosthetic groups, a Rieske-type [Fe_2S_2] center and a mononuclear iron, are associated with the α-subunit in the (αβ)*n*-type enzymes (Parales et al., 1998). Being a conserved region of ARHDs and presented the catalytic activity in the recognition of substrate (Nam et al., 2001), the α-subunit reflects the phylogenetic affiliation between dioxygenases. In such case, the α-subunit is used as a target in phylogeny and diversity studies of ARHDs.

The last classification of ARHDs (Nam et al., 2001) was based on the homology of the amino acid sequences of the α-subunit of the enzymes, and identified four distinct groups for 54 oxygenases (Figure 15.2). Group I consists of homo-multimer oxygenases and has only the α-subunit, while groups II, III, and IV include heterodimer oxygenases with α- and β-subunits. The group II contains Benzoate/Toluate dioxygenases, the group III Naphthalene/Polycyclic Aromatic Hydrocarbon (PAH) dioxygenases, and group IV benzene/toluene/biphenyl dioxygenases. Group I includes oxygenases sharing a low homology with each other and having various

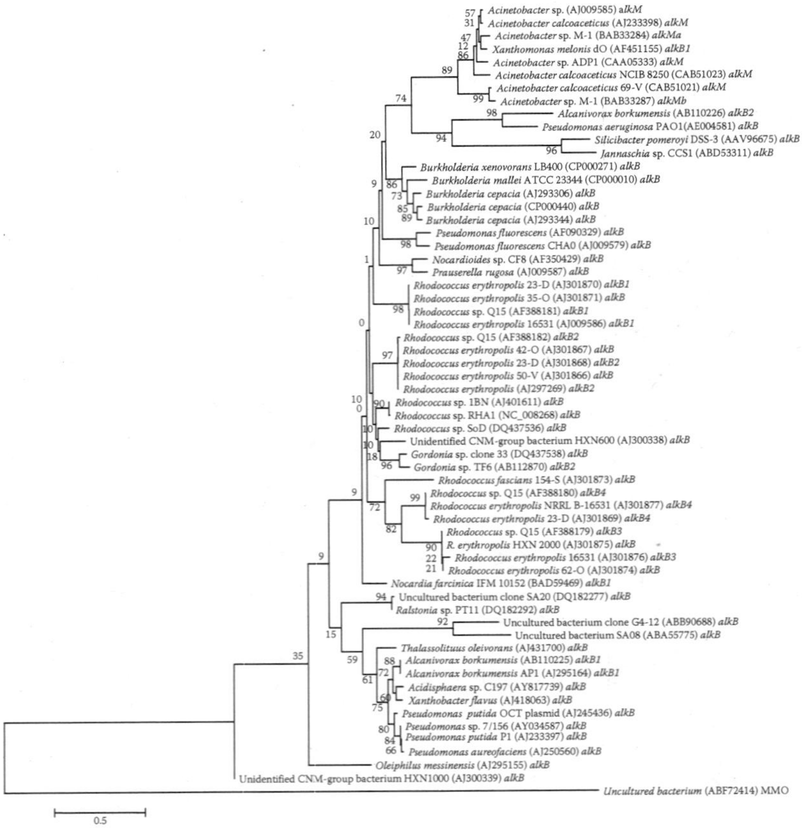

FIGURE 15.1 Phylogenetic tree based on the alignment of amino acid sequences of membrane-bound alkane monooxygenases from several α-, β-, and γ-Proteobacteria and Actinobacteria. The tree was constructed by using the Neighbor-Joining (NJ) method in MEGA3 program with the p-distance model and pairwise deletion of gaps/missing data, substitution model PAM Matrix (Dayhoff), and bootstrap of 1000 repetition. The scale bar denotes divergence percentage between sequences. Outgroup: methane monooxygenase.

lengths of 329–446 amino acid residues, while other dioxygenases classified into groups II, III, and IV have approximately 450 residues.

15.2.2 Genes Described in Hydrocarbon-Degrading Bacteria

The genes encoding monooxygenases and dioxygenases have been characterized in *P. putida* (formerly *P. oleovorans*) ATCC 29347 (Pp *alkB*), *Rhodococcus* sp. strain Q15 (Rh *alkB1*, Rh *alkB2*), *Acinetobacter* sp. ADP-1 (Ac *alkM*), *P. putida* ATCC 17484 (*ndoB*), *P. putida* F1 (*todC1*), *P. putida* ATCC 33015 (*xylE*, *cat23*), and *Pseudomonas pseudoalcaligenes* KF 707 (*bphA*). Pp *alkB*, located on the OCT plasmid, a member of the IncP-2 family of plasmids, encodes alkane hydroxylase, a membrane-bound monooxygenase that is the first enzyme of the short-chain alkane

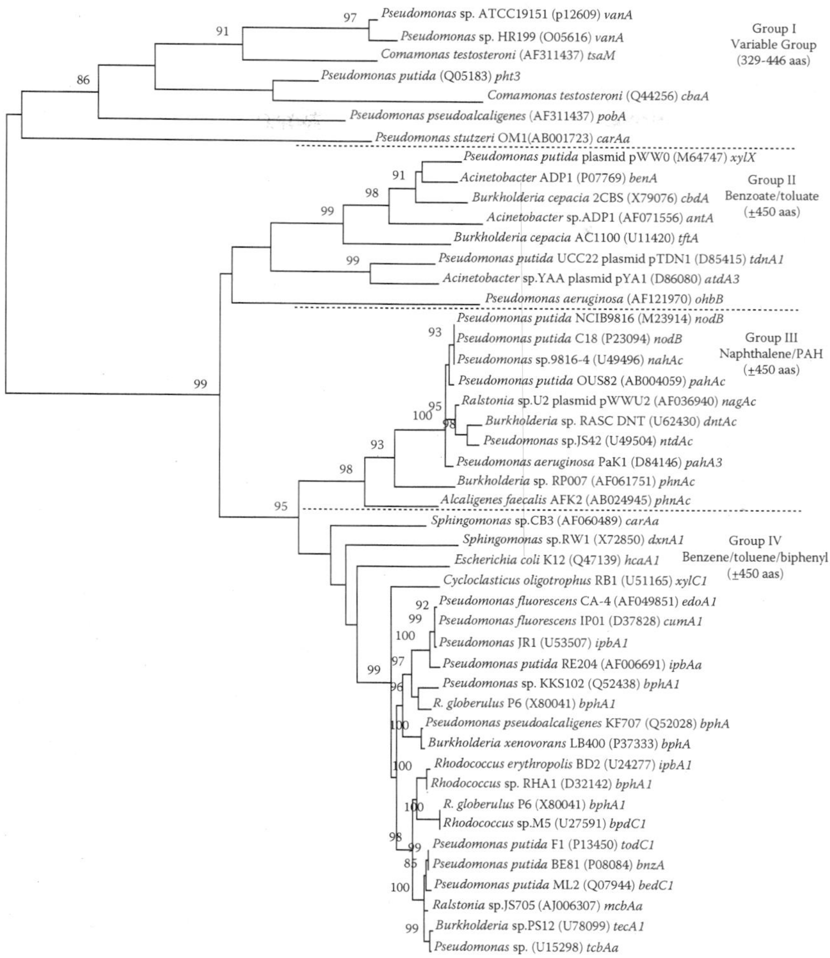

FIGURE 15.2 Phylogenetic trees based on the alignment of amino acid sequences of α-subunit from dioxygenases from the four groups of ARHDs, according to the Nam classification (Nam et al., 2001). Neighbor-Joining tree, substitution model: PAM Matrix (Dayhoff), and bootstrap of 1000 repetition. The scale bar denotes divergence percentage between sequences.

(C_6–C_{12}) degradation pathway (Kok et al., 1989; van Beilen et al., 1994). Recently, highly similar multiple alkane hydroxylase systems have been isolated and characterized in *R. erythropolis* NRRL B-16531 and *Rhodococcus* sp. strain Q15 (Whyte et al., 2002b). This genus and other closely related high %G+C actinomycetes are considered important members of hydrocarbon-degrading microbial populations in contaminated soils (Warhurst and Fewson, 1994). Both organisms are psychrotrophs that are able to degrade a wide range of *n*-alkanes (C_8–C_{32}) and contain at least four alkane monooxygenase homologues (Rh *alkB1*, Rh *alkB2*, Rh *alkB3*, and Rh *alkB4*); when the rhodococcal *alkB2* was cloned into *P. fluorescens* KOB2Δ1, the recombinant strain was able to mineralize and grow on C_{12}–C_{16} *n*-alkanes (Whyte et al., 2002b). Ac *alkM*, one of the five essential genes encoding alkane utilization by *Acinetobacter* sp. ADP-1, is required for the degradation of C_{12}–C_{18} *n*-alkanes (Ratajczak et al., 1998). The natural diversity of the alkane hydroxylase system has recently been presented by van Beilen et al. (2003). The *ndoB* gene, located on the NAH7 plasmid, is one of the three genes that encode naphthalene dioxygenase, the first enzyme in the *P. putida* ATCC 17484 naphthalene degradation pathway (Kurkela et al., 1988). The *todC1* gene, located on the chromosome of *P. putida* F1, encodes for the α-subunit of terminal dioxygenase, one of the three components of toluene dioxygenase, the first enzyme involved in the toluene degradation pathway (Zylstra and Gibson, 1989). The *xylE* and *cat23* genes, located on the TOL plasmid of *P. putida* ATCC 33015, encode for catechol 2,3-dioxygenase, a key enzyme involved in the lower degradation pathway of aromatic compounds, such as toluene and xylene (Nakai et al., 1983; Moon et al., 1995; Okuta et al., 1998; Laramee et al., 2000). The *bphA* gene, located on the chromosome of the polychlorinated biphenyl-degrading strain *P. pseudoalcaligenes* KF707, encodes for biphenyl dioxygenase, a multicomponent enzyme responsible for the catalysis of the initial oxidation of biphenyl and chlorobiphenyls (Furukawa and Arimura, 1987; Asturias et al., 1995).

15.3 BIOGEOGRAPHY AND SURVEY FOR *ALK* AND ARHD GENES IN POLAR SOILS

Several studies have been conducted in cold environments in Alaska, the Arctic (Lindstrom et al., 1991; Button et al., 1992; Pritchard et al., 1992; Master and Mohn, 1998; Whyte et al., 1999a) and the Antarctic (Delille et al., 1997; Mac Cormack and Fraile, 1997; Aislabie et al., 1999; Bej et al., 2000; Delille and Delille, 2000; Aislabie et al., 2006) to specifically examine petroleum hydrocarbon and polychlorinated compound degradations in different ecosystems, but relatively little genotypic information is available for Antarctic soils. For the bioremediation of oil-contaminated Antarctic soils, indigenous microbial populations that degrade alkanes and aromatic compounds are necessary, since the Antarctic Treaty prohibits the introduction of foreign organisms (Aislabie et al., 1998).

In a survey for indigenous microbial hydrocarbon degradation genes in Antarctic soils reported by Luz et al. (2004) and Whyte et al. (2002a), PCR and hybridization analyses of DNA extracts from the 16 Antarctic soil samples, using specific primers (Table 15.1), resulted in the amplification of DNA fragments of three of the screened alkane monooxygenase genes (Pp *alkB*, Rh *alkB1*, Rh *alkB2*). Overall, the rhodococcal *alkB1* and *alkB2* genes were more frequently detected among the samples than the Pp *alkB* gene. The Pp *alkB* genotype was detected more frequently in the contaminated soils around the diesel fuel storage tanks at Antarctic Station Comandante Ferraz (CF) than in the control soils. However, *Acinetobacter alkM* was not detected by PCR or hybridization analyses.

The *ndoB* genotype was detected in 69% of the Antarctic soils. The *xylE* PCR amplification products were detected mainly in control soils, different from the *cat23* genotype that was detected only in contaminated soils around CF.

TABLE 15.1
Oligonucleotide Sequences of Primers, Probes, and Reference Strains

Catabolic Gene	Sequence (5′- to -3′)	PCR Fragment Size (bp)	Reference Strains (GenBank Accession No.)	References
	Alkane Monooxygenase			
(Pp)*alkB*b-F	TGGCCGGCTACTCCGATGATCGGAATCTGG	870	*P. putida* ATCC 29347	van Beilen et al. (2001)
(Pp)*alkB*e-R	CGCGTGGTGATCCGAGTGCCGCTGAAGGTG		(X65936)	
Rh *alkB1*-F2	ATCTGGGCGCGTTGGGATTTGAGCG	642	*Rhodococcus* sp. Q15	Whyte et al. (2002b)
Rh *alkB1*-R1	CGCATGGTGATCGCTGTGCCGCTGC		(AF388181)	
Rh *alkB2*-F1	ACTCTGGCGCAGTCGTTTTACGGCC	552	*Rhodococcus* sp. Q15	Whyte et al. (2002b)
Rh *alkB2*-R1	CCCACTGGGCAGGTTGGGCGCACCG		(AF388182)	
(Ac)*alkM*-F	CCTGTCTCATTTGGCGCTCGTTCCTACAGG	496	*Acinetobacter* sp. ADP-1	Ratajczak et al. (1998)
(Ac)*alkM*-R	CCAAAGTGGCGGAATCATAGCAGGC		(AJ002316)	
	Naphthalene Dioxygenase			
*ndoB*b-F	CACTCATGATAGCCTGATTCCTGCCCCCGGCG	642	*P. putida* ATCC 17484	Kurkela et al. (1988)
*ndoB*e-R	CCGTCCCACAACACACCCATGCCGCTGCCG		(M23914)	
	Toluene Dioxygenase			
*todC*1-F	CGGGTGGGCTTACGACACCGCCGGCAATCT	560	*P. putida* F1 (J04996)	Zylstra and Gibson (1989)
*todC*1-R	TCGAGCCGCGCTCCACGCTACCCAGACGTT			

(*continued*)

TABLE 15.1 (continued)
Oligonucleotide Sequences of Primers, Probes, and Reference Strains

Catabolic Gene	Sequence (5′- to -3′)	PCR Fragment Size (bp)	Reference Strains (GenBank Accession No.)	References
	Catechol-2,3-Dioxygenase			
*xylE*b-F	GTGCAGCTGCGTGTACTGGACATGAGCAAG	834	*P. putida* ATCC 33015 (M65205)	Nakai et al. (1983)
*XylE*e-R	GCCCAGCTGGTCGGTGGTCCAGGTCACCGG			
cat2,3 1a-F	AGGTGCTCGGTTTCTACCTGGCCG	405–408	*P. putida* ATCC 33015 (M64747)	Laramee et al. (2000)
cat2,3 6a-R	ACGGTCATGAATCGTTCGTTGAG			
	Biphenyl Dioxygenase			
*bphA*1-F	TCACCTGCAGCTATCACGGCTGG	830	*P. pseudoalcaligenes* KF707 (M83673)	Furukawa et al. (1987)
*bphA*1-R	GGATCTCCACCCAGTTCTCGCCATCGTCCTG			

When biogeography is considered, PCR and hybridization analyses revealed that the bacterial communities in both Antarctic and subtropical Brazilian soils possessed genes with homology to the *ndoB*, *todC1*, *xylE*, *cat23*, and *bphA1* genes (Luz et al., 2004). A similar geographical distribution of these genes was found in Brazil, Puerto Rico, and the United States (Pellizari et al., 1996), and in the Canadian Arctic (Whyte et al., 1996, 1997, 1999a). The most common genotypes found in contaminated and control soils were *todC1* and *bphA1* from Antarctica and *todC1* and *cat23* from Brazil; these genes were involved in the degradation of toluene and catechol. It is difficult to predict the origin of these genes in Antarctic microbial community. However, Fulthorpe et al. (1998), suggested that a high level of genotypic diversity within a population suggests that the ability to catabolize a specific compound is associated with natural adaptation, and that the horizontal transfer of novel genes between geographically distant but related strains is a common event.

Among the nine different catabolic genes examined, seven were detected in nonimpacted areas from the Antarctica. The presence of this genotypic diversity in uncontaminated ecosystems could be related to biogenic and/or geochemical sources of hydrocarbon compounds in these areas. Also, the quantification of these genes in the Antarctic environment can enhance understanding of their presence in the pristine areas. For these reasons, qPCR for catabolic genes was applied to four Antarctic soil samples where *bph* genes were previously detected. A set of degenerated primers (Bellicanta, 2004) designed based on the highly conserved regions of amino acids sequences that embody the catalytic center [2Fe-2S]-Rieske of ARHDs α-subunit was used.

Copy numbers of ARHD genes in soil samples were 4.0 copies of gene/ng of genomic DNA in sample 1A, 17.5 copies of gene/ng of genomic DNA in sample 1B, 3.3 copies/ng of genomic DNA to sample 2A, and the lower number of copies, 2.3 copies/ng of genomic DNA was observed to the sample 2B. These sample numbers refer to four soil samples from *in situ* microcosms located near the diesel fuel storage tanks around the Brazilian Antarctic Station Comandante Ferraz.

The lower numbers of ARHD genes when compared with the higher numbers reported in the contaminated samples of temperate soil (Yergeau et al., 2007) can be possibly justified for the presence of low hydrocarbon concentration in the soils analyzed.

Once the dispersion of catabolic genes was consistently detected in pristine and contaminated soil samples from Antarctica, and the genotypes described appear to be ubiquitous in the environment, the knowledge about the biodiversity of these genes can provide insight into the origins of hydrocarbon-degrading genes' in polar environments.

15.4 DIVERSITY OF *ALK* AND ARHD GENES IN ANTARCTIC ENVIRONMENTS

Recent data concerning the diversity of Alks and ARHDs in marine Antarctic sediment from Antarctic Peninsula, using degenerate primers clone libraries, reveal different distributions of the hydrocarbon-degrading genes in the sites analyzed. The sediments were sampled in two different sites in the Admiralty Bay, King George

Island (Figure 15.2). One site is located in front of the Brazilian scientific station Comandante Ferraz (CF) (62°05.12′S and 58°23.07′W), affected by human activity. The second one is a pristine site located across the bay, 3.2 km away, identified as Botany Point (BP) (62°05.97′S and 58°20.49′W) (Kuhn et al., 2009).

The results clearly showed the difference between CF and BP *alk* genes communities with high diversity indices found in BP and two predominant Operational Protein Families (OPFs) (Schloss and Handelsman, 2008) in CF: S1 (related to *Silicibacter pomeroyi* DSS-3, *alkB*) and S2 (related to *Acinetobacter* sp. ADP1, *alkM*). On the other hand, the diversity and distribution of ARHD genes in the same communities were similar, with both communities having the same predominant groups: D1 (related to *Rhodococcus* sp. RHA1 plasmid pRHL1, *bphl*) and D2 (*Burkholderia xenovorans* LB400, *bphA1*) (Figure 15.3).

In *alk* gene libraries, new genes related to *alkB* previously described in α-, β-, and γ-Proteobacteria (*S. pomeroyi, Acidisphaera* sp., *Alcanivorax borkumensis, P. putida*), Actinobacteria (*Gordonia* sp., *Prauserella rugosa, Nocardioides* sp., *Rhodococcus* sp., *Nocardia farcinica*) and environmental clones (S7, S11, and S14) were detected. The *alkM* related to *alk* genes described in γ-Proteobacteria (*Acinetobacter* sp.) were also detected (99.88% identity) with a 35% frequency. In general, the *alk* sequences ranged from 53.10% to 69.60% identity to reference sequences by nucleotide level, and 50.90%–73.40% identity to reference sequences by amino acid level. With the exception of OPF S2, *alk* environmental sequences showed neither significance nor identity with nucleotide database sequences. However, protein alignments showed that all of these PCR fragments encode peptides with a protein sequence identity to Alk enzymes. Moreover, all sequences showed the presence of the two internal conserved regions of alkane monooxygenases (motif B, EHXXGHH; motif C; NYXEHYG) (Kuhn et al., 2009). These suggest that some cold-adapted Antarctic marine communities bear *alk* genes not yet described and that these genes are phylogenetically distinct from mesophilic bacteria communities (Figure 15.4).

ARHD genes found in the same sites were related to biphenyl, benzene, chlorobenzene, isopropylbenzene, toluene, and dibenzothiophene dioxygenases (dioxygenases from Group IV according Nam et al., 2001) already described (Figures 15.3 and 15.5). Besides the detection of known dioxygenases genes, ARHD genes related to mesophilic uncultured clones were also detected. The ARHD sequences ranged from 62.76% to 99.37% identity to reference sequences by nucleotide level, and 53.38%–98.82% identity to reference sequences by amino acid level.

Many dioxygenase sequences were associated to dioxygenases described in plasmids and transposons as OPFs D1 and D11. OPF D1 presented 96.56% nucleotide identity to *bphA1* from *Rhodococcus* sp. RHA1 pRHL1 plasmid and OPF D11 presented 98% nucleotide identity to *bphA1* from *Ralstonia oxalatica* transposon TN4371 and *bphA1* from *Burkholderia oxalatica* 2A plasmid Pijbl. It was not possible to detect phylogenetic distinct groups between ARHD genes as observed in *alk* genes distribution. However, Antarctic ARHD clone sequences were related to five distinct substrate groups (Figure 15.5) previously described in gram positive (purple color) and gram negative (pink color) bacteria. The radial tree in Figure 15.5 shows the clustering of ARHDs according to the substrates oxidized.

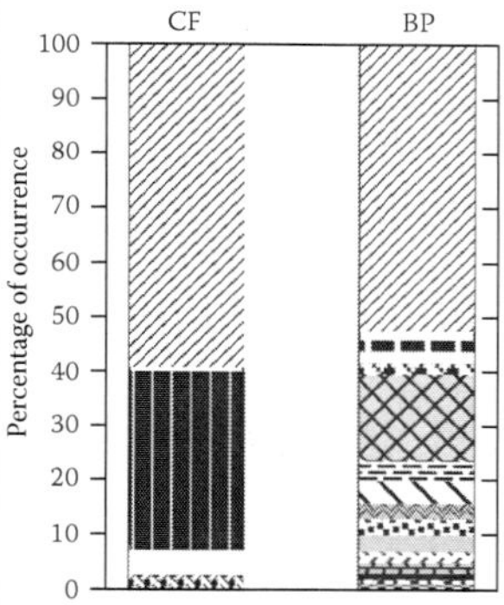

Alks

OPF	Gene	Microorganism	Average of identity (%) nt	aa
S1	*alkB*	*Silicibacter pomeroyi* DSS-3	64.34	63.73
S2	*alkM*	*Acinetobacter* sp. ADP1	99.88	99.80
S3	*alkB2*	*Gordonia sp.* TF6	57.27	58.32
S4	*alkB*	*Prauserella rugosa/ nocardioides* sp.	65.45	71.85
S5	*alkB*	*Acidisphaera* sp. C197	58.63	65.72
S6	*alkB*	*Silicibacter pomeroyi* DSS-3	59.90	58.50
S7	*alkB*	Uncultured bacterium SA08	59.55	59.45
S8	*alkB1*	*Alcanivorax borkumensis*	56.46	56.60
S9	*alkB*	*Rhodococcus* sp. SoD	60.00	60.20
S10	*alkB1*	*Nocardia farcinica* IFM	60.40	61.00
S11	*alkB*	Uncultured bacterium G4–12	53.10	50.90
S12	*alkB*	*Pseudomonas putida* P1	54.40	52.80
S13	*alkB*	*Rhodococcus* sp. 1BN	61.00	60.20
S14	*alkB*	Uncultured bacterium G4–12	63.40	65.20
S15	*alkM*	*Acinetobacter* sp. M-1	69.60	73.40

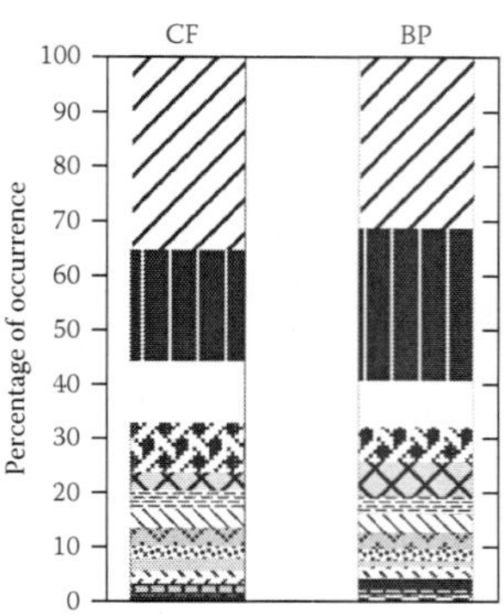

ARHDs

OPF	Gene	Microorganism	Average of identity (%) nt	aa
D1	*bphA1*	*Rhodococcus* sp. RHA1 plasmid pRHL1	95.02	96.56
D2	*bphA1*	*Burkholderia xenovorans* LB 400	97.93	97.91
D3	*dbdCa*	*Xanthobacter polyaromaticivorans*	77.52	83.49
D4	*todC1*	*Pseudomonas putida*	99.37	98.82
D5	ARHD	Uncultured bacterium clone ado 6	64.13	53.38
D6	ARHD	*Mycobacterium vanbaalenii* PYR-1	80.35	78.85
D7	*dbdCa*	*Xanthobacter polyaromaticivorans*	74.53	76.60
D8	ARHD	Uncultured bacterium clone tar91–67	69.15	64.98
D9	ARHD	Uncultured bacterium clone tar91–67	63.30	64.00
D10	ARHD	Uncultured bacterium clone tar91–67	65.10	68.50
D11	*bphA1*	*Comamonas testosteroni*	95.52	98.37
D12	*bphA1*	*Burkholderia xenovorans* LB400	82.90	79.20
D13	ARHD	Uncultured bacterium clone tar91–67	62.76	67.40
D14	*bphA1*	*Burkholderia* sp. JB1	78.30	86.10
D15	ARHD	Uncultured bacterium clone tar91–67	63.30	63.30

FIGURE 15.3 Comparison between Comandante Ferraz (CF) and Botany Point (BP) clone libraries. Photo: indicates the sample sites (marine sediment from 20 m depth). Graphics: *alks* and ARHDs genes percentage of occurrence and the average of identity between clone sequences and GenBank reference sequences: (nt) percentage of identity by nucleotide level; (aa) percentage of identity by amino acid level. S1 to S15 are OPFs of *alk* genes found in the libraries. S1 to S4 was detected in both sample sites, and S5 to S15 was found only in BP site. D1 to D15 are OPFs of ARHDs genes found in the libraries. D1 to D11 and D13 are OPFs detected in both sites. D12 was detected only in CF site, and D14 and D15 are OPFs found only in BP site.

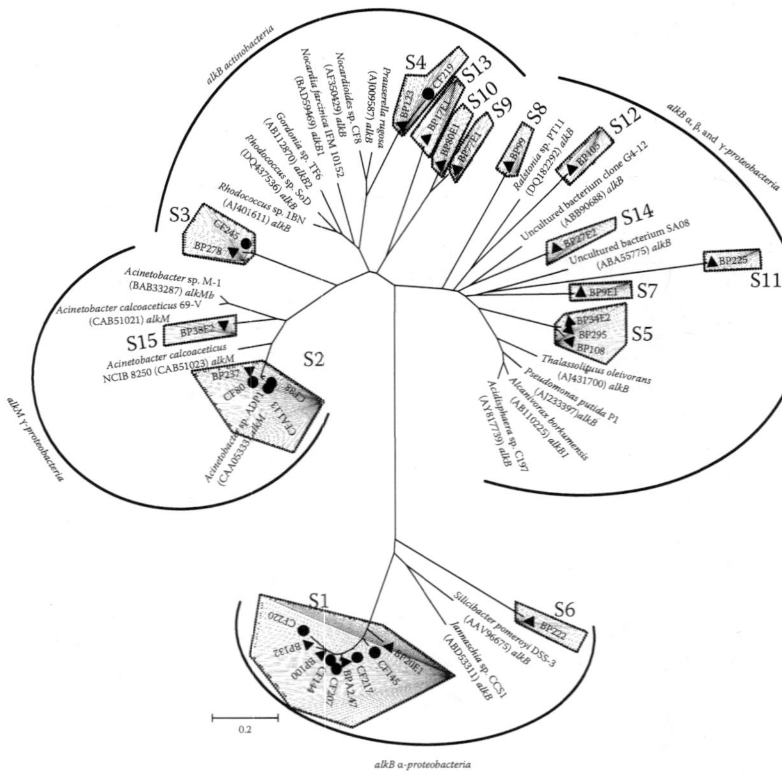

FIGURE 15.4 Phylogenetic tree based on the alignment of amino acid sequences of membrane-bound alkane monooxygenases from clones obtained in this study and reference strains (comprising 162–169 amino acids, corresponding to position 143–310 of *alkB* from *P. putida* GPo1 (van Beilen et al., 2005)). The tree was constructed by using the Neighbor-Joining (NJ) method in MEGA3 program with the p-distance model and pairwise deletion of gaps/missing data, substitution model PAM Matrix (Dayhoff), and bootstrap of 1000 repetition. The scale bar denotes divergence percentage between sequences. Symbols: clones from Comandante Ferraz (•); clones from Botany Point (▲). Access numbers of *alk* genes from GenBank database reference strains are in parentheses. (From Kuhn, E. et al., *Environ. Microbiol.*, 11, 669, 2009. With permission.)

The detection of Group IV dioxygenases in the Antarctic marine sediment are consistent with the results found by Nastasi (2006) and Luz et al. (2004). Nastasi (2006) detected the presence of *bphA1* and *todC1* in diesel-degrading consortia of marine sediment samples from South Shetlands Island using specific primers for hydrocarbon-degrading genes (Table 15.1). The *bphA1* gene was amplified from a biphenyl-degrading consortium from the same CF sample site from where biphenyl-degrading genes were analyzed in sediment samples by degenerate primers, showing that the genes found in environment could be active in microbial consortia.

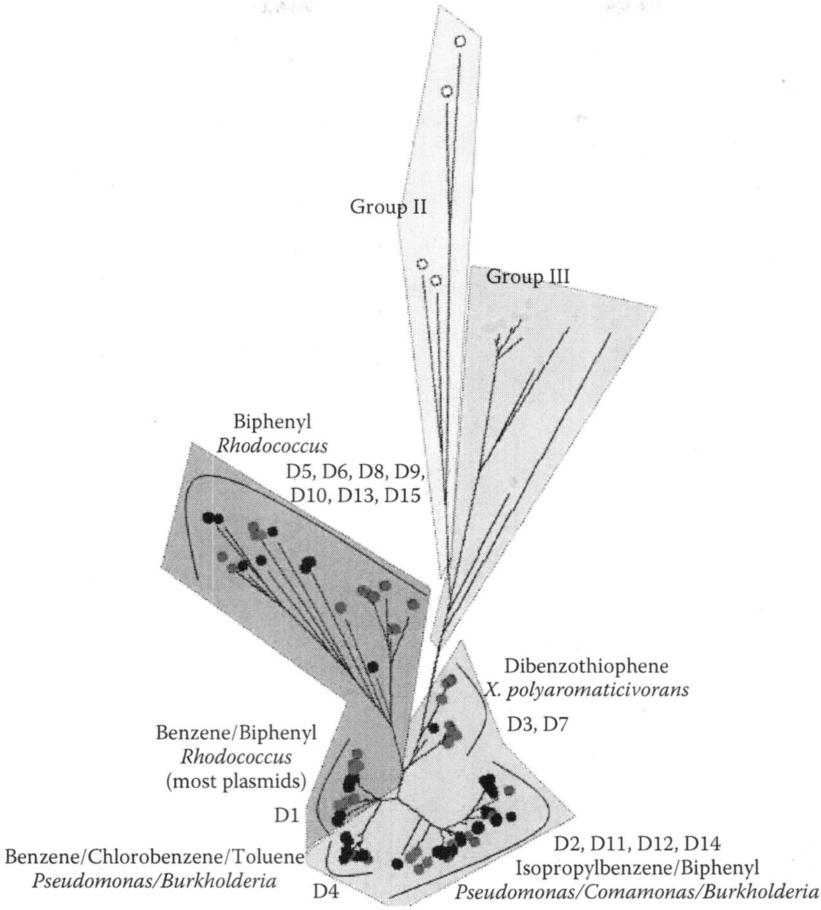

FIGURE 15.5 (See color insert following page 276.) ARHDs radial tree showing the clustering according to the substrates oxidized by the members of Group IV dioxygenases. Antarctic clone sequences were related to five distinct substrate groups previously described in Gram Positive (purple color) bacteria and Gram Negative (pink color) bacteria such as *Rhodococcus*, *Pseudomonas*, *Comamonas*, *Burkholderia*, and *Xanthobacter*. Group I was not included in this tree because of the low homology of the sequences related to the other three Groups of dioxygenases.

15.5 SUMMARY

Hydrocarbon-degrading genes appear to be widespread in the environment, including the poles.

Despite the prevalence of few groups of new genes having been reported, the detection of novel genes showed that the biodiversity of these genes in polar environments remains scarce.

In contrast to the new alkane-degrading genes present in the cold-tolerant bacteria described above, the genes used for aromatic degradation by psychrotolerant and psychrophilic communities do not appear to differ significantly from those identified in mesophilic isolates. In fact, Whyte et al. (1996) found that catabolic genes from several aromatic-degrading psychrotolerant strains had homology to those described in mesophilic bacteria, although other isolates appeared to have novel genes. Thus, additional molecular analyses need to be conducted in order to test for the presence of catabolic genes specifically adapted to polar regions.

The presence of ARHD and *alk* genes in pristine Antarctic sites brings new questions related to the origin of these genes in Antarctica. The search for the presence and diversity of these genes in permafrost samples or soil exposed after a glacier retreat can help us reveal how recently these genes were introduced in the environment.

The significance of the findings discussed in this chapter suggests a need for new studies on gene expression to elucidate potential links between microbial community structures and their functions in bioremediation processes in polar environments. New techniques, such as functional microarray analysis (Yergeau et al., 2007) and the Stable Isotope Probing–DNA (SIP) (Leigh et al., 2007) applied to polar environment, are promising tools to predict which genes and microorganisms are responsible for metabolizing organic pollutants in poles.

REFERENCES

Aislabie, J., M. Balks, N. Astori, G. Stevenson, and R. Symons. 1999. Polycyclic aromatic hydrocarbons in fuel-oil contaminated soils, Antarctica. *Chemosphere* 39:2201–2207.

Aislabie, J., M. McLeod, and R. Fraser. 1998. Potential for biodegradation of hydrocarbons in soil from the Ross Dependency, Antarctica. *Appl. Microbiol. Biotechnol.* 49:210–214.

Aislabie, J., D.J. Saul, and J.M. Foght. 2006. Bioremediation of hydrocarbon-contaminated polar soils. *Extremophiles* 10:171–179.

Asturias, J.A., E. Diaz, and K.N. Timmis. 1995. The evolutionary relationship of biphenyl dioxygenase from gram-positive *Rhodococcus globerulus* P6 to multicomponent dioxygenases from gram-negative bacteria. *Gene* 156:11–18.

Bej, A.K., D. Saul, and J. Aislabie. 2000. Cold-tolerant alkane-degrading *Rhodococcus* species from Antarctica. *Polar Biol.* 23:100–105.

Bellicanta, G.S. 2004. Diversity of catabolic genes in sediment samples of Santos and São Vicente Estuarine Systems, SP. PhD diss., University of São Paulo, Brazil.

Button, D.K., B.R. Robertson, D. McIntosh, and F. Juttner. 1992. Interactions between marine bacteria and dissolved-phase and beached hydrocarbons after the Exxon Valdez oil spill. *Appl. Environ. Microbiol.* 58:243–251.

Coleman, N.V., N.B. Bui, and A.J. Holmes. 2006. Soluble di-iron monooxygenase gene diversity in soils, sediments and ethene enrichments. *Environ. Microbiol.* 8:1228–1239.

Cripps, G.C. and J. Priddle, 1991. Hydrocarbons in the Antarctic marine environment. *Antarc. Sci.* 3:233–250.

Delille, D. and B. Delille. 2000. Field observations on the variability of crude oil impact on indigenous hydrocarbon-degrading bacteria from sub-Antarctic intertidal sediments. *Mar. Environ. Res.* 49:403–417.

Delille, D., A. Basseres, and A. Dessommes. 1997. Seasonal variation of bacteria in sea ice contaminated by diesel fuel and dispersed crude oil. *Microb. Ecol.* 33:97–105.

Fulthorpe, R.R., A.N. Rhodes, and J.M. Tiedje. 1998. High levels of endemicity of 3-chlorobenzoate-degrading soil bacteria. *Appl. Environ. Microbiol.* 64:1620–1627.

Furukawa, K. and N. Arimura. 1987. Purification and properties of 2,3-dihydroxybiphenyl dioxygenase from polychlorinated biphenyl-degrading *Pseudomonas pseudoalcaligenes* and *Pseudomonas aeruginosa* carrying the cloned *bphC* gene. *J. Bacteriol.* 169:924–927.

Head, I.M., D.M. Jones, and W.F. Roling. 2006. Marine microorganisms make a meal of oil. *Nat. Rev. Microbiol.* 4:173–182.

Kok, M., R. Oldenhuis, M.P. van der Linden, P. Raatjes, J. Kingma, P.H. van Lelyveld, and B. Witholt. 1989. The *Pseudomonas oleovorans* alkane hydroxylase gene. Sequence and expression. *J. Biol. Chem.* 264:5435–5441.

Kuhn, E., G.S. Bellicanta, and V.H. Pellizari. 2009. New *alk* genes detected in Antarctic marine sediments. *Environ. Microbiol.* 11:669–673.

Kurkela, S., H. Lehvaslaiho, E.T. Palva, and T.H. Teeri. 1988. Cloning, nucleotide sequence and characterization of genes encoding naphthalene dioxygenase of *Pseudomonas putida* strain NCIB9816. *Gene* 73:355–362.

Laramee, L., J.R. Lawrence, and C.W. Greer. 2000. Molecular analysis and development of 16S rRNA oligonucleotide probes to characterize a diclofop-methyl-degrading biofilm consortium. *Can. J. Microbiol.* 46:133–142.

Leigh, M.B., V.H. Pellizari, O. Uhlik, R. Sutka, J. Rodrigues, N.E. Ostrom, J. Zhou, and J.M. Tiedje. 2007. Biphenyl-utilizing bacteria and their functional genes in a pine root zone contaminated with polychlorinated biphenyls (PCBs). *ISME J.* 1:134–148.

Lindstrom, J.E., R.C. Prince, J.C. Clark, M.J. Grossman, T.R. Yeager, J.F. Braddock, and E.J. Brown. 1991. Microbial populations and hydrocarbon biodegradation potentials in fertilized shoreline sediments affected by the T/V Exxon Valdez oil spill. *Appl. Environ. Microbiol.* 57:2514–2522.

Luz, A.P., E.M.P. Ciapina, R.C. Gamba, M.S. Lauretto, E.W.C. Farias, M.C. Bicego, S. Taniguchi, R.C. Montone, and V.H. Pellizari. 2006. Potential for bioremediation of hydrocarbon polluted soils in the Maritime Antarctic. *Antarc. Sci.* 18:335–343.

Luz, A.P., V.H. Pellizari, L.G. Whyte, and C.W. Greer. 2004. A survey of indigenous microbial hydrocarbon degradation genes in soils from Antarctica and Brazil. *Can. J. Microbiol.* 50:323–333.

MacCormack, W.P. and E.R. Fraile. 1997. Characterization of a hydrocarbon degrading psychrotrophic Antarctic bacterium. *Antarc. Sci.* 9:150–155.

Margesin, R. and F. Schinner. 1999. Biodegradation of diesel oil by cold-adapted microorganisms in presence of sodium dodecyl sulfate. *Chemosphere* 38:3463–3472.

Margesin, R., D. Labbe, F. Schinner, C.W. Greer, and L.G. Whyte. 2003. Characterization of hydrocarbon-degrading microbial populations in contaminated and pristine Alpine soils. *Appl. Environ. Microbiol.* 69:3085–3092.

Master, E.R. and W.W. Mohn. 1998. Psychrotolerant bacteria isolated from arctic soil that degrade polychlorinated biphenyls at low temperatures. *Appl. Environ. Microbiol.* 64:4823–4829.

Milcic-Terzic, J., Y. Lopez-Vidal, M.M. Vrvic, and S. Saval. 2001. Detection of catabolic genes in indigenous microbial consortia isolated from a diesel-contaminated soil. *Bioresour. Technol.* 78:47–54.

Mishra, V., L. Rup, and C. Srinivasan. 2001. Enzymes and operons mediating xenobiotic degradation in bacteria. *Crit. Rev. Microbiol.* 27:133–166.

Moon, J., H. Chang, K.R. Min, and Y. Kim. 1995. Cloning and sequencing of the catechol 2,3-dioxygenase gene of *Alcaligenes* sp. KF711. *Biochem. Biophys. Res. Commun.* 208:943–949.

Nakai, C., H. Kagamiyama, M. Nozaki, T. Nakazawa, S. Inouye, Y. Ebina, and A. Nakazawa. 1983. Complete nucleotide sequence of the metapyrocatechase gene on the TOI plasmid of *Pseudomonas putida* mt-2. *J. Biol. Chem.* 258:2923–2928.

Nam, J.W., H. Nojiri, T. Yoshida, H. Habe, H. Yamane, and T. Omori. 2001. New classification system for oxygenase components involved in ring-hydroxylating oxygenations. *Biosci. Biotechnol. Biochem.* 65:254–263.

Nastasi, F.R. 2006. Diversity characterization of xenobiotic compounds (hydrocarbons and PCB's) degrading-consortia in marine sediment from Brazil and Antarctica. Master diss., University of São Paulo, Brazil.

Okuta, A., K. Ohnishi, and S. Harayama. 1998. PCR isolation of catechol 2,3-dioxygenase gene fragments from environmental samples and their assembly into functional genes. *Gene* 212:221–228.

Parales, J.V., R.E. Parales, S.M. Resnick, and D.T. Gibson. 1998. Enzyme specificity of 2-nitrotoluene 2,3-dioxygenase from *Pseudomonas* sp. strain JS42 is determined by the C-terminal region of the alpha subunit of the oxygenase component. *J. Bacteriol.* 180:1194–1199.

Pellizari, V.H., S. Bezborodnikov, J.F. Quensen III, and J.M. Tiedje. 1996. Evaluation of strains isolated by growth on naphthalene and biphenyl for hybridization of genes to dioxygenase probes and polychlorinated biphenyl-degrading ability. *Appl. Environ. Microbiol.* 62:2053–2058.

Peressutti, S.R., A.M. Alvarez, and O.H. Pucci. 2003. Dynamics of hydrocarbon-degrading bacteriocenosis of an experimental oil pollution in Patagonian soil. *Int. Biodeterior. Biodegrad.* 52:21–30.

Pritchard, P.H., J.G. Mueller, J.C. Roger, F.V. Kremer, and J.A. Glaser. 1992. Oil spill bioremediation: Experiences, lessons and results from the Exxon Valdez oil spill in Alaska. *Biodegradation* 3:315–335.

Ratajczak, A., W. Geissdorfer, and W. Hillen. 1998. Alkane hydroxylase from *Acinetobacter* sp. strain ADP1 is encoded by *alkM* and belongs to a new family of bacterial integral-membrane hydrocarbon hydroxylases. *Appl. Environ. Microbiol.* 64:1175–1179.

Ruberto, L.A.M., S.C. Vazquez, A. Curtosi, M.C. Mestre, E. Pelletier, and W. Mac Cormack. 2006. Phenanthrene biodegradation in soils using an Antarctic Bacterial Consortium. *Bioremediat. J.* 10:191–201.

Schloss, P.D. and J. Handelsman. 2008. A statistical toolbox for metagenomics: Assessing functional diversity in microbial communities. *BMC Bioinform.* 9:34.

Shanklin, J. and E. Whittle. 2003. Evidence linking the *Pseudomonas oleovorans* alkane omega-hydroxylase, an integral membrane diiron enzyme, and the fatty acid desaturase family. *FEBS Lett.* 545:188–192.

Shanklin, J., E. Whittle, and B.G. Fox. 1994. Eight histidine residues are catalytically essential in a membrane-associated iron enzyme, stearoyl-CoA desaturase, and are conserved in alkane hydroxylase and xylene monooxygenase. *Biochemistry* 33:12787–12794.

Smits, T.H., S.B. Balada, B. Witholt, and J.B. van Beilen. 2002. Functional analysis of alkane hydroxylases from gram-negative and gram-positive bacteria. *J. Bacteriol.* 184:1733–1742.

Smits, T.H., M. Rothlisberger, B. Witholt, and J.B. van Beilen. 1999. Molecular screening for alkane hydroxylase genes in Gram-negative and Gram-positive strains. *Environ. Microbiol.* 1:307–317.

Sotsky, J.B., C.W. Greer, and R.M. Atlas. 1994. Frequency of genes in aromatic and aliphatic hydrocarbon biodegradation pathways within bacterial populations from Alaskan sediments. *Can. J. Microbiol.* 40:981–985.

Staley, J.T. 1997. Biodiversity: Are microbial species threatened? *Curr. Opin. Biotechnol.* 8:340–345.

Staley, J.T. and J.J. Gosink. 1999. Poles apart: Biodiversity and biogeography of sea ice bacteria. *Annu. Rev. Microbiol.* 53:189–215.

Stallwood, B., J. Shears, P.A. Williams, and K.A. Hughes. 2005. Low temperature bioremediation of oil-contaminated soil using biostimulation and bioaugmentation with a *Pseudomonas* sp. from maritime Antarctica. *J. Appl. Microbiol.* 99:794–802.

van Beilen, J.B. and E.G. Funhoff. 2007. Alkane hydroxylases involved in microbial alkane degradation. *Appl. Microbiol. Biotechnol.* 74:13–21.

van Beilen, J.B., Z. Li, W.A. Duetz, T.H. Smits, and B. Witholt. 2003. Diversity of alkane hydroxylase systems in the environment. *Oil Gas Sci. Technol.* 58:427–440.

van Beilen, J.B., S. Panke, S. Lucchini, A.G. Franchini, M. Röthlisberger, and B. Witholt. 2001. Analysis of *Pseudomonas putida* alkane degradation gene clusters and flanking insertion sequences: Evolution and regulation of the *alk*-genes. *Microbiology* 147:1621–1630.

van Beilen, J.B., T.H. Smits, F.F. Roos, T. Brunner, S.B. Balada, M. Rothlisberger, and B. Witholt. 2005. Identification of an amino acid position that determines the substrate range of integral membrane alkane hydroxylases. *J. Bacteriol.* 187:85–91.

van Beilen, J.B., T.H. Smits, L.G. Whyte, S. Schorcht, M. Rothlisberger, T. Plaggemeier, K.H. Engesser, and B. Witholt. 2002. Alkane hydroxylase homologues in Gram-positive strains. *Environ. Microbiol.* 4:676–682.

van Beilen, J.B., M.G. Wubbolts, and B. Witholt. 1994. Genetics of alkane oxidation by *Pseudomonas oleovorans. Biodegradation* 5:161–174.

Volkman, J.K., D.G. Holdsworth, G.P. Neill, and H.J. Bavor Jr. 1992. Identification of natural, anthropogenic and petroleum hydrocarbons in aquatic sediments. *Sci. Total Environ.* 112:203–219.

Warhurst, A.M. and C.A. Fewson. 1994. Biotransformations catalyzed by the genus *Rhodococcus. Crit Rev. Biotechnol.* 14:29–73.

Watkinson, R.J. and P. Morgan. 1990. Physiology of aliphatic hydrocarbon-degrading microorganisms. *Biodegradation* 1:79–92.

Whyte, L.G., L. Bourbonniere, and C.W. Greer. 1997. Biodegradation of petroleum hydrocarbons by psychrotrophic *Pseudomonas* strains possessing both alkane (*alk*) and naphthalene (*nah*) catabolic pathways. *Appl. Environ. Microbiol.* 63:3719–3723.

Whyte, L.G., C.W. Greer, and W.E. Inniss. 1996. Assessment of the biodegradation potential of psychrotrophic microorganisms. *Can. J. Microbiol.* 42:99–106.

Whyte, L.G., L. Bourbonniere, C. Bellerose, and C.W. Greer. 1999a. Bioremediation assessment of hydrocarbon-contaminated soils from the high Arctic. *Bioremediat. J.* 3:69–80.

Whyte, L.G., S.J. Slagman, F. Pietrantonio, L. Bourbonniere, S.F. Koval, J.R. Lawrence, W.E. Inniss, and C.W. Greer. 1999b. Physiological adaptations involved in alkane assimilation at a low temperature by *Rhodococcus* sp. strain Q15. *Appl. Environ. Microbiol.* 65:2961–2968.

Whyte, L.G., A. Schultz, J.B. van Beilen, A.P. Luz, V.H. Pellizari, D. Labbe, and C.W. Greer. 2002a. Prevalence of alkane monooxygenase genes in Arctic and Antarctic hydrocarbon-contaminated and pristine soils. *FEMS Microbiol. Ecol.* 41:141–150.

Whyte, L.G., T.H. Smits, D. Labbe, B. Witholt, C.W. Greer, and J.B. van Beilen. 2002b. Gene cloning and characterization of multiple alkane hydroxylase systems in *Rhodococcus* strains Q15 and NRRL B-16531. *Appl. Environ. Microbiol.* 68:5933–5942.

Wyatt, T.M. 1984. The microbial degradation of hydrocarbons. *Trends Biochem. Sci.* 9:20–23.

Yergeau, E., S. Kang, Z. He, J. Zhou, and G.A. Kowalchuk. 2007. Functional microarray analysis of nitrogen and carbon cycling genes across an Antarctic latitudinal transect. *ISME J.* 1:163–179.

Zylstra, G.J. and D.T. Gibson. 1989. Toluene degradation by *Pseudomonas putida* F1. Nucleotide sequence of the *todC1C2BADE* genes and their expression in Escherichia coli. *J. Biol. Chem.* 264:14940–14946.

16 Potential Use of Real-Time PCR to Assess Changes in the Microbial Population Structure and Function during Bioremediation of Polar Soils

Shane M. Powell

CONTENTS

16.1 INTRODUCTION

Real-time PCR is a powerful tool for exploring microbial community structure and function. It allows the quantitative measurement of specific genes from DNA or RNA extracted from cultures or environmental samples. It is possible to target groups of bacteria on a phylogenetic basis using the 16S rRNA gene, or genes for particular functions, such as sulfate reduction or hydrocarbon degradation. The target of the assay is determined by the primer and/or probe pair. The specificity can be broad (genus or group level), or narrow (a particular species, strain, or genotype) allowing insight into microbial community structure and function in much greater detail than before. Rather than only being able to detect the presence or absence of a particular species or gene, as is possible with standard PCR, we are now able to follow changes in the abundance of a gene in response to outside factors. For example, Ben-Dov et al. (2007) followed the community dynamics of sulfate-reducing bacteria in wastewater using the dissimilatory sulfate reductase (*dsrA*) and adenosine-5′-phosphosulfate reductase (*aspA*) genes. Another example is the use of the ammonium monooxygenase gene (*amoA*) to follow the effect of fertilization with ammonium sulfate in both microcosms and a field trial (Okano et al., 2004). In bioremediation studies, the concentration of catabolic genes such as naphthalene dioxygenase (*ndoB*) or alkane monooxygenase (*alkB*) has been correlated with the soil hydrocarbon concentration (Powell et al., 2006a; Baldwin et al., 2008), providing a link between the microbial process and its effect on its environment.

The ability to measure specific genes provides a better understanding of the microbial processes that occur in a particular system. The polar regions are host to microbial communities that have a range of environmental factors to contend with in addition to adapting to utilizing pollutants. These conditions can include extreme cold, large temperature fluctuations, low water availability, low nutrient availability (more so in Antarctica than the Arctic), and the effect of repeated freeze–thaw cycles. Real-time PCR is a tool to help to increase our understanding of all these processes and how they interact with biodegradation. An improved understanding of microbial community dynamics will lead to improvements in the design of bioremediation strategies. This is especially important in polar regions where implementing bioremediation programs is made more complicated by the climate (short summer season, severe weather events) and remote location (increased transport costs, limited facilities).

16.2 REAL-TIME PCR TECHNOLOGY

The development of the polymerase chain reaction (PCR) in the 1980s as a way of accessing and analyzing nucleic acids from cells and the environment started a revolution in microbial ecology. However, standard PCR is only suitable for qualitative, not quantitative, analyses: there is no way of determining how many copies of a particular gene are present in a sample. Methods such as most-probable-number PCR (Toranzos et al., 1993; van Elsas et al., 1997) and competitive PCR (Simon et al., 1992; Telenti et al., 1992) were developed as quantitative PCR methods. Although these methods work well, they are labor intensive and are not cost effective for large sample numbers. In the 1990s, real-time PCR was introduced in medical research as a means of quantifying the number of copies of a particular gene in specific tissues. As with other methods originally developed for the field of medical molecular biology, environmental microbiologists soon adapted real-time PCR for their own needs. This allowed a shift in focus from "what genes are present in this ecosystem" to "how abundant are these genes" and "what affects their expression."

Only a brief explanation of the mechanics of real-time PCR is presented here as excellent reviews are available (see, e.g., Heid et al., 1996 or MacKay, 2004).

Real-time PCR works on the basic principle that with every cycle of PCR, the number of copies of the target amplicon doubles until a reagent, usually either the nucleotides or the primers, becomes limiting and a plateau is reached. The more the copies of the template DNA are present in the sample at the beginning of the PCR, the sooner the plateau is reached (see Figure 16.1). That is, the threshold cycle (C_T), the cycle where an arbitrary amount of double-stranded DNA is present, is smaller when greater amounts of template DNA are present in the initial sample. The relationship is a log-linear one described by

$$C_T = M \times \log(\text{template concentration [copies template/}\mu\text{L]}) + B$$

For a perfectly efficient reaction (doubling of PCR products every cycle), $M = -3.322$ and B is 1. In practical terms, M is the slope of the standard curve (plotted as threshold cycle against the number of copies in template) and B is the y-intercept (Figure 16.2).

The innovation in the development of real-time PCR was the use of fluorophores to measure the amount of double-stranded DNA present, which made it possible to follow the progress of the reaction and determine the threshold cycle. Real-time PCR machines measure the fluorescence in each reaction tube every cycle and as the amount of double-stranded DNA increases, so does the fluorescence. Three types of chemistries (ways of linking the fluorophore to the double-stranded DNA) have been developed with many variations commercially available. The simplest are the intercalating dyes that fluoresce when bound to double-stranded DNA. Hydrolysis probes and hybridization chemistries both rely on the interaction of a reporter and a quencher fluorophore attached to a probe (see Figure 16.3). When the reporter and the quencher are in close proximity, the quencher absorbs the emitted energy from

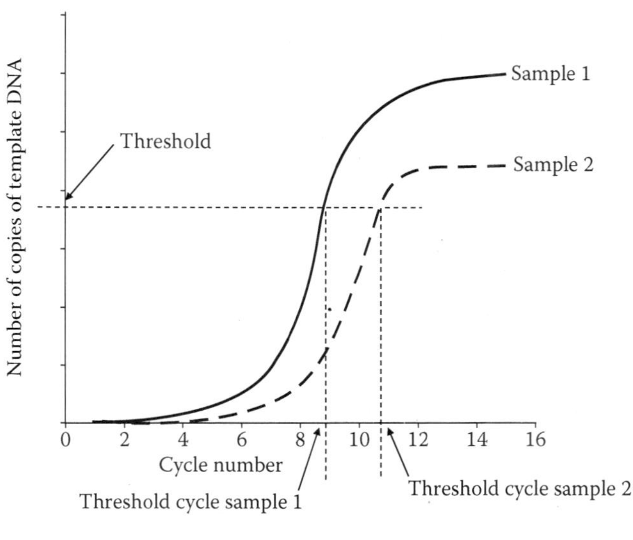

FIGURE 16.1 Progression of real-time PCR reactions and the determination of threshold cycle.

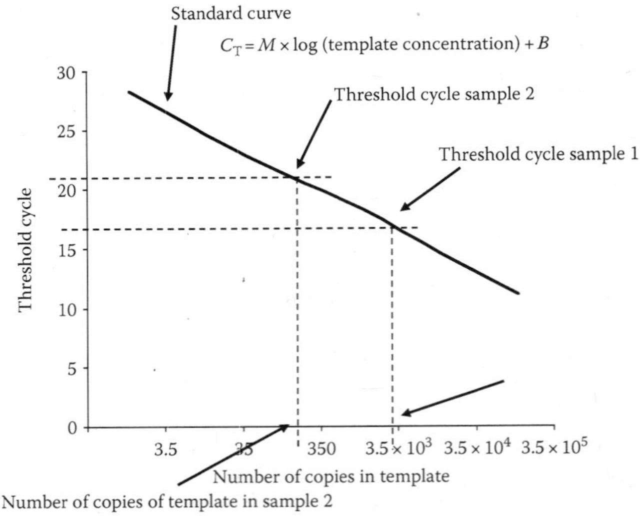

FIGURE 16.2 Real-time PCR standard curve showing relationship between threshold cycle and number of copies of template in sample.

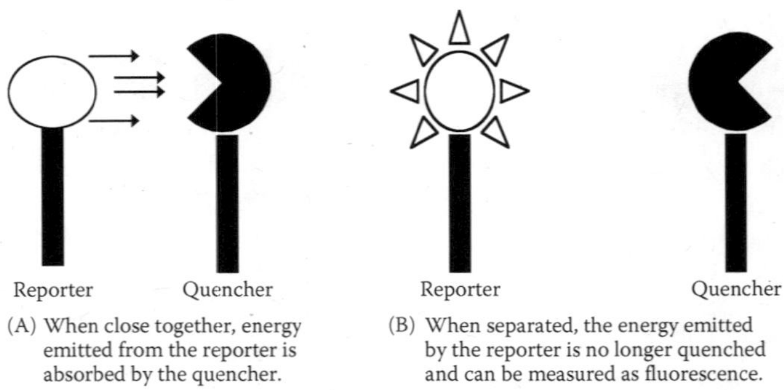

FIGURE 16.3 Interaction between reporter and quencher fluorophores in hybridization and hydrolysis probes.

the reporter and there is no signal. However, when the two are separated there is no quenching of the fluorescence and a signal can be measured.

SYBR Green I is the most common intercalator dye used in real-time PCR. As the rounds of amplification continue, the amount of double-stranded DNA increases (doubles in a perfectly efficient reaction), and hence the fluorescence also increases. However, the drawback to this method is that SYBR Green binds to all double-stranded DNA, including primer dimers (primer artifacts) that are formed when primers anneal to each other. Careful experimental design and analysis of the PCR products must be carried out to ensure that nonspecific products are not causing an increase in the fluorescence and hence an overestimation of the gene copy number.

The use of hydrolysis probes developed from the 5′ nuclease or Taqman assays that originally used radiolabeled probes (see MacKay, 2004). This system utilizes the 5′–3′ exonuclease activity of DNA polymerase to separate the reporter and the quencher fluorophores. The probe molecule, with a reporter attached to the 5′ end and a quencher attached to the 3′ end, binds to a target molecule between the two primer sites. As the target molecule is amplified, the 5′ nuclease activity of DNA polymerase cleaves the probe separating the reporter and the quencher thus increasing the fluorescence. With each round of PCR there are subsequently more target molecules for the probe to bind to. It is then cleaved by the DNA polymerase separating the reporter and the quencher and hence fluorescence is emitted.

Hybridization probes, also known as "hairpin oligoprobes," have complementary 5′ and 3′ ends that bind to each other to form a hairpin structure. The most common of these are called molecular beacons. The probes are designed to bind to a region of the target DNA between the two primer sites. A reporter fluorophore is attached to the 5′ end and a quencher to the 3′ end of the probe. When the probe is not bound to the target, the 5′ and 3′ ends bind to each other and the fluorescence is quenched. However, when the probe binds to the target the hairpin structure is lost as the two ends are separated from each other. The fluorescence from the reporter is no longer quenched, and a signal can be measured. As the PCR continues and more amplicon molecules are present, more probe molecules bind to the target that increases the emitted fluorescence.

16.3 ADVANTAGES AND LIMITATIONS OF REAL-TIME PCR

Real-time PCR has the same advantages and limitations of other PCR-based methods compared to culture-based methods. It avoids the bias of the noncultivable majority but introduces its own biases, such as preferential amplification in mixed template PCR reactions. All methods have some limitations, and therefore it is important to choose the most appropriate method for the hypothesis and to interpret the data taking into account the limitations of the methodology. Guidelines have recently been published (Bustin et al., 2009) outlining the reporting requirements for real-time PCR experiments. It is expected that this will not only improve the quality of published real-time PCR but also enable reviewers (and readers) to evaluate real-time PCR experiments and make comparisons of different studies more easily.

No special handling, such as chemical preservation, is required for samples that are to be used for real-time PCR. If the analysis is to be based on DNA, then simply freezing the samples at −18°C is adequate. If the analysis is to be based on RNA then freezing at −80°C, prompt RNA extraction or freezing with the addition of an RNA stabilization product is more suitable. This simplifies fieldwork as samples can be collected and frozen without the need for specialized processing, which may be difficult in the field.

In practical terms, real-time PCR is fast and less labor-intensive than many other methods, particularly if robotic systems are utilized for setting up a large number of reactions. There are no requirements for postreaction processing, and the data analysis is simplified by the system software. This is the biggest advantage that real-time PCR has to offer microbial ecology because it means that once an assay has been designed and optimized it is relatively easy to process large numbers of samples. Most systems have 96-well plate or 72-tube configurations with a 2 h turnaround, so with standard quality control protocols including the duplicates of samples, standard curves, and negative controls it is possible to process over a 100 samples per day. The most time-consuming part of the process is the DNA or RNA extractions. The ability to easily analyze large numbers of samples improves experimental and sampling design by increasing the statistical power of an experiment meaning that subtle differences between groups (e.g., different treatments or times) are more likely to be detected.

Unlike many other analytical methods, the specificity of real-time PCR assays can be controlled through the design of primers and/or probes. Assays can be designed to be highly specific, e.g., for a gene carried by a particular strain (see Kohno et al. (2002) in which specific primers were designed to detect the variation in alkane monooxygenase that degrades *n*-alkanes with more than 12 carbons), or more broadly specific such as detecting all strains from a particular genus (e.g., detection of *Desulfotomaculum* by Stubner (2002)). This flexibility can be very useful. For example, it is possible to use universal primers for the 16S rRNA gene to measure the total number of copies of 16S rRNA and then use group or species specific primers/probes to measure the number of copies of a particular phylogenetic group, thus obtaining a realistic picture of the proportion of the microbial community belonging to that group. The use of a ratio of specific:total 16S rRNA genes overcomes the problem of the multiple copies of the 16S rRNA gene that occurs when this gene is measured alone (see below).

Real-time PCR is also very sensitive and can detect in the order of 100 copies of a gene per reaction. If it is assumed that there is one copy of the gene of interest per cell, and that 1 g of sample was extracted to obtain 50 μL of a DNA suspension and 2 μL of the extract was used in the reaction, then 100 copies per reaction would equate to 2500 copies per g of sample. Given that 1 g of soil may contain over 1×10^6 cells, then 100 copies per reaction represents 0.25% of the population. This level of sensitivity means that contamination can easily become a problem, and it is imperative that negative controls are always included and standard precautions are always taken when setting up reactions. Several studies (Gruntzig et al., 2001; Okano et al., 2004; Powell et al., 2006a) have reported that the limit of detection is around 300 copies per reaction; below this level standard curves become nonlinear and unreliable.

The success of real-time PCR is dependent on the primer and probe designs. To design primers that have the desired specificity requires knowledge of the gene sequence, and this is often lacking for genes other than the 16S rRNA gene. For example, in June 2008 there were 49 sequences for the benzylsuccinate synthase gene, a key gene in the anaerobic degradation of aromatic hydrocarbons (Beller et al., 2002). Of these, only 11 are from isolates representing 10 different species from 7 genera. It is unlikely that this number represents the total diversity of anaerobic aromatic-degrading organisms. The situation with naphthalene dioxygenase is better although again unlikely to cover the real genetic diversity of this enzyme. There are currently 214 sequences for naphthalene dioxygenase in GenBank (excluding putative genes). Eighty-three of these were from uncultured organisms, only 131 are from hydrocarbon-degrading isolates. These isolates come from 18 different species from 15 genera (although many were identified only to genus level so the species diversity could be higher).

The complexity of the ecosystem under investigation can add another dimension to the task of developing real-time PCR assays. When working with a pure culture *in vitro*, designing a real-time PCR assay to measure the activity of a specific gene is relatively straightforward. However, when working with environmental samples that contain an unknown number of different species with an unknown number of variations of a gene, it is difficult to know how much of that diversity is being captured. Some processes are carried out by more than one enzyme, some enzymes have conserved genes and others divergent genes. The alkane monooxygenase gene is one such example: there are variants that are responsible for the degradation of only short-chain or only long-chain alkanes (Sei et al., 2003), and there are published primer sets that distinguish between the *alkB* gene from *Pseudomonas putida* and that from *Rhodococcus erythropolis* (Luz et al., 2004). Designing an assay to quantitatively measure all alkane monooxygenase genes in a sample is very difficult. In these situations, it is often best to tailor the assay around the diversity present in the sample set by, for example, first creating a clone library of the gene of interest (Powell et al., 2006a).

After designing an appropriate primer and probe, optimizing reaction conditions is the next step in developing and validating a real-time PCR assay. As with standard PCR, many factors can be manipulated including primer, probe, and template concentrations, annealing temperature and the length of each step. Ideally, in every round of PCR, the number of copies of the gene of interest doubles resulting in a reaction efficiency of 100%. This can be calculated from the slope of the standard curve and is automatically done with most real-time PCR software platforms. In practice, few reactions consistently display 100% efficiency. A generally accepted rule of thumb is that reaction efficiencies between 90% and 110% are acceptable (e.g., Ambion TechNotes). A variation in the reaction efficiency affects the calculation of the number of copies of the gene from the C_T. One cause of different reaction efficiencies is differences in either the template or primer sequences. Degenerate primers, which are a mixture of nearly identical primers with one or more nucleotide differences, may amplify the template with different efficiencies. However, this is a controversial point as few studies have been carried out on the effect of degenerate primers on reaction efficiency. One study on woodchuck hepatitis virus (Huang and Buckwold, 2005) found that the degenerate primers amplified all forms of the gene with equal efficiency.

As there is a log–linear relationship between the C_T (raw signal) and the number of copies of the gene of interest, small differences in the C_T result in large differences in copy number. Although this means that real-time PCR is very sensitive to small changes in gene copy numbers (useful for applications such as gene expression), it also means that the reproducibility of the method can be quite low. The assay coefficient of variation (CV) can be based on C_T (raw data) or on the calculated gene copy number (processed data). In studies that have looked at both (Dionisi et al., 2003; Powell et al., 2006a; Smith et al., 2006), the CV based on C_T was always within acceptable limits (less than 5%) whereas that based on copy number was up to 20%.

A disadvantage of real-time PCR when used to look at processes via functional genes is that if DNA is used as the template then there is no way to distinguish between dead cells, live cells, and live but dormant (viable but not culturable, VBNC) cells. Using mRNA as the template provides an estimate of active genes as Lee et al. (2008) showed by measuring different dehalogenase genes to follow the biodegradation of trichlorethene. However, the difficulty of obtaining mRNA from environmental samples, particularly soils and sediments, is reflected in the lack of bioremediation studies using mRNA-based techniques. Recently methods utilizing the ability of propidium monoazide bromide (PMA) or ethidium monoazide bromide (EMA) to bind to DNA have been developed. Monoazides are only able to enter membrane-compromised cells, where, after photoactivation (exposure to light), they bind to DNA preventing the DNA from being amplified in subsequent PCR (Rudi et al., 2005). In this way, only DNA from cells that were viable at the time of the monoazide addition will be amplified and detected. This method has only been used with simple matrices such as biofilms so far; there are no reports of its use in microcosm experiments or bioremediation field trials. However, it is a promising development that is being applied to different types of environmental samples (Pisz et al., 2007; Luo et al., 2008).

16.4 REAL-TIME PCR IN BIOREMEDIATION

Real-time PCR assays have been published for many genes that are involved in hydrocarbon biodegradation including monooxygenases, dioxygenases, and genes for anaerobic processes such as benzoyl CoA reductase and dehalogenases. The published examples of real-time PCR assays are given in Table 16.1. Some of these assays were designed with primers that have broad targets while others have targeted catabolic genes that are specific to a study site or a particular species/group of organisms. In one case (Hristova et al., 2001), the 16S rRNA gene was measured to follow the population dynamics of a particular strain (PM1) throughout a bioaugmentation trial.

Catabolic genes often have quite divergent sequences. This can be useful in studies looking at the evolution and transmission of catabolic genes through the environment. However, it makes quantification by real-time PCR difficult if a conserved region of the gene suitable for primer sites cannot be found. Primer sets have been published in which there are degenerate primers, however, this is less than ideal as some templates will amplify with a greater efficiency than others. Other approaches are to target very specific subsets of a gene or to use multiple primer/probe sets to capture all the diversity present. DeBruyn et al. (2007) designed real-time PCR primer and probe sets for aromatic dioxygenases found in different groups of bacteria. They were able to distinguish between the pyrene dioxygenase of *Mycobacterium* (*nidA*) and the naphthalene dioxygenases found in Beta-proteobacteria (*nagAc*) and Gamma-proteobacteria (*nahAc*). In this case, the authors concluded that the relationship between polyaromatic hydrocarbon (PAH) degradation and the levels of these three genes made *nidA* and *nagAc*, but not *nahAc*, suitable biomarkers for biodegradation.

Assays that target the 16S rRNA gene can also be useful in the context of biodegradation and bioremediation. Where particular groups of bacteria are known to be important, real-time PCR can be used to follow their population dynamics.

TABLE 16.1
Examples of Published Real-Time PCR Assays Useful in Bioremediation

Enzyme	Gene	Target Details	Study Site	Reference
Ribosomal RNA	16S rRNA	MTBE-degrading strain PM1	Sediment in microcosms	Hristova et al. (2001)
Ribosomal RNA	16S rRNA	Ammonia-oxidizing bacteria	Petroleum waste landfarm, Finland	Kurola et al. (2005)
Alkane monooxygenase	*alkB*	Alkane-degrading soil microbes	Contaminated Antarctic soil	Powell et al. (2006a)
Naphthalene dioxygenase	*nahAC*, *ndo*, and *pahAC*	BTEX-degrading microbes	Gasoline-contaminated groundwater	Baldwin et al. (2008)
Toluene dioxygenase	*TodC1*	BTEX-degrading microbes	Gasoline-contaminated groundwater	Baldwin et al. (2008)
Xylene monooxygenase	*xylM*	BTEX-degrading microbes	Gasoline-contaminated groundwater	Baldwin et al. (2008)
Pyrene dioxygenase	*nidA*	*Mycobacterium* spp.	Coal tar-contaminated soil	DeBruyn et al. (2007)
Naphthalene dioxygenases	*nagAc* and *nahAc*	Beta-proteobacteria and Gamma-proteobacteria	Coal tar-contaminated soil	DeBruyn et al. (2007)
Benzylsuccinate synthase	*bss*	Anaerobic aromatic degrading microbes	Underground fuel tanks	Beller et al. (2002)
Benzoyl CoA reductase	BCR	Aromatic-degrading, denitrifying microbes	Gasoline-contaminated groundwater	Hosoda et al. (2005)
Benzoyl CoA reductase	*bcr* or *bzd*	Aromatic-degrading, denitrifying microbes	Estuarine sediment	Song and Ward (2005)
Reductive dehalogenase	*tceA*, *bvcA*, and *vcrA*	Dehalogenating, *Dehalococcoides*	Pure cultures, contaminated groundwater mRNA from groundwater	Ritalahti et al. (2006) and Lee et al. (2008)

This is also useful for following the progress of bioaugmentation where a degrading organism is added to the contaminated site. The 16S rRNA gene has also been used to estimate total numbers of bacterial cells. However, the number of copies of this gene per cell varies between species, which makes it unsuitable for estimating total cell numbers in a mixed culture. For example, *Desulfurococcus mucosis* has one copy of the 16S rRNA gene per cell whereas *Bacillus cereus* has 10–12 copies per cell (Fogel et al., 1999). This means that for a sample with 10 *D. mucosis* cells and 2 *B. cereus* cells (a total of 12 cells), approximately 30 copies of the 16S rRNA gene will be detected. If another sample has 2 *D. mucosis* cells and 10 *B. cereus* cells (again a total of 12 cells), it will have 102 copies of the 16S rRNA gene and will appear to have over three times the number of cells as the first sample. Other genes such as the ribosomal polymerase B (*rpoB*) gene that are also found in all cells are conserved genes and appear to have only one copy per cell (Dahllof et al., 2000), which makes the *rpoB* gene more suitable for monitoring the total number of bacteria.

16.5 REAL-TIME PCR IN POLAR MICROBIOLOGY

Very little has been published using real-time PCR to investigate the biodegradation and bioremediation processes in polar regions. There have been surveys of catabolic genes in the Arctic and Antarctic using PCR-based methods (e.g., Whyte et al., 2002; Luz et al., 2004) but surprisingly few quantitative studies. Real-time PCR has been used to investigate polar ecosystems but not often for biodegradation studies. For example, Abell and Bowman (2005) studied the distribution of the Flavobacteria in the Southern Ocean using targeted real-time PCR. Yergeau et al. (2007) correlated population structure, measured with real-time PCR of the 16S rRNA gene, to particular environmental variables to determine which had the greatest effect on community structure. At the current time, there are only two published studies that use real-time PCR to follow microbial populations during a bioremediation trial in the polar region (Powell et al., 2006a,b).

16.6 CASE STUDY: FERTILIZATION FIELD TRIAL EXPERIMENT AT OLD CASEY STATION, ANTARCTICA

Old Casey station was occupied by Australia's Antarctic program between 1969 and 1988. A spill of around 36,000L of Special Antarctic Blend (SAB) diesel occurred in 1982. As part of a project looking at ways to clean up the old station, a field trial looking at the effects of nutrient and water addition to the soil (biostimulation) was conducted. Soil was collected, sieved to 4.75mm, placed in metal containers and reburied in the ground to simulate natural conditions. Two treatments were applied: low controlled release nutrients (100mg PO_4^{3-} kg^{-1} 200mg NO_3^- kg^{-1}, 300mg NH_4^+ kg^{-1}) and high controlled release nutrients (400mg PO_4^{3-} kg^{-1}, 700mg NO_3^- kg^{-1}, 900mg NH_4^+ kg^{-1}). These were compared to a control treatment, which had no addition of fertilizer. There were four replicates for each treatment. Surface samples were taken approximately 1 year later, and after 5 years the containers were removed from the ground and samples taken at the top (0–35mm), middle (35–70mm), and bottom (70–100mm) depths in the soil profile.

Initially the aerobic biodegradation of hydrocarbons in the surface samples was investigated (Powell et al., 2006a). Most probable number counts were performed to estimate the numbers of total heterotrophic, alkane-degrading, and SAB-degrading bacteria 1 year and 5 years after the fertilization. A real-time PCR assay was developed to target the alkane monooxygenase gene *alkB*. Degenerate primers were used to create a clone library, and the clone sequences were used to design primers that would capture the most common forms of the *alkB* gene in these samples. The primer set for the *rpoB* gene used by Dahllof et al. (2000) was used to estimate the total number of bacteria in the soil, and the proportion of microbes containing the *alkB* gene was calculated as the ratio of the number of copies of *alkB*: number of copies of *rpoB*. There was a strong positive correlation ($r = 0.87$) between the concentration of *n*-alkanes in the soil and this ratio. As the *n*-alkanes were degraded, the proportion of microbes carrying the *alkB* gene decreased. Culture-based methods, used on the same samples, showed that the number of diesel degrading bacteria remained high. This suggests that as the alkanes were degraded, microbes that were able to utilize other components of the diesel became a dominant part of the community at the expense of the alkane degraders. This was a positive result for the bioremediation of the soil as it indicated that after the easily degraded alkanes were lost from the soil, other components of the SAB diesel were being used as carbon sources.

Degradation processes at three depths in the soil profile in the control and high nutrient treatment were then explored (Powell et al., 2006b). It was hypothesized that because of the excess nutrients added to the soil, hydrocarbon degradation coupled to denitrification may be occurring where the soil oxygen availability was low, i.e., in the deeper regions of the soil profile. Real-time PCR results showed that the increase in the levels of *nosZ*, the gene for nitrous oxide reductase that is part of the denitrification process, was significantly higher in the fertilized samples than in the control samples. Using the primers described by Song and Ward (2005) for benzoyl Co-A reductase, a key gene in the anaerobic degradation pathway for aromatic hydrocarbons, we were also able to show that anaerobic aromatic hydrocarbon degrading bacteria were much more prevalent in the fertilized samples compared to the control samples from the same depth. We then conducted a microcosm experiment to demonstrate that hydrocarbon degradation under denitrifying conditions occurred in Antarctic soils. This was the first report of this process in Antarctic soils. These results had implications for the design of bioremediation strategies in terms of oxygen requirements and the form of nitrogen addition to the soil. More work is required to determine the significance of anaerobic processes in the Antarctic soil ecosystem.

16.7 POTENTIAL ADVANCEMENTS IN REAL TIME PCR

As our knowledge of the enzymes and genes involved in biodegradation increases, so will the possibilities for using real-time PCR to quantitatively measure these genes in the environment. Three areas offer the most promise for developing the use of real-time PCR to monitor biodegradation processes in the environment:

1. Expanding the knowledge base of gene sequences for enzymes involved in biodegradation processes
2. Developing the ability to differentiate between live and dead cells to produce a more representative picture of active processes
3. Improving multiplexing capabilities which will increase the efficiency and productivity of this method

One of the disadvantages of using DNA-based methods to investigate microbial communities is that it is not possible to distinguish between DNA from active cells and DNA from inactive cells. Real-time PCR was developed as a method to measure the gene expression in tissues by measuring the amount of mRNA present. This results in a clear measure of active cells. However, the extraction of quality mRNA from environmental samples such as soils is extremely difficult as it is a small molecule with a short half life. It is possible to isolate rRNA from the environment; but this only allows the exploration of community structure via the 16S rRNA gene rather than metabolic genes and community function. The use of the monoazides (PMA or EMA) to target live cells for DNA extraction currently holds the most promise for elucidating active microbial processes in environmental samples.

Multiplex real-time PCR reactions are those where more than one gene in measured in a single reaction. This is done by utilizing different fluorophores for the different genes. As the available technology has improved, so have the possibilities for multiplex reactions with some manufacturers now claiming that up to four targets can be measured per reaction. In reality it is difficult, although sometime possible (for example Ward and Bej, 2006)) to produce four primer and probe sets that have similar optimal PCR conditions and do not interfere with each other. However, the ability to measure even two genes simultaneously would increase the efficiency and decrease the cost of real-time PCR analyses.

16.8 SUMMARY

The design of real-time PCR assays is dependent on quality sequence information, and further use of real-time PCR in bioremediation will be dependent on continual research into the genetics of degrading strains. This is currently the biggest limitation of real-time PCR: it is only possible to measure what we know exists.

However, real-time PCR can be an important part of the design of bioremediation treatments by elucidating what parts of the microbial community are active at different stages and what is limiting the degradation process. This knowledge is required to decide whether natural attenuation or bioremediation is an appropriate approach for a particular contaminated site. Real-time PCR can also be used to monitor changes in microbial community structure, which provides information on the effect of the contamination on the ecosystem and allows treatment end points to be set. This is particularly pertinent for the Antarctic as the microbial ecosystem is more easily monitored than the invertebrate or transient faunal populations. The application of real-time PCR to biodegradative microorganisms will contribute significantly to our understanding of the role of microbial biodiversity and functional biodiversity in the Antarctic environment.

Real-time PCR is a potentially powerful tool for gaining a deeper understanding of microbial processes *in situ*, which is, after all, the aim of microbial ecology research.

REFERENCES

Abell, G. C. J. and J. P. Bowman. 2005. Ecological and biogeographic relationships of class Flavobacteria in the Southern Ocean. *FEMS Microbiology Ecology* 51 (2):265–277.

Baldwin, B. R., C. H. Nakatsu, and L. Nies. 2008. Enumeration of aromatic oxygenase genes to evaluate monitored natural attenuation at gasoline-contaminated sites. *Water Research* 42 (3):723–731.

Beller, H. R., S. R. Kane, T. C. Legler, and P. J. J. Alvarez. 2002. A real-time polymerase chain reaction method for monitoring anaerobic, hydrocarbon degrading bacteria based on a catabolic gene. *Environmental Science and Technology* 36:3977–3984.

Ben-Dov, E., A. Brenner, and A. Kushmaro. 2007. Quantification of sulfate-reducing bacteria in industrial wastewater, by real-time polymerase chain reaction (PCR) using dsrA and apsA genes. *Microbial Ecology* 54 (3):439–451.

Bustin, S. A. V. Benes, J. A. Garson, J. Hellemans, J. Huggett, M. Kubista, R. Mueller, T. Nolan, M. W. Pfaffl, G. L. Shipley, J. Vandesompele, and C. T. Wittwer. 2009. The MIQE guidelines: Minimum information for publication of quantitative real-time PCR experiments. *Clinical Chemistry* 55:611–622.

Dahllof, I., H. Baillie, and S. Kjelleberg. 2000. rpoB-based microbial community analysis avoids limitations inherent in 16S rRNA gene intraspecies heterogeneity. *Applied and Environmental Microbiology* 66:3376–3380.

DeBruyn, J. M., C. S. Chewning, and G. S. Sayler. 2007. Comparative quantitative prevalence of Mycobacteria and functionally abundant nidA, nahAc and nagAc dioxygenase genes in coal tar contaminated sediments. *Environmental Science and Technology* 41(15): 5426–5432.

Dionisi, H. M., G. Harms, A. C. Layton, I. R. Gregory, J. Parker, S. A. Hawkins, K. G. Robinson, and G. S. Sayler. 2003. Power analysis for real-time PCR quantification of genes in activated sludge and analysis of variability introduced by DNA extraction. *Applied and Environmental Microbiology* 69:6597–6604.

Fogel, G. B., C. R. Collins, J. Li, and C. F. Brunk. 1999. Prokaryotic genome size and SSU rDNA copy number: Estimation of microbial relative abundance from a mixed population. *Microbial Ecology* 38:93–113.

Gruntzig, V., S. C. Nold, J. Zhou, and J. M. Tiedje. 2001. *Pseudomonas stutzeri* nitrite reductase gene abundance in environmental samples measured by real-time PCR. *Applied and Environmental Microbiology* 67:760–768.

Heid, C. A., J. Stevens, K. J. Livak, and P. M. Williams. 1996. Real time quantitative PCR. *Genome Research* 6:986–994.

Hosoda, A., Y. Kasai, N. Hamamura, Y. Takahata, and K. Watanabe. 2005. Development of a PCR method for the detection and quantification of benzoyl-CoA reductase genes and its application to monitored natural attenuation. *Biodegradation* 16 (6):591–601.

Hristova, K. R., C. M. Lutenegger, and K. M. Scow. 2001. Detection and quantification of methyl tert-butyl ether degrading strain PM1 by real time TaqMan PCR. *Applied and Environmental Microbiology* 67:5154–5160.

Huang, Z. H. and V. E. Buckwold. 2005. A TaqMan PCR assay using degenerate primers for the quantitative detection of woodchuck hepatitis virus DNA of multiple genotypes. *Molecular and Cellular Probes* 19 (4):282–289.

Kohno, T., Y. Sugimoto, K. Sei, and K. Mori. 2002. Design of PCR primers and gene probes for general detection of alkane degrading bacteria. *Microbes and Environments* 17:114–121.

Kurola, J., M. Salkinoja-Salonen, T. Aarnio, J. Hultman, and M. Romantschuk. 2005. Activity, diversity and population size of ammonia-oxidising bacteria in oil-contaminated landfarming soil. *FEMS Microbiology Letters* 250 (1):33–38.

Lee, P. K. H., T. W. Macbeth, K. S. Sorenson Jr., R. A. Deeb, and L. Alvarez-Cohen. 2008. Quantifying genes and transcripts to assess the in situ physiology of "*Dehalococcoides*" spp. in a trichloroethene-contaminated groundwater site. *Applied and Environmental Microbiology* 74:2728–2739.

Luo, L. X., C. Walters, H. Bolkan, X. L. Liu, and J. Q. Li. 2008. Quantification of viable cells of *Clavibacter michiganensis* subsp. michiganensis using a DNA binding dye and a real-time PCR assay. *Plant Pathology* 57 (2):332–337.

Luz, A. P., V. H. Pellizari, L. G. Whyte, and C. W. Greer. 2004. A survey of indigenous microbial hydrocarbon degradation genes in soils from Antarctica and Brazil. *Canadian Journal of Microbiology* 50 (5):323–333.

MacKay, I. M. 2004. Real-time PCR in the microbiology laboratory. *Clinical Microbiology and Infection* 10:190–212.

Okano, Y., K. R. Hristova, C. M. Leutenegger, L. E. Jackson, R. F. Denison, B. Gebreyesus, D. Lebauer, and K. M. Scow. 2004. Application of real-time PCR to study effects of ammonium on population size of ammonia-oxidizing bacteria in soil. *Applied and Environmental Microbiology* 70:1008–1016.

Pisz, J. M., J. R. Lawrence, A. N. Schafer, and S. D. Siciliano. 2007. Differentiation of genes extracted from non-viable versus viable micro-organisms in environmental samples using ethidium monoazide bromide. *Journal of Microbiological Methods* 71:312–318.

Powell, S. M., S. H. Ferguson, J. P. Bowman, and I. Snape. 2006a. Using real-time PCR to assess changes in the hydrocarbon-degrading microbial community in Antarctic soil during bioremediation. *Microbial Ecology* 52 (3):523–532.

Powell, S. M., S. H. Ferguson, I. Snape, and S. D. Siciliano. 2006b. Fertilization stimulates anaerobic fuel degradation of Antarctic soils by denitrifying microorganisms. *Environmental Science & Technology* 40:2011–2017.

Ritalahti, K. M., B. K. Amos, Y. Sung, Q. Z. Wu, S. S. Koenigsberg, and F. E. Loffler. 2006. Quantitative PCR targeting 16S rRNA and reductive dehalogenase genes simultaneously monitors multiple *Dehalococcoides* strains. *Applied and Environmental Microbiology* 72 (4):2765–2774.

Rudi, K., B. Moen, S. M. Dromtorp, and A. L. Holck. 2005. Use of ethidium monoazide and PCR in combination for quantification of viable and dead cells in complex samples. *Applied and Environmental Microbiology* 71 (2):1018–1024.

Sei, K., Y. Sugimoto, K. Mori, H. Maki, and T. Kohno. 2003. Monitoring of alkane-degrading bacteria in a sea-water microcosm during crude oil degradation by polymerase chain reaction based on alkane-catabolic genes. *Environmental Microbiology* 5:517–522.

Simon, L., R. C. Levesque, and M. Lalonde. 1992. Rapid quantitation by PCR of endomycorrhizal fungi colonizing roots. *PCR Methods and Applications* 2 (1):76–80.

Smith, C. J., D. B. Nedwell, L. F. Dong, and A. M. Osborn. 2006. Evaluation of quantitative polymerase chain reaction based approaches for determining gene copy and gene transcript numbers in environmental samples. *Environmental Microbiology* 8:804–815.

Song, B. and B. B. Ward. 2005. Genetic diversity of benzoyl coenzyme A reductase genes detected in denitrifying isolates and Estuarine sediment communities. *Applied and Environmental Microbiology* 71 (4):2036–2045.

Stubner, S. 2002. Enumeration of 16S rDNA of *Desulfotomaculum* lineage 1 in rice field soil by real-time PCR with SybrGreen detection. *Journal of Microbiological Methods* 50:155–164.

Telenti, A., P. Imboden, and D. Germann. 1992. Competitive polymerase chain-reaction using an internal standard—Application to the quantitation of Viral-DNA. *Journal of Virological Methods* 39 (3):259–268.

Toranzos, G. A., A. J. Alvarez, and E. A. Dvorsky. 1993. Application of the polymerase chain-reaction technique to the detection of pathogens in water. *Water Science and Technology* 27 (3–4):207–210.

van Elsas, J. D., V. Mantynen, and A. C. Wolters. 1997. Soil DNA extraction and assessment of the fate of *Mycobacterium chlorophenolicum* strain PCP-1 in different soils by 16S ribosomal RNA gene sequence based most-probable-number PCR and immunofluorescence. *Biology and Fertility of Soils* 24 (2):188–195.

Ward, L. N. and A. K. Bej. 2006. Detection of *Vibrio parahaemolyticus* in shellfish by use of multiplexed real-time PCR with Taqman flourescent probes. *Applied Environmental Microbiology* 72:2031–2042.

Whyte, L. G., A. Schultz, J. B. van Beilen, A. P. Luz, V. Pellizari, D. Labbe, and C. W. Greer. 2002. Prevalence of alkane monooxygenase genes in Arctic and Antarctic hydrocarbon-contaminated and pristine soils. *FEMS Microbiology Ecology* 41 (2):141–150.

Yergeau, E., S. Bokhorst, A. H. L. Huiskes, H. T. S. Boschker, R. Aerts, and G. A. Kowalchuk. 2007. Size and structure of bacterial, fungal and nematode communities along an Antarctic environmental gradient. *FEMS Microbiology Ecology* 59 (2):436–451.

17 Microbial Bioremediation in Polar Environments: Current Status and Future Directions

Ronald M. Atlas

CONTENTS

17.1 INTRODUCTION

Although bioremediation is a green technology that can be advantageous for application to polar regions where natural processes and pristine conditions are highly valued (see Atlas and Philp, 2005 for an overview of the applicability of bioremediation for real world cleanup efforts), it should only be relied upon if it can achieve acceptable end points of decontamination. Put another way, bioremediation can be a cost effective way of lowering the concentrations of pollutants to levels that can be accommodated in the environment without causing any ecological harm, but bioremediation has limitations that must be taken into account in considering its potential applicability to this remediation of contaminated polar environments. In some cases, bioremediation through biostimulation (e.g., fertilizer addition) is the best technology to use; in other cases it is appropriate to rely upon natural attenuation (intrinsic bioremediation), which may be the only option in the tool box with a positive benefit-to-cost ratio due to expense and the ecological damage of digging up and transporting the contaminated soils or sediments (see Atlas and Bragg, 2009 for a discussion on when and when not to consider active bioremediation).

In particular, when considering the applicability of bioremediation, it is critical to establish the performance parameters that must be achieved in order to determine if bioremediation can be an effective technology. How fast must the pollutant be removed? What are the acceptable end points? Will the end points be set based upon ecological toxicity or specific pollutant concentrations? Will natural biodegradative processes be adequate, or will interventions be necessary, such as biostimulation through nutrient addition or tilling, to reduce the concentrations of contaminants and if so can such bioremediation achieve the desired cleanup end points?

When considering the potential for using bioremediation in the future in Arctic and Antarctic ecosystems it is important to take into account which pollutants are most likely to contaminate specific environments and whether the indigenous microorganisms have the metabolic capacity to attack those compounds. Due to the extraction of petroleum in the Arctic and the use of petroleum-based fuels for transportation and generation of heat and electricity in both the Arctic and the Antarctic, hydrocarbons are the most likely contaminants in polar environments that might be amenable to bioremediation (Snape et al., 2001). Other contaminants, such as PCBs and pesticides, are detectable in polar regions as a result of long distance atmospheric transport from temperate regions, but are generally disseminated so widely and at concentrations that would preclude any active cleanup. Some contained sites contaminated with pollutants, such as PCBs at the abandoned Distant Early Warning (DEW line) radar stations across the United States, Canadian, and Greenland Arctic, are probably best treated using physical means, including burial at hazardous waste disposal sites.

With regard to hydrocarbon pollutants, there already have been several accidental oil and fuel spills that have contaminated Arctic and Antarctic ecosystems (see, e.g., Aislabie et al., 2004; Cripps and Shears, 1997; Delille and Pelletier, 2002; Jaraulaa et al., 2008; McCarthy et al., 2004; Mohn et al., 2001; Poland et al., 2003; Stallwood et al., 2005; Whyte et al., 1999). Despite attention to the prevention of additional accidental oil and fuel spillages, they are likely to occur and probably will become more frequent as human polar scientific and commercial activities increase. Thus, developing appropriate cleanup technologies for accidental crude oil and refined petroleum products spills of varying magnitudes is a significant challenge, and it is within that context that the potential uses of bioremediation need to be critically evaluated.

As documented throughout this book, microorganisms in Arctic and Antarctic ecosystems have been found to have sufficient capacities for degrading hydrocarbons for bioremediation to be considered as a potential cleanup technology for polar regions. A number of studies (see, e.g., Aislabie et al., 1998, 2008; Delille, 2000; Delille et al., 2007; Rike et al., 2001; Sexstone and Atlas, 1977) have shown that both Antarctic and Arctic soil microbial communities are able to degrade a range of potentially contaminating aliphatic and aromatic hydrocarbons, even in regions that have not previously experienced oil contamination. Studies that have been conducted *in situ* where oil has contaminated the environment (see, e.g., Powell et al., 2007; Prince et al., 2003) indicate that both Antarctic and Arctic marine sediment microbial communities are also capable of adapting to utilize hydrocarbons from spilt fuel and oil present at environmentally realistic concentrations. Thus, indigenous microbes should be capable of degrading many of the petroleum hydrocarbons that may contaminate polar terrestrial and aquatic ecosystems.

But different microbial populations have specific metabolic capabilities and cannot utilize all hydrocarbon pollutants. Powell et al. (2007), for example, showed that the biodegradation of petroleum hydrocarbons occurs in Antarctic marine sediments, but the longevity of hydrocarbons in Antarctic sediment is influenced by location across relatively small distances. She and her colleagues demonstrated that faster degradation occurs in the Antarctic at sites with a prior history of oil contamination suggesting that adaptation is important to early term biodegradation. Diverse populations of microorganisms likely will be necessary to attack complex mixtures of hydrocarbons that occur when crude oil and refined fuels contaminate the environment. In the case of crude oils, some recalcitrant compounds will not be utilized by microbes and will remain in the environment. Such residues, like asphaltic pavements, generally are not bioavailable and usually are not toxic or dangerous to organisms and ecosystem functions. In fact, at least one Arctic oil spill that occurred along the trans-Alaska pipeline was treated by converting it to an asphalt car parking area.

Although highly weathered residues generally are not bioavailable, the overall nature of the oil, including the polar components and by-products of microbial degradation, must be considered in determining whether or not to rely upon bioremediation as a remediation treatment for the cleanup of pollutants in polar regions. Studies in temperate regions show that as oil weathers and the viscosity increases both the bioavailability and the rates of biodegradation decline; more viscous oils are degraded at increasingly slower rates (McMillen et al., 1995). Also the more highly substituted and higher molecular weight compounds tend to be far less water soluble than low molecular weight aromatics and to be more resistant to microbial attack (Atlas and Bragg, 2009). Consequently it is not unusual to find asphalt pavements years after an oil spill, as occurred following the *Metula* spill in Tierra del Fuego and the *Arrow* spill in Nova Scotia (Lee et al., 2003; Owens et al., 2007).

When considering bioremediation for treating oil-contaminated environments it also is necessary to recognize that the rates of biodegradation are relatively slow compared to physical cleanup methods; bioremediation is not an instantaneous solution and it may take many years to reduce hydrocarbon concentrations to acceptable levels in contaminated polar ecosystems. Even when high numbers of hydrocarbon degraders are present the actual rates of hydrocarbon biodegradation may be severely limited by environmental factors, and the physical location and state of the oil. Hence, bioremediation is often used only after the physical removal of surface oil. Also since the biodegradation of hydrocarbons follows first order kinetics, as the concentration of the pollutant decreases so do the rates of further biodegradative removal. Thus, reaching near zero concentrations of even readily biodegradable hydrocarbons often takes extended periods of time before the concentrations of hydrocarbon pollutants reach acceptable levels. The expectation of achieving zero concentrations of polluting hydrocarbons through biodegradation *in situ* is not reasonable.

17.2 ARCTIC ENVIRONMENTS

The Arctic has a much higher probability of petroleum (crude oil) pollution than the Antarctic. This is because petroleum reserves are being exploited throughout the Arctic, leading to a high potential for oil contamination of the environment. Petroleum is being extracted at various Arctic locations and the spillages of crude oil are likely at these sites and along transport routes. Given the complexity of the hydrocarbon mixtures found in crude oils, this represents a major challenge to the metabolic capabilities of microbial communities in Arctic ecosystems. Human settlements in the Arctic, with their reliance on gasoline and fuel oils, also contribute to the likelihood of the hydrocarbon contamination of Arctic ecosystems. There are numerous towns and villages across the Arctic where refined oils are stored and used for transportation, heating, and power. The spillages of these fuels endanger potable water supplies as well as ecosystems supporting wildlife.

According to the U.S. Geological Survey (USGS) (Bird et al., 2008; U.S. Geological Survey, 2008a; U.S. Geological Survey, 2008b): "The Arctic is an area of high petroleum resource potential, low data density, high geologic uncertainty, and sensitive environmental conditions. A large portion of the remaining global endowment of oil and gas resources has long been thought to exist in the high northern latitudes of Russia, Norway, Greenland, United States, and Canada. However, the quality, quantity, and distribution of these resources are poorly understood." Owing to the potential importance of Arctic petroleum reserves for meeting global energy needs, the Circum-Arctic Resource Appraisal (CARA) was undertaken by the USGS to produce a comprehensive and unbiased probabilistic estimate of undiscovered petroleum resources in the high northern latitudes.

The USGS has estimated that areas north of the Arctic Circle have 90 billion barrels of undiscovered, technically recoverable oil. This represents 13% of the undiscovered oil in the world. Of the estimated totals, more than half of the undiscovered oil resources are estimated to occur in just three geologic provinces—Arctic Alaska, the Amerasia Basin, and the East Greenland Rift Basins. More than 70% of the potential Arctic oil resources, 84% of which occur offshore, are estimated to occur in Arctic Alaska, Amerasia Basin, East Greenland Rift Basins, East Barents Basins, and West Greenland–East Canada. Four basins are likely to be the focus of the petroleum industry: the Kronprins Christian Basin, which is likely to have large reserves, the southwest Greenland basin, due to its proximity to markets, and the more oil-prone basins of Laptev and Baffin Bay.

Extensive drilling done in the Canadian Arctic during the 1970s and 1980s found an estimated 1.9 billion barrels of oil which were deemed to be insufficient to justify development at that time. A more recent assessment concluded that the Canadian Arctic has immense untapped energy resources consisting of oil, natural gas, condensate, oil sands, oil shales, and gas hydrates located in both the Northwest Territories and Nunavut (Hogg and Enachescu, 2008). These are thought to be commercially viable to supply growing global needs for petroleum. Greenland is believed to have some of the world's largest remaining oil resources. According to the USGS, waters off northeastern Greenland (north and south of the Arctic Circle) could contain up to 110 billion barrels of oil. In the United States, oil has been produced along the North Slope of the Brooks Range Mountains.

Oil currently flows across Arctic tundra from Prudhoe Bay through the trans-Alaska pipeline to the sub-Arctic port of Valdez where it is shipped by oil tankers to the lower 48 states. Much controversy centers on whether oil development should be permitted in the Alaskan Arctic Wildlife Refuge where there are believed to be extensive oil reserves but also where wildlife, including a very large herd of caribou, could be endangered by petroleum development. Whether or not this particular region is developed, it is clear that petroleum development in many Arctic regions will present a threat for spillages that would require remediation. It is also clear that standards for the cleanup of any major oil spills in the Arctic Wildlife Refuge would have to protect wildlife and be based upon restoring ecological functions as quickly as possible.

Other Arctic regions may soon become favorable for petroleum development. Global warming will likely increase this threat as new regions become ice free, making drilling and transport of oil more feasible. The reduction of polar sea ice and the increasing worldwide demand for energy will probably result in a dramatic increase in the number of vessels carrying petroleum in Arctic waters. It is possible that sufficient sea ice will disappear to make the Arctic seas a major transport route for shipping between Asia and Europe. Thus, the likelihood of a significant contamination of Arctic ecosystems with petroleum hydrocarbons is on an upward trajectory, making it especially important to consider what role bioremediation would play in future cleanup efforts. A recent report concluded that "the existing infrastructure for responding to maritime accidents in the Arctic is limited and more needs to be done to enhance emergency response capacity as Arctic sea ice declines and ship traffic in the region increases" (University of New Hampshire and the National Oceanic and Atmospheric Administration, 2009).

Several Arctic Oil spills have already occurred. One of the worst Arctic oil spills occurred in Russia in 1994. This spill, which may have been eight times bigger than the *Exxon Valdez* oil spill, occurred near the town of Usinsk. A pipeline just south of the Arctic Circle began leaking in February but the oil was contained within a dike until October when the dike collapsed. About 102,000 tonnes of oil contaminated the Siberian tundra. The spill reached the Kolva River, a tributary of the Pechora River, which flows into the Barents Sea. Another spill occurred in 2006 when about 267,000 gallons (1 million liters) of oil leaked onto the tundra of Alaska's North Slope at Prudhoe Bay. The spill occurred in February, and it took 5 days to discover. The spill, which covered about 2 acres (0.8 ha), occurred in an area where pipes are buried and covered with gravel to allow passage by animals. The pipe which was laid in the early 1970s had corroded. Prior to this spill, the largest in the North Slope was a 38,850-gallon (147,063-L) spill in 1989. In each of these spills, physical means were used for the cleanup, and there was no active bioremediation. Oil that was not recovered is still undergoing natural biodegradation in the impacted ecosystems.

17.3 ANTARCTIC ENVIRONMENTS

The Antarctic differs significantly from the Arctic in that it is more pristine and has not been subject to the same human impact and pollution. The Antarctic remains mainly a natural habitat. The only people currently living in the Antarctic are scientists. A limited number of tourists, albeit a growing number, visit the Antarctic Peninsula but do not move far from the immediate coastal shorelines. Although several fuel spills have occurred during early exploration and more recently near scientific research stations, the major impact on the Antarctic is coming from air pollution produced in temperate regions. Global warming is causing significant ice melting and the injection of chemicals into the atmosphere has resulted in a depletion of ozone and hence exposure to more ultraviolet radiation in the Antarctic due to the creation of an ozone hole in the atmosphere over the Antarctic Continent. Unlike the Arctic where there is a rush to extract petroleum reserves, there is no petroleum or other mineral exploitation in the Antarctic and this will be the case as long as the provisions of the Antarctic Treaty remain in force.

The Protocol on Environmental Protection to the Antarctic Treaty, also known as the Antarctic-Environmental Protocol, which is a part of the Antarctic Treaty System, provides for comprehensive environmental protection of the Antarctic environment and dependent and associated ecosystems. The Parties to the Treaty have committed themselves to the comprehensive protection of the Antarctic environment and dependent and associated ecosystems, and have designated Antarctica as a natural reserve devoted to peace and science. Any activity relating to mineral resources, other than scientific research, is prohibited by the Treaty. The Treaty limits or prohibits the use of certain chemicals, e.g., PCBs, and requires the development of emergency response action plans for incidents, e.g., fuel spills, with potential adverse effects on the Antarctic environment or dependent and associated ecosystems.

As discussed in other chapters of this book there have been several fuel oil spills in the Antarctic, some of which have raised serious concerns about potential ecological harm. An accidental soil contamination by diesel fuel, for example, occurred in the Crozet Island between July and November 1997 near the "Alfred Faure" scientific station where more than 20,000 L of diesel fuel was spilled in the vicinity of the power station (Delille and Pelletier, 2002). A much larger spillage occurred on January 28, 1989 when the *Bahia Paraiso* ran aground near Janus and DeLaca islands less than two miles from Palmer Station. As a result of that ship accident, an estimated 600,000 L of diesel fuel and other petroleum products leaked into the coastal Antarctic waters. As vessels continue to sail in the ice-laden waters of Antarctica more such spills can be anticipated. It can be expected that most fuel oils spilled into coastal waters around Antarctica will be subject to natural weathering, including biodegradation by indigenous microorganisms as was the case following the *Bahia Paraiso* spillage.

Other spills in the Antarctic have largely been terrestrial and primarily associated with scientific research bases. Most hydrocarbon contamination in the McMurdo Sound area, where the United States has its major Antarctic research station, has occurred around landfills, helipads, and refueling areas, and from leaks in fuel tanks and fuel lines. Relatively high levels of hydrocarbon contamination are also found at research stations in other parts of Antarctica (Saul et al., 2005). Much of this contamination has occurred in soils adjacent to the stations where human activities are focused (Aislabie et al., 2004). In each case the spillage has been of refined oil, e.g., diesel fuel or gasoline, that is a far less complex mixture than a crude oil. Hence, with regard to the likely contaminants bioremediation in the Antarctic is less challenging to microbial communities than the Arctic with regard to the range of hydrocarbons that would need to be degraded. Also as long as bioremediation can be carried out *in situ* without disturbing the environment it may be preferred over other methods, e.g., digging up soil that may cause slumping and the loss of permafrost would not be favored under the Antarctic Protocol.

17.4 WHAT DO WE KNOW AND WHAT DO WE NEED TO KNOW FOR BIOREMEDIATION TO BE SUCCESSFUL IN POLAR REGIONS?

Many studies have shown that hydrocarbon-degrading microorganisms are ubiquitous in polar regions. A number of studies have characterized these microorganisms. The result is that, as documented in the earlier chapters in this book, we know that hydrocarbon-degrading microorganisms are likely to be present in sufficient numbers and with sufficient metabolic capabilities to attack many different hydrocarbons in soils and waters in Arctic and Antarctic ecosystems that are, or may become, contaminated with petroleum hydrocarbons. The only exceptions may be on the surfaces of ice and in low carbon soils of the Antarctic Dry Valleys where microbial populations may be limited in numbers and diversities. In most polar ecosystems though, the indigenous microbial communities should respond to the presence of hydrocarbons and changes in community structure should be expected to result in increased numbers of hydrocarbon-degrading microorganisms that may persist even after the disappearance of the contaminating hydrocarbons. Following the hydrocarbon contamination of Antarctic soils, for example a significant shift was observed to organisms more capable of degrading diesel fuel (Saul et al., 2005).

Ice represents a special environment that is prevalent in polar regions and must be considered in developing contingency plans for Arctic and Antarctic spillages of crude and refined oils. As an example of the fate of spilled fuel on polar ice, evaporation was the main mechanism for the loss of hydrocarbons from an approximately 730L aviation diesel fuel spill that contaminated ice covering Lake Fryxell (McMurdo Dry Valleys, East Antarctica) following a January 2003 helicopter crash. Biodegradation was restricted to the regions of sediment-bearing ice where alkane and aromatic hydrocarbon degraders were able to utilize the spilled fuel (Jaraulaa et al., 2008). In areas where sediments are present on ice, an ecologically and physiologically complex microbial consortium develops from the relatively nutrient- and carbon-enriched-mixture of sediment and water. In the Arctic microbial communities associated with fjord ice have been shown to be capable of degrading hydrocarbons; however, the evidence suggests that hydrocarbon biodegradation will occur only very slowly and negligibly in the upper layers of contaminated ice (Brakstad et al., 2008).

Even though hydrocarbon-degrading microorganisms are indigenous and readily found in soil and aquatic Arctic and Antarctic ecosystems, as stated by Westlake (1982), no single microbial species has the enzymatic ability to metabolize all of the types of compounds found in a crude oil; thus, a community of microbes is responsible for the natural biodegradation of the complex hydrocarbon mixtures found in oil. Much more information is needed about the diversity of hydrocarbon-degrading microorganisms in Antarctic and Arctic regions if a predictive capability that could aid in decisions about the applicability and potential performance of bioremediation in polar regions is to be achieved.

Regardless of the valuable additional scientific information that can be generated about the indigenous microorganisms that are capable of degrading hydrocarbons and the other compounds in fuels, e.g., anti-icing compounds, in various Arctic and Antarctic ecosystems, seeding with a specialized exogenous microorganism, or a mixture of such organisms, should not be necessary to treat oil spillages that contaminate Arctic or Antarctic water, coastal soil, and shorelines since bioaugmentation *in situ* has not proven successful for treating oil spills even in temperate environments. This is fortunate since the introduction of exogenous microorganisms would be prohibited by the international Antarctic Treaty and almost certainly would be blocked by the legal action from being used in the Arctic. An exception though might be some regions of ice, the soils of the Dry Valleys, and the edge of the polar plateaus where there may be limited numbers and diversity of hydrocarbon-degrading microorganisms.

The lack of realistic prospects for developing a superbug to treat oil spills has not dissuaded a number of investigators from undertaking research to find microbes that could be used for *in situ* bioaugmentation. From the mid-1980s up to the late 1990s numerous attempts were made to design genetically modified microorganisms for environmental release as agents for the bioremediation of various pollutants (Urgun-Demirtas et al., 2006). Yet, "the field eventually came to a standstill because of regulatory restrictions and after multiple failures to program bacteria to behave in a predictable fashion in real world settings" (Cases and de Lorenzo, 2005). According to de Lorenzo (2008), part of the problem can be traced to the multiscale complexity associated with bioremediation (much beyond improving one new enzyme or pathway), and the need to take aboard design principles for complex circuits that are routine in systems engineering. de Lorenzo (2008) considers bioremediation as the exposure of a whole mixture of chemical structures to an intricate multispecies metabolic network present in a polluted scenario. He points out that the complexity involved in such events is growingly amenable to the conceptual frame and the tools of systems biology because (1) "the availability of genes, genomes, and metagenomes of biodegradative microorganisms makes it possible to model and even predict the fate of chemicals through the global metabolic network that results from connecting all known biochemical transactions" and (2) "Recent computational resources increasingly help in the design of superior biocatalysts for the biodegradation and biotransformations of desired chemicals."

Given the lack of success in bioaugmentation for treating spilled oil, the search for specialized microorganisms that can degrade hydrocarbons as potential seed organisms for *in situ* bioremediation applications does not a appear to be a fruitful line of research. Additionally, for the Antarctic, the introduction of foreign microorganisms into the environment is prohibited. This does not suggest that studies should not be conducted to characterize microbial populations that can degrade polluting hydrocarbons. Such studies, particularly those that employ molecular methods to characterize the metabolic capabilities of the indigenous microbial populations, are potentially quite useful for predicting and monitoring the fate of pollutants. In addition, specialized microorganisms or microbial communities with appropriate genetic capabilities might be used in contained bioreactors, e.g., for the treatment of contaminated water or in contained biopiles to treat contaminated soils.

Although there have been relatively few studies performed on the potential use of bioremediation in diverse polar environments, the studies that have been performed have shown that fertilizer application can stimulate biodegradation in Arctic and Antarctic ecosystems suggesting that bioremediation may be a successful strategy for cleanup of oil and fuel spills.

de Lorenzo (2008) points out that the factors at play in bioremediation scenarios include more elements than just the biological catalysts and the contaminants. Their dynamic interactions occur in concrete abiotic settings, which are defined by a whole of physicochemical conditions: O_2 tension, electron acceptors, water, temperature, granulation, and others, many of which change over time and the course of the catalysis and determine the species composition of the endogenous microbial communities as much as (or more than) the availability of given chemical species as C and energy source. de Lorenzo (2008) proposes that "there are three dimensions to the effectiveness of any bioremediation process, only one of them (the catabolic landscape) being biological. The chemical landscape of the place, including nutrients-to-be, electron donors/acceptors and stressors has a dynamic interplay with the biological vector of the system on the abiotic background imposed by the micro/macrogeography of the location at stake. This includes humidity, conductivity, temperature, pressure, texture, matric conditions, redox (O_2) status, etc." In fact, it is the engineering of these latter environmental conditions that is key to successful bioremediation.

Swannell et al. (1996) critically examined the applications of bioremediation in field trials of both experimental and accidental oil spillages, including those in Arctic and Antarctic coastal waters and shorelines. They concluded that the potential for bioremediation for open polar marine waters has not been demonstrated and that there is little chance that bioremediation would be used except when shorelines or terrestrial ecosystems are impacted. Sirvins and Angles (1986) did perform a sea trial in Antarctica with the oleophilic fertilizer Inipol EAP 22 to treat Arabian light crude oil that indicated the possible stimulation of biodegradation. But the limited data were inadequate to establish the effectiveness of bioremediation (Swannell et al., 1996). Sveum and Ladousse (1989), who also carried out a field trial with Inipol EAP 22 in a lagoon system on the coast of Kings Bay, Spitsbergen, Norway, found that there was no increase in the rate of biodegradation over an 80 day period following fertilizer addition. They hypothesized that the photooxidation of the oil on the sea surface as a result of 24-h exposure to the Arctic sun occurred faster than biodegradation. It is also possible that the fertilizer dispersed before it could stimulate the microbial biodegradation of the open-water oil slick. The conclusion was that nutrient addition was ineffective for the treatment of open-water oil slicks in the Arctic.

In contrast to these unsuccessful open-water bioremediation trials, nutrient additions to soils and shorelines have proven more promising for stimulating microbial oil degradation. Braddock et al. (1997) conducted treatability studies for the potential treatment of contaminated soils near Barrow Alaska. Fertilizer application enhanced microbial hydrocarbon-degrading activity, and both aromatic and aliphatic hydrocarbons were degraded at increased rates from these soils following fertilizer application (Braddock et al., 1999).

In a field trial carried out in 1976 at Spitsbergen, Norway, Sendstad et al. (1984) also found that fertilizer application stimulated oil biodegradation. This study supported the ability of bioremediation to accelerate the rate of oil biodegradation on Arctic shorelines but not the extent of biodegradation. The results are consistent with the sub-Arctic Alaskan *Exxon Valdez* experience where fertilizer application was demonstrated to increase the rates of oil biodegradation by up to five times shortly after the spill (Bragg et al., 1994). However, in the study conducted at Spitzbergen, within 4 years natural attenuation caught up, and the extent of degradation was similar on fertilized and unfertilized shoreline plots (Sendstad et al., 1984).

The Baffin Island Oil Spill Project, which was an extensive field study carried out between 1980 and 1983 in Canada's eastern Arctic at Cape Hatt on the northern end of Baffin Island, included an assessment of bioremediation (Prince et al., 2002; Sendstad, 1980; Sendstad et al., 1982; Sergy and Blackall, 1987). In this study, nutrients were not well retained in the sediments. Nevertheless, the addition of nutrients resulted in an increase in the number of oil-degrading bacteria and respiratory carbon dioxide production in comparison to untreated controls. Although little difference in oil concentration was noted between the nutrient-treated and control plots, the results suggest that bioremediation may have some potential, even in an Arctic environment. The enhancement of oil biodegradation rates by fertilizer addition was evident 2 years after the experiment began as indicated by changes in the alkane/isoprenoid ratio in the residual oil. A combination of fertilizer application and tilling of oil and fertilizer into the sediment supported the highest rates of biodegradation.

The ability of fertilizer application to stimulate biodegradation was tested following an accidental spillage of 88,000 L of gasoline from storage tanks into Kings Bay, Spitsbergen, in November 1985 (Sveum, 1987; Sveum and Ladousse, 1989). The oil was stranded on coarse sediments and had weathered for 2 months before any fertilizer was applied. It was estimated that fertilizer (Inipol EAP 22) addition resulted in two- to threefold increase in the biodegradation of the alkanes in the oil. Based upon their Arctic field studies, Sveum and Bech (1994) suggested that bioremediation should be targeted primarily at oil that remains in beach sediments after the initial weathering and physical removal of surface oil.

The application of slow-release and soluble fertilizers proved to be an effective and environmentally benign way of stimulating oil biodegradation on an Arctic shoreline (Prince et al., 2003). These experiments were conducted on shorelines near Sveagruva, Spitzbergen in the summers of 1997 and 1998. Fertilizer application to the surface of the beach delivered nutrients to the oiled sediment beneath the beach surface. The addition of fertilizer supported a greater biodegradation of oil than on nonfertilized plots. The rate of oil biodegradation was approximately doubled over the course of a year by fertilizer applications in the first 2 months after the spill. It is noteworthy that fertilizer applications over a relatively short period had a pronounced effect on the chemistry of the residual oil 1 year later.

Several bioremediation experiments have demonstrated the potential efficacy of bioremediation in Antarctica to treat hydrocarbon-contaminated ecosystems (Aislabie et al., 2001, 2004; Coulon et al., 2004; Delille, 2000; Delille et al., 2007; Kerry, 1993, Snape et al., 2001; Stallwood et al., 2005). The addition of nutrients was found to stimulate the biodegradation of oil on a number of contaminated subpolar shorelines (Delille et al., 2002; Pelletier et al., 2004). In a field study begun in December 2000 in the vicinity of the scientific research station of Port-aux-Français (sub-Antarctic Kerguelen Archipelago), the addition of the slow release fertilizer Inipol EAP-22 significantly improved the rates of hydrocarbon biodegradation. However, even after 4 years, the toxicity of oiled soils as determined by Microtox solid phase tests showed a persistent response in spite of an apparent significant degradation of alkanes and aromatics. This raises the question as to whether bioremediation can achieve the necessary performance for environmental restoration even when microbial degradation can reduce significantly the concentrations of the polluting hydrocarbons.

Because *in situ* conditions are most likely to be suboptimal for biodegradation in polar soils, *ex situ* bioremediation may turn out to be the method of choice for ameliorating and controlling the factors limiting microbial activity, i.e., low and fluctuating soil temperatures, low levels of nutrients, and possible alkalinity and low moisture (Aislabie et al., 2006). Bioremediation experiments conducted on-site in the Arctic indicate that land farming and biopiles may be useful approaches for the bioremediation of polar soils. *Ex situ* biopiles have been used successfully for the bioremediation of diesel-contaminated Arctic soils using biostimulation (heating, nutrients, and aeration) (Filler et al., 2001; Mohn et al., 2001). Nutrient amendment in biopiles has consistently improved the bioremediation of hydrocarbon-contaminated soils in the Arctic (Mohn et al., 2001; Thomassin-Lacroix et al., 2001). In one study, Braddock et al. (1999) found that the loss of linear alkanes was more strongly affected by the addition of fertilizers than branched alkanes. In contrast to fertilizer addition, the bioaugmentation of the biopiles with enrichments of cold-adapted microbes has yielded variable results in the Arctic (Mohn et al., 2001; Thomassin-Lacroix et al., 2001). Landfarming also has been demonstrated to be a potentially useful *ex situ* bioremediation method for polar regions (McCarthy et al., 2004). A field scale landfarming experiment conducted at Barrow, Alaska found that hydrocarbons in a diesel fuel could be reduced to target levels (McCarthy et al., 2004).

Two relevant questions that need to be considered when evaluating the potential use of bioremediation are what is the ultimate standard that should be used for measuring success and how should bioremediation (e.g., hydrocarbon biodegradation) be measured? Many different standards and methodologies are currently employed and these may change over the course of treatment, making it very difficult to carry out in advance the necessary research. As a result both laboratory and field demonstration projects generally have proven to be necessary for real-world applications before bioremediation could be used for treating major oil spills, e.g., the sub-Arctic *Exxon Valdez* oil spill (see Bragg et al., 1992 for a discussion on the testing and application of bioremediation in Prince William Sound Alaska).

Much research has been done on how to assess the biodegradation of complex hydrocarbon mixtures, e.g., crude oil spills, in actual field settings. Because of environmental variability (patchiness of distribution), the extraction of hydrocarbons and measurement of total extractable petroleum hydrocarbons (sometimes referred to as TPH or TEH) has proven to be inadequate for assessing rates and extent of pollutant removal achievable by biodegradation. Yet the concentration of extractable hydrocarbons is often used as the measure of residual pollutant for determining if the environment is clean enough.

Measuring only total extractable hydrocarbons, however, overlooks the presence of naturally occurring hydrocarbons in the environment, e.g., from extensive peat deposits in the Arctic and also fails to take into account the fact that toxicity is associated with specific hydrocarbons, such as some aromatic compounds. The need to consider individual hydrocarbons, as opposed to the total mixture, complicates the assessment of a suitable treatment end point unless very detailed chemical analyses or ecotoxicity tests are performed. Various measures have been proposed for determining the appropriate ecotoxicological end points of hydrocarbon biodegradation, but particular measures and specific cleanup levels derived from temperate zones may be too liberal for soil contamination in sub-Antarctic islands and other polar regions (see, e.g., Schafer et al., 2007). Clearly caution must be used as commonly used tests, e.g., Microtox or CYP1A (hepatic cytochrome P450 1A) induction, may not provide the necessary data for making appropriate decisions about the actual environmental impact of a pollutant.

With regard to the evaluation of oil biodegradation and the efficacy of bioremediation, this often is a difficult task, especially in the field because of the complexity of weathering processes and the heterogeneity of contaminated sites; oil contaminated sites are often highly heterogeneous. Oil concentrations can vary greatly within a small area, and consequently, variability associated with field studies can be so high as to preclude or interfere with the ability to discern significant treatment differences. It is possible though to reduce the spatial variability of the data on biodegradation by normalizing to a conserved chemical species within the mixture (Douglas et al., 1994). Early studies on oil biodegradation—conducted at a time when gas chromatography was widely employed as the state of the art analytical methodology for tracking the fate of individual aliphatic and aromatic compounds in residual oil—often employed pristane and phytane as presumed conserved internal standards (Atlas and Bartha, 1973). A number of studies on Arctic and Antarctic hydrocarbon biodegradations discussed in this book also have relied upon heptadecane to pristane ratios for assessing biodegradation. While comparisons of the relative rates of the disappearance of linear alkanes and light aromatics to pristane or phytane are useful in laboratory studies and the early stages of oil biodegradation, both pristane and phytane are subject to extensive biodegradation that severely limits their usefulness for monitoring biodegradation and the extent of the weathering of complex hydrocarbon mixtures in the environment over extended periods.

Various other compounds present in crude oils are far more resistant to biodegradation and make better biomarkers (Atlas and Bragg, 2007; Douglas et al., 1996; Wang et al., 2001). For example, after 22 years of weathering most paraffins and polycyclic aromatic hydrocarbons (PAHs) in oiled sediments from Chedabucto Bay were lost, but multiring alkyl PAH homologues such as chrysenes and biomarker compounds (triterpanes and steranes) remained, making these candidate biomarkers for use in monitoring the weathering of petroleum in the environment (Lee et al., 2003).

Hopanes have proven to be useful conserved internal markers for following the biodegradation of crude oil contamination (Prince at al., 1994), trimethylphenanthrenes for following the biodegradation of diesel fuel (Douglas et al., 1992), and 2,2,3,3-tetramethylbutane and 1,1,3-trimethylcyclopentane for following the biodegradation of gasoline and condensate (Townsend et al., 2004). Hopanes have been frequently used as the biomarker for monitoring the weathering of petroleum following major marine oil spills, including for example the *Amoco Cadiz* spill in Brittany France (Atlas et al., 1981; Gundlach et al., 1983), the *Exxon Valdez* spill in Prince William Sound Alaska (Bragg et al., 1992, 1994; Prince, 1993; Prince and Douglas, 2005; Prince et al., 1993), and the *Metula* spill in Chile at the Straight of Megellan (Wang et al., 2001). In cases where hopane has not been measured, the data can be normalized to another constituent such as C2-, C3-, or C4-chrysene, that is relatively resistant to biodegradation, albeit not as resistant as hopanes or stearanes (Zhu et al., 2001). Chrysene has been used by Delille et al. (2007) for measuring the biodegradation of diesel fuel in sub-Antarctic soils; they also used hopane as a conserved biomarker for measuring oil biodegradation.

Wang et al. (2001) performed an extensive study on biomarker resistance to biodegradation, finding that the degree of biodegradation of biomarker compounds was strongly correlated with their molecular structures and that even the most refractory biomarker compounds showed some degree of biodegradation. The norhapanes and C29-$\alpha\beta\beta$-steranes exhibited more resistance to biodegradation than the diasteranes, C27 steranes, tricyclic terpanes, and pentacyclic terpanes. The C29-18α(H), 21β(H)-30-norneohopane and C29-$\alpha\beta\beta$-stigmastanes (20R and 20S) appeared to be the most biodegradation-resistant terpane and sterane compounds among the biomarkers examined in this study. Based on the results of analyses of approximately 100 oil samples collected in Prince William Sound over the period from 1999 to 2006 (10–17 years after the *Exxon Valdez* oil spill), Atlas and Bragg (2007) reported that the most stable biomarker compound to be used for quantifying the loss of PAH, or other oil components, is C29R-stigmastane. C29$\beta\beta R$-stigmastane and C27S-cholestane were almost as stable as C29R-stigmastane. C30-hopane was subject to minor biodegradation, resulting in about a 4% underestimation of PAH depletion. Using chrysenes underestimated PAH depletion by about 9%–10%. These results are consistent with those of prior field studies monitoring oil over extended periods (Wang et al., 2001).

Ultimately we need to return to the critical questions that need to be addressed before determining whether bioremediation should be considered as a potential cleanup method: (1) how clean is clean? and (2) can bioremediation achieve that level of cleanup within an acceptable time period? At this point, decisions can only be made on a case-by-case basis with limited certainties as to whether relying upon microbial biodegradation will be adequate. More research is needed to produce data that would support reliable assessments of the potential benefits of bioremediation, which performance criteria should be employed for assessing necessary standards for acceptable cleanup levels, and whether bioremediation can achieve the necessary end points for the successful restoration of contaminated polar environments.

17.5 CONCLUDING REMARKS

The following represent some of the key messages regarding our current understanding of polar microbiology and future needs relative to the potential for bioremediation. Indigenous hydrocarbon-degrading microorganisms are present in polar marine and terrestrial ecosystems. It can be anticipated that if there are coastal or terrestrial spills of crude or refined oils in the Arctic or refined oils in Antarctic coastal or terrestrial ecosystems, then microorganisms that can biodegrade the contaminating hydrocarbons will be present. It is not clear that this will be true for ice where microbial populations are lower, less diverse, and concentrated in brine channels. More needs to be known about the fate of oil under ice, especially for the Arctic where petroleum extraction and shipping heighten the possibility of a major oil spill that could contaminate ice surfaces. Further studies are also needed to characterize the hydrocarbon degradation capabilities of indigenous microorganisms so as to develop a predictive capability that can be used in developing strategies for treating spilled oil and fuels in polar regions. These studies should focus on the diversity of hydrocarbon-degrading capabilities of communities that are present. In this regard, the molecular characterization of both the diversity of microorganisms based upon rRNA or other genetic analyses and the diversity of genes involved in hydrocarbon metabolism will be needed to provide sufficient scientific knowledge to predict the abilities of indigenous microbial communities to degrade complex hydrocarbon mixtures.

With regard to bioremediation as a technology for hydrocarbon cleanup in polar regions, the focus should be on the biostimulation of indigenous microorganisms and not on the bioaugmentation which has not been demonstrated to be an effective approach for the *in situ* treatment of petroleum pollutants. It is important to understand the environmental constraints that limit the rates of hydrocarbon biodegradation in specific environments and how bioremediation solutions can be engineered. In some cases, it may be necessary to transfer contaminated soils to biopiles, landfarms, or bioreactors to facilitate the optimization of environmental conditions so that hydrocarbon biodegradation can proceed optimally. The development of effective *ex situ* treatments needs much greater attention with regard to the potential applicability for decontaminating polar soils.

Finally, while a number of studies have demonstrated that bioremediation should be applicable in polar regions, more field studies are needed to establish that bioremediation can meet performance criteria for the successful cleanup of oil contamination. Pollution prevention remains the key for protecting pristine Arctic and Antarctic ecosystems. In situations where accidental pollution occurs, bioremediation should be considered as a potential cleanup technology. When doing so it is necessary to determine if a particular bioremediation strategy can achieve the desired end points for decontamination within an acceptable time frame to minimize environmental impact and ecological harm. Bioremediation is a useful technology but it has limitations. Optimistically further research efforts can reduce the uncertainties currently associated with using a biological approach to environmental cleanup, and bioremediation engineering solutions will be developed that can be used for the treatment of contaminated polar ecosystems.

REFERENCES

Aislabie, J., J.M. Balks, J.M. Foght, and E.J. Waterhouse. 2004. Hydrocarbon spills on Antarctic soil: Effect and management. *Environmental Science and Technology* 38:1265–1274.

Aislabie, J., R. Fraser, S. Duncan, and R.L. Farrell. 2001. Effects of oil spills on microbial heterotrophs in Antarctic soils. *Polar Biology* 24:308–313.

Aislabie, J., M. McLeod, and R. Fraser. 1998. Potential of biodegradation of hydrocarbons in soil from the Ross Dependency, Antarctica. *Applied Microbiology and Biotechnology* 49:210–214.

Aislabie, J., J. Ryburn, and A. Sarmur. 2008. Hexadecane mineralization activity in ornithogenic soil from Seabee Hook, Cape Hallett, Antarctica. *Polar Biology* 31:421–428.

Aislabie J., D.J. Saul, and J.M. Foght. 2006. Bioremediation of hydrocarbon-contaminated polar soils. *Extremophiles* 10:171–179.

Atlas, R.M. and R. Bartha. 1973. Fate and effects of oil pollution in the marine environment. *Residue Reviews* 49:49–85.

Atlas, R.M. and J. Bragg. 2007. Assessing the long-term weathering of petroleum on shorelines: Uses of conserved components for calibrating loss and bioremediation potential. In *Proceedings of the 29th Arctic and Marine Oilspill Program (AMOP)*, vol. 1. June 5–7, Edmonton, Alberta, Canada, pp. 263–290.

Atlas, R.M. and J.R. Bragg. 2009. Bioremediation of marine oil spills: When and when not—The *Exxon Valdez* experience. *Microbial Biotechnology* 2:213–221.

Atlas, R.M. and J. Philp. 2005. *Bioremediation: Applied Microbial Solutions for Real-World Environmental Cleanup*. ASM Press, Washington, D.C.

Atlas, R.M., P.D. Boehm, and J.A. Calder. 1981. Chemical and biological weathering of oil, from the Amoco Cadiz Spillage within the Littoral Zone. *Estuarine and Coastal Marine Science* 12:589–608.

Bird, K.J., R.R. Charpentier, D.L. Gautier, D.W. Houseknecht, T.R. Klett, J.K. Pitman, T.E. Moore, C.J. Schenk, M.E. Tennyson, and C.J. Wandrey. 2008. *U.S. Geological Survey Fact Sheet 2008–3049—Circum-Arctic Resource Appraisal: Estimates of Undiscovered Oil and Gas North of the Arctic Circle*. U.S. Geological Survey, Denver. http://pubs.usgs.gov/fs/2008/3049/.

Braddock, J.F., M.L. Ruth, P.H. Catterall, J.L. Walworth, and K.A. McCarthy. 1997. Enhancement and inhibition of microbial activity in hydrocarbons-contaminated Arctic soils: Implications for nutrient-amended bioremediation. *Environmental Science Technology* 31:2078–2084.

Braddock, J.F., J.L. Walworth, and K.A. McCarthy. 1999. Biodegradation of aliphatic vs. aromatic hydrocarbons in fertilized arctic soils. *Bioremediation Journal* 3:105–116.

Bragg, J.R., R.C. Prince, and R.M. Atlas. 1994. Effectiveness of bioremediation for oiled intertidal shorelines. *Nature* 368:413–418.

Bragg, J.R., R.C. Prince, J.B. Wilkinson, and R.M. Atlas. 1992. *Bioremediation for Shoreline Cleanup Following the 1989 Alaskan Oil Spill*. Exxon, Houston, TX.

Brakstad, O.G., I. Nonstad. L.-G. Faksness, and P.J. Brandvik. 2008. Responses of microbial communities in Arctic sea ice after contamination by crude petroleum oil. *Microbial Ecology* 55:540–552.

Cases, I. and V. de Lorenzo. 2005. Genetically modified organisms for the environment: Stories of success and failure and what we have learned from them. *International Microbiology* 8:213–222.

Cripps, G.C. and J. Shears. 1997. The fate in the marine environment of a minor diesel fuel spill from an Antarctic research station. *Environmental Monitoring Assessment* 46:221–232.

Coulon, F., E. Pelletier, R. St. Louis, L. Gourhant, and D. Delille. 2004. Degradation of petroleum hydrocarbons in two sub-Antarctic soils: Influence of an oleophilic fertilizer. *Environmental Toxicology Chemistry* 23:1893–1901.

Delille, D. 2000. Response of Antarctic soil bacterial assemblages to contamination by diesel fuel and crude oil. *Microbial Ecology* 40:159–168.

Delille, D. and E. Pelletier. 2002. Natural attenuation of diesel-oil contamination in a subantarctic soil (Crozet Island). *Polar Biology* 25:682–687.

Delille, D., F. Coulon, and E. Pelletier. 2007. Long-term changes of bacterial abundance, hydrocarbon concentration and toxicity during a biostimulation treatment of oil-amended organic and mineral sub-Antarctic soils. *Polar Biology* 30:925–933.

Delille, D., B. Delille, and E. Pelletier. 2002. Effectiveness of bioremediation of crude oil contaminated subAntarctic intertidal sediment: The microbial response. *Microbioal Ecology* 44:118–126.

de Lorenzo, V. 2008. Systems biology approaches to bioremediation. *Current Opinion in Biotechnology* 19:1–11.

Douglas, G.S., A.E. Bence, R.C. Prince, S.J. McMillen, and E.L. Butler. 1996. Environmental stability of selected petroleum source and weathering ratios. *Environmental Science and Technology* 30:2332–2339.

Douglas, G.S., K.J. McCarthy, D.T. Dahlen, J.A. Seavey. W.G. Steinhauer, R.C. Prince, and D.L. Elmendorf. 1992. The use of hydrocarbon analyses for environmental assessment and remediation. *Journal of Soil Contamination* 1:197–216.

Douglas, G.S., R.C. Prince, E.L. Butler, and W.G. Steinhauer. 1994. The use of internal chemical indicators in petroleum and refined products to evaluate the extent of biodegradation. In R.E. Hinchee, B.C. Alleman, R.E. Hoeppel, and R.N. Miller (eds.), *Hydrocarbon Bioremediation*, Chap. 20. Lewis Publishers, Boca Raton, FL, pp. 219–236.

Filler, D.M., J.E. Lindstrom, J.F. Braddock, R.A. Johnson, and R. Nickalaski. 2001. Integral biopile components for successful bioremediation in the Arctic. *Cold Regions Science and Technology* 32:143–156.

Gundlach, E.R., P.D. Boehm, M. Marchand, R.M. Atlas, D.M. Ward, and D.A. Wolfe. 1983. The fate of Amoco-Cadiz oil. *Science* 221:122–129.

Hogg, J. and M. Enachescu. 2008. The Canadian Arctic Petroleum Potential: What does the Next Decade Hold for Canada's Northern Resources? http://74.125.95.132/search?q=cache:x2wKwblVVokJ:www.cspg.org/conventions/abstracts/2008abstracts/280.pdf+The+Canadian+Arctic+Petroleum+Potential:+What+does+the+Next+Decade+Hold+for+Canada%E2%80%99s+Northern+Resources%3F&hl=en&ct=clnk&cd=1&gl=us.

Jaraulaa, C.M.B., F. Keniga, P.T. Dorana, J.C. Priscub, and K.A. Welch. 2008. SPME-GCMS study of the natural attenuation of aviation dieselspilled on the perennial ice cover of Lake Fryxell, Antarctica. *Science of the Total Environment* 407:250–262.

Kerry, E. 1993. Bioremediation of experimental petroleum spills on mineral soils in the Vesfold Hills, Antarctica. *Polar Biology* 13:163–170.

Lee, K., R.C. Prince, C.W. Greer, K.G. Doe, J.E.H. Wilson, S.E. Cobanli, G.D. Wohlgeschaffen, D. Alroumi, T. King, and G.H. Tremblay. 2003. Composition and toxicity of residual Bunker C fuel oil in intertidal sediments after 30 years. *Spill Science Techology Bulletin* 8:187–199.

McCarthy, K., L. Walker, L. Vigoren, and J. Bartel. 2004. Remediation of spilled hydrocarbons by *in situ* landfarming at an arctic site. *Cold Regions Science and Technology* 40:31–39.

McMillen, S.J., A.G. Requejo, G.N. Young, P.S. Davis, P.D. Cook, J.M. Kerr, and N.R. Gray. 1995. Bioremediation potential of crude oil spilled on soil 3. In R.E. Hinchee et al. (eds.), *Microbial Processes for Bioremediation*, Battelle Press, Columbus, OH, pp. 91–99.

Mohn, W.W., C.Z. Radziminski, M.-C. Fortin, and K.J. Reimer. 2001. On site bioremediation of hydrocarbon-contaminated Arctic tundra soils in inoculated biopiles. *Applied Microbiology and Biotechnology* 57:242–247.

Owens, E.H., E. Elliott Taylor, and B. Humphrey. 2007. The persistence and character of stranded oil on coarse-sediment beaches. *Marine Pollution Bulletin* 56:14–26.

Pelletier, E., D. Delille, and B. Delille. 2004. Crude oil bioremediation in sub-Antarctic intertidal sediments: Chemistry and toxicity of oiled residues. *Marine Environmental Research* 57:311–327.

Poland, J.S., M.J. Riddle, and B.A. Zeeb. 2003. Contaminants in the Arctic and the Antarctic: A comparison of sources, impacts, and remediation options. *Polar Record* 39:369–384.

Powell, S.M., P.McA. Harvey, J.S. Stark, I. Snape, and M.J. Riddle. 2007. Biodegradation of petroleum products in experimental plots in Antarctic marine sediments is location dependent. *Marine Pollution Bulletin* 54:434–440.

Prince, R.C. 1993. Petroleum spill bioremediation in marine environments. *Critical Reviews in Microbiology* 19:217–242.

Prince, R.C. and G.S. Douglas. 2005. Quantification of hydrocarbon biodegradation using internal markers. In R. Margesin and F. Shinner (eds.), *Soil Biology*, vol. 5. Springer-Verlag, Berlin Heidelberg, Germany, pp. 179–188.

Prince, R.C., R. Bare, R.M. Garrett, M.J. Grossman, C.F. Haith, L.G. Keim, K. Lee, G.J. Holtoms, P. Lambert, G.A. Sergy, E.H. Owens, and C.C. Guenette. 2003. Bioeremediation of stranded oil on an Arctic shoreline. *Spill Science and Technology Bulletin* 8:303–312.

Prince, R.C., J.R. Clark, J.E. Lindstrom, E.L. Butler, E.J. Brown, G. Winter, M.J. Grossman, P.R. Parrish, R.E. Bare, J.F. Braddock, W.G. Steinhauer, G.S. Douglas, J.M. Kennedy, P.J. Barter, J.R. Bragg, E.J. Harner, and R.M. Atlas. 1993. Bioremediation of the Exxon Valdez oil spill: Monitoring safety and efficacy. In R.E. Hinchee, B.C. Alleman, R.E. Hoeppel, and R.N. Miller (eds.), *Hydrocarbon Bioremediation*, Chap. 11. Lewis Publishers, Boca Raton, FL, pp. 107–124.

Prince, R.C., D.L. Elmendorf, J.R. Lute, C.S. Hsu, C.E. Haith, J.D. Senius, G.J. Dechert, G.S. Douglas, and E.L. Butler. 1994. 17α(H),21β(H)-hopane as a conserved internal marker for estimating the biodegradation of crude oil. *Environmental Science and Technology* 28:142–145.

Prince, R.C., E.H. Owens, and G.A. Sergy. 2002. Weathering of an Arctic oil spill over 20 years: The BIOS experiment revisited. *Marine Pollution Bulletin* 44:1236–1242.

Rike, A.G., M. Børessen, and A. Instanes. 2001. Response of cold-adapted microbial populations in permafrost profile to hydrocarbon contaminants. *Polar Record* 37:239–248.

Saul, D.J., J.M. Aislabie, C.E. Brown, L. Harris, and J.M. Foght. 2005. Hydrocarbon contamination changes the bacterial diversity of soil from around Scott Base, Antarctica. *FEMS Microbiology Ecology* 53:141–155.

Schafer, A.N., I. Snape, and S.D. Siciliano. 2007. Soil biogeochemical toxicity end points for sub-Antarctic islands contaminated with petroleum hydrocarbons. *Environmental Toxicology and Chemistry* 26:890–897.

Sexstone, A.J. and R.M. Atlas. 1977. Response of microbial populations in Arctic tundra soils to crude oil. *Canadian Journal of Microbiology* 23:1327–1333.

Sendstad, E. 1980. Accelerated biodegradation of crude oil on Arctic shorelines. In *Proceedings of the Third Arctic and Marine Oilspill Program Technical Seminar*. Environment Canada, Ottawa, Canada, pp. 402–416.

Sendstad, E., T. Hoddo, P. Sveum, K. Eimhjellen, K. Josefson, O. Nilsen, and T. Sommer. 1982. Enhanced oil biodegradation on an Arctic shoreline. In *Proceedings of the Fifth Arctic Marine Oilspill Program Technical Seminar*. Environment Canada, Ottawa, Canada, pp. 331–340.

Sendstad, E., P. Sveum, L.J. Endal, Y. Brattbakk, and O. Ronning. 1984. Studies on a seven year old seashore crude oil spill on Spitsbergen. In *Proceedings of the Seventh Arctic Marine Oilspill Program Technical Seminar*. Environment Canada, Ottawa, Canada, pp. 60–74.

Sergy, G.A. and P.J. Blackall. 1987. Design and conclusions of the Baffin Island Oil Spill Project. *Arctic* 40:1–19.

Sirvins, A. and M. Angles. 1986. Development and effects on marine environment of a nutrient formula to control pollution by petroleum hydrocarbons. *NATO ASI Series* G9:357–404.

Snape, I., M.J. Riddle, J.S. Stark, C.M. Cole, C.K. King, S. Duquesne, and D.B. Gore. 2001. Management and remediation of contaminated sites at Casey Station, Antarctica. *Polar Record* 37:199–214.

Stallwood, B., J. Shears, P.A. Williams, and K.A. Hugues. 2005. Low temperature bioremediation of oil-contaminated soil using biostimulation and bioaugmentation with a *Pseudomonas* sp. from maritime Antarctica. *Journal of Applied Microbiology* 99:794–802.

Sveum, P. 1987. Accidentally spilled gas-oil in a shoreline sediment on Spitsbergen: Natural fate and enhancement of biodegradation. In *Proceedings of the 10th Arctic and Marine Oilspill Program Technical Seminar.* Environment Canada, Ottawa, Canada, pp. 177–192.

Sveum, P. and C. Bech. 1994. Bioremediation and physical removal of oil on shore. In R.E. Hinchee, B.C. Alleman, R.E. Hoeppel, and R.N. Miller (eds.), *Hydrocarbon Bioremediation.* Lewis Publishers, Boca Raton, FL, pp. 311–317.

Sveum, P. and A. Ladousse. 1989. Biodegradation of oil in the Arctic: Enhancement by oil-soluble fertilizer application. In *Proceedings of the 1989 Oil Spill Conference.* American Petroleum Institute, Washington, D.C., pp. 439–446.

Swannell, R.P.J., K. Lee, and M. McDonagh. 1996. Field evaluations of marine oil spill bioremediation. *Microbiological Reviews* 60:342–365.

Thomassin-Lacroix, E.J.M., Z. Yu, M. Eriksson, K.J. Reimer, and W.W. Mohn. 2001. DNA-based and culture-based characterization of hydrocarbon-degrading consortium enriched from Arctic soil. *Canadian Journal of Microbiology* 47:1107–1115.

Townsend, G.T., R.C. Prince, and J.M. Suflita. 2004. Anaerobic biodegradation of alicyclic constituents of gasoline and natural gas condensate by bacteria from an anoxic aquifer. *FEMS Microbiology Ecology* 49:129–135.

U.S. Geological Survey. 2008a. *Circum-Arctic Resource Appraisal.* U.S. Geological Survey, Menlo Park, CA. http://energy.usgs.gov/arctic/.

U.S. Geological Survey, (USGS). 2008b. *90 Billion Barrels of Oil and 1,670 Trillion Cubic Feet of Natural Gas Assessed in the Arctic.* U.S. Geological Survey, Reston, VA. http://www.usgs.gov/newsroom/article.asp?ID=1980.

University of New Hampshire/National Oceanic Atmospheric Administration. 2009. Opening the Arctic Seas: Envisioning Disaster & Framing Solutions—Workshop Report. www.crrc.unh.edu/workshops/arctic_spill_summit/index.html.

Urgun-Demirtas, M., B. Stark, and K. Pagilla. 2006. Use of genetically engineered microorganisms (GEMs) for the bioremediation of contaminants. *Critical Reviews in Biotechnology* 26:145–164.

Wang, Z., M. Fingas, E.H. Owens, L. Sigouin, and C.E. Brown. 2001. Long-term fate and persistence of the spilled Metula Oil in a marine salt marsh environment: Degradation of petroleum biomarkers. *Journal Chromatography A*, 926:275–290.

Westlake, D.W.S. 1982. Micro-organisms and the degradation of oil under northern marine conditions. In J.B. Sprague, J.H. Vandermeulen, and P.G. Wells (eds.), *Oil and Dispersants in Canadian Seas—Research Appraisal and Recommendations.* Publication EPS-3-EC-82-2. Environmental Protection Service Canada, Ottawa, Canada, pp. 47–50.

Whyte, L.G., L. Bourbonniere, C. Bellerose, and C.W. Greer. 1999. Bioremediation assessment of hydrocarbon-contaminated soils from the High Arctic. *Bioremediation Journal* 3:69–79.

Zhu, X., A.D. Venosa, M.T. Suidan, and K. Lee. 2001. *Guidelines for the Bioremediation of Marine Shorelines and Freshwater Wetlands.* U.S. EPA, Office of Research and Development, National Risk Management Research Laboratory, Land Remediation and Pollution Control Division, Cincinnati, OH.

Index

A

N

O

P

Q